Coding for Optical Channels

Ivan Djordjevic • William Ryan • Bane Vasic

Coding for Optical Channels

with Chapter 8 contributed by
Anantha Raman Krishnan and Shiva K. Planjery

Springer

Ivan Djordjevic
Department of Electrical
& Computer Engineering
University of Arizona
Tucson, AZ 85721
USA
ivan@ece.arizona.edu

William Ryan
Department of Electrical
& Computer Engineering
University of Arizona
Tucson, AZ 85721
USA
ryan@ece.arizona.edu

Bane Vasic
Department of Electrical
& Computer Engineering
University of Arizona
Tucson, AZ 85721
USA
vasic@ece.arizona.edu

ISBN 978-1-4419-5568-5 e-ISBN 978-1-4419-5569-2
DOI 10.1007/978-1-4419-5569-2
Springer New York Dordrecht Heidelberg London

Library of Congress Control Number: 2009943537

Printed on acid-free paper

Springer is part of Springer Science+Business Media (www.springer.com)

To my parents (Blagoje and Verica), brother Slavisa, and to Milena.
Ivan Djordjevic
To Stephanie, Faith, Grant, Luke, and the Bandii.
William Ryan

Preface

Recent years have witnessed a dramatic resurgence of interest in channel coding within the optical communications community, as evidenced by the increase of the number of publications, and many eye-catching implementations and experimental demonstrations presented at major conferences. The main drivers behind the emergence of channel coding in optical communications are: (1) high demands in bandwidth thanks to the recent growth of Internet usage, IPTV, VoIP, and YouTube; and (2) rapid advance of silicon signal processing capability. In recent years, with the rapid growth of data-centric services and the general deployment of broadband access networks, there has been a strong demand driving the dense wavelength division multiplexing (DWDM) network upgrade from 10 Gb/s per channel to more spectrally efficient 40 Gb/s or 100 Gb/s per channel, and beyond. The 100 Gb/s Ethernet (100 GbE) is currently under standardization for both local area networks (LANs) and wide area networks (WANs). The 400 Gb/s and 1 Tb/s are regarded to be the next steps after 100 Gb/s and have started already attracting research community interests. Migrating to higher transmission rates comes along with numerous challenges such as degradation in the signal quality due to different linear and nonlinear channel impairments and increased installation costs. To deal with those channel impairments novel advanced techniques in modulation and detection, coding, and signal processing should be developed. Such topics will be described in detail in this book.

The introduction of sophisticated electronic digital signal processing (DSP), coherent detection, and coding could fundamentally alter the optical networks as we see them today. DSP has played a vital role in wireless communication and has enabled so-called software-defined radio (SDR). Thanks to the recent resurgence of coherent detection and the recent drive toward dynamically reconfigurable optical networks with transmission speeds beyond 100 Gb/s, DSP and forward error correction (FEC) are becoming increasingly important. Regardless of the data destination, an optical transport system (OTS) must provide the predefined bit-error rate (BER) performance. To achieve a target BER regardless of the data destination, the future OTS should be able to adjust the FEC strength according to the optical channel conditions. Such an approach leads us toward the *software-defined optical transmission* (SDOT) in which the transponder can be adapted or reconfigured to multiple standards, multiple modulation formats, or code rates, a concept very similar to SDR.

Although channel coding for optical channels has gained prominence and emerged as the leading ultra-high-speed optical transmission enabling technology, FEC seems to be rather alien to many optical engineers. The optical engineers are aware that FEC potentially hold the keys to solving many major problems for today's "fragile" and "rigid" optical networks, but feel intimidated by sophisticated coding terminology. This book is intended to give a coherent, self-contained, and comprehensive introduction to the fundamentals of channel coding and DSP for optical communications. It is designed for three diverse groups of researchers: (1) optical communication engineers who are proficient in the optical science and are interested in applying coding theory and DSP, but not familiar with basic coding concepts, (2) wireless communication engineers who are very much adequate with their DSP and coding skill sets, but are disoriented by the perceived huge gap between optical and RF communications worlds, and (3) coding experts interested in entering the world of optical communications. An attempt has been made to make the individual chapters self-contained as much as possible while maintaining the flow and connection between them.

This book is organized into 11 chapters, and treats topics related to modulation, DSP and coding for optical channels starting from the fundamentals of optical communication and major channel impairments and noise sources, through DSP and coding, to various applications, such as single-mode fiber transmission, multimode fiber transmission, free space-optical systems, and optical access networks. This book presents interesting research problems in the emerging field of channel coding, constrained coding, coded-modulation, and turbo equalization and touches on the intriguing issue related to future research topics in coding for optical channels. The main purpose of this book is: (1) to describe the FEC schemes currently in use in optical communications, (2) to describe different classes of codes on graphs of high interest for next-generation high-speed optical transport, (3) to describe how to combine multilevel modulation and channel coding optimally, and (4) to describe how to perform equalization and soft decoding jointly, in a so-called turbo equalization fashion.

The authors would like to thank their colleagues and students, in particular, H. G. Batshon, L. L. Minkov, M. Arabaci, A. R. Krishnan, S. K. Planjery, L. Xu, T. Wang, M. Cvijetic, F. Kueppers, S. Denic, M. Ivkovic, M. A. Neifeld, J. Anguita, T. Mizuochi, X. Liu, I. Gabitov, N. Alic, G. T. Djordjevic, and S. K. Turitsyn, whose collaboration or support on the subject of coding for optical channels contributed directly and indirectly to the completion of this book. The authors would like to acknowledge National Science Foundation (NSF), NEC Laboratories America, and Opnext for their support of coding theory and coding for optical channels research activities.

Finally, special thanks are extended to Charles Glaser and Amanda Davis of Springer US for their tremendous effort in organizing the logistics of this book including editing and promotion, which is indispensible to make this book happen.

Tucson, AZ

Ivan Djordjevic
William Ryan
Bane Vasic

Contents

About the Authors

Ivan B. Djordjevic is an Assistant Professor in the Department of Electrical and Computer Engineering at the University of Arizona, Tucson, where he is directing the Optical Communications Systems Laboratory (OCSL). He is the author or coauthor of more than 100 journal publications and more than 100 conference papers. Dr. Djordjevic is also the author, together with Dr. William Shieh, of the book *OFDM for Optical Communications*, Elsevier, 2009. His current research interests include optical networks, error control coding, constrained coding, coded modulation, turbo equalization, OFDM applications, and quantum error correction. He is an Associate Editor for the *Research Letters in Optics* and the *International Journal of Optics.*

William E. Ryan is a Professor in the Department of Electrical and Computer Engineering at the University of Arizona. He has published numerous papers in the leading conferences and journals in the area of communication theory and channel coding. He is also the author (with Shu Lin) of the graduate textbook *Channel Codes: Classical and Modern*, Cambridge University Press, 2009. His research interests are in coding and signal processing with applications to magnetic data storage and data communications.

Dr. Bane Vasic received his B.Sc., M.Sc., and Ph.D. from the University of Nis, Serbia. He is currently a Professor of Electrical and Computer Engineering and Mathematics at the University of Arizona, Tucson. Prior to this appointment, he was at Bell Laboratories. He is a Member of the Editorial Board of the IEEE Transactions on Magnetics, and was a chair or technical program chair for several workshops and conferences including: IEEE CTW 2003 and 2007, DIMACS Workgroup and Workshop on Theoretical Advances in Information Recording, 2004, LANL Workshop on Applications of Statistical Physics to Coding Theory, 2004, Communication Theory Symposium within ICC 2006. He has authored a number of journal and conference articles, book chapters and edited three books, and his patents are implemented in Bell Labs chips. His research interests include coding theory, communication theory, constrained systems, and digital communications and recording.

Chapter 1
Introduction

We live in a time officially proclaimed as the information era, which is closely related to Internet technology and characterized by never-ending demands for higher information capacity [1]. Optical transmission links are established around the globe, and the optical fiber connection extends from the global backbone to access networks, all the way down to the curb, building, home, and desk [1–9]. Despite of the Internet "bubble" occurred in the early 2000s, the Internet traffic has been growing at astonishing rate ranging from 75 to 125% per year [6]. Given the recent growth of Internet usage, IPTV, and VoIP, it has become clear that current 10-Gb/s Ethernet rate is insufficient to satisfy the bandwidth demands in near future. For example, Internet2 has announced 2 years ago a capacity upgrade of its next-generation IP network from 10 Gb/s to 100 Gb/s [7]. According to some industry experts, 100-Gb/s transmission is needed by the end of 2009, while 1 Tb/s should be standardized by the year 2012–2013 [7].

The performance of fiber-optic communication systems operating at these high rates is degraded significantly due to several channel impairments including intra- and interchannel fiber nonlinearities, the nonlinear phase noise, and polarization mode dispersion (PMD). These effects constitute the current limiting factors in efforts to accommodate demands for higher capacities/speeds, longer link lengths, and more flexible wavelength switching and routing capabilities in optical networks. To deal with those channel impairments, novel advanced techniques in modulation and detection, coding, and signal processing should be developed, and some important approaches will be described in this book. The book represents a coherent and comprehensive introduction to the fundamentals of optical communications, digital signal processing (DSP), and coding for optical channels.

In this chapter, we provide a historical overview on optical communications (Sect. 1.1), introduce basics of optical communication and networking (Sect. 1.2), describe the current optical communication trends (Sect. 1.3), and explain why the coding in optical channels is of utmost importance (Sect. 1.4). In Sect. 1.5, we describe the organization of the book.

I. Djordjevic et al., *Coding for Optical Channels*,
DOI 10.1007/978-1-4419-5569-2_1,

1.1 Historical Perspective of Optical Communications

The earliest optical communications systems consisted of fire or smoke signals, signaling lamps, and semaphore flags to convey a single piece of information [3]. For example, a relatively sophisticated ancient communication system, along the Great Wall of China, was composed of countless beacon towers. In this ancient communication system, the number of lanterns or the color of smoke was used as a means to inform the size of an invading enemy, which represents a crude form of multilevel signaling. By using the beacon towers, with the guards in each tower, positioned at regular distances along the Great Wall, a message could be transmitted from one end of the Great Wall to the other, more than 7,300 km, in slightly more than 1 h [9]. Therefore, this ancient communication system has many similarities with today's relay or regeneration systems, in which the beacon towers can be considered as relays. Relay or regeneration systems were further studied by Claude Chappe in 1792 to transmit coded messages over a distance of 100 km [3].

Thanks to the success of telegraphy, telephony, and radio communications in the first half of twentieth century, the optical communication systems were actually forgotten. However, in the late twentieth century, different communication systems came to saturation in terms of reach and capacity. For instance, a typical coaxial cable-based transport system operating at 155 Mb/s requires the regeneration at approximately every 1 km, which is costly to operate and maintain. The natural step was to study the optical communication systems, which can dramatically increase the total capacity. The research in optical communication was boosted upon demonstration of a laser principle [10]. The first step was to fabricate an appropriate optical transmission medium. Kao and Hockman [11] proposed to use the optical fiber as the medium, although at the time it had unacceptable fiber loss. Their argument was that attenuation mostly was coming from impurities, rather than any fundamental physical effect such as Rayleigh scattering, which could be reduced by improving the fabrication process. Their prediction was that an optical fiber with attenuation of 20 dB/km should be sufficient for telecom applications, which surprisingly was developed within 5 years since initial proposal, by researchers from Cornell. This invention opens up opportunities for development of fiber-optic communication systems. Several generations of optical communication systems were developed since then. The *first generation* appeared in 1980s, and the operating wavelength was 0.8 μm with 45 Mb/s data rate. Repeater spacing was 10 km, which was much greater than that for comparable coax systems. Lower installation and maintenance costs resulted from fewer repeaters.

The *second generation*, which was focused on a transmission near 1.3 μm to take advantage of the low attenuation (<1 dB/km) and low dispersion, was deployed during the early 1980s. Sources and detectors were developed that use InGaAsP semiconductor. The bit rate of these systems was limited to <100 Mb/s due to dispersion in multimode fibers (MMFs). *Single-mode fiber* (SMF) was then incorporated. By 1987 the second generation systems were operating at 1.7 Gb/s at 1.3 μm with repeater spacing of 50 km.

The *third generation* systems were based on the use of 1.55-μm sources and detectors. At this wavelength the attenuation of fused silica fiber is minimal. The deployment of these systems was delayed, however, due to the relatively large dispersion at this wavelength. Two approaches were proposed to solve the dispersion problem. The first approach was to develop single-mode lasers and the second was to develop dispersion shifted fiber (DSF) at 1.55 μm. In 1990, 1.55-μm systems operating at 2.5 Gb/s were commercially available and were capable of operating at 10 Gb/s for distances of 100 km [1–3]. The best performance was achieved with DSFs in conjunction with single-mode lasers. A *drawback* of these systems was the need for *electronic regeneration* with repeaters typically spaced every 60–70 km. Coherent detection methods were investigated in late 1980s and early 1990s to increase receiver sensitivity. However, this approach was super ceded by the development of the optical amplifier.

The *fourth generation* systems are based on the use of *optical amplifiers* to increase repeater spacing and *wavelength division multiplexing* (WDM) to increase the aggregate bit rate. Erbium-doped fiber amplifiers (EDFAs) were developed to amplify signals without electronic regeneration during the 1980s [1–3]. In 1991, signals could be transmitted 14,300 km at 5 Gb/s without electronic regeneration [1–3]. The first transpacific commercial system went into operation sending signals over 11,300 km at 5 Gb/s and other systems are being deployed [1–3]. System capacity is increased through use of WDM. Multiple wavelengths can be amplified with the same optical amplifier. In 1996, 20 $\times$ 5 Gb/s signals were transmitted over 9,100 km providing a total bit rate of 100 Gb/s and a bandwidth–length (B–L) product of 910 (Tb/s) km. [1–3] In these broad band systems, dispersion becomes an important issue to be addressed.

In the *fifth generation* systems, the effort is primarily concerned with the fiber dispersion problem. Optical amplifiers solve the loss problem but increase the dispersion problem since dispersion effects accumulate over multiple amplification stages. An ultimate solution is based on the novel concept of *optical solitons* [1–3]. These are pulses that preserve their shape during propagation in a lossless fiber by counteracting the effect of dispersion through fiber nonlinearity. Experiments using stimulated Raman scattering (SRS) as the nonlinearity to compensate for both loss and dispersion were effective in transmitting signals over 4,000 km [1–3]. EDFAs were first used to amplify solitons in 1989 [1–3]. By 1994 a demonstration of soliton transmission over 9,400 km was performed at a bit rate of 70 Gb/s by multiplexing seven 10-Gb/s channels [1–3]. In parallel, dispersion compensating fibers (DCFs) were invented to deal with chromatic dispersion, and various dispersion maps were proposed [1–3] (see Sect. 1.3.4 for more details).

In *sixth generation* systems, the efforts have been directed toward realizing greater capacity of fiber systems by multiplexing a large number of wavelengths. These systems are referred to as dense wavelength division multiplexing (DWDM) systems. Systems with wavelength separation of 0.8 nm are currently in operation and efforts are pushing to reduce this to <0.5 nm. Controlling wavelength stability and the development of wavelength demultiplexing devices are critical to this effort. Systems are currently operating at 10 Gb/s and 40 Gb/s.

The current research focus is related to 100 Gb/s per wavelength optical transmission and beyond, by employing various multilevel modulation and coding schemes, polarization-multiplexing, DSP, and coherent detection. The orthogonal frequency division multiplexing (OFDM) appears to be an excellent candidate to deal with chromatic dispersion and PMD, but is sensitive to four-wave mixing (FWM) between subcarriers due to fiber nonlinearities [9].

1.2 Optical Transmission and Optical Networking

An exemplary WDM optical network, which can be used to identify the key optical components, concepts, and system parameters is shown in Fig. 1.1. The end-to-end optical transmission involves both electrical and optical signal paths. To perform conversion from electrical to optical domain the optical transmitters are used, while to perform conversion in opposite direction (optical to electrical conversion) the optical receivers are used. The SMF serves as a foundation of an optical transmission system because the optical fiber is used as medium to transport the optical signals from source to destination. The optical fibers attenuate the signal during transmission, and someone has to use optical amplifiers, such as EDFAs, Raman amplifiers, or parametric amplifiers, to restore the signal quality. Unfortunately, the amplification process is accompanied with the noise addition. For better exploitation of enormous bandwidth of SMF, the WDM concept is introduced, which corresponds to the scheme with multiple optical carriers at different wavelengths that are modulated by using independent electrical bit streams, as shown in Fig. 1.1, and then transmitted over the same SMF. During transmission of WDM signals, occasionally several wavelengths have to be added/dropped, which is performed by the optical add–drop multiplexer (OADM), as shown in Fig. 1.1. The optical networks require

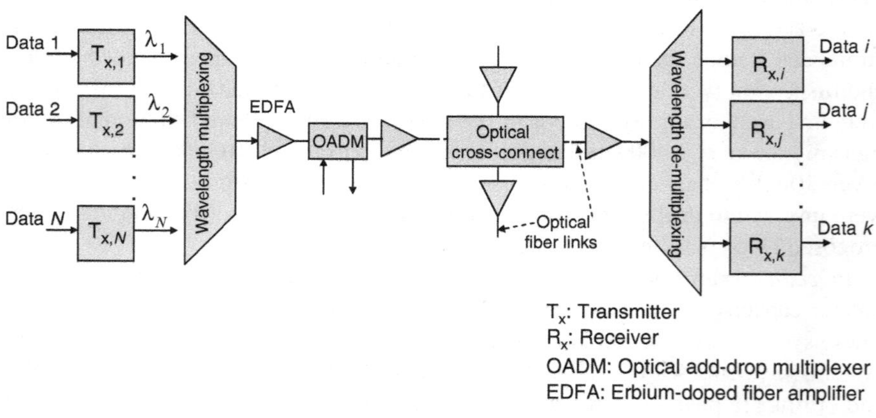

Fig. 1.1 An exemplary WDM optical network identifying key optical components, concepts, and parameters

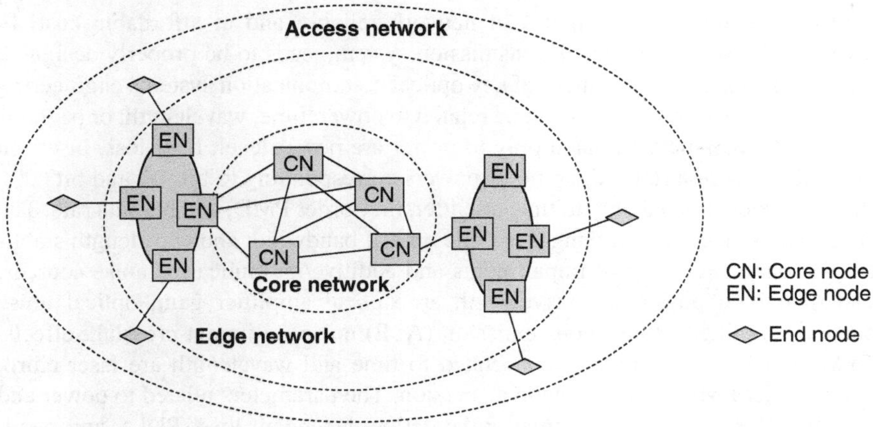

Fig. 1.2 A typical optical networking architecture

the switching of information among different fibers, which is performed by the optical cross-connect (OXS). To combine several distinct wavelength channels into composite channel the wavelength multiplexers are used. On the other hand, to split the composite WDM channel into distinct wavelength channels, the wavelength demultiplexers is used. To impose the information signal (be in digital or analog form) optical modulators are used. The optical modulators are commonly used in combination with semiconductor lasers.

To provide a global picture, we describe a typical optical network shown in Fig. 1.2. We can identify three ellipses representing the *core network*, the *edge network*, and the *access network* [1]. The long-haul core network interconnects big cities, major communications hubs, and even different continents by means of submarine transmission systems. The core networks are often called the wide area networks (WANs) or interchange carrier networks. The edge optical networks are deployed within smaller geographical areas and are commonly recognized as metropolitan area networks (MANs) or local exchange carrier networks. The access networks represent peripheral part of optical network and provide the last-mile access or the bandwidth distribution to the individual end-users.

The optical transmission systems can be *classified* according to different criterions. If *transmission length* is used for classification, we can identify very short reach (hundreds of meters), short reach (several kilometers), long reach (tens and hundreds of kilometers), and ultra-long reach (thousands of kilometers) optical transmission systems. When *bit rate* is used as classification criteria, the optical transmission systems can be classified as low-speed (tens of Mb/s), medium-speed (hundreds Mb/s), and high-speed (Gb/s). Finally, from *application perspective* point of view the systems can be either power budget (loss) limited or bandwidth (transmission speed) limited.

The ultimate goal of an optical signal transmission system is usually defined as achieving desired bit error rate (BER) performance between two end users or between two intermediate nodes in network reliably and at affordable cost. In order to achieve so, an optical transmission system needs to be properly designed, which includes the management of key optical communication systems engineering parameters. These parameters can be related to power, time, wavelength, or be interrelated. The parameters related only to power are power level, fiber loss, insertion loss, and extinction ratio (the ratio of powers corresponding to bit "1" and bit "0"). The parameters related only to time are jitter, first-order PMD, and bit/data rate. The parameters related to wavelength include optical bandwidth and wavelength stability. The parameters, signal impairments and additive/multiplicative noise sources, related to both power and wavelength are optical amplifier gain, optical noise [such as amplified-spontaneous emission (ASE) noise], different crosstalk effects, FWM, and SRS. The parameters related to time and wavelength are laser chirp, second-order PMD, and chromatic dispersion. The parameters related to power and time are BER, modulation format, polarization-dependent loss (PDL), and quantum noise. Finally, the channel impairments related to time, power, and wavelength simultaneously are self-phase modulation (SPM), cross-phase modulation (CPM), and stimulated Brillouin scattering (SBS). Any detection scenario must include electronic noise, such as thermal noise, which is associated with receiver design. These different parameters, noise sources, and channel impairments are subject of Chaps. 2 and 3.

Different high-speed optical transmission *enabling technologies* can either be related to the usage of novel/better *devices*, such as Raman and parametric amplifiers, PMD and chromatic dispersion compensators, and modulators, or be related to the novel *methods*, such as advanced modulation formats (various multilevel modulation schemes with both direct and coherent detections and OFDM), forward error correction (FEC), coded modulation, constrained (modulation/line) coding, advanced detection schemes [maximum likelihood sequence detection/estimation (MLSD/E) and maximum a posteriori probability (MAP) detection (Bahl–Cocke–Jelinek–Raviv (BCJR)-algorithm-based equalizers)], and various multiplexing schemes [polarization-multiplexing, optical time division multiplexing (OTDM), subcarrier multiplexing (SCM), code division multiplexing (CDM), and OFDM]. These various enabling technologies will be described in the following chapters.

An important concept to be introduced here is related to the so-called *lightwave path*, which can be defined as the trace that optical signal passes between the source and destination without experiencing any opto-electrical-opto (O-E-O) conversion [1]. Generally speaking, the lightwave paths may differ in lengths and in the information capacity that is carried along and can traverse though different portions of an optical network. The lightwave path can be considered as bandwidth wrapper for lower speed transmission channels, which form virtual circuit services [1]. The time division multiplexing (TDM) technique is applied to aggregate the bandwidth of virtual circuits before it is wrapped in the lightwave path. TDM of virtual circuits can be either *fixed* (each circuit receives a guaranteed amount of the bandwidth – a bandwidth pipe) or *statistical* (in packet-switching the data content is divided into data packets, which can be handled independently). The fixed multiplexing of

Table 1.1 Bit rates for different synchronous/asynchronous optical channels

Synchronous (TDM) channels	Bit rate	Asynchronous (data) channels	Bit rate
DS-1	1.544 Mb/s	10-BaseT Ethernet	10 Mb/s
E-1	2.048 Mb/s	100-BaseT Ethernet	100 Mb/s
OC-1	51.84 Mb/s	FDDI	100 Mb/s
OC-3=STM-1	155.52 Mb/s	ESCON	200 Mb/s
		Fiber Channel-I	200 Mb/s
OC-12=STM-4	602.08 Mb/s	Fiber Channel-II	400 Mb/s
		Fiber Channel-III	800 Mb/s
OC-48=STM-16	2.488 Gb/s	Gb Ethernet	1 Gb/s
OC-192=STM-64	9.953 Gb/s	10-Gb Ethernet	10 Gb/s
OC-768=STM-256	39.813 Gb/s	40-Gb Ethernet	40 Gb/s

Table 1.2 The wavelength bands for fiber-optics communications

Wavelength band	Descriptor	Wavelength range (nm)
O-band	Original	1,260–1,360
E-band	Extended	1,360–1,460
S-band	Short	1,460–1,530
C-band	Conventional	1,530–1,565
L-band	Long	1,565–1,625
U-band	Ultra-long	1,625–1,675

virtual circuits is defined by SONET/SDH standards. Bit-rates of different bandwidth channels, for both synchronous and asynchronous transmission, are given in Table 1.1, due to Cvijetic [1].

Optical fiber is the key point of an optical transmission system because it has much wider available bandwidth, lower signal attenuation, and smaller signal distortions compared with any other wired or free-space physical media. The total bandwidth is approximately 400 nm, or around 50 THz, when related to the wavelength region with fiber attenuation being bellow 0.5 dB/km. The usable optical bandwidth is commonly split into several wavelength bands, as shown in Table 1.2 due to Ramaswami and Sivarajan [2]. The bands around the minimum attenuation point, usually referred to as C and L bands, are the most suitable for high channel count DWDM transmission. The wavelength region around 1,300 nm is less favorable for optical signal transmission because signal attenuation is higher than attenuation in S, C, and L bands. On the other hand, it is quite suitable for CATV signals, and the course-WDM (CWDM) technique is usually employed in this region.

The key optical components, which will be described in Chap. 2, can be classified as follows (1) semiconductor light sources [light-emitting diodes (LEDs) and semiconductor lasers: Fabry–Perot (FP), distributed feedback (DFB), distributed Bragg reflector (DBR), vertical cavity surface emitting (VCSEL), and tunable lasers (external cavity laser, mutilaser chip, three-section tunable)]; (2) optical modulators [direct optical modulators and external modulators: Mach–Zehnder modulator

(MZM) and electroabsorption modulator]; (3) optical fibers (MMFs and SMFs); (4) optical amplifiers [semiconductor optical amplifier (SOA), EDFA, Raman amplifiers, and parametric amplifiers]; (5) photodiodes [PIN, avalanche photodiodes (APDs), and metal–semiconductor–metal (MSM) photodetectors]; and (6) various optical components [optical isolators, optical circulators, optical filters, optical couplers, optical switches, and optical multiplexers/demultiplexers].

A monochromatic electromagnetic wave, which is commonly used as a signal carrier, can be represented through its electric field as $\boldsymbol{E}(t) = \boldsymbol{p}A\cos(\omega t + \phi)$ (A – amplitude, ω – frequency, ϕ – phase, $\boldsymbol{p}$ – polarization orientation), for which each parameter can be used to impose the message signal. If the message signal is analog, the corresponding modulation formats are amplitude modulation (AM), frequency modulation (FM), phase modulation (PM), and polarization modulation (PolM). On the other hand, when the modulating signal is digital then the carrier signal duration is limited to symbol duration, and corresponding modulation formats are amplitude shift keying (ASK), frequency shift keying (FSK), phase shift keying (PSK), and polarization shift keying (PolSK).

In order to better utilize the enormous bandwidth of the optical fiber, we have to transmit simultaneously many channels over the same bandwidth through *multiplexing*. The commonly used methods of multiplexing in optical communications are given below as follows:

- *Wavelength-division multiplexing* (WDM) is already introduced in Fig. 1.1.
- *Time-division multiplexing* (TDM), in which many lower-speed signals are time-interleaved to generate a high-speed signal. The multiplexing can be performed either in electrical domain, when is known as electrical TDM (ETDM), or in optical domain, when is known as optical TDM (OTDM).
- *Frequency-division multiplexing* (FDM), in which continuous-wave (CW) modulation is used to translate the spectrum of the message signal into a specific frequency slot of the passband of optical channel. The optical version of FDM is commonly referred to as WDM.
- *Orthogonal frequency-division multiplexing* (OFDM) is a particular version of FDM in which the orthogonality among subcarrier is obtained by providing that each subcarrier has exactly an integer number of cycles in the symbol interval. The number of cycles between adjacent subcarriers differs by exactly one.
- *Subcarrier multiplexing* (SCM) is again a particular version of FDM in which different independent data streams are first microwave multiplexed and then transmitted using the same wavelength carrier.
- *Code-division multiplexing* (CDM), in which each message signal is identified by a unique *signature sequence* ("code"), with signature sequences being orthogonal to each other.

During the transmission over an optical fiber, the transmitted signal is impaired by various noise sources and channel impairments. The noise sources can be additive in nature (dark current noise, thermal noise, ASE noise, and crosstalk noise) or be multiplicative in nature [mode partition noise (MPN), laser intensity noise (RIN), modal noise, quantum shot noise, and avalanche shot noise]. Different channel impairments can be related to fiber attenuation, insertion loss, dispersion effects, or fiber nonlinearities. Fiber attenuation originates from material absorption, which can be intrinsic (ultraviolet, infrared) or extrinsic (water vapor, Fe, Cu, Co, Ni, Mn, Cr, various dopants: GeO_2, P_2O_5, B_2O_3), Rayleigh scattering, and waveguide imperfections (Mie scattering, bending losses, etc.). The dispersion effects can originate from intermodal (multimode) dispersion (in MMFs), chromatic dispersion (material and waveguide dispersion effects present in SMF), PMD, and PDL. The fiber nonlinearities can originate from nonelastic scattering effects (SBS, SRS) or Kerr nonlinearities (SPM, XPM, FWM). Various noise sources and optical channel impairments are described in Chap. 3.

1.3 Optical Communications Trends

The invention of Internet has fundamentally changed the underlying information communication infrastructure and has lead to the world-wide telecom boom in the late 1990s and early 2000s [9]. This development coincided with the development and deployment of WDM optical amplified systems. Surprisingly, the Internet traffic has continued its rapid growth despite the so-called "Internet bubble" in the equity market. Namely, some new applications, such as IPTV and YouTube, have emerged, which continued to drive the high bandwidth demands so that the growth of the Internet does not seem to saturate any soon. Moreover, the CISCO's projection of the Internet traffic growth, up to 2012, shows an exponential dependence from 2002 to 2012 [9]. This exponential Internet traffic growth places an enormous pressure to the underlying information infrastructure at every level, from core to access networks. In the rest of this section, we describe several trends in optical communications arising from rapid IP traffic growth.

1.3.1 Migration to 100 Gb/s Ethernet and Beyond

Over the past decades the Ethernet (IEEE 802.3) has expanded from original share-medium LAN technology to a reliable standard across all level of the networks [9]. Ethernet has become the undisputed communication technology of choice in terms of cost and reliability. Because the IP backbones have grown so quickly that some large ISPs already reported router-to-router trunk connectivity exceeding 100 Gb/s in 2007; some industry experts believe that the 100-Gb/s Ethernet (100 GbE) standard is too late, while 1-Tb/s Ethernet standard should be available by 2012–2013 [8]. The migration of the line rate from 10 Gb/s to 100 Gb/s is expected to help in reduction of capital and operational costs. Since the migration to 100 GbE leads to fewer pipes, but of larger bandwidth among the IP routers, it is expected to simplify the traffic management.

The deployment of transmission systems based on data rates of 100 Gb/s is advantageous for many reasons [12] (1) smaller number of high-speed channels for a given bandwidth results in better spectral efficiency because no bandwidth is wasted on separating the channels; (2) a single high-speed transponder may replace many low-speed ones resulting in reduction of the number of optoelectronic devices and easier monitoring; and (3) a reduced number of channels for a given bandwidth permits using simpler optical switching devices and simpler routing algorithms. However, increasing data rates of fiber-optics communication systems is associated with numerous technological obstacles [10–15] such as increased sensitivity to fiber nonlinearities, high sensitivity to PMD, and increased demands in dispersion accuracy.

1.3.2 Dynamically Reconfigurable Optical Networks

The explosive growth of bandwidth-rich Internet video applications places tremendous strain on the traditional communication networks [9]. Although the link capacity can be enhanced by migration to 40 Gb/s or 100 Gb/s or by employing more WDM channels, such simplistic approach works very well in only a point-to-point communication. In order to be able to adjust to ever changing requirements for the bandwidth demand, the optical networks must be able to dynamically add, drop, and route the wavelength channels at individual nodes. If this operation is performed in optical domain, the transponder cost related to optical-to-electrical and electrical-to-optical conversions will be eliminated. This type of wavelength management in optical domain is performed by reconfigurable optical add–drop multiplexer (ROADM) [16].

Current limitations of photonics-enabled networks also result from the heterogeneity of the infrastructure and consequential bottlenecks at different boundaries and interfaces. In optically routed networks, neighboring DWDM channels carry random traffic patterns in which different lightwave paths experience different penalties due to the deployment of ROADMs and wavelength cross-connects (WXCs). Different wavelength channels carrying the traffic to different destinations can have quite different signal-to-noise ratios (SNRs) and spectral distortions due to cascaded filtering effects, as illustrated in Fig. 1.3a. The Internet of the future should be able to support a wide range of services containing a large amount of multimedia over different network types at high speed.

The ROADMs are commonly used to provide interconnection of many distribution networks (see Fig. 1.3b). Since their introduction in 2003, ROADMs have become a mainstream for use in core networks. Unfortunately, the use of reconfigurable networks with the transport speed exceeding 100 Gb/s imposes big challenges to the network designers, because at such high speeds the transmitted signal is very sensitive to PMD, residual chromatic dispersion, ROADMs concatenation filtering effect, and imperfections in electrical and optical components. It is therefore mandatory to perform per channel optical dispersion compensation. For long-haul

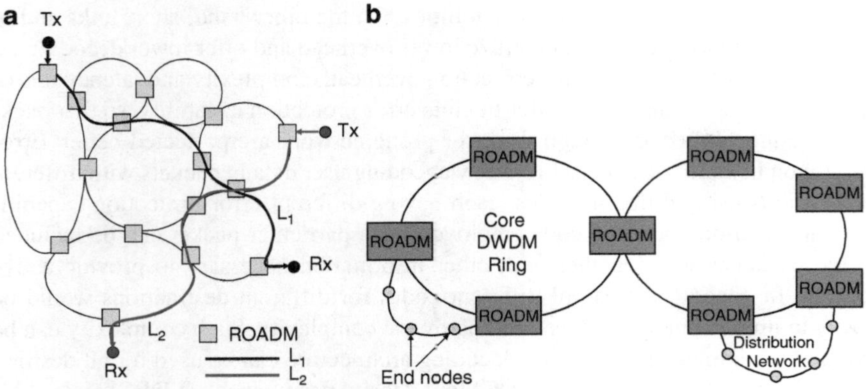

Fig. 1.3 (**a**) Different lightwave paths (L_1 and L_2) in an optically routed network; (**b**) optical distribution networks connected by ROADMs

transmission, it is also necessary to perform PMD compensation at 100 Gb/s. The optical PMD compensators usually have large footprint and are highly lossy and expensive. It is quite clear that such a "brute-force" migration from 10 GbE to 100 GbE cannot achieve at all the original goal of the cost saving.

1.3.3 Software-Defined Optical Transmission

In response to the emergence of analog and digital standards in wireless communications during 1980s, the concept of software-defined ratio (SDR) has been introduced [17–19]. SDRs are capable of changing scrambling and encryption coding in an ad hoc manner. Additionally, they can change their modulation format, data rate, channel coding, and voice codecs, while providing flexible and interoperable communication. The flexibility to adapt to major transmission parameters in accordance to the existing channel/traffic conditions is another key benefit. When channel conditions are favorable, SDRs can increase the signal constellation size or decrease it when channel conditions become poor, therefore giving SDRs the advantage of improved noise immunity. Channel coding schemes of SDRs can also be adapted to better deal with the effects of fading and interference.

As expected, similar challenges arise in modern optical communications [9]. A number of various "advanced" modulation formats have been proposed for the next generation 100-Gb/s Ethernet optical transport [20–26]. We envision an optical network in which all packets are error protected at edge nodes and can provide a target BER performance regardless of the packet destination. Because in this network the encoding and decoding is performed in edge nodes, it is compatible with existing modulation and multiplexing and switching techniques and can be used in different network environments. To enable the long-haul transmission, the employed codes must have good error correction capabilities. On the other hand, short links and/or highfidelity lightwave paths can utilize lower overhead and offer lower decoding latency. In order to optimize error correction overhead, complexity and latency it is of crucial importance that codes offer flexible error protection capability, so that packets traveling further or through the error prone network are protected better. Error protection flexibility can be achieved by encoding user data in packets with different destinations using different codes, each having different error protection capabilities. Information about the code employed for a particular packet can be included in the packet header together with other information necessary to provide active routing. In general, the use of different codes for different destinations would be costly to implement due to increased hardware complexity. Such complexity can be avoided if a unified encoding and decoding architecture can be used for all destinations. The structured quasicyclic (QC) low-density parity-check (LDPC) codes [27] provide us with this unique feature. Such an approach leads toward the *software-defined optical transmission* (SDOT) in which the transponder can be adapted or reconfigured to multiple standards, multiple modulation formats, or code rates, the concept very similar to SDR. In addition, the SDOT system should be able to (1) dynamically set up the physical link without any human intervention, (2) assign an optimal line rate and signal constellation size in accordance with the optical link conditions, (3) select between multicarrier mode and single-carrier mode, (4) chose an optimum code rate, and (5) accurately report various channel parameters (optical SNR, chromatic dispersion, PMD, electrical SNR) so to predict the fault and alarm before it causes the traffic interruption.

The concept of SDOT is illustrated in Fig. 1.4a. A key difference with respect to a conventional optical transmission system is the presence of DAC/ADC and DSP in the architecture of the SDOT. The SDOT promotes the migration from analog to digital domain to enhance the optical transmissions by providing a dynamic adaptation to the optical channel conditions and reconfiguration to an appropriate modulation format and code rate.

The electronic dispersion compensation (EDC) is a key point important for the success of SDOT concept. The very early approaches of EDC are essentially hardware based including feed-forward equalizer (FFE) and decision feedback equalizer (DFE), with limited performance improvement [28]. The EDC by DSP means has shown much better tolerance to various channel impairments. SDOT shown in Fig. 1.4 provides a generic architecture applicable to various EDCs by DSP techniques. For example, for conventional optical front ends and intensity modulation/direct detection (IM/DD) systems, the MLSD/E or turbo equalization can be used [13, 15]. The precompensation can be used together with optical in-phase/quadrature (IQ) modulator and direction detection, which is illustrated in Fig. 1.4b. For a coherent detection front end, digital phase estimation can be used instead of conventional optical phase-locked loop (OPLL). For coherent detection and optical IQ modulator (see Fig. 1.4b), coherent optical OFDM can be used to compensate for chromatic dispersion and PMD. Although the front ends in these examples are different, they all take the advantage of DSP to improve the chromatic dispersion and PMD tolerance.

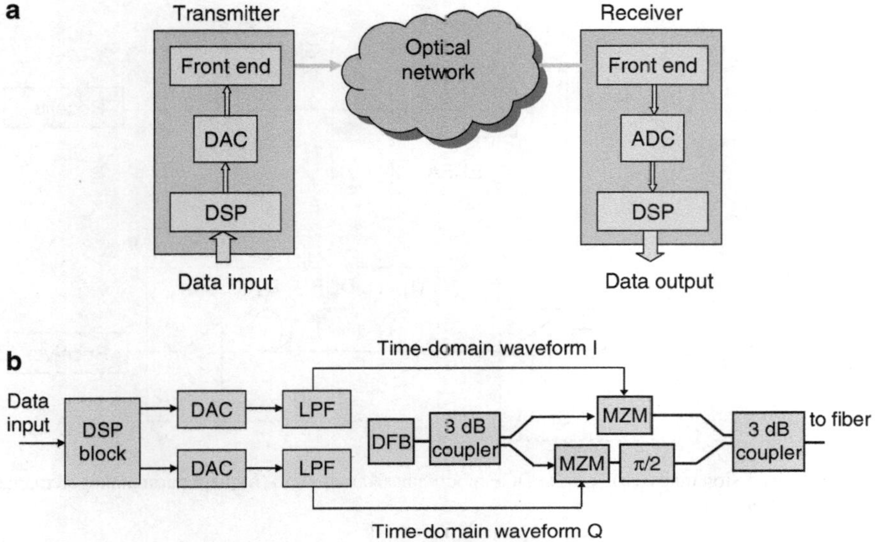

Fig. 1.4 (**a**) Conceptual diagram of software-defined optical transmission concept; (**b**) a typical transmitter front-end architecture (for single polarization) applicable to both direct and coherent detections. *DAC* Digital-to-analog converter, *ADC* analog-to-digital converter, *MZM* Mach–Zehnder modulator

1.3.4 Digital Signal Processing and Coherent Detection

The conventional optical systems employ a dispersion map to deal with accumulated chromatic dispersion of SMF. In dispersion maps, the DCF modules are deployed at the amplifier stage to compensate for the accumulated dispersion of the transmission link. The DCF could be placed at the optical amplifier site within double-stage amplifier, which is typical for terrestrial applications (see Fig. 1.5a), or be distributed in a dispersion mapped span, which is typical for submarine systems (see Fig. 1.5b). Such dispersion maps work very well for transmission systems operating at 10 Gb/s and below, but become extremely tedious at 40 Gb/s and beyond, requiring both the dispersion and dispersion slope of the DCF to be matched precisely. Any residual mismatched dispersion needs to be compensated using a fixed or tunable optical dispersion module, and this is performed on per channel basis. On the other hand, the electron equalizer has the advantages of lower cost, small footprint, and ease of adaption and is suitable for employment in SDOT scenario [9]. The electronic equalizers employ the classical equalization approaches such as FFE, DFE, Viterbi equalizer, and turbo equalizer. The early stage electrical equalizers mostly utilized SiGe or InP/AlGaAs technology with the channel memory limited to 2 or 3 bits and have successfully been commercial deployed [29]. The major breakthrough in electronic signal processing took place when researchers from Nortel promoted their predistortion equalizer supporting 5,000-km transmission over SMF

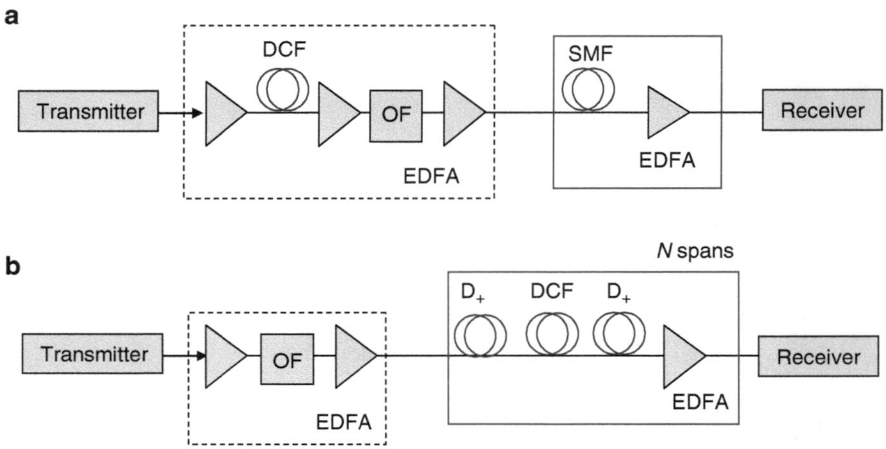

Fig. 1.5 Dispersion maps: (**a**) discrete DCF module based maps and (**b**) dispersion managed maps. *OF* Optical filter

without any optical dispersion compensator [30]. This work raised an interesting and fundamental question to the optical community whether it is necessary at all to use the dispersion maps, such as those shown in Fig. 1.5. The "era" of electronic DSP-assisted optical transmission has arrived, followed by the subsequent dramatic revival of the coherent optical communications [9].

1.3.5 OFDM for Optical Communications

OFDM is an efficient approach to deal with intersymbol interference (ISI) due to chromatic dispersion and PMD. By providing that the guard interval is larger than the combined delay spread due to chromatic dispersion and maximum differential group delay (DGD), the ISI can be eliminated successfully [9]. In the early of the 2006, three groups independently proposed two flavors of optical OFDM for long-haul application that were also aimed at eliminating the need for dispersion management in optical transmission systems [9], including direct-detection optical OFDM [31, 32] and coherent optical OFDM (CO-OFDM) [33]. CO-OFDM provides the superior performance in terms of spectral efficiency, receiver sensitivity, and polarization-dispersion resilience, but requires a little bit higher implementation complexity compared to that of direct detection OFDM [9]. The transmission experiments of CO-OFDM conducted in the research laboratories have reached 100-Gb/s transmission over 1,000 km of SMF [34–36].

The block diagram of an OFDM system with direct detection for transmission over MMF links is shown in Fig. 1.6, as an illustration. The demultiplexer converts incoming information data stream into parallel form. Constellation mapper maps

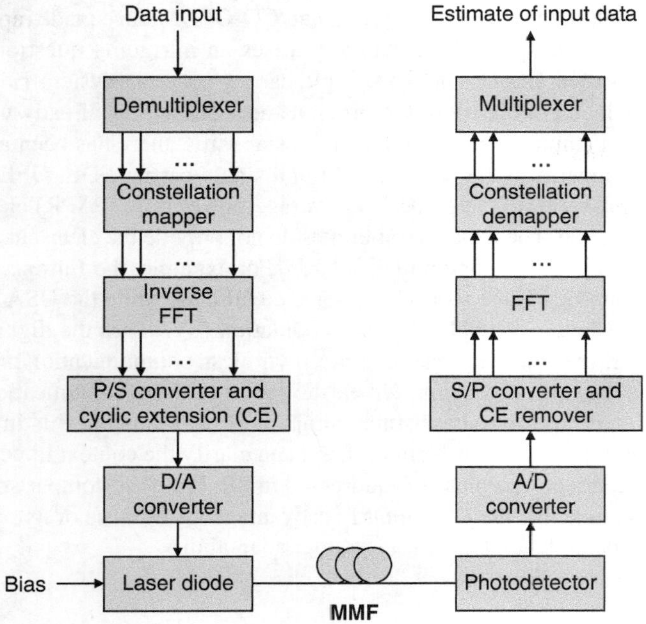

Fig. 1.6 The block diagram of an OFDM system with direct detection suitable for transmission over MMF links. *P/S* Parallel-to-serial, *S/P* serial-to-parallel

parallel data into N subcarriers using quadrature amplitude modulation (QAM). The modulation is performed by applying the inverse fast Fourier transform (IFFT), which transforms frequency-domain parallel data into time-domain parallel data. The digital-to-analog (D/A) converter performs the conversion from digital to analog domain. Typically, D/A converter contains a transmit filter. As shown in Fig. 1.6, to reduce the system cost direct modulation of laser diode can be used. Because the negative signals cannot be transmitted over an intensity modulation with direct detection, the bias voltage is used to convert the negative portion of OFDM signal to positive. Since this approach is power inefficient, the clipping can be used. At the receiver side, upon optical-to-electrical conversion by photodetector, DC bias blocking, analog-to-digital (A/D) conversion, cyclic removal, and serial-to-parallel (S/P) conversion, the demodulation is performed by the FFT. With sufficient number of subcarriers, the OFDM symbol duration can be made arbitrarily small increasing therefore the tolerance to ISI due to chromatic dispersion and PMD. By providing that the cyclic extension guard interval (see Fig. 1.6) is longer than total delay spread due to chromatic dispersion and maximum value of DGD, chromatic dispersion and PMD can be completely (at least in principle) compensated for. The cyclic extension can be performed by inserting the last $N_G/2$ samples of the IFFT effective portion of symbol at the beginning of the OFDM symbol and appending the first $N_G/2$ samples at the end.

As both single-carrier coherent system and CO-OFDM have made rapid progress toward 100-GbE transport, there naturally arises an intriguing question about the superiority between CO-OFDM and single-carrier coherent system [9]. Although OFDM has gained popularity in the previous decade and has already widely been implemented in a number of communication standards, there has been an on-going debate on the superiority between OFDM and single-carrier FDE. OFDM has two fundamental problems (1) large peak-to-average-power ratio (PAPR) and (2) sensitivity to phase noise. The debates sometimes do not provide the clear-cut answer and the result can be a split decision in standards. For example, the Europe, Japan, and most of the countries in the world have selected OFDM, while the USA has chosen single-carrier 8-level vestigial sideband modulation (8VSB) as the digital TV standard. However, the same arguments used in wireless communication are not quite applicable to the optical domain. Given the fact that the communication channel, devices, and systems are quite distinct compared to RF domain, it is imperative to understand thoroughly the problems at hand and clarify the context in which the debate is being conducted, which was addressed in [9]. From the comparison provided in [9], we conclude that the CO-OFDM is advantageous because of its scalability to the ever increasing data rate and transponder adaptability.

1.4 Forward Error Correction for Optical Communications and Networks

The state-of-the-art fiber-optics communication systems standardized by the ITU employ concatenated BCH/RS codes [37, 38]. The RS(255,239) in particular has been used in a broad range of long-haul communication systems, and it is commonly considered as the first-generation of FEC [41, 42]. The elementary FEC schemes (BCH, RS, or convolutional codes) may be combined to design more powerful FEC schemes, e.g., RS(255,239) + RS(255,233). Several classes of concatenation codes are listed in ITU-T G975.1. Different concatenation schemes, such as the concatenation of two RS codes or the concatenation of RS and convolutional codes, are commonly considered as second generation of FEC [41]. In recent years, iteratively decodable codes, like turbo codes [38–41] and LDPC codes [13–15, 42, 44–50], have generated significant research attention. In [40], Sab and Lemarie proposed an FEC scheme based on block turbo code for long-haul DWDM optical transmission systems. In recent papers [13–15, 44–48], we have shown that iteratively decodable LDPC codes outperform turbo product codes (TPCs) in BER performance. The decoder complexity of these codes is comparable (or lower) to that of TPCs and is significantly lower than that of serial/parallel concatenated turbo codes. For reasons mentioned above, LDPC code is a viable and attractive choice for the FEC scheme for 100-Gb/s optical transmission systems. The soft iteratively decodable codes (turbo and LDPC codes), also known as codes on graphs, in optical community are commonly referred to as the *third generation* of FEC [41, 43].

Codes on graphs have revolutionized communications and are becoming standard in many applications. LDPC codes, invented by Gallager [42] in 1960s, are linear block codes for which the parity check matrix has low density of ones. LDPC codes have generated great interests in the coding community recently, and this has resulted in a great deal of understanding of the different aspects of LDPC codes and their decoding process. An iterative LDPC decoder based on the *sum–product algorithm* (SPA) has been shown to achieve a performance as close as 0.0045 dB to the Shannon limit [49]. The inherent low complexity of this decoder opens up avenues for its use in different high-speed applications, including optical communications.

In Fig. 1.7, we show the recent progress in FEC for high-speed optical communications; the figure is adopted and modified from [51]. The horizontal axis denotes the year, while the vertical axis denotes the net coding gain (NCG). The first generation FEC schemes appeared during 1987–1993 and RS(255,239) of code rate 0.93 (overhead 7%) shows the NCG of 5.8 dB. The second generation FEC was developed during 2000–2004, with the best performing concatenated code showing the NCG of 9.4 dB for code rate of 0.8 (25% of redundancy) [51]. Focus, since then was on codes on graphs, turbo and LDPC codes, with potential NCG above 10 dB. The codes on graphs are commonly referred to third generation FEC for optical communications, as indicated above.

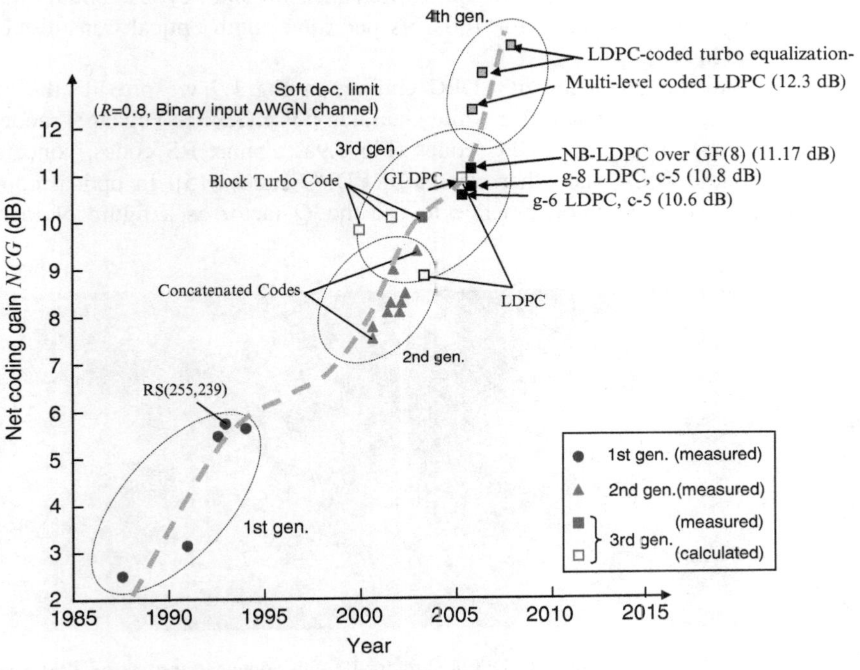

Fig. 1.7 Recent progress in FEC for optical communication systems. (Modified from ref. [51]; © IEEE 2006; reprinted with permission.)

The new emerging applications such as IPTV and YouTube have continued to drive the bandwidth demand. The traffic in the IP backbones has grown so quickly that some large ISPs already reported router-to-router trunk connectivity over 100 Gb/s in 2007. As the response, the network operators consider 100 Gb/s per wavelength optical transport. The multilevel modulation schemes are being studied as an efficient way, which in combination with channel coding, in so-called coded modulation [14, 15], can provide a solution to high bandwidth demands. However, at such high speed the signal quality is degraded significantly due to intrachannel nonlinearities and PMD. The turbo equalization [13, 15] has been proposed to simultaneously deal with various linear and nonlinear channel impairments. Different coded modulation schemes and different turbo equalization schemes can be considered as the fourth generation channel coding for next generation high-speed optical transport.

As an illustration of recent progress in FEC for high-speed optical transmission, in Fig. 1.8 we show an FPGA evaluation board due to Mizuochi et al. [52] for study of feasibility of FEC for 100-Gb/s optical transmission. The authors in [52] have demonstrated a 2-bit soft-decision-based LDPC (9216,7936) and RS (992,956) FEC coding, with four iterations using live FPGA emulation boards and demonstrated the NCG of 9.0 dB at 31.3 Gb/s (at output BER of 10^{-13}). In combination with polarization-multiplexed QPSK, this FEC scheme represents the first potential candidate for 100 Gb/s per wavelength optical transmission and beyond. Some other LDPC code proposals suitable for 100 Gb/s per wavelength optical transmission may be found in [15].

To illustrate the efficiency of LDPC coding, in Fig. 1.9 we provide the results of simulations for an additive white Gaussian noise (AWGN) channel model. We compare the large girth LDPC codes (Fig. 1.9a) against RS codes, concatenated RS codes, TPCs, and other classes of LDPC codes [15]. In optical communications, it is a common practice to use the Q-factor as a figure of merit

Fig. 1.8 An FPGA board for study of LDPC coding for 100-Gb/s optical transmission. (After ref. [52]; © IEEE 2009; reprinted with permission.)

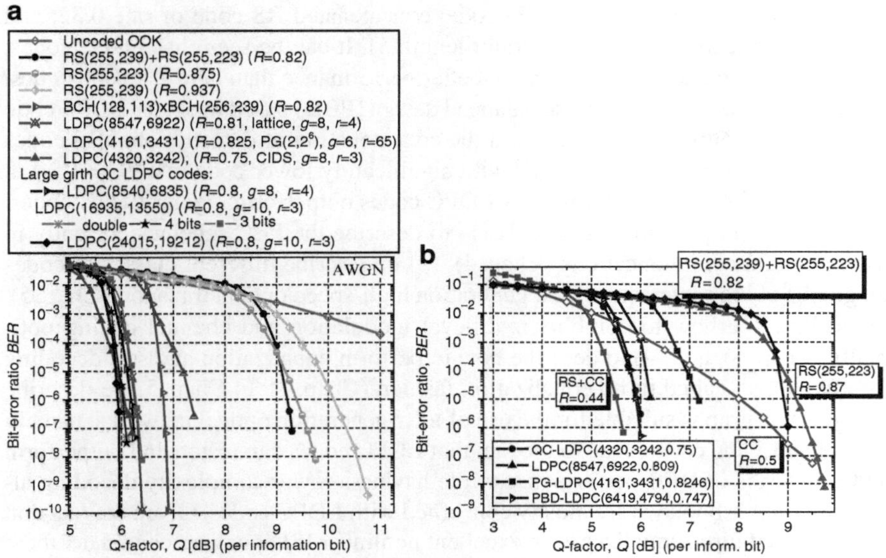

Fig. 1.9 (**a**) Large girth QC LDPC codes against RS codes, concatenated RS codes, TPCs, and previously proposed LDPC codes on an AWGN channel model and (**b**) LDPC codes vs. convolutional, concatenated RS, and concatenation of convolutional and RS codes on an AWGN channel. Number of iterations in sum–product-with-correction-term algorithm was set to 25. (After ref. [15]; © IEEE 2009; reprinted with permission.)

of binary modulation schemes instead of SNR.[1] In all simulation results in this figure, we maintained the double precision. For the LDPC(16935,13550) code, we also provided 3- and 4-bit fixed-point simulation results (see Fig. 1.9a). Our results indicate that the 4-bit representation performs comparable to the double-precision representation whereas the 3-bit representation performs 0.27 dB worse than the double-precision representation at the BER of 2×10^{-8}. The large-girth[2] LDPC(24015,19212) code of rate 0.8 outperforms the concatenation RS(255,239) + RS(255,223) (of rate 0.82) by 3.35 dB and RS(255,239) by 4.75 dB both at BER of 10^{-7}. The same LDPC code outperforms projective geometry (PG) $(2, 2^6)$ based LDPC(4161,3431) (of rate 0.825) by 1.49 dB at BER of 10^{-7} and outperforms CIDS-based LDPC(4320,3242) of rate 0.75 [47] by 0.25 dB. At BER of 10^{-10}, it outperforms lattice-based LDPC(8547,6922) of rate 0.81 (designed as described in [45]) by 0.44 dB and BCH(128,113) × BCH(256,239) TPC of rate 0.82 by 0.95 dB. The NCG at BER of 10^{-12} is 10.95 dB. In Fig. 1.9b, different LDPC codes

[1] The Q-factor is defined as $Q = (\mu_1 - \mu_0)/(\sigma_1 + \sigma_0)$, where μ_j and $\sigma_j (j = 0, 1)$ represent the mean and the standard deviation corresponding to the bits $j = 0, 1$.

[2] The girth represents the shortest cycle in corresponding bipartite graph representation of a parity-check matrix.

are compared against RS (255,223) code, concatenated RS code of rate 0.82, and convolutional code (CC) (of constraint length 5). It can be seen that LDPC codes, both regular and irregular, offer much better performance than hard-decision codes. It should be noticed that pairwise balanced design (PBD) based irregular LDPC code of rate 0.75 is only 0.4 dB away from the concatenation of convolutional RS codes (denoted in Fig. 1.9b as RS + CC) with significantly lower code rate $R = 0.44$ at BER of 10^{-6}. As expected, irregular LDPC codes outperform regular LDPC codes.

The main purpose of this book is (1) to describe the FEC schemes currently in use in optical communications (Chap. 4), (2) to describe different classes of codes on graphs of high interest for next generation high-speed optical transport (Chap. 5), (3) to describe how to combine multilevel modulation and channel coding optimally (Chap. 6), and (4) to describe how to perform equalization and soft decoding jointly, in a so-called turbo equalization fashion (Chap. 7). In Chap. 5, we describe the codes on graphs suitable for the use of next-generation optical networks, namely TPCs and LDPC codes. Due to the fact that LDPC codes can match and outperform TPCs in terms of BER performance while having a lower complexity decoding algorithm, in this book we are mostly concerned with LDPC codes. Given the fact that LDPC-coded turbo equalizer is an excellent nonlinear ISI equalizer candidate, there naturally arises the question about fundamental limits on channel capacity of coded-modulation schemes, which is addressed in Chap. 10. After this brief description of the book, in Sect. 1.5 we provide detailed description of every chapter in the book.

1.5 Organization of the Book

This chapter is devoted to the overview/organization of the book. This book contains 11 chapters covering a number of topics related to coding and DSP for optical channels.

In Chap. 2, we describe the key optical components used in contemporary optical communication systems, basic modulation formats, and direct detection and coherent detection principles.

In Chap. 3, we describe basic signal and noise parameters; major channel impairments including chromatic dispersion, PMD, and fiber nonlinearities; and system engineering process.

Chapter 4 is devoted to various hard decision-coding schemes currently in use in optical communications, such as BCH codes, RS codes, concatenated codes, and interleaving process. For completeness of presentation, the convolution codes are also described.

In Chap. 5, we describe codes on graphs, such as turbo codes, TPCs, and LDPC codes, that are currently in research phase. We describe their performance for thermal-noise-dominated scenario.

Chapter 6 is devoted to the optimization of multilevel modulation and coding process to achieve the best possible BER performance. Several coded-modulation schemes are described (1) multilevel coding [14, 53], (2) multidimensional coded modulation [24,25], and (3) coded-OFDM [9,22,31]. Using this approach, modulation, coding, and multiplexing are performed in a unified fashion so that all related signal processing is performed on symbol level. At this level, dealing with various nonlinear effects and PMD is more manageable, while the aggregate rate is maintained above 100 Gb/s.

Chapter 7 is devoted to the compensation of different linear and nonlinear channel impairments by turbo equalization. We describe a *multilevel* ($M > 2$) MAP turbo equalization scheme based on multilevel BCJR algorithm [54] based equalizer (called here multilevel BCJR equalizer) [15]. When used in combination with large girth LDPC codes as channel codes, this scheme represents a universal equalizer scheme for *simultaneous* suppression of fiber nonlinearities, for chromatic dispersion compensation, and for PMD compensation, applicable to both direct and coherent detections. To further improve the overall BER performance, we perform the iteration of extrinsic LLRs between LDPC decoder and multilevel BCJR equalizer. We use the extrinsic information transfer (EXIT) chart approach due to ten Brink [55] to match the LDPC decoders and multilevel BCJR equalizer. We further show how to combine this scheme with multilevel-coded modulation schemes with coherent detection.

In Chap. 8, we describe the use of constrained coding to deal with intrachannel nonlinearities. The role of a constrained coding is to impose certain constraints on a transmitted sequence in order to avoid those waveforms that are most likely to be incorrectly received. A constrained encoder translates an arbitrary user bit stream into a bit stream that satisfies the constraints of the channel. In this chapter we describe (1) time-domain constraints, such as runlength constraints, special forbidden pattern, distance enhancing codes, and self-synchronized codes; (2) spectral constraints, such as DC free constraint, spectral null constraint, higher order spectral zeros, and discrete spectral components; and (3) composite constraints. We further describe the ghost-pulse-constrained codes suitable to deal with intrachannel nonlinearities. Finally, we describe how to optimally combine constrained and channel codes, such as reverse concatenation with soft iterative decoding.

In Chap. 9, we describe free-space optical (FSO) systems that are robust in the presence of atmospheric turbulence such as (1) coded-OFDM, (2) coded MIMO concept, and (3) raptor-code-based FSO systems. We further describe different schemes suitable to solve the incompatibility problem between wireless/RF and optical channels: M-ary QAM and OFDM, M-ary pulse-amplitude modulation (PAM), and M-ary pulse-position modulation (PPM). We also describe different power-efficient coding schemes. We further discuss different adaptive modulation and coding techniques. We conclude the chapter with different coded modulation schemes suitable for hybrid optical communication systems.

Given the fact that turbo equalization is an excellent approach to deal with nonlinear ISI due to chromatic dispersion, PMD, and fiber nonlinearities, there naturally arises the question about fundamental limits on channel capacity of fiber-optics channels. In Chap. 10, we describe how to determine the channel capacity under different scenarios. The channel capacity evaluation of both single-carrier and multicarrier (in particular OFDM) is described. We describe how to determine the channel capacity when the nonlinear interaction of Kerr nonlinearities and ASE noise is taken into account. The channel capacity of hybrid optical-wireless communication systems is addressed as well.

In Chap. 11, we illustrate and emphasize why coding for optical channels is a vibrant and fast progressing field that provides great potentials for both practical development and research endeavor in the field of optical communications. We also give out our views on what are the important problems in coding for optical channels. The main problem in decoder implementation for large girth *binary* LDPC codes is the excessive codeword length, and a fully parallel implementation on a single FPGA is quite a challenging problem. To solve this problem, in Chap. 11, we also consider large-girth *nonbinary* LDPC codes over GF(2^m). By designing codes over higher-order fields, we aim to achieve the coding gains comparable to binary LDPC codes but for shorter codeword lengths.

References

1. Cvijetic M (2004) Optical transmission systems engineering. Artech House, Boston, MA
2. Ramaswami R, Sivarajan K (2002) Optical networks: a practical perspective, 2nd edn. Morgan Kaufman, San Fransisco, CA
3. Agrawal GP (2002) Fiber-optic communication systems, 3rd edn. Wiley, New York
4. Agrawal GP (2004) Lightwave technology: components and devices. Wiley, New York
5. Agrawal GP (2005) Lightwave technology: telecommunication systems. Wiley, New York
6. Melle S, Jaeger J, Perkins D, Vusirikala V (2007) Market drivers and implementation options for 100-GbE transport over the WAN. IEEE Appl Pract 45(11):18–24
7. Internet2 and Level 3 communications to deploy next generation nationwide research network Internet2 press release Available at http://news.internet2.edu/index.cfm
8. McDonough J (2007) Moving standards to 100 GbE and beyond. IEEE Commun Mag 45(11):6–9
9. Shieh W, Djordjevic I (2009) OFDM for optical communications. Elsevier, Amsterdam
10. Maiman TH (1960) Stimulated optical radiation in ruby. Nature 187(4736):493–494
11. Kao KC, Hockman GA (1966) Dielectric-fiber surface waveguides for optical frequencies. Proc. IEEE 113:1151–1158
12. Essiambre E-J, Raybon G, Mikkelsen B (2002) Pseudo-linear transmission of high-speed TDM signals at 40 and 160 Gb/s. In: Kaminow IP, Li T (eds) Optical fiber telecommunications IVB. Academic, San Diego, CA, pp 233–304
13. Djordjevic IB, Minkov LL, Batshon HG (2008) Mitigation of linear and nonlinear impairments in high-speed optical networks by using LDPC-coded turbo equalization. IEEE J Sel Areas Commun, Optical Commun Netw 26(6):73–83
14. Djordjevic IB, Cvijetic M, Xu L, Wang T (2007) Using LDPC-coded modulation and coherent detection for ultra high-speed optical transmission. IEEE/OSA J Lightwave Technol 25: 3619–3625
15. Djordjevic IB, Arabaci M, Minkov L (2009) Next generation FEC for high-capacity communication in optical transport networks. IEEE/OSA J Lightwave Technol 27(16):3518–3530 (Invited Paper)
16. Basch EB, Egorov R, Gringeri S, Elby S (2006) Architectural tradeoffs for reconfigurable dense wavelength division multiplexing systems. IEEE J Sel Top Quantum Electron 12:615–626
17. Burns P (2003) Software defined radio for 3G. Artech House, Boston, MA
18. Kenningotn PB (2005) RF and baseband techniques for software defined radio. Artech House, Boston, MA

19. Mitola J (1995) The software radio architecture. IEEE Commun Mag 33(5):26–38
20. Winzer PJ, Raybon G, Duelk M (2005) 107-Gb/s optical ETDM transmitter for 100 G Ethernet transport In: Proceedings of European conference on optical communication, Paper no. Th4.1.1, Glasgow, Scotland
21. Sun H, Wu KT, Roberts K (2008) Real-time measurements of a 40 Gb/s coherent system. Opt Express 16:873–879
22. Djordjevic IB, Vasic B (2006) 100 Gb/s transmission using orthogonal frequency division multiplexing. IEEE Photon Technol Lett 18(15):1576–1578
23. Djordjevic IB, Cvijetic M, Xu L, Wang T (2007) Proposal for beyond 100 Gb/s optical transmission based on bit-interleaved LDPC-coded modulation. IEEE Photon Technol Lett 19(12):874–876
24. Batshon HG, Djordjevic IB, Minkov LL, Xu L, Wang T, Cvijetic M (2008) Proposal to achieve 1 Tb/s per wavelength transmission using 3-dimensional LDPC-coded modulation. IEEE Photon. Technol Lett 20(9):721–723
25. Batshon HG, Djordjevic IB, Xu L, Wang T (2009) Multidimensional LDPC-coded modulation for beyond 400 Gb/s per wavelength transmission. IEEE Photon Technol Lett 21(16): 1139–1141
26. Fludger CRS, Duthel T, Van Den Borne D et al (2008) Coherent equalization and POLMUX-RZ-DQPSK for robust 100-GE transmission. IEEE/OSA J Lightwave Technol 26:64–72
27. Djordjevic IB (2009) Adaptive LDPC-coded multilevel modulation with coherent detection for high-speed optical transmission In: Proceedings of IEEE photonics society summer topicals 20–22 July 2009, Newport Beach, CA, Paper no. WC1. 2
28. Buchali F, Bülow H (2004) Adaptive PMD compensation by electrical and optical techniques. IEEE/OSA J Lightwave Technol 22:1116–1126
29. Elbers JP, Wernz H, Griesser H et al (2005) Measurement of the dispersion tolerance of optical duobinary with an MLSE-receiver at 10.7 Gb/s In: Proceedings of optical fiber communication conference, Los Angeles, CA, Paper no. OThJ4
30. McGhan D, Laperle C, Savchenko A, Li C, Mark G, O'Sullivan M (2006) 5120-km RZ-DPSK transmission over G.652 fiber at 10 Gb/s without optical dispersion compensation. IEEE Photon Technol Lett 18:400–402
31. Djordjevic IB, Vasic B (2006) Orthogonal frequency division multiplexing for high-speed optical transmission. Opt Express 14:3767–3775
32. Lowery AJ, Du L, Armstrong J (2006) Orthogonal frequency division multiplexing for adaptive dispersion compensation in long haul WDM systems In: Proceedings of optical fiber communication conference, Anaheim, CA, Paper no. PDP 39
33. Shieh W, Athaudage C (2006) Coherent optical orthogonal frequency division multiplexing. Electron Lett 42:587–589
34. Shieh W, Yang Q, Ma Y (2008) 107 Gb/s coherent optical OFDM transmission over 1000-km SSMF fiber using orthogonal band multiplexing. Opt Express 16:6378–6386
35. Jansen SL, Morita I, Tanaka H (2008) 10x121.9-Gb/s PDM-OFDM transmission with 2-b/s/Hz spectral efficiency over 1000 km of SSMF In: Proceedings of optical fiber communication conference, San Diego, CA, Paper no. PDP2
36. Kobayash T, Sano A, Yamada E (2008) Electro-optically subcarrier multiplexed 110 Gb/s OFDM signal transmission over 80 km SMF without dispersion compensation. Electron Lett 44:225–226
37. ITU, Telecommunication Standardization Sector. Forward error correction for submarine systems. Technical recommendation G.975/G709
38. Sab OA (2001) FEC techniques in submarine transmission systems. In: Proceedings of optical fiber communication conference, vol 2, pp TuF1-1–TuF1-3
39. Pyndiah RM (1998) Near optimum decoding of product codes. IEEE Trans Commun 46: 1003–1010
40. Sab OA, Lemarie V (2001) Block turbo code performances for long-haul DWDM optical transmission systems. OFC 3:280–282

41. Mizuochi T et al (2004) Forward error correction based on block turbo code with 3-bit soft decision for 10 Gb/s optical communication systems. IEEE J Sel Top Quantum Electron 10(2):376–386
42. Gallager RG (1963) Low density parity check codes. MIT, Cambridge, MA
43. Mizuochi T et al (2003) Next generation FEC for optical transmission systems. In: Proceedings of optical fiber communication conference (OFC 2003), vol 2, pp 527–528
44. Djordjevic IB, Milenkovic O, Vasic B (2005) Generalized low-density parity-check codes for optical communication systems. IEEE/OSA J. Lightwave Technol 23:1939–1946
45. Vasic B, Djordjevic IB, Kostuk R (2003) Low-density parity check codes and iterative decoding for long haul optical communication systems. IEEE/OSA J Lightwave Technol 21:438–446
46. Djordjevic IB et al (2004) Projective plane iteratively decodable block codes for WDM high-speed long-haul transmission systems. IEEE/OSA J Lightwave Technol 22:695–702
47. Milenkovic O, Djordjevic IB, Vasic B (2004) Block-circulant low-density parity-check codes for optical communication systems IEEE/LEOS J Sel Top Quantum Electron 10:294–299
48. Vasic B, Djordjevic IB (2002) Low-density parity check codes for long haul optical communications systems. IEEE Photon Technol Lett 14:1208–1210
49. Chung S et al (2001) On the design of low-density parity-check codes within 0.0045 dB of the Shannon limit. IEEE Commun Lett 5:58–60
50. Davey MC, MacKay DJC (1998) Low-density parity check codes over GF(q). IEEE Commun Lett 2:165–167
51. Mizuochi T (2006) Recent progress in forward error correction and its interplay with transmission impairments. IEEE Sel Top Quantum Electron 12(4):544–554
52. Mizuochi T, Konishi Y, Miyata Y, Inoue T, Onohara K, Kametani S, Sugihara T, Kubo K, Yoshida H, Kobayashi T, Ichikawa T (2009) Experimental demonstration of concatenated LDPC and RS codes by FPGAs emulation. IEEE Photon Technol Lett 21(18):1302–1304
53. Djordjevic IB, Vasic B (2006) Multilevel coding in M-ary DPSK/differential QAM high-speed optical transmission with direct detection. IEEE/OSA J Lightwave Technol 24:420–428
54. Bahl LR, Cocke J, Jelinek F, Raviv J (1974) Optimal decoding of linear codes for minimizing symbol error rate. IEEE Trans Inform Theory IT-20(2):284–287
55. ten Brink S (2001) Convergence behavior of iteratively decoded parallel concatenated codes. IEEE Trans Commun 40:1727–1737

Chapter 2
Fundamentals of Optical Communication

The ultimate goal of the optical signal transmission is to achieve the predetermined bit-error ratio (BER) between any two nodes in an optical network. The optical transmission system has to be properly designed in order to provide the reliable operation during its lifetime, which includes the management of key engineering parameters.

In this chapter, we describe the key optical components used in contemporary optical communication systems; basic direct detection modulation schemes; and basic coherent detection schemes. This chapter is based on [1–24].

The chapter is organized as follows. After the brief introduction, in Sect. 2.2 the key optical components are identified and described at a level sufficient to understand other chapters in the book without having any background in optical communications. In Sect. 2.3, different modulation formats with direct detection are described. Section 2.4 is devoted to different modulation schemes with coherent detection.

2.1 Introduction

The optical transmission system design [1–5] involves accounting for different effects that may degrade the signal during modulation, propagation, and detection processes. The transmission quality is assessed by the received *signal-to-noise ratio* (SNR), which is the ratio between signal power and noise power at the decision point. The SNR is related to the receiver *sensitivity*, the minimum received optical power needed to keep SNR at the specified level.

In digital optical communications, bit-error rate (BER), defined as the ratio of bits in error to total number of transmitted bit at the decision point, is commonly used as a figure of merit. In that sense, the *receiver sensitivity* is defined as the minimum required received optical power to keep BER below a given value. The three *types of parameters* important *from the system engineering point of view* include (1) optical signal parameters that determine the signal level, (2) the optical noise parameters that determine the BER, and (3) the impairment parameters that determine the power margin to be allocated to compensate for their impact. The optical

I. Djordjevic et al., *Coding for Optical Channels*,
DOI 10.1007/978-1-4419-5569-2_2,

signal parameters defining the signal level include optical transmitter output power, extinction ratio, optical amplification gain, and photodiode responsivity. The total noise is a stochastic process composed of both additive noise components and multiplicative (nonadditive) noise components. There exist a number of impairments that deteriorate the signal quality during transmission such as fiber attenuation, chromatic dispersion, polarization mode dispersion (PMD), polarization-dependent loss (PDL), fiber nonlinearities, insertion loss, and frequency chirp; just to mention a few. A proper design process involves different steps to provide a prespecified transmission system quality and to balance different system parameters. The systems parameter can be related to power, time, wavelength, or combination of them. Given this general description of different signal and noise parameters, we turn our attention to the key optical components.

2.2 Key Optical Components

This section describes the basic optical components used in an optical transmission system. An exemplary optical network identifying the key optical components is shown in Fig. 2.1. The end-to-end optical transmission involves both electrical and optical signal paths. To perform conversion from electrical to optical domain, the optical transmitters are used, while to perform conversion in opposite direction (optical to electrical conversion), the optical receivers are used. The optical fibers serve as foundation of an optical transmission system because the optical fiber is used as a medium to transport the optical signals from source to destination. The optical fibers attenuate the signal during transmission, and someone has to use optical amplifiers, such as erbium-doped fiber amplifiers (EDFAs), Raman amplifiers, or parametric amplifiers, to restore the signal quality [1–5]. However, the process of amplification is accompanied with noise addition. The simplest optical transmission system employs only one wavelength. The wavelength division multiplexing (WDM) can

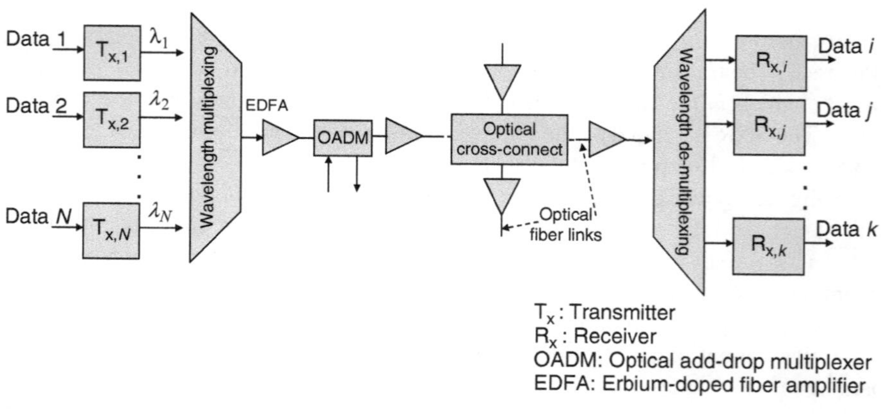

Fig. 2.1 An exemplary optical network identifying key optical components

be considered as an upgrade of the single-wavelength system. WDM corresponds to the scheme in which multiple optical carriers at different wavelengths are modulated by using independent electrical bit streams, as shown in Fig. 2.1, and then transmitted over the same fiber. WDM has potential of exploiting the enormous bandwidth offered by the optical fiber. During transmission of WDM signals, occasionally one or several wavelengths are to be added or dropped, which is performed by the optical component known as optical add–drop multiplexer (OADM), as illustrated in Fig. 2.1. The optical networks require the switching of information among different fibers, which is performed in optical cross-connect (OXS). To combine several distinct wavelength channels into composite channel, the wavelength multiplexers are used. On the other hand, to split the composite WDM channel into distinct wavelength channels, the wavelength demultiplexers are used. To impose the information signal, optical modulators are used. The optical modulators are commonly used in combination with semiconductor lasers.

The typical receiver configuration with direct detection is shown in Fig. 2.2 [1–5]. The main purpose of the optical receiver, terminating the lightwave path, is to convert the signal coming from single-mode fiber from optical to electrical domain and process appropriately such obtained electrical signal to recover the data being transmitted. The incoming optical signal may be preamplified by an optical amplifier and further processed by an optical filter to reduce the level of amplified spontaneous emission (ASE) noise or by wavelength demultiplexer to select a desired wavelength channel. The optical signal is converted into electrical domain by using a photodetector, followed by an electrical postamplifier. To deal with residual intersymbol interference (ISI), an equalizer may be used. The main purpose of clock recovery circuit is to provide timing for decision circuit by extracting the clock from the received signal. The clock recovery circuit is most commonly implemented using the phase-lock loop (PLL). Finally, the purpose of decision circuit is to provide the binary sequence being transmitted by comparing the sampled signal to a predetermined threshold. Whenever the received sample is larger than the threshold, the decision circuit decides in favor of bit 1, otherwise in favor of bit 0.

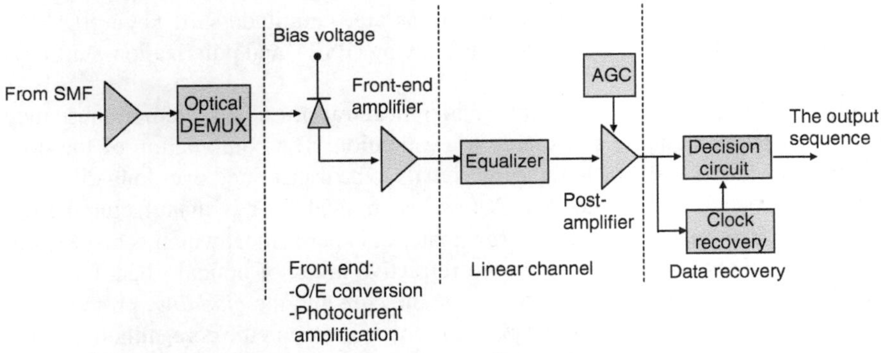

Fig. 2.2 A typical direct detection receiver architecture. *O/E* optical to electrical and *AGC* automatic gain control

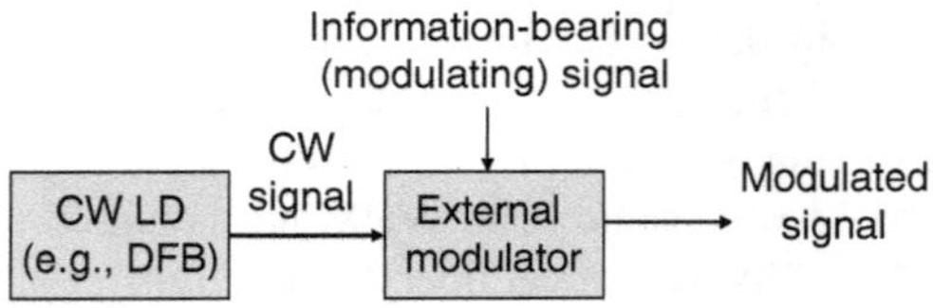

Fig. 2.3 An illustration of external modulation principle. *DFB* distributed feedback laser

The optical signal generated by semiconductor laser has to be modulated by information signal before being transmitted over the optical fiber. This can be achieved by *directly modulating* the bias current of semiconductor laser, which can be done even at high speed (even up to 40 Gb/s in certain lasers). Unfortunately, this concept although conceptually simple is rarely used in practice because of the frequency chirp introduced by direct modulation, nonuniform frequency response, and large current swing needed to provide operation. For transmitters operating at 10 Gb/s and above, instead, the semiconductor laser diode (LD) is commonly biased at constant current to provide continuous wave (CW) output, and *external modulators* are used to impose the information signal to be transmitted. The most popular modulators are electro-optic optical modulators, such as Mach–Zehnder modulators, and electroabsorption modulators. The principle of the external modulator is illustrated in Fig. 2.3. Through the external modulation process, a certain parameter of the CW signal, used as a signal carrier, is varied in accordance with the information-bearing signal. For example, a monochromatic electromagnetic wave is commonly used as a carrier, and its electrical field $\boldsymbol{E}(t)$ can be represented by

$$\boldsymbol{E}(t) = \boldsymbol{p}A\cos(\omega t + \varphi), \tag{2.1}$$

where A, ω, and φ are amplitude, frequency, and phase, respectively; while $\boldsymbol{p}$ denotes the polarization orientation. Each of those parameters can be used to carry information, and the information-bearing signal can be either CW or discrete. If the information-bearing signal is CW, corresponding modulation formats are amplitude modulation (AM), frequency modulation (FM), phase modulation (PM), and polarization modulation (PolM). On the other hand, if the information-bearing signal is digital, the corresponding modulations are: amplitude-shift keying (ASK), frequency-shift keying (FSK), phase-shift keying (PSK), and polarization-shift keying (PolSK).

Optical fibers serve as foundation of an optical transmission system because they transport optical signals from source to destination. The combination of low-loss and large bandwidth allows high-speed signals to be transmitted over long distances before the regeneration is needed. A low-loss optical fiber is manufactured from several different materials; the base row material is pure silica, which is mixed with different dopants in order to adjust the refractive index of optical fiber. The optical fiber consists of two waveguide layers, the *core* and the *cladding*, protected by buffer coating. The majority of the power is concentrated in the core, although some portion can spread to the cladding. There exists a difference in refractive indices between the core and cladding, which is achieved by mixing dopants, commonly added to the fiber core. There exist two types of optical fibers: multimode fiber (MMF) and single-mode fiber (SMF). Multimode optical fibers transfer the light through a collection of spatial transversal modes. Each mode, defined through a specified combination of electrical and magnetic components, occupies a different cross section of the optical fiber core and takes a slightly distinguished path along the optical fiber. The difference in mode path lengths in multimode optical fibers produces a difference in arrival times at the receiving point. This phenomenon is known as *multimode dispersion* (or *intermodal dispersion*) and causes signal distortion and imposes the limitations in signal bandwidth. The second type of optical fibers, SMFs, effectively eliminates multimode dispersion by limiting the number of propagating modes to a *fundamental* one. SMFs, however, introduce another signal impairment known as the chromatic dispersion. *Chromatic dispersion* is caused by the difference in velocities among different spectral components within the same mode.

The attenuation of signal propagating through optical fiber is low compared to that through other transmission media, such as copper cables or free space. Nevertheless, we have to amplify the attenuated signal from time to time, to restore the signal level, without any conversion into electrical domain. This can be done in optical amplifiers, through the process of *stimulated emission*. The main ingredient of an optical amplifier is the *optical gain* realized through the amplifier pumping (being either electrical or optical) to achieve the so-called *population inversion*. The common types of optical amplifiers are semiconductor optical amplifiers (SOAs), EDFAs, and Raman amplifiers. The amplification process is commonly followed by the noise process, not related to the signal, which occurs due to spontaneous emission. The amplification process degrades the SNR, because of ASE added to the signal in every amplifier stage.

Before providing more details about basic building blocks identified in this section, let us give more global picture by describing a typical optical network shown in Fig. 2.4. We can identify three ellipses representing the *core network*, the *edge*

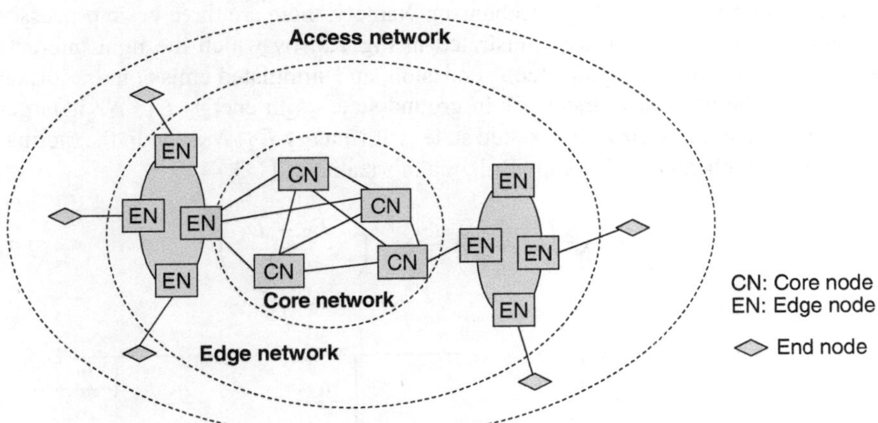

Fig. 2.4 A typical optical networking architecture

network, and the *access network* [1]. The long-haul core network interconnects big cities, major communications hubs, and even different continents by means of submarine transmission systems. The core networks are often called the wide area networks (WANs) or interchange carrier networks. The edge optical networks are deployed within smaller geographical areas and are commonly recognized as metropolitan area networks (MANs) or local exchange carrier networks. The access networks represent peripheral part of optical network and provide the last-mile access or the bandwidth distribution to the individual end-users. The common access networks are local area networks (LANs) and distribution networks. The common physical network topologies are *mesh* network (often present in core networks), *ring* network (in edge networks), and *star* networks (commonly used in access networks).

Given this general description of key optical components in the rest of this section, we provide more details about basic building blocks: optical transmitters are described in Sect. 2.2.1, optical receivers in Sect. 2.2.2, optical amplifiers in Sect. 2.2.3, optical fibers in Sect. 2.2.4, and other optical building blocks, such as multiplexers/demultiplexers, optical filters, OADMs, optical switches, couplers, etc., are described in Sect. 2.2.5.

2.2.1 Optical Transmitters

The role of the optical transmitter is to generate the optical signal, impose the information-bearing signal, and launch the modulated signal into the optical fiber. The semiconductor light sources are commonly used in state-of-the-art optical communication systems. The light generation process occurs in certain semiconductor materials due to recombination of electrons and holes in p–n junctions, under direct biasing. Depending on the nature of the recombination process, we can classify different semiconductor light sources as either light-emitting diodes (LEDs) in which spontaneous recombination dominates or semiconductor lasers in which the stimulated emission is a dominating mechanism. Namely, there are three basic processes in semiconductor materials, as illustrated in Fig. 2.5, by which the light interacts with matter: absorption, spontaneous emission, and stimulated emission. In normal conditions, the number of electrons in ground state (with energy E_1) N_1 is larger than the number of electrons in excited state (with energy E_2) N_2, and in the thermal equilibrium their ratio follows the Boltzmann's statistics [1,3,4]

$$\frac{N_2}{N_1} = \exp\left(-\frac{h\nu}{k_B T}\right) = \exp\left(-\frac{E_2 - E_1}{k_B T}\right), \tag{2.2}$$

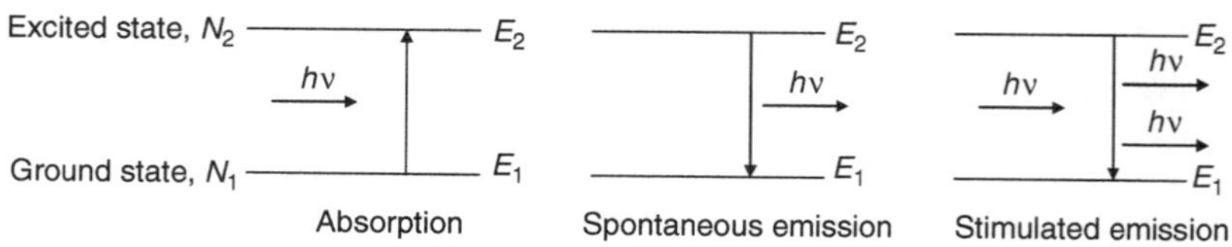

Fig. 2.5 Illustrating the interaction of the light with the matter

where $h\nu$ is a photon energy (h is the Plank's constant and ν is the optical frequency proportional to the energy difference between the energy levels $E_2 - E_1$), k_B is the Boltzmann's constant, and T is the absolute temperature. In the same regime, the spontaneous emission rate $dN_{2,\mathrm{spon}}/dt = A_{21}N_2$($A_{21}$ denotes the spontaneous emission coefficient) and the stimulated emission rate $dN_{2,\mathrm{stim}}/dt = B_{21}\rho(\nu)N_2$ (B_{21} denotes the stimulated emission coefficient and $\rho(\nu)$ denotes the spectral density of electromagnetic energy) are equalized with absorption rate $dN_{1,\mathrm{abs}}/dt = A_{12}\rho(\nu)N_1$ (A_{12} denotes the absorption coefficient):

$$A_{21}N_2 + B_{21}\rho(\nu)N_2 = B_{12}\rho(\nu)N_1. \tag{2.3}$$

In the visible or near-infrared region ($h\nu \sim 1\,\mathrm{eV}$), the spontaneous emission always dominates over stimulated emission in thermal equilibrium at room temperature ($k_B T \approx 25\,\mathrm{meV}$) [see (2.2)]. The stimulated emission rate can exceed absorption rate only when $N_2 > N_1$; the condition is referred to as *population inversion* and can never be realized for systems being in thermal equilibrium. The population inversion is a prerequisite for laser operation; and in atomic system, it is achieved by using *three- and four-levels pumping schemes* (an external energy source raises the atomic population from ground to an excited state). There are three basic components required to sustain stimulated emission and to form useful laser output: the pump source, the active medium, and the feedback mirrors. The active medium can be solid (such as in semiconductor lasers), gaseous, or liquid in nature. The pump can be electrical (e.g., semiconductor lasers), optical, or chemical. The purpose of the pump is to achieve the population inversion. The basic structure of semiconductor laser of Fabry–Perot type is shown in Fig. 2.6a, together with the equivalent model. The injection (bias) current flows through the p–n junction and stimulates the recombination of electrons and holes, leading to the generation of photons.

For the lasing action to be sustainable, the gain and phase matching condition should be satisfied. In the active medium, both *gain/absorption* described by $\gamma(\nu)$ and *scattering* described by α_s are present. The intensity inside the cavity can be described by the following dependence $I(z) = I_0 \exp[(\gamma(\nu) - \alpha_s)z]$. The lasing is possible when collective gain is larger than the loss after a round trip pass through the cavity:

$$I(2L) = I_0 R_1 R_2 \exp\left[2L(\gamma(\nu) - \alpha_s)\right] = I_0, \tag{2.4}$$

where R_1 and R_2 are facet reflectivities (see Fig. 2.6), L is the length of active medium, and I_0 and $I(2L)$ correspond to initial and round-trip intensities. The gain threshold is obtained by solving (2.4) per γ:

$$\gamma_{\mathrm{th}} = \alpha_s + \frac{1}{2L}\ln\left(\frac{1}{R_1 R_2}\right) = \alpha_{\mathrm{int}} + \alpha_{\mathrm{mir}}, \tag{2.5}$$

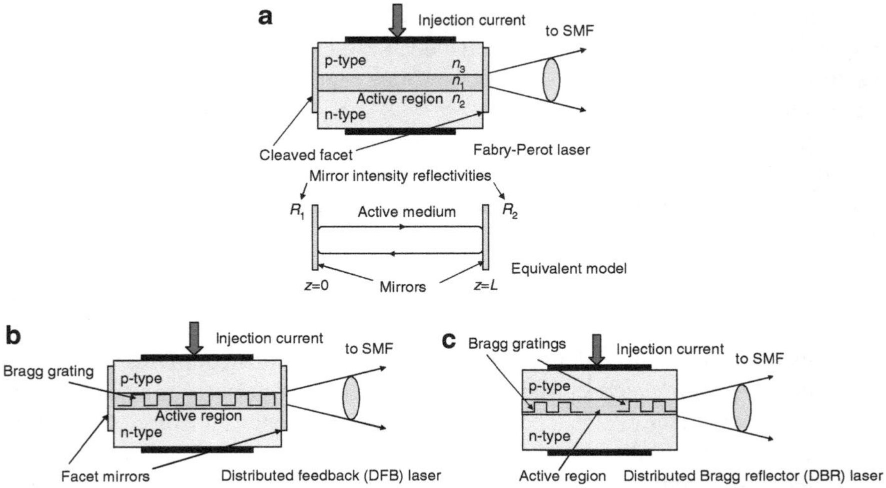

Fig. 2.6 Semiconductor lasers (**a**) Fabry–Perot semiconductor laser, (**b**) DFB laser, and (**c**) DBR laser

where the internal losses (corresponding to α_s) and mirror losses ($(1/2L)\ln(1/R_1R_2)$) are denoted by α_{int} and α_{mir}, respectively. After the round trip the resultant phase must be equal to the initial phase, leading to the phase-matching condition:

$$\exp[-j2\beta L] = 1 \Rightarrow 2\beta L = q2\pi, \beta = 2\pi n/\lambda, \tag{2.6}$$

where β denotes the propagation constant, n is the refractive index of active medium, and λ is the free-space wavelength. The phase-matching condition can be satisfied for many different integers q, representing different longitudinal modes of frequency $\nu_q = qc/(2nL)$. The separation between neighboring longitudinal modes is known as the *free-spectral range*:

$$\Delta\nu = \nu_q - \nu_{q-1} = q\frac{c}{2nL} - (q-1)\frac{c}{2nL} = \frac{c}{2nL}. \tag{2.7}$$

Because of the presence of longitudinal modes, the Fabry–Perot laser belongs to the class of multimode lasers. To improve the coherence of output light and the laser modulation speed, distributed feedback (DFB) lasers, shown in Fig. 2.6b, are used. The key idea of this laser is to effectively select one of the longitudinal modes while suppressing the remaining ones. This is achieved by introducing the Bragg grating inside of the laser cavity. The wavelength of selected longitudinal mode can be determined from the Bragg condition:

$$2\Lambda = m\frac{\lambda}{n_{av}} \Rightarrow \lambda_B = \frac{2n_{av}\Lambda}{m}, \tag{2.8}$$

where Λ is the grating period, n_{av} is the average refractive index of a waveguide mode, and λ/n_{av} is the average wavelength of the light in the waveguide mode. If the grating element is put outside of the active region or instead of the facet mirrors, the distributed Bragg reflector (DBR) laser is obtained, which is illustrated in Fig. 2.6c. Both DFB and DBR lasers belong to the class of single-mode lasers. Different semiconductor lasers shown in Fig. 2.6 are edge-emitting lasers.

Another important semiconductor laser type is vertical cavity surface emitting laser (VCSEL), which emits the light vertical to the active layer plane [1–4]. The VCSELs are usually based on In-GaAs-P layers acting as Bragg reflectors and provide the positive feedback leading to the stimulated emission.

The spectral curve of the single-mode lasers is a result of transition between discrete energy levels and can often be represented using the Lorentzian shape [1–4]:

$$g(\nu) = \frac{\Delta\nu}{2\pi\left[(\nu-\nu_0)^2 + \left(\frac{\Delta\nu}{2}\right)^2\right]}, \tag{2.9}$$

where ν_0 is the central optical frequency and $\Delta\nu$ represents the laser linewidth [1]:

$$\Delta\nu = \frac{n_{sp}G\left(1+\alpha_{chirp}^2\right)}{4\pi P}, \tag{2.10}$$

where n_{sp} is the spontaneous emission factor, G is the net rate of stimulated emission, P denotes the output power, and α_{chirp} is the chirp factor (representing the amplitude–phase coupling parameter).

The small-signal frequency response of the semiconductor laser is determined by [3, 4]

$$H(\omega) = \frac{\Omega_R^2 + \Gamma_R^2}{(\Omega_R + \omega - j\Gamma_R)(\Omega_R - \omega + j\Gamma_R)}, \tag{2.11}$$

where Γ_R is the damping factor and Ω_R is the relaxation frequency.

$$\Omega_R^2 \approx \frac{G_N P_b}{\tau_p} \quad (\Gamma_R \ll \Omega_R)$$

with G_N being the net rate of stimulated emission, P_b being the output power corresponding to the bias current, and τ_p being the photon lifetime related to the excited energy level. The modulation bandwidth (defined as 3-dB bandwidth) is therefore determined by the relaxation frequency

$$\omega_{3\,dB} = \sqrt{1+\sqrt{2}}\,\Omega_R,$$

and for fast semiconductor lasers it can be 30 GHz. Unfortunately, the direct modulation of semiconductor lasers leads to frequency chirp, which can be described by the instantaneous frequency shift from steady-state frequency ν_0 as follows [1]:

$$\delta v(t) = \frac{\alpha_{\text{chirp}}}{4\pi}\left[\frac{d}{dt}\ln P(t) + \chi P(t)\right] = C_{\text{dyn}}\frac{d}{dt}P(t) + C_{\text{ad}}P(t), \tag{2.12}$$

where $P(t)$ is the time variation of the output power, χ is the constant (varying from zero to several tens) related to the material and design parameters, and α_{chirp} is the chirp factor defined as the ratio between the refractive index n change and gain G change with respect to the number of carriers N: $\alpha_{\text{chirp}} = (dn/dN)/(dG/dN)$. The first term on the right-hand side of (2.12) represents dynamic (transient or instantaneous) chirp and the second term the adiabatic (steady-state) frequency chirp. The random fluctuation in carrier density due to spontaneous emission also leads to the linewidth enhancement proportional to $\left(1+\alpha_{\text{chirp}}^2\right)$. To avoid the chirp problem, the external modulation is used; while the semiconductor lasers are biased by a dc voltage to produce a continuous wave operation.

There are two types of external modulators commonly used in practice: Mach–Zehnder modulator (MZM) and electroabsorption modulator (EAM), whose operational principle is illustrated in Fig. 2.7. The MZM is based on electro-optic effect, the effect that in certain materials (such as $LiNbO_3$) where the refractive index n changes with respect to the voltage V applied across electrodes [4]:

$$\Delta n = -\frac{1}{2}\Gamma n^3 r_{33}(V/d_e) \Rightarrow \Delta\phi = \frac{2\pi}{\lambda}\Delta n L, \tag{2.13}$$

where Δn denotes the refractive index change, $\Delta\phi$ is corresponding phase change, r_{33} is the electro-optic coefficient (~30.9 pm/V in $LiNbO_3$), d_e is the separation of electrodes, L is the electrode length, and λ is the wavelength of the light. The MZM (see Fig. 2.7a) is a planar waveguide structure deposited on the substrate, with two pairs of electrodes (1) for high-speed ac voltage representing the modulation data (RF) signal and (2) for dc bias voltage. Let $V_1(t)$ and $V_2(t)$ denote the electrical drive signals on the upper and lower electrodes, respectively. The output electrical field $E_{\text{out}}(t)$ of the second Y-branch can be related to the input electrical filed E_{in} by

$$E_{\text{out}}(t) = \frac{1}{2}\left[\exp\left(j\frac{\pi}{V_\pi}V_1(t)\right) + \exp\left(j\frac{\pi}{V_\pi}V_2(t)\right)\right]E_{\text{in}}, \tag{2.14}$$

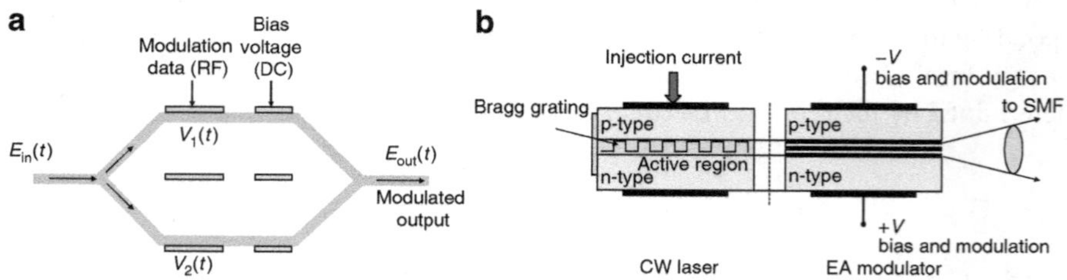

Fig. 2.7 External optical modulators (**a**) Mach–Zehnder modulator and (**b**) electroabsorption modulator

where V_π is differential drive voltage ($V_1 - V_2 = V_\pi$) resulting in differential phase shift of π rad between two waveguides. Possible modulation formats that can be used with this MZM include: on–off keying (OOK) with zero/nonzero chirp, binary phase-shift keying (BPSK), differential phase-shift keying (DPSK), quadrature phase-shift keying (QPSK), differential QPSK (DQPSK), and return-to-zero (RZ) with duty cycle 33%, 50%, or 67%. For example, for zero-chirp OOK or BPSK the following complementary drive signals should be applied:

$$V_1(t) = V(t) - V_\pi/2, V_2(t) = -V(t) + V_\pi/2 \Rightarrow \frac{E_{\text{out}}(t)}{E_{\text{in}}} = \sin\left(\frac{\pi V(t)}{V_\pi}\right).$$

The electroabsorption modulator (EAM) is a semiconductor-based planar waveguide composed of multiple p-type and n-type layers that form multiple quantum wells (MQWs). The basic design of EAM is similar to that of semiconductor lasers. The MQW is used to support the quantum-confined Stark effect (the absorption spectrum being a function of applied field) more effectively. Because of similarities of EAMs and semiconductor lasers design it is possible to fabricate them on the same substrate (see Fig. 2.7b), providing that EAM and laser are electrically isolated. Bandgap of quantum wells is larger than photon energy, so that the light is completely transmitted in the absence of bias, which corresponds to the ON state. When the reverse bias is applied the input signal is absorbed, which corresponds to the OFF state. The modulation speed of EAMs is typically comparable to the modulation speed of MZMs. However, the extinction ratio (the ratio of average powers corresponding to symbol 1 and symbol 0) is lower.

2.2.2 *Optical Receivers*

The purpose of the optical receiver is to convert the optical signal into electrical domain and to recover the transmitted data. The typical OOK receiver configuration is already given in Fig. 2.2. We can identify three different stages: front-end stage, the linear channel stage, and data recovery stage. The front-end stage is composed of a photodetector and a preamplifier. The most commonly used front-end stages are high-impedance front-end and transimpedance front-end, both shown in Fig. 2.8. High-impedance front end (Fig. 2.8a) employs a large value load resistance to reduce the level of thermal noise and has a good receiver sensitivity. However, the bandwidth of this scheme is low because the RC constant is large. To achieve both the high receiver sensitivity and large bandwidth, the transimpedance front-end scheme, shown in Fig. 2.8b, is used. Even though the load resistance is high, the negative feedback reduces the effective input resistance by a factor of $G - 1$, where G is the front-end amplifier gain. The bandwidth is increased for the same factor compared to high-impedance front-end scheme.

The photodiode is an integral part of both front-end stage schemes. The key role of the photodiode is to absorb photons in incoming optical signal and convert

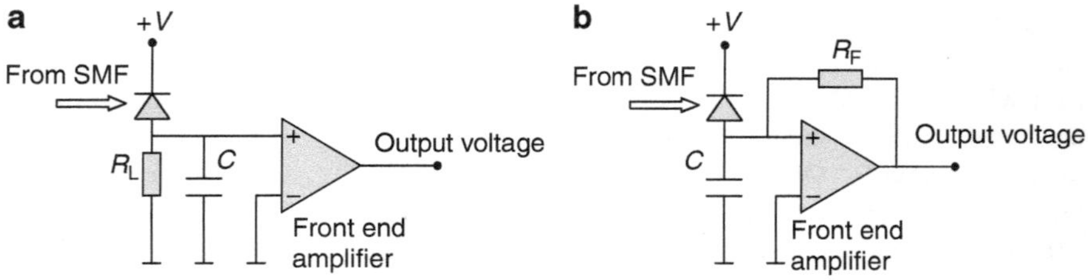

Fig. 2.8 Optical receiver front-end stage schemes (**a**) high-impedance front-end and (**b**) transimpedance front-end

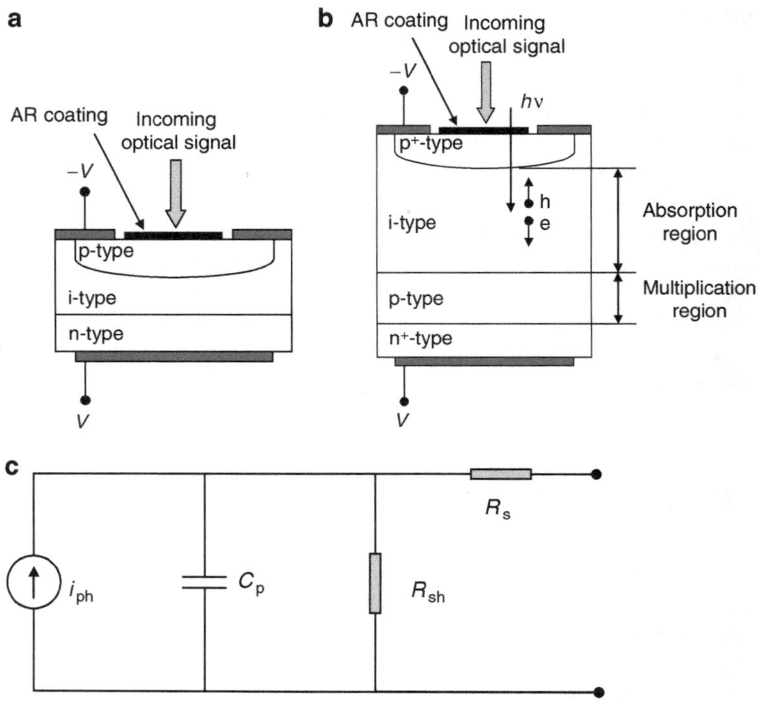

Fig. 2.9 The semiconductor photodiodes (**a**) p–i–n photodiode and (**b**) avalanche photodiode. (**c**) The equivalent p–i–n photodiode model

them back to the electrical level through the process opposite to the one taking place in semiconductor lasers. The common photodiodes are p–n photodiode, p–i–n photodiode, avalanche photodiode (APD), and metal–semiconductor–metal (MSM) photodetectors [3, 4]. *p-n photodiode* is based on a reverse-biased p–n junction. The thickness of the depletion region is often less than the absorption depth for incident light, and the photons are absorbed outside of the depletion region, leading to the slow response speed. *p–i–n photodiode* consists of an intrinsic region sandwiched between p- and n-type layers, as shown in Fig. 2.9a. Under the reverse bias, the depletion depth can be made sufficiently thick to absorb most of the incident photons. *Avalanche photodiode*, shown in Fig. 2.9b, is a modified p–i–n photodiode that is operated at very high reverse bias. Under high-field conditions, photogenerated carriers induce generation of *secondary electron–hole pairs* by the process of *impact ionization*, and this process leads to internal electrical gain. *MSM photodetectors* employ interdigitated Schottky barrier contacts on one face of the device and are compatible with planar processing and optoelectronic integration. Depending on the device design, the device is illuminated through the p- or n-type contact. In Si, Ge, or GaAs diodes, the substrate is absorbing so that the device has to be illuminated through the top contact, as shown in Fig. 2.9a. On the other hand, in InGaAs or InGaAsP, the substrate is transparent, and the device can be designed to be illuminated either through the substrate or through the top contact. In order to increase the depletion region and to minimize the diffusion current component, an intrinsic layer (i-type) is introduced to the p–i–n photodiode structure. The p–i–n photodiode is reverse biased and has very high internal impedance, meaning that it acts as a current source generating the photocurrent proportional to the incoming optical signal power. The equivalent scheme of p–i–n photodiode is shown in Fig. 2.9c. Typically the internal series resistance R_s is low, while the internal shunt resistance is high, so that the junction capacitance C_p dominates and can be determined by

$$C_p = \varepsilon_s \frac{A}{w} = \left[\frac{\varepsilon_s N_A N_D}{2(N_A - N_D)(V_0 - V_A)} \right]^{1/2}, \tag{2.15}$$

where ε_s is the semiconductor permittivity, A is the area of the space charge region (SCR), w is the width of SCR, N_A and N_D denote dopant (acceptor and donor) densities, V_0 is the built-in potential across the junction, and V_A is applied negative voltage. The photocurrent $i_{ph}(t)$ is proportional to the power of incident light $P(t)$, that is $i_{ph}(t) = RP(t)$, where R [A/W] is the photodiode responsivity. The photodiode responsivity is related to the quantum efficiency η, defined as the ratio of number of generated electrons and the number of incident photons, by $R = \eta q/h\nu$, where q is an electron charge and $h\nu$ is a photon energy. Using this model we can determine the 3-dB bandwidth of high-impedance front-end scheme as $B_{3\,dB} = 1/(2\pi R_L C_p)$ and the 3-dB bandwidth of transimpedance front-end scheme as $B_{3\,dB} = (G+1)/(2\pi R_F C_p)$, which is G times higher than bandwidth of high-impedance front-end scheme.

2.2.3 Optical Fibers

Optical fibers serve as foundation of an optical transmission system because they transport optical signals from source to destination. The combination of low-loss and extremely large bandwidth allows high-speed signals to be transmitted over long distances before the regeneration becomes necessary. A low-loss optical fiber is manufactured from several different materials; the base row material is pure silica,

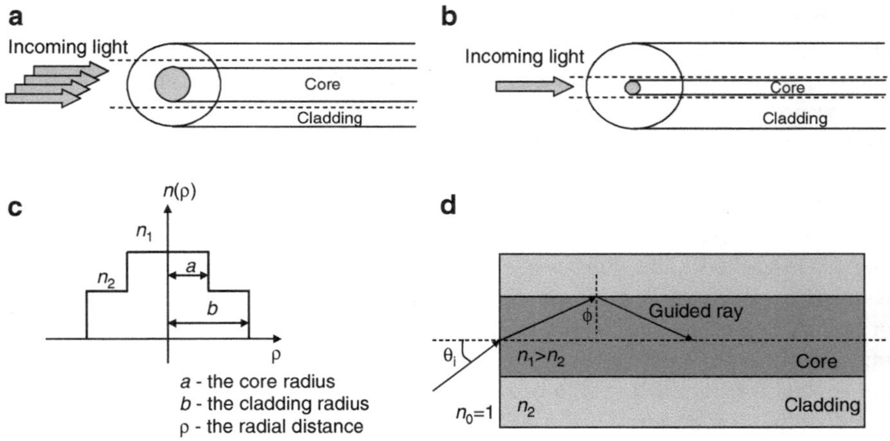

Fig. 2.10 Optical fibers (**a**) multimode optical fiber, (**b**) single-mode optical fiber, (**c**) refractive-index profile for step-index fiber, and (**d**) the light confinement in step-index fibers through the total internal reflection

which is mixed with different dopants in order to adjust the refractive index of optical fiber. The optical fiber, shown in Fig. 2.10, consists of two waveguide layers, the *core* (of refractive index n_1) and the *cladding* (of refractive index n_2), protected by the jacket (the buffer coating). The majority of the power is concentrated in the core, although some portion can spread to the cladding. There is a difference in refractive indices between the core and cladding ($n_1 > n_2$), which is achieved by a mix of dopants commonly added to the fiber core. The refractive-index profile for step-index fiber is shown in Fig. 2.10c, while the illustration of light confinement by the total internal reflection is shown in Fig. 2.10d. The ray will be totally reflected from the core–cladding interface (a guided ray) if the following condition is satisfied:

$$n_0 \sin \theta_\mathrm{i} < \sqrt{n_1^2 - n_2^2},$$

where θ_i is the angle of incidence. $\max(n_0 \sin \theta_\mathrm{i})$ defines the light gathering capacity of an optical fiber and it is called the *numerical aperture* (NA):

$$\mathrm{NA} = \sqrt{n_1^2 - n_2^2} \approx n_1 \sqrt{2\Delta}, \quad \Delta = 1,$$

where Δ is the *normalized index difference* defined as $\Delta = (n_1 - n_2)/n_1$. Therefore, from the geometrical optics point of view, light propagates in optical fiber due to series of total internal reflections that occur at the core–cladding interface. The smallest angle of incidence ϕ (see Fig. 2.10d) for which the total internal reflection occurs is called the *critical angle* and equals $\sin^{-1} n_2/n_1$.

There exist two types of optical fibers: MMF (shown in Fig. 2.10a) and SMF (shown in Fig. 2.10b). Multimode optical fibers transfer the light through a collection of spatial transversal modes. Each mode, defined through a specified combination of electrical and magnetic components, occupies a different cross section of the optical fiber core and takes a slightly distinguished path along the optical fiber. The difference in mode path lengths in multimode optical fibers produces a difference in arrival times at the receiving point. This phenomenon is known as *multimode dispersion* (or *intermodal dispersion*) and causes signal distortion and imposes the limitations in signal bandwidth. The second type of optical fibers, SMFs, effectively eliminates multimode dispersion by limiting the number of propagating modes to a *fundamental* one. The *fundamental mode* occupies the central portion of the optical fiber and has an energy maximum at the axis of the optical fiber core. Its radial distribution can be approximated by Gaussian curve. The number of modes (M) that can effectively propagate through an optical fiber is determined by the *normalized frequency* (*V parameter* or V number): $M \approx V^2/2$, when V is large. The normalized frequency is defined by

$$V = \frac{2\pi a}{\lambda}\sqrt{n_1^2 - n_2^2}, \tag{2.16}$$

where a is the fiber core radius, λ is the carrier wavelength, and n_1 and n_2 are refractive indices related to the fiber core and the fiber cladding, respectively.

Each mode propagating through the fiber is characterized by its own propagation constant β. The dependence of the electric and magnetic fields on axial coordinate z is expressed through the factor $\exp(-j\beta z)$. The propagation constant must satisfy the following condition:

$$2\pi n_2/\lambda < \beta < 2\pi n_1/\lambda. \tag{2.17}$$

In order to evaluate the transmission characteristics of the optical fiber, the functional dependence of the mode propagation constant on the optical signal wavelength has to be known. The *normalized propagation constant b* is defined for that purpose:

$$b = \frac{\beta^2 - (2\pi n_2/\lambda)^2}{(2\pi n_1/\lambda)^2 - (2\pi n_2/\lambda)^2}. \tag{2.18}$$

The normalized propagation constant is related to the normalized frequency V by [1,3]

$$b(V) \approx (1.1428 - 0.9960/V)^2, \quad 1.5 \le V \le 2.5. \tag{2.19}$$

The multimode dispersion can effectively be eliminated by limiting the number of propagating modes to a fundamental one: $V \le V_c = 2.405$ with V_c being the cutoff frequency. The cutoff frequency is controlled by keeping the core radius small and the normalized index difference $\Delta = (n_1 - n_2)/n_1$ between 0.2 and 0.3%.

2.2.4 Optical Amplifiers

The purpose of an optical amplifier is to restore the signal power level, reduced due to losses during propagation, without any optical to electrical conversion. The general form of an optical amplifier is given in Fig. 2.11a. Most optical amplifiers amplify incident light through the *stimulated emission*, the same mechanism that is used in lasers, but without the feedback mechanism. The main ingredient is the *optical gain* realized through the amplifier pumping (electrical or optical) to achieve the *population inversion*. The optical gain, generally speaking, is not only a function of frequency, but also a function of local beam intensity. To illustrate the basic concepts, we consider the case in which the gain medium is modeled as *two-level system*, as shown in Fig. 2.11b. The *amplification factor* G is defined as the ratio of amplifier output P_{out} and input P_{in} powers $G = P_{\text{out}}/P_{\text{in}}$. The amplification factor can be determined by knowing the dependence of evolution of power through the gain media [3]:

$$\frac{\mathrm{d}P}{\mathrm{d}z} = gP, \quad g(\omega) = \frac{g_0}{1 + (\omega - \omega_0)^2 T_2^2 + P/P_{\mathrm{S}}}, \tag{2.20}$$

g is the gain coefficient, g_0 is the gain peak value, ω_0 is the atomic transition frequency, T_2 is the dipole relaxation time (<1 ps), ω is the optical frequency of incident signal, P is the incident signal power, and P_{S} is the saturation power.

In the unsaturated regime ($P \ll P_{\mathrm{S}}$), the differential equation (2.20) can be solved by separation of variables to get the following dependence of power $P(z) = P(0)\exp(gz)$, so that the amplification factor can be obtained by

$$G(\omega) = \exp[g(\omega)L], \tag{2.21}$$

and corresponding full-width half-maximum (FWHM) bandwidth is determined by

$$\Delta\nu_{\mathrm{A}} = \Delta\nu_{\mathrm{g}}\sqrt{\frac{\ln 2}{\ln(G_0/2)}}, \quad G_0 = \exp(g_0 L), \tag{2.22}$$

where $\Delta\nu_{\mathrm{g}}$ is the FWHM gain coefficient bandwidth.

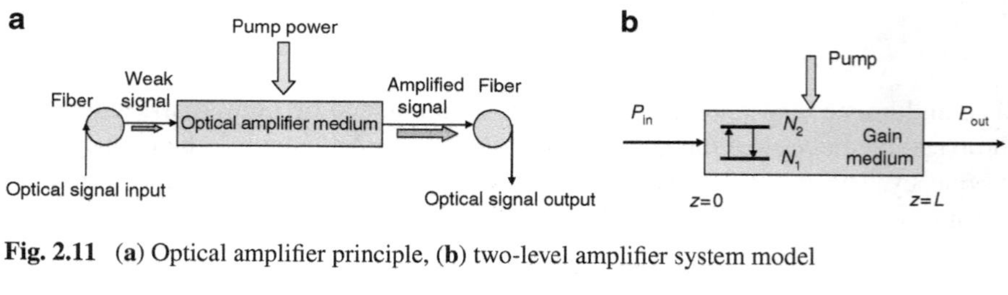

Fig. 2.11 (**a**) Optical amplifier principle, (**b**) two-level amplifier system model

The *gain saturation* comes from the power dependence of the gain coefficient (2.20). The coefficient is reduced when the incident power P becomes comparable with the saturation power P_S. Let us assume that the incident frequency is tuned to the peak gain ($\omega = \omega_0$), then from (2.20) we obtain

$$\frac{dP}{dz} = \frac{g_0 P}{1 + P/P_S}. \tag{2.23}$$

By solving the differential equation (2.23) with respect to the *boundary conditions*, $P(0) = P_{in}$, and $P(L) = P_{out} = GP_{in}$, we get

$$G = G_0 \exp\left[-\frac{G-1}{G}\frac{P_{out}}{P_S}\right]. \tag{2.24}$$

From (2.24) we can determine another important optical amplifier parameter, the *output saturation power* as being the optical power at which the gain G is reduced to $G_0/2$ (3 dB down):

$$P_{out}^{sat} = \frac{G_0 \ln 2}{G_0 - 2} P_S \approx (\ln 2) P_S \approx 0.69 P_S (G_0 > 20\,\text{dB}). \tag{2.25}$$

Three common applications of optical amplifiers are (1) power boosters (of transmitters), (2) in-line amplifiers, (3) optical preamplifiers, which is illustrated in Fig. 2.12. The booster (power) amplifiers are placed at the optical transmitter side to enhance the transmitted power level or to compensate for the losses of optical elements between the laser and optical fibers, such as optical coupler, WDM multiplexers, and external optical modulators. The in-line amplifiers are placed along the transmission link to compensate for the losses incurred during propagation of optical signal. The optical preamplifiers are used to increase the signal level before the photodetection takes place, improving therefore the receiver sensitivity.

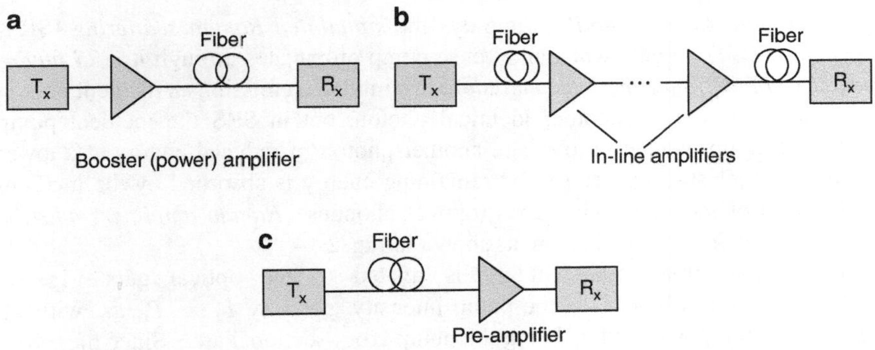

Fig. 2.12 Possible application of optical amplifiers (**a**) booster amplifier, (**b**) in-line amplifiers, and (**c**) preamplifier

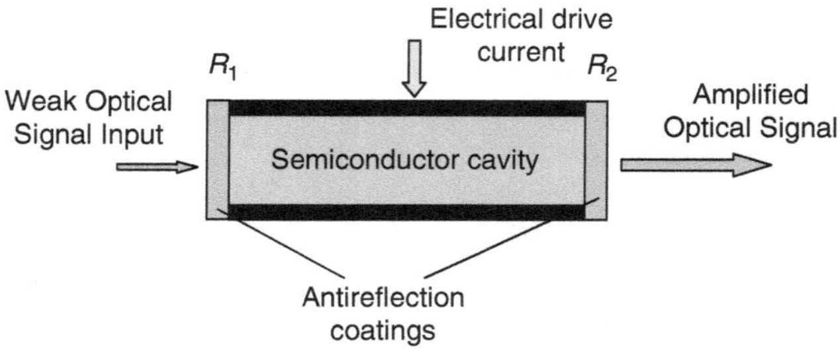

Fig. 2.13 The SOA operation principle

Several types of optical amplifiers have been introduced so far: *SOAs*, *fiber Raman* (and Brillouin) *amplifiers*, rare-earth-doped fiber amplifiers (erbium-doped *EDFA* operating at 1,500 nm and praseodymium-doped PDFA operating at 1,300 nm), and *parametric amplifiers*. Semiconductor lasers act as amplifiers before reaching the threshold. To prevent the lasing, *antireflection* (*AR*) *coatings* are used in SOAs, as shown in Fig. 2.13. Even with AR coating, the multiple reflections must be included when considering the Fabry–Perot (FP) cavity. The gain of FP amplifiers is given by [3]

$$G_{\mathrm{FP}}(\nu) = \frac{(1-R_1)(1-R_2)G(\nu)}{\left(1-G(\nu)\sqrt{R_1R_2}\right)^2 + 4G\sqrt{R_1R_2}\sin^2\left[\pi(\nu-\nu_{\mathrm{m}})/\Delta\nu_{\mathrm{L}}\right]}, \tag{2.26}$$

where R_1 and R_2 denote the facet reflectivities, $G(\nu)$ is the single-pass amplification factor, ν_{m} is the cavity resonance frequency, and $\Delta\nu_{\mathrm{L}}$ is the free-spectral range.

FP amplifier bandwidth can be determined as follows [3]:

$$\Delta\nu_{\mathrm{A}} = \frac{2\Delta\nu_{\mathrm{L}}}{\pi}\sin^{-1}\left[\frac{1-G\sqrt{R_1R_2}}{\left(4G\sqrt{R_1R_2}\right)^{1/2}}\right]. \tag{2.27}$$

A *fiber-based Raman amplifier* employs the *stimulated Raman scattering* (SRS) occurring in silica fibers when an intense pump propagates through it. *SRS fundamentally differs from stimulated emission*: in stimulated emission an incident photon stimulates emission of another identical photon, but in SRS the incident pump photon gives up its energy to create another photon of reduced energy at a lower frequency (inelastic scattering); the remaining energy is absorbed by the medium in the form of molecular vibrations (optical phonons). *Raman amplifiers must be pumped optically* to provide gain, as shown in Fig. 2.14.

The Raman-gain coefficient g_{R} is related to the optical gain $g(z)$ as $g(z) = g_{\mathrm{R}}I_{\mathrm{p}}(z)$, where I_{p} is the pump intensity given by $I_{\mathrm{p}} = P_{\mathrm{p}}/a_{\mathrm{p}}$, with P_{p} being the pump power and a_{p} being the pump cross-sectional area. Since the cross-sectional area is different for different types of fibers, the ratio $g_{\mathrm{R}}/a_{\mathrm{p}}$ is the measure of the Raman-gain efficiency. The DCF efficiency can even be eight times better

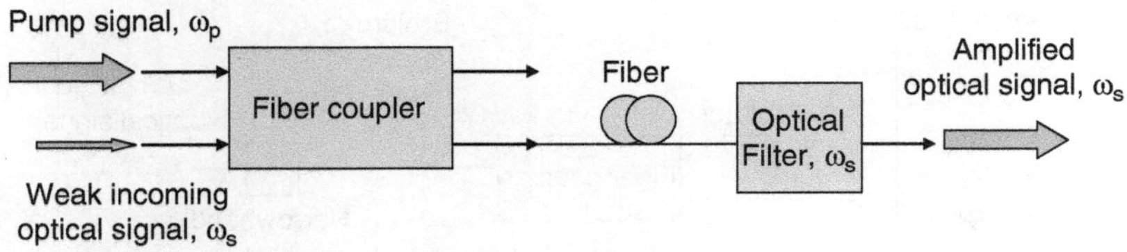

Fig. 2.14 The Raman amplifier operation principle in forward-pumping configuration

than that of a standard SMF, as shown in [3]. The evolution of the pump P_p and signal P_s powers (in distance z) can be studied by solving the system of *coupled differential equations* below [1,3]:

$$\frac{dP_s}{d_z} = -\alpha_s P_s + \frac{g_R}{a_p} P_p P_s, \quad \frac{dP_p}{dz} = -\alpha_p P_p - \frac{\omega_p}{\omega_s}\frac{g_R}{a_p} P_p P_s, \tag{2.28}$$

where a_s denotes the signal cross-sectional area and ω_p and ω_s denote the pump and signal frequency, respectively, while other parameters are already introduced above.

In *small-signal amplification regime* (when the pump depletion can be neglected), the pump power evolution is exponential, $P_p(z) = P_p(0)\exp[-\alpha_p z]$, so that the *Raman amplifier gain* is found to be

$$G_A = \frac{P_s(0)\exp(g_R P_p(0) L_{eff}/a_p - \alpha_s L)}{P_s(0)\exp(-\alpha_s L)} = \exp(g_0 L),$$
$$g_0 = g_R \frac{P_p(0)}{a_p}\frac{L_{eff}}{L} \approx \frac{g_R P_p(0)}{a_p \alpha_p L}\ (\alpha_p L \ll 1). \tag{2.29}$$

The origin of saturation in Raman amplifiers is pump power depletion, which is quite different from that in SOAs. *Saturated amplifier gain* G_S can be determined (assuming $\alpha_p = \alpha_s$) by [3]

$$G_S = \frac{1 + r_0}{r_0 + 1/G_A^{1+r_0}}, \quad r_0 = \frac{\omega_p}{\omega_s}\frac{P_s(0)}{P_p(0)}. \tag{2.30}$$

The amplifier gain is reduced down by 3 dB when $G_A r_0 \approx 1$, the condition that is satisfied when the amplified signal power becomes comparable to the input pump power $P_s(L) = P_p(0)$. Typically $P_0 \sim 1$ W, and channel powers in a WDM systems are around 1 mW, meaning that *Raman amplifier operates in unsaturated or linear regime*.

The *rare-earth doped fiber amplifiers* are finding increasing importance in optical communication systems. The most important class is EDFAs due to their ability to amplify in 1.55-μm wavelength range. The active medium consists of 10–30 m length of optical fiber highly doped with a rare-earth element, such as erbium (Er), ytterbium (Yb), neodymium (Nd), or praseodymium (Pr). The host fiber material

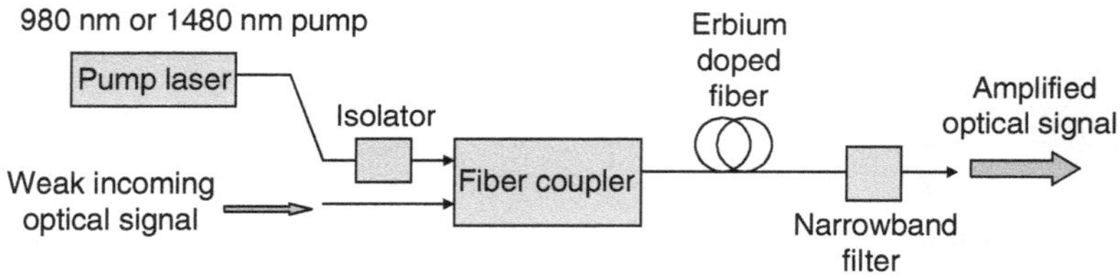

Fig. 2.15 The general EDFA configuration

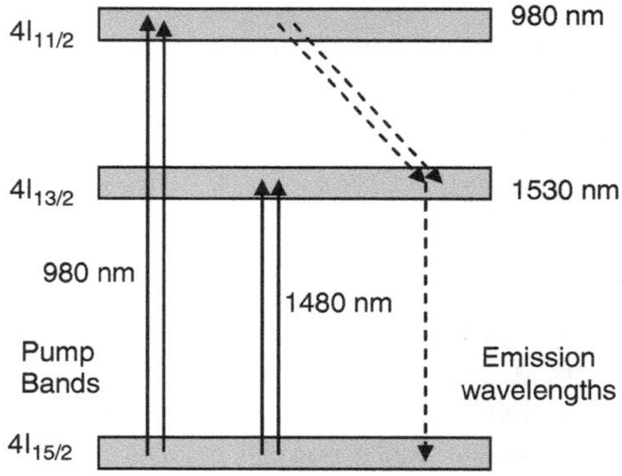

Fig. 2.16 The EDFA energy diagram

can be pure silica, a fluoride-based glass, or a multicomponent glass. General EDFA configuration is shown in Fig. 2.15.

The pumping at a suitable wavelength provides gain through *population inversion*. The gain spectrum depends on the pumping scheme as well as on the presence of other dopants, such as Ge or Al within the core. The amorphous nature of silica broadens the energy levels of Er^{3+} into the bands, as shown in Fig. 2.16.

The pumping is primarily done in optical domain with the *primary pump wavelengths* at 1.48 μm and 0.98-μm. The atoms pumped to the $4I_{11/2}$ level (with 980 nm pump) decay to the primary emission transition band. The pumping with 1.48 μm is directly excited to the upper transition levels of the emission band.

EDFAs can be designed to operate in such a way that the pump and signal travel in opposite directions; this configuration is commonly referred to as *backward pumping*. In *bidirectional pumping*, the amplifier is pumped in both directions simultaneously by using two semiconductor lasers located at both fiber ends.

2.2.5 *Other Optical Components*

Different optical components can be classified, depending on whether they can operate without an external electric power source or not, into two broad categories: passive or active. Important active components are lasers, external modulators, optical amplifiers, photodiodes, optical switches, and wavelength converters. On the

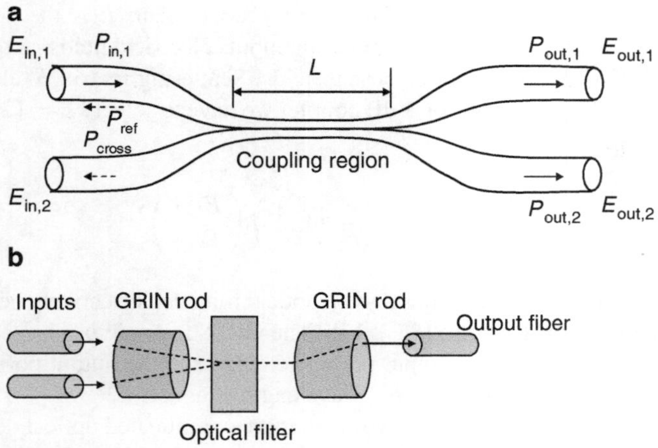

Fig. 2.17 Optical couplers (**a**) fiber fusing based and (**b**) GRIN rod based

other hand, important passive components are optical couplers, isolators, multiplexers/demultiplexers, and filters. Some components, such as optical filters, can be either passive or active depending on operational principle. In this section, we will briefly explain some important optical components not being described in previous sections.

The 2×2 optical coupler is a fundamental device that can either be implemented using the fiber fusing or be based on graded-index (GRIN) rods and optical filters, as shown in Fig. 2.17. The fused optical couplers (shown in Fig. 2.17a) are obtained when the cladding of two optical fibers are removed, the cores are brought together, and then heated and stretched. The obtained waveguide structure can exchange energy in the coupling region between the branches. If both inputs are used 2×2 coupler is obtained, if only one input is used 1×2 coupler is obtained. The optical couplers are recognized either as optical tap (1×2) couplers or directional (2×2) couplers. The power coupler splitting ratio depending on purpose can be different with typical values being 50%/50%, 10%/90%, 5%/95%, and 1%/99%. Directional coupler parameters (defined when only input 1 is active) are splitting ratio $P_{out,1}/(P_{out,1} + P_{out,2})$, excess loss $10\log_{10}[P_{in,1}/(P_{out,1} + P_{out,2})]$, insertion loss $10\log_{10}(P_{in,i}/P_{out,j})$, and crosstalk $10\log_{10}(P_{cross}/P_{in,1})$. The operation principle of directional coupler can be explained using coupled mode theory [15] or simple scattering (propagation) matrix $\boldsymbol{S}$ approach, assuming that a coupler is lossless and reciprocal device:

$$\begin{pmatrix} E_{out,1} \\ E_{out,2} \end{pmatrix} = \boldsymbol{S} \begin{pmatrix} E_{in,1} \\ E_{in,2} \end{pmatrix} = \begin{pmatrix} s_{11} & s_{12} \\ s_{21} & s_{22} \end{pmatrix} \begin{pmatrix} E_{in,1} \\ E_{in,2} \end{pmatrix}$$
$$= e^{-j\beta L} \begin{pmatrix} \cos(kL) & j\sin(kL) \\ j\sin(kL) & \cos(kL) \end{pmatrix} \begin{pmatrix} E_{in,1} \\ E_{in,2} \end{pmatrix}, \tag{2.31}$$

where β is propagation constant, k is coupling coefficient, L is the coupling region length, $E_{\text{in},1}$ and $E_{\text{in},2}$ are corresponding inputs electrical fields, and $E_{\text{out},1}$ and $E_{\text{out},2}$ are corresponding output electrical fields. Scattering matrix $\boldsymbol{S}$ elements are denoted with s_{ij}. For example, for 3-dB coupler we have to select $kL = (2m+1)\pi/4$ (m is a positive integer) to get

$$\begin{pmatrix} E_{\text{out},1} \\ E_{\text{out},2} \end{pmatrix} = \frac{1}{\sqrt{2}} \begin{pmatrix} 1 & \mathrm{j} \\ \mathrm{j} & 1 \end{pmatrix} \begin{pmatrix} E_{\text{in},1} \\ E_{\text{in},2} \end{pmatrix}. \tag{2.32}$$

The combination of two GRIN rods and an optical filter can effectively be used as an optical coupler, as illustrated in Fig. 2.17b. The GRIN rods are used as collimators, to collimate the light from two input ports and deliver to the output port, while the optical filter is used to select a desired wavelength channel.

The optical couplers can be used to create more complicated optical devices such as $M \times N$ optical stars, directional optical switches, different optical filters, multiplexers, etc.

An *optical filter* modifies the spectrum of incoming light and can mathematically be described by corresponding transfer function $H_{\text{of}}(\omega)$:

$$E_{\text{out}}(t) = \frac{1}{2\pi} \int_{-\infty}^{\infty} \tilde{E}_{\text{in}}(\omega) H_{\text{of}}(\omega) \mathrm{e}^{\mathrm{j}\omega t} \mathrm{d}\omega, \tag{2.33}$$

where $E_{\text{in}}(t)$ and $E_{\text{out}}(t)$ denote the input and output electrical field, respectively, and we used $\sim$ to denote the FT as before. Depending on the operational principle, the optical filters can be classified into two broad categories as *diffraction* or *interference* filters. The important class of optical filters is tunable optical filters, which are able to dynamically change the operating frequency to the desired wavelength channel. The *basic tunable optical filter types* include tunable 2×2 directional couplers, Fabry–Perot (FP) filters, Mach–Zehnder (MZ) interferometer filters, Michelson filters, and acousto-optical filters. Two basic optical filters, FP filter and MZ interferometer filter, are shown in Fig. 2.18. An FP filter is in fact a cavity between two high-reflectivity mirrors. It can act as a tunable optical filter if the cavity length is controlled for example by using a piezoelectric transducer. Tunable FP filters can also be made by using liquid crystals, dielectric thin films,

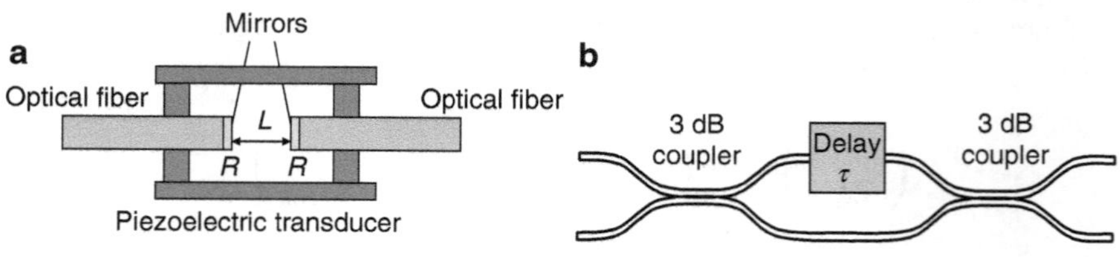

Fig. 2.18 Basic optical filters (**a**) Fabry–Perot filter and (**b**) Mach–Zehnder interferometer

semiconductor waveguides, etc. A *transfer function of an FP filter* whose mirrors have the same reflectivity R, and the cavity length L can be written as [4]

$$H_{FP}(\omega) = \frac{(1 - R)e^{j\pi}}{1 - Re^{j\omega\tau}}, \quad \tau = 2L/v_g, \tag{2.34}$$

where τ is the round-trip time within the cavity and v_g is the group velocity. The transfer function of FP filter is periodic with period being the free spectral range (FSR) $\Delta\nu = v_g/(2L)$. Another FP filter important parameter is the *finesse* defined as [1,4]

$$F = \Delta\nu/\Delta\nu_{FP} \cong \frac{\pi\sqrt{R}}{1 - R}, \tag{2.35}$$

where $\Delta\nu_{FP}$ is the FP transmission peak width.

Mach–Zehnder interferometer filters, shown in Fig. 2.18b, can also be used as tunable optical filters. The first coupler splits the signal into two equal parts, which acquire different phase shifts before they interfere at the second coupler. Several MZ interferometers can be cascaded to create an optical filter. When cross output of 3-dB coupler is used, the square magnitude of transfer function is $|H_c(\omega)| = \cos^2(\omega\tau/2)$ [3], so that the transfer function of M-stage MZ filter based on 3-dB couplers can be written as

$$|H_{MZ}(\omega)|^2 = \prod_{m=1}^{M} \cos^2(\omega\tau_m/2), \tag{2.36}$$

where τ_m is the adjustable delay of mth ($m = 1, 2, \ldots, M$) cascade.

Multiplexers and demultiplexers are basic devices of a WDM system. Demultiplexers contain a *wavelength-selective element* to separate the channels of a WDM signal. Based on underlying physical principle, different demultiplexer devices can be classified as *diffraction-based demultiplexers* (based on a diffraction grating) and *interference-based demultiplexers* (based on optical filters and directional couplers).

Diffraction-based demultiplexers are based on Bragg diffraction effect and use an angular dispersive element, such as the diffraction grating. The incoming composite light signal is reflected from the grating and dispersed spatially into different wavelength components, as shown in Fig. 2.19a. Different wavelength components are focused by lenses and sent to individual optical fibers. The same device can be used as multiplexer by switching the roles of input and output ports. This device can be implemented using either conventional or GRIN lenses. To simplify design, the concave grating can be used.

The second group of optical multiplexers is based on interference effect and employs the optical couplers and filters to combine different wavelength channels into a composite WDM signal. The multiplexers employing the interference effect include thin-film filters multiplexers, and the array waveguide grating (AWG) [1–4,8,9,16], which is shown in Fig. 2.19b. The AWG is highly versatile WDM device, because it can be used as a multiplexer, a demultiplexer, a drop-and-insert element, or even as a wavelength router. It consists of M_{input} and M_{output} slab waveguides and two identical focusing planar star couplers connected by N uncoupled waveguides with a

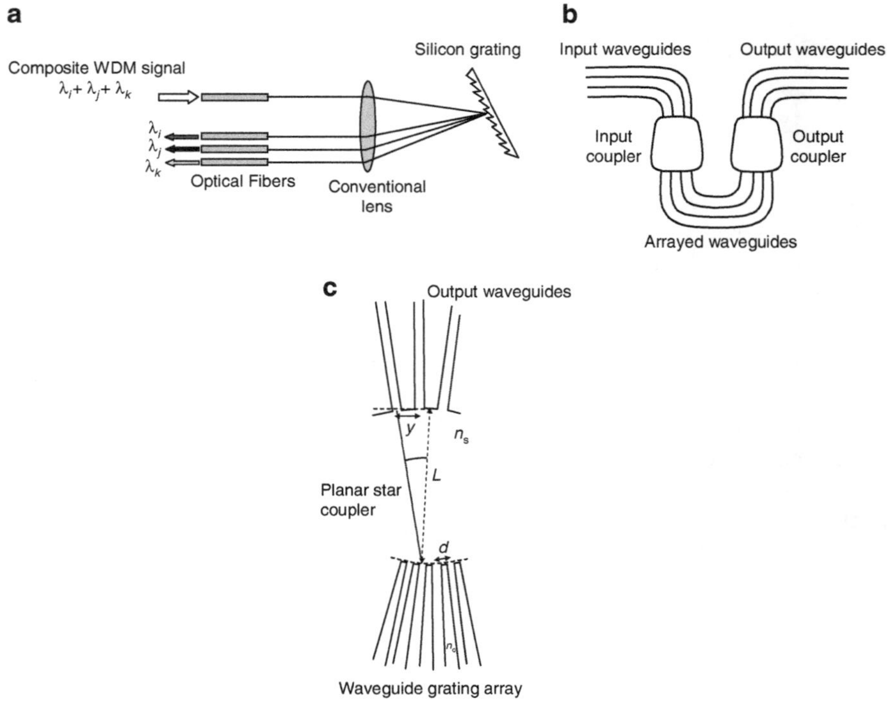

Fig. 2.19 Optical multiplexers/demultiplexers (**a**) grating-based demultiplexer, (**b**) array waveguide grating (AWG) demultiplexer, and (**c**) illustrating the operating principle of AWG

propagation constant β. The length of adjacent waveguides in the central region differ by a constant value ΔL, with corresponding phase difference being $2\pi n_c \Delta L/\lambda$, where n_c is the refractive index of arrayed waveguides. Based on the phase-matching condition (see Fig. 2.19c) and knowing that the focusing is achieved when the path length difference ΔL between adjacent array waveguides be an integer multiple of the central design wavelength λ_c, that is $n_c \Delta L = m\lambda_c$, we derive the following expression for the channel spacing [16]:

$$\Delta\nu = \frac{y}{L}\frac{n_s c d}{m\lambda^2}\frac{n_c}{n_g}, \tag{2.37}$$

where d is the spacing between the grating array waveguides, y is the spacing between the centers of output ports, L is the separation between center of arrayed waveguides and center of output waveguides, n_c is the refractive index of waveguides in grating array, n_s is the refractive index of star coupler, n_g the group index, and m is the diffraction order. The FSR can be obtained by [16]

$$\Delta\nu_{\mathrm{FSR}} = \frac{c}{n_{\mathrm{g}}(\Delta L + d \ \sin \Theta_{\mathrm{in},i} + d \ \sin \Theta_{\mathrm{out},j})}, \tag{2.38}$$

where the diffraction angles from the ith input $\Theta_{\mathrm{in},i}$ and jth output $\Theta_{\mathrm{out},j}$ ports (measured from the center of the array) can be found as $\Theta_{\mathrm{in},i} = iy/L$, and $\Theta_{\mathrm{out},j} = jy/L$, respectively.

Given this description of basic building blocks used in the state-of-the-art optical communication systems, we turn our attention to the description of different direct detection schemes.

2.3 Direct Detection Modulation Schemes

Basic optical modulation formats can be categorized as follows (1) On–Off Keying (OOK), where the 1 is represented by the presence of the pulse while the 0 by the absence of a pulse; (2) Amplitude-shift keying (ASK), where the information is embedded in the amplitude of the sinusoidal pulse; (3) Phase-shift keying (PSK), where the information is embedded in the phase; (4) Frequency-shift keying (FSK), where the information is embedded in the frequency; and (5) Polarization-shift keying (PolSK), where the information is embedded in the polarization. In this section, we will discuss the modulation formats with direct detection, namely (1) non-return-to-zero (NRZ), (2) return-to-zero (RZ), (3) alternate mark inversion (AMI), (4) duobinary modulation, (5) carrier-suppressed RZ, (6) NRZ-differential phase-shift keying (NRZ-DPSK), and (7) RZ-differential phase-shift keying (RZ-DPSK).

2.3.1 *Non-Return-to-Zero*

NRZ is considered to be the simplest modulation format. In early modulation phases, it was used due to its immunity to laser phase noise and for its low bandwidth requirements relative to other modulation formats. Recently, these issues are not the main concern with the advanced technology and high speed, and so, NRZ is merely used for comparison purposes with other formats.

The block diagram of an NRZ transmitter is shown in Fig. 2.20, and it is composed of a laser source and an external modulator. In our case, the modulator is chosen to be a Mach–Zehnder modulator (MZM). The output of the NRZ transmitter is shown in Fig. 2.21.

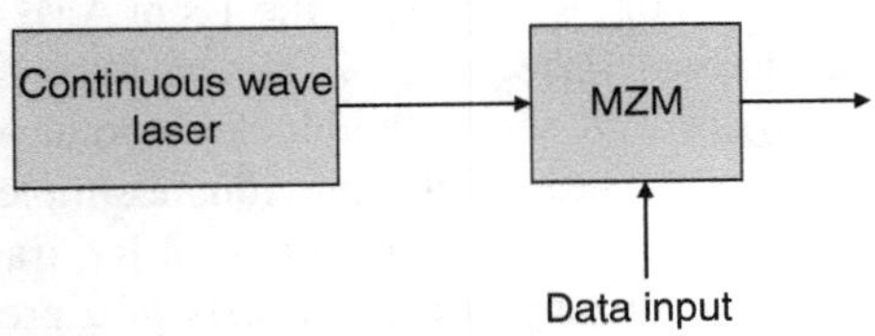

Fig. 2.20 Non-return-to-zero transmitter. *MZM* Mach–Zehnder modulator

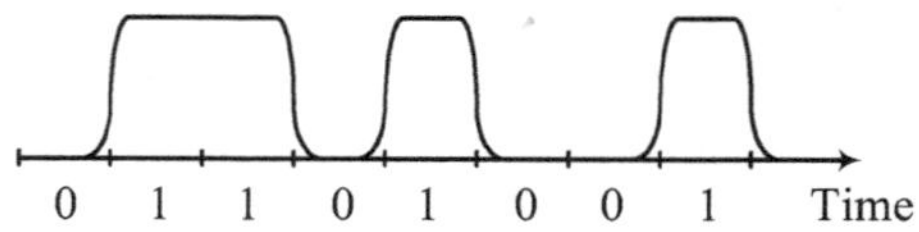

Fig. 2.21 NRZ signal

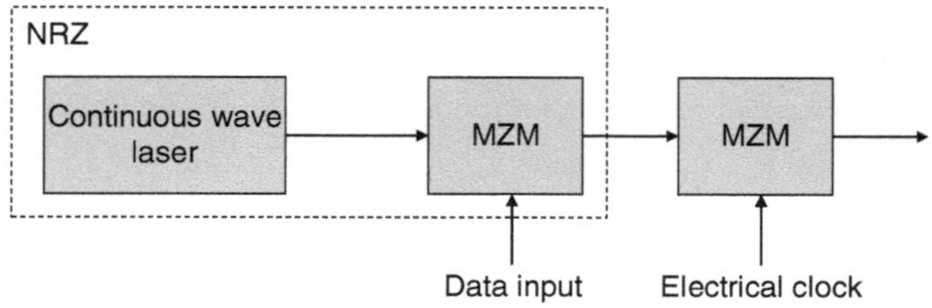

Fig. 2.22 Return-to-zero transmitter

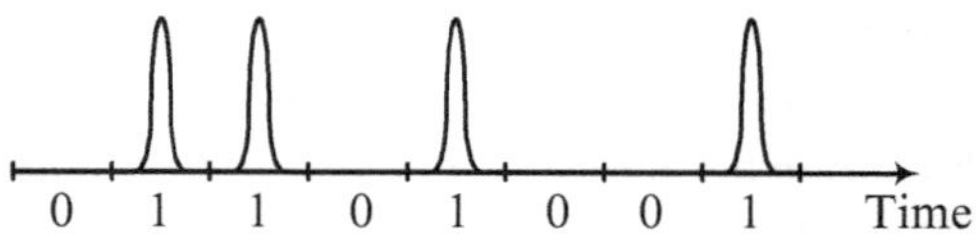

Fig. 2.23 RZ signal

2.3.2 Return-to-Zero

RZ is one of the basic subcategories of the OOK modulation format. Its bandwidth is higher than that of the corresponding NRZ at the same data rate due to the fact that in each "1" bit period T, the signal amplitude reaches the amplitude specified for the "1" bit and goes down to zero. RZ is characterized by the duty cycle, which is the ratio of the pulse width at $1/\sqrt{2}$ of the maximum amplitude to the bit period T.

Figure 2.22 shows the block diagram of the RZ transmitter which is basically an NRZ transmitter with an extra external modulator driven by an electrical clock that can be achieved by a sinusoidal signal at the half data rate. Figure 2.23 shows the RZ signal output of the modulator.

2.3.3 Alternate Mark Inversion

AMI utilizes amplitude to transmit information, the amplitude levels are either "on" or "off," while the optical phase levels for the "on" state are either "0" or "π," i.e., for a "0" bit, the output of the AMI modulator is "0", while for a "1" bit, the output is either "1" or "−1." In fact, the 1's in AMI are represented in an alternating fashion, i.e., no two consecutive "1" bits have the same sign.

Figure 2.24 shows the block diagram of the AMI transmitter with the same binary input used for the previous example for NRZ and RZ (i.e., 01101001). As noticed, the AMI transmitter is an RZ transmitter that is driven by AMI-encoded data. The driving circuit consists of a precoder and an encoder that utilize delay

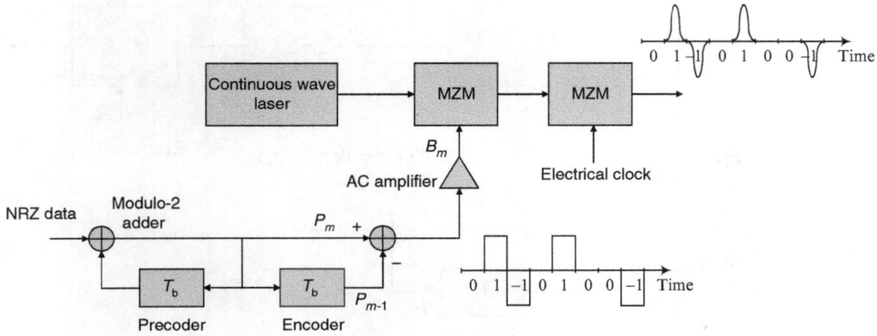

Fig. 2.24 The block diagram of the AMI transmitter

Table 2.1 AMI modulated signal generation and detection

m	0	1	2	3	4	5	6	7	8
D_m		0	1	1	0	1	0	0	1
P_m	0	0	1	0	0	1	1	1	0
B_m		0	1	−1	0	1	0	0	−1
$\|B_m\|$		0	1	1	0	1	0	0	1

lines. The precoder gives out P_m by adding the input bit to the value stored in the delay line by a modulo-2 adder, while the encoder gives out B_m by subtracting P_{m-1} from P_m.

Table 2.1 shows an example of calculating P_m and B_m from the data input D_m. We assume that the initial values in the delay lines are set to "0," so at $m = 0$, $P_0 = 0$, then we calculate P_m by adding P_{m-1} to D_m. B_m on the other hand is calculated by subtracting P_{m-1} from P_m. The original sequence is recovered by photodetector, which removes the phase information.

2.3.4 *Duobinary Modulation Format*

The optical duobinary signal has two intensity levels "on" and "off." The "on" state signal can have one of the two optical phases, 0 and π. The two "on" states correspond to the logic states "1" and "−1" of the duobinary encoded signal, and the "off" state corresponds to the logic state "0" of the duobinary encoded signal. According to the duobinary encoding rule, the logic states "−1," "0," and "1" of the duobinary encoded signal correspond to the logic states "0," "1," and "0" of the original binary signal, respectively. Therefore, the original signal can be recovered by inverting the directly detected signal.

The schematic diagram of the transmitter for the optical duobinary signal format is shown in Fig. 2.25. The transmitter configuration of the optical duobinary signal format is identical to that of the RZ signal format. However, the signal that drives the data modulator is not the original NRZ data, but the duobinary-encoded data

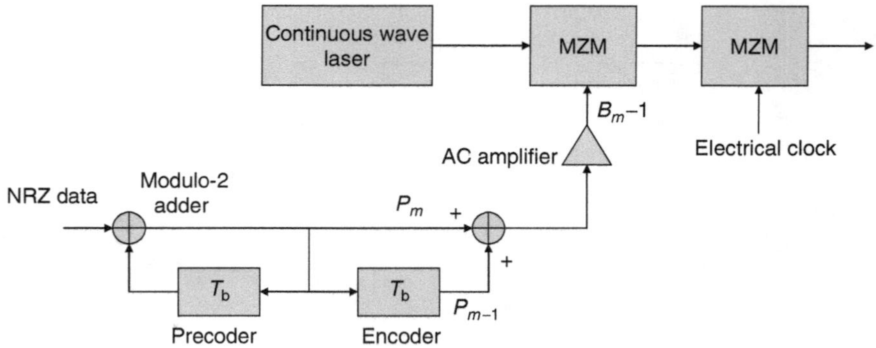

Fig. 2.25 The block diagram of the duobinary transmitter

Table 2.2 Duobinary signal generation and detection

m	0	1	2	3	4	5	6	7	8
D_m		0	1	1	0	1	0	0	1
P_m	0	0	1	0	0	1	1	1	0
B_m		0	1	1	0	1	2	2	1
$B_m - 1$		−1	0	0	−1	0	1	1	0
$\|B_m - 1\|$		1	0	0	1	0	1	1	0
$\overline{\|B_m - 1\|}$		0	1	1	0	1	0	0	1

sequence. The first step in the modulation process is to feed the original NRZ data sequence D_m to the differential precoder in order to avoid error propagation at the receiver caused by the preceding received data not being recovered correctly. The precoder gives out P_m by adding the input bit to the value stored in the delay line by a modulo-2 adder, while the encoder gives out B_m by subtracting P_{m-1} from P_m. Table 2.2 shows an example of generating P_m and B_m from the data input D_m, as well as recovery of the transmitted sequence on receiver side by photodetector and NOT gate.

2.3.5 Carrier-Suppressed Return-to-Zero

Carrier-suppressed return-to-zero (CS-RZ) modulation format was proposed by Sano and Miyamoto [25]. The major difference between a CS-RZ and a conventional RZ is that in CS-RZ optical signal there is a π phase shift between adjacent bits, as shown in Fig. 2.26. Therefore, the average optical field in a CS-RZ signal is zero (there is no DC component). In the frequency domain, this translates into a carrier suppression in the optical spectrum. In general, the generation of a CS-RZ optical signal requires two electro-optic modulators. The first intensity modulator generates a conventional chirp-free RZ optical signal and the second electro-optic phase modulator produces a π optical phase shift between adjacent

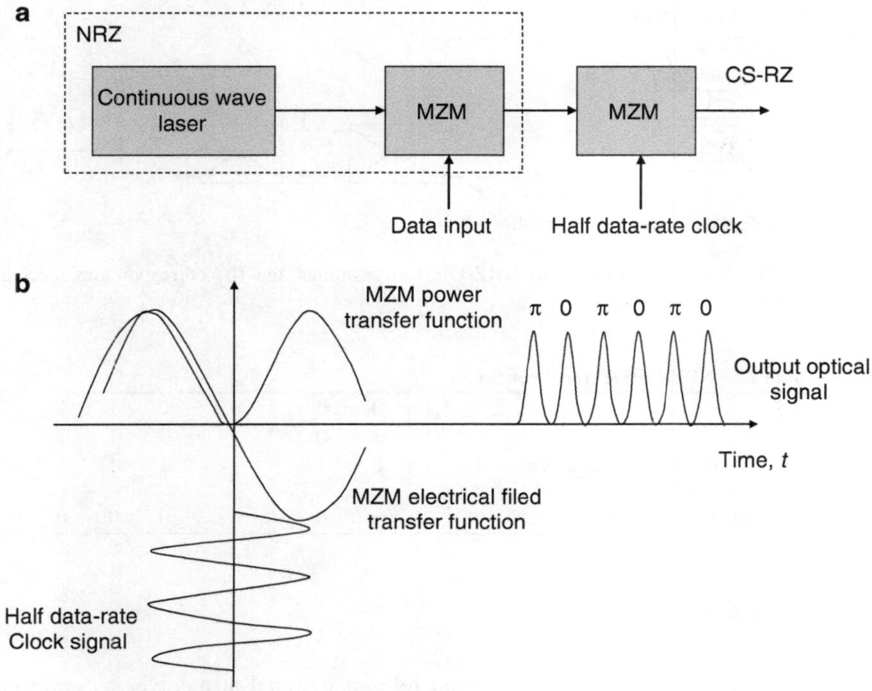

Fig. 2.26 (**a**) The block diagram of the CS-RZ transmitter and (**b**) the illustration of generation of CS-RZ sequence

bits. In this transmitter configuration, the modulation bandwidths of both of these two electro-optic modulators have to be in the same level as the data rate. The figure below shows one possible version of CS-RZ transmitter. In this configuration, the first intensity modulator encodes the NRZ data. The second Mach–Zehnder type intensity modulator is biased at the minimum power transmission point and driven by a sinusoid at the half data rate. The MZ intensity modulator biased at this condition performs frequency doubling for the modulating signal, and the output pulse train is phase alternated between adjacent bits. This configuration reduces the bandwidth requirement for the electro-optical modulators. In fact, the bandwidth required for the second modulator is only a half of the signal data rate.

CS-RZ is another modulation format considered to have better tolerance to fiber nonlinearity and residual chromatic dispersion. Phase alternating between adjacent bit slots reduces the fundamental frequency components to half of the data rate and regular RZ intensity bit pattern makes it easy to find the optimum dispersion compensation. In addition, carrier suppression reduces the efficiency of four-wave mixing (FWM) in WDM systems.

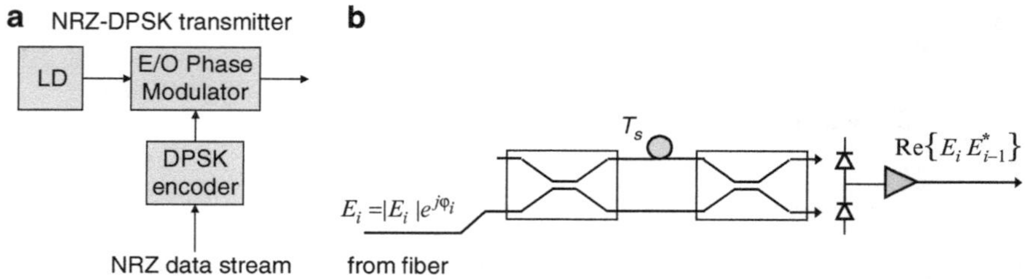

Fig. 2.27 The block diagram of (**a**) NRZ-DPSK transmitter and (**b**) corresponding receiver configuration. *LD* Laser diode

Table 2.3 NRZ-DPSK signal generation

$\{b_k\}$		1	0	0	1	0	0	1	1
$\{d_{k-1}\}$		1	1	0	1	1	0	1	1
Differentially encoded sequence $\{d_k\}$, $d_k = \overline{b_k \oplus d_{k-1}}$	1	1	0	1	1	0	1	1	1
Transmitted phase (rad)	0	0	π	0	0	π	0	0	0

2.3.6 NRZ-DPSK

DPSK eliminates the need for a coherent reference signal at receiver by combining two basic operations at transmitter (1) differential encoding and (2) phase-shift keying. To send symbol 0, we phase advance the current signal waveform by 180°, and to send a symbol 1, we leave the phase unchanged. The transmitter and receiver block diagrams are shown in Fig. 2.27a. The receiver "measures" the relative phase difference received during two successive bit intervals. The following differentially encoded rule applied: If the incoming binary symbol b_k is 1, leave the symbol d_k unchanged with respect to previous bit; if the incoming binary symbol b_k is 0, change the symbol d_k with respect to the previous bit (see Table 2.3 for an illustrative example).

At a DPSK optical receiver, shown in Fig. 2.27b, a one-bit-delay Mach–Zehnder interferometer (MZI), correlates each bit with previous bit and makes the phase-to-intensity conversion. When the two consecutive bits are in-phase, they are added constructively in the MZI and results in a high signal level. Otherwise, if there is a π phase difference between the two bits, they cancel each other in the MZI and results in a low signal level. In a practical DPSK receiver, the MZI has two balanced output ports (constructive port and destructive port).

2.3.7 RZ-DPSK

In order to improve system tolerance to nonlinear distortion and to achieve a longer transmission distance, return-to-zero DPSK (RZ-DPSK) has been proposed. In this modulation format, an optical pulse appears in each bit slot, with the binary data

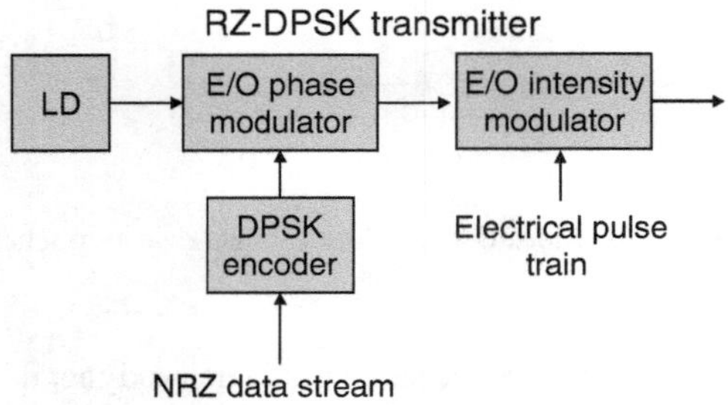

Fig. 2.28 The block diagram of RZ-DPSK transmitter

encoded as either a "0" or a "π" phase shift between adjacent bits. In general, the width of the optical pulses is narrower than the bit slot and therefore, the signal optical power returns to zero at the edge of each bit slot. In order to generate the RZDPSK optical signal, one more intensity modulator has to be used compared to the generation of NRZ-DPSK. The block diagram of a RZ-DPSK transmitter is shown in Fig. 2.28.

First, an electro-optic phase modulator generates a conventional NRZ-DPSK optical signal, and then, this NRZ-DPSK optical signal is sampled by a periodic pulse train at the clock rate through an electro-optic intensity modulator. In this modulation format, the signal optical intensity is no longer constant; this will probably introduce the sensitivity to self-phase modulation (SPM). In addition, due to the narrow pulse intensity sampling, the optical spectrum of RZ-DPSK is wider than a conventional NRZ-DPSK. Intuitively, this wide optical spectrum would make the system more susceptible to chromatic dispersion. However, in long distance optical systems, periodic dispersion compensation is often used and RZ modulation format makes it easy to find the optimum dispersion compensation because of its regular bit patterns.

2.4 Coherent Detection Modulation Schemes

In order to exploit the enormous bandwidth potential of the optical fiber, different multiplexing techniques (OTDMA, WDMA, CDMA, SCMA), modulation formats (OOK, ASK, PSK, FSK, PolSK, CPFSK, DPSK, etc.), demodulation schemes (DD or coherent), and technologies were developed; and some important aspects are discussed in this section. The coherent detection offers several important advantages compared to direct detection (1) improved receiver sensitivity, (2) better frequency selectivity, (3) possibility of using constant amplitude modulation formats (FSK, PSK), (4) tunable optical receivers similar to RF receivers are possible, and (5) with coherent detection the chromatic dispersion and PMD can easier be mitigated.

Based on whether local laser operating frequency coincides with incoming optical signal, the coherent reception can identified as (1) homodyne, in which the

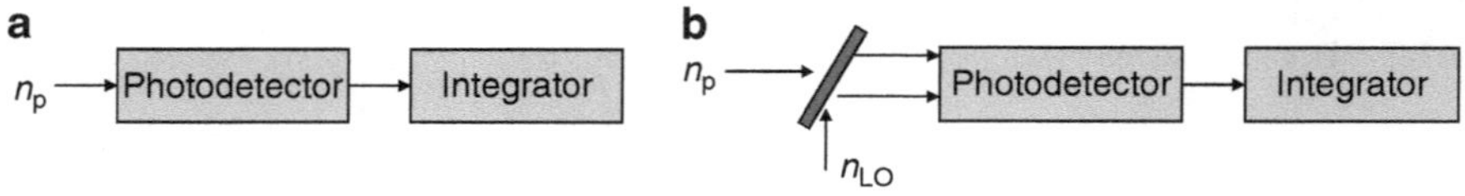

Fig. 2.29 Block diagrams of (**a**) direct detection scheme and (**b**) coherent detection scheme

frequencies are the same or (2) heterodyne, in which the frequencies are different so that all related signal processing upon photodetection is performed at suitable intermediate frequency (IF). Different coherent detection can be classified into following categories (1) synchronous (PSK, FSK, ASK), (2) asynchronous (FSK, ASK), (3) differential detection (CPFSK, DPSK), (4) phase diversity receivers, (5) polarization diversity receivers, and (6) polarization multiplexing schemes. Synchronous detection schemes can further be categorized as residual carrier or suppressed carrier [Costas loop/decision-driven loop (DDL)] based.

In Fig. 2.29, we show the basic difference between direct detection (Fig. 2.29a) and coherent detection (Fig. 2.29b) schemes. Coherent detection, in addition to the photodetector and integrator already in use for direct detection, employs a local laser whose frequency and polarization is matched to the frequency and polarization of incoming optical signal. The incoming optical signal and local laser output signal are mixed in optical domain (a mirror is used in this illustrative example, in practice an optical hybrid is used instead).

Because the photodetector generates the photocurrent in proportion to the input optical power, the direct detection integrator output signal can be written as follows:

$$\begin{aligned} s_0'(t) &= 0; \quad 0 \le t \le T, \\ s_1'(t) &= RP_s; \quad 0 \le t \le T, \end{aligned} \tag{2.39}$$

where P_s is the input optical signal power, R is the photodiode responsivity introduced earlier, and T is the bit duration. The subscripts 0 and 1 are used to denote the transmitted bits (zero and one). It is very common in optical communications to express the photocurrent output signal in terms of number of photons per bit $n_p = RP_sT/q$ (q is an electron charge) as given below:

$$\begin{aligned} s_0(t) &= 0; \quad 0 \le t \le T, \\ s_1(t) &= \frac{n_p}{T}; \quad 0 \le t \le T. \end{aligned} \tag{2.40}$$

The corresponding coherent detector integrator output signal, for homodyne synchronous detection, can be written as

$$s_{0,1}(t) = \frac{1}{2T}\left(\mp\sqrt{2n_p} + \sqrt{2n_{LO}}\right)^2; \quad 0 \le t \le T, \tag{2.41}$$

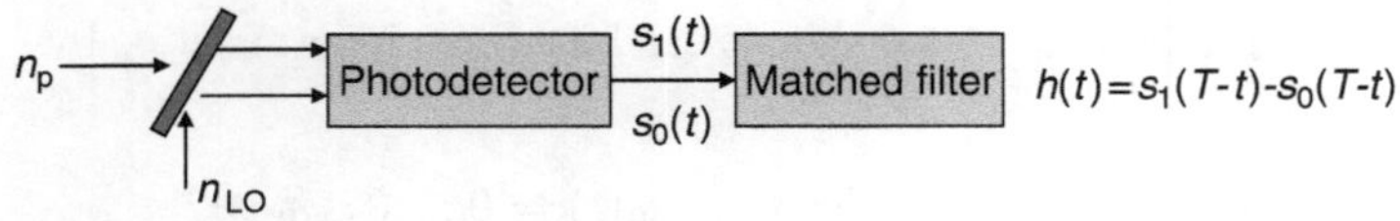

Fig. 2.30 The synchronous matched filter detector configuration

where n_{LO} denotes the average number of local laser photons per bit, sign "–" corresponds to transmitted zero, and sign "+" to transmitted one bits. The optimum coherent detection receiver (minimizing the bit error probability P_b), shown in Fig. 2.30, is the matched filter (or correlation) receiver with impulse response given by $h(t) = s_1(T - t) - s_0(T - t)$ [26].

General expression for matched filter output signal, applicable to different modulation formats (such as ASK, FSK, PSK) and shot-noise-dominated scenario, can be written as follows:

$$s'_{0,1}(t) = 2R\sqrt{P_S P_{LO}}\cos(\omega_{0,1}(t) + \theta_{0,1}(t)) + n_{sh}(t); \quad 0 \le t \le T, \tag{2.42}$$

where P_{LO} denotes the local laser output signal, while $\omega_{0,1}$ and $\theta_{0,1}$ represent frequency and phase corresponding to bit 0 and 1. n_{sh}(t) denotes the shot noise process, which commonly can be modeled as zero-mean Gaussian with power spectral density $N'_0 = 2RqP_{LO}$. Because the bit energy can be determined by $E' = E'_{0,1} = \int_0^T s'^2_{0,1}(t)\mathrm{d}t \cong 2R^2 P_S P_{LO} T$, the corresponding SNR is related to number of photons per bit as given below:

$$\frac{E'}{N'_0} = \frac{2R^2 P_s P_{LO} T}{2RqP_{LO}} = \frac{RP_s T}{q} = n_p. \tag{2.43}$$

Therefore, the general expression (2.42) can also be written in terms of n_p:

$$s_{0,1}(t) = \sqrt{\frac{2n_p}{T}}\cos(\omega_{0,1}(t) + \theta_{0,1}(t)); \quad 0 \le t \le T. \tag{2.44}$$

Correlation coefficient ρ and Euclidean distance d between transmitted signals (s_0 and s_1) are defined as

$$\rho = \frac{1}{\sqrt{E_0 E_1}}\int_0^T s_0(t)s_1(t)\mathrm{d}t; \quad d^2 = E_0 + E_1 - 2\rho\sqrt{E_0 E_1}, \tag{2.45}$$

where E_i is the energy of ith ($i = 0, 1$) bit. The probability of error is related to the Euclidean distance as follows:

$$P_b = \frac{1}{2}\mathrm{erfc}\left(\frac{d}{2\sqrt{N_0}}\right). \tag{2.46}$$

For ASK systems, the transmitted signals can be written in terms of n_p by

$$s_1(t) = \sqrt{\frac{2n_\mathrm{p}}{T}}\cos(\omega_1 t); \quad s_0(t) = 0; \quad 0 \le t \le T \tag{2.47}$$

By using (2.45), the Euclidean distance squared is found to be $d^2 = n_\mathrm{p}$ so that corresponding expression for bit error probability is obtained from (2.46) as follows:

$$P_\mathrm{b} = \frac{1}{2}\mathrm{erfc}\left(\frac{\sqrt{n_\mathrm{p}}}{2}\right). \tag{2.48}$$

The bit error probability of 10^{-9} is achieved for 72 photons per bit.

For continuous phase FSK (CPFSK) systems, the transmitted symbols can be represented by

$$\begin{aligned} s_1(t) &= \sqrt{\frac{2n_\mathrm{p}}{T}}\cos(\omega_1 t); \quad 0 \le t \le T, \\ s_0(t) &= \sqrt{\frac{2n_\mathrm{p}}{T}}\cos(\omega_0 t); \quad 0 \le t \le T. \end{aligned} \tag{2.49}$$

The corresponding correlation coefficient can be found by

$$\rho = \frac{2}{T}\int_0^T \cos\,\omega_0 t\,\cos\,\omega_1 t\mathrm{d}t \cong \frac{\sin 2\pi m}{2\pi m}, \qquad m = \frac{|\omega_1 - \omega_0|}{2\pi/T}, \tag{2.50}$$

where m is the modulation index. The corresponding bit error probability is obtained by substituting (2.50) in (2.46) by

$$P_\mathrm{b} = \frac{1}{2}\mathrm{erfc}\left(\sqrt{\frac{n_\mathrm{p}}{2}\left(1 - \frac{\sin 2\pi m}{2\pi m}\right)}\right). \tag{2.51}$$

For $m = 0.5p$ ($p = 1, 2, \ldots$), the number of required photons per bit to achieve P_b of 10^{-9} is 36, while the minimum $n_\mathrm{p} = 29.6$ is obtained for $m = 0.715$.

For direct modulation PSK (DM-PSK) systems, the transmitted symbols can be represented by

$$\begin{aligned} s_1(t) &= \sqrt{\frac{2n_\mathrm{p}}{T}}\cos(\theta_1(t)); \quad 0 \le t \le T; \quad \theta_1(t) = \begin{cases} \omega_\mathrm{IF} t + \frac{\pi m}{T} t; & 0 \le t \le T/(2m) \\ \omega_\mathrm{IF} t + \pi/2; & T/(2m) \le t \le T \end{cases}, \\ s_0(t) &= \sqrt{\frac{2n_\mathrm{p}}{T}}\cos(\theta_0(t)); \quad 0 \le t \le T; \quad \theta_0(t) = \begin{cases} \omega_\mathrm{IF} t - \frac{\pi m}{T} t; & 0 \le t \le T/(2m) \\ \omega_\mathrm{IF} t - \pi/2; & T/(2m) \le t \le T \end{cases}, \end{aligned} \tag{2.52}$$

where $\omega_{\rm IF} = |\omega_{\rm s} - \omega_{\rm LO}|$ is the intermediate frequency. The correlation coefficient is obtained from (2.45) as $\rho \cong 1/(2m) - 1$, and corresponding probability of error is given by

$$P_{\rm b} = \frac{1}{2}{\rm erfc}\left(\sqrt{n_{\rm p}\left(1 - \frac{1}{4m}\right)}\right). \tag{2.53}$$

For $m = 1/2$, the required number of photons per bit to achieve $P_{\rm b}$ of 10^{-9} is 36 (that is the same as in FSK systems), while when $m \to \infty$ the required $n_{\rm p}$ is 18.

For FSK systems, with two oscillators the transmitted signals are represented by

$$\begin{aligned} s_1(t) &= \sqrt{\frac{2n_{\rm p}}{T}}\cos(\omega_1 t + \theta); \quad 0 \le t \le T, \\ s_0(t) &= \sqrt{\frac{2n_{\rm p}}{T}}\cos(\omega_0 t + \theta); \quad 0 \le t \le T, \\ \theta &= \begin{cases} [\pi - \pi m]_\pi; & 2p \le m \le 2p+1 \\ [-\pi m]_\pi; & 2p+1 \le m \le 2(p+1) \end{cases} \quad p = 0, 1, 2, \ldots. \end{aligned} \tag{2.54}$$

(We use $[x]_\pi$ to denote mod π operation.) The correlation coefficient is obtained by substituting (2.54) in (2.45) $\rho \cong (\sin(2\pi m + \theta) - \sin\theta)/(2\pi m)$, while probability of error expression is obtained by

$$P_{\rm e} = \frac{1}{2}{\rm erfc}\left(\sqrt{\frac{n_{\rm p}}{2}\left(1 + \frac{|\sin \pi m|}{\pi m}\right)}\right). \tag{2.55}$$

The bit error probability $P_{\rm b}$ of 10^{-9} is achieved for $n_{\rm p} = 36/(1 + |\sin \pi m|/\pi m)$. For example, $n_{\rm p} = 18$ when $m \to 0$, while $n_{\rm p} = 36$ when $m = 1, 2, \ldots$.. In Fig. 2.31, we provide the comparison of different FSK modulation formats in terms of required $n_{\rm p}$ to achieve $P_{\rm b}$ of 10^{-9} vs. modulation index m.

For PSK systems, the transmitted signal can be written by

$$\begin{aligned} s_1(t) &= \sqrt{\frac{2n_{\rm p}}{T}}\cos(\omega_1 t); \quad 0 \le t \le T, \\ s_0(t) &= \sqrt{\frac{2n_{\rm p}}{T}}\cos(\omega_1 t + \pi); \quad 0 \le t \le T. \end{aligned} \tag{2.56}$$

The corresponding probability of error is given by

$$P_{\rm e} = \frac{1}{2}{\rm erfc}\left(\sqrt{\frac{n_{\rm p}}{2}}\right). \tag{2.57}$$

The bit error probability $P_{\rm b}$ of 10^{-9} is achieved for $n_{\rm p} = 18$ in heterodyne case and $n_{\rm p} = 9$ in homodyne case. In Fig. 2.32, we provide the bit error probabilities

Fig. 2.31 Comparison of FSK systems

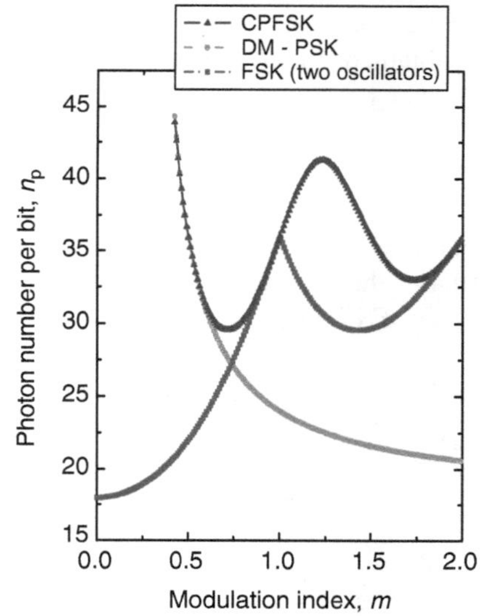

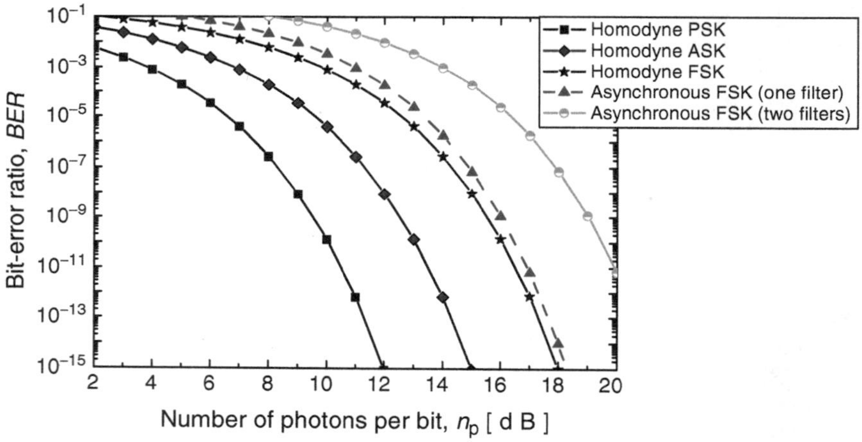

Fig. 2.32 BER performance comparison

for different modulation schemes discussed above. Similarly in conventional digital communications systems, the synchronous coherent detection PSK scheme performs the best. The comparison of different modulation formats in terms of receiver sensitivity, defined as required number of photons per bit n_p to achieve BER of 10^{-9}, is given in Table 2.4, which is adopted from [27]. For more details on various modulation schemes for coherent detection, an interested reader is referred to an excellent book due to Jacobsen [27].

Table 2.4 Receiver sensitivities in terms of required n_p to achieve BER of 10^{-9}

System		Coherent heterodyne	Coherent homodyne	IM/DD
Quantum limit				20
Super quantum limit			5	
ASK	Matched filter	72	36	
	Asynchronous	76		
CPFSK	$m = 0.5, 1, 1.5, \ldots$	36		
Matched filter	$m = 0.715$	29.6		
FSK	$m = 0$	18	9	
Two oscillators	$m = 1, 2, \ldots$	36		
Matched filter	$m = 1.43$	29.6		
CPFSK	$m = 0.5$	61.9	30.9	
Delay-detection	$m = 0.8$	36.2		
	$m = 1$	43.6		
FSK	$m = 1, 2, \ldots$	40		
Asynchronous	$m = 1.5$	47		
DM-PSK	$m = 0.5$	36	18	
Matched filter	$m = 1$	24	12	
	$m = 2$	20.6	10.3	
DM-DPSK	$m = 0.5$	61.9	30.9	
Delay-detection	$m = 1$	30.6	15.3	
	$m = 2$	24.4	12.2	
PolSK		36	18	
Matched filter				
PolSK (and FSK)	$m = 0.5, 1, 1.5, \ldots$	72		
Matched filter	$m = 0.715$	59.2		
	$m = 1.43$	59.2		
PolSK asynchronous		40		
PSK		18	9	
Matched filter				
DPSK		20	10	
DPSK IF $= 1/(2T)$		24.7		

2.4.1 Optical Hybrids and Balanced Receivers

So far we used an optical mirror to perform optical mixing before photodetection takes place. In practice, this operation is performed by four-port device known as optical hybrid, which is shown in Fig. 2.33.

Electrical fields at output ports E_{1o} and E_{2o} are related to the electrical fields at input ports E_{1i} and E_{2i} as follows:

$$E_{1o} = (E_{1i} + E_{2i})\sqrt{1-k},$$
$$E_{2o} = (E_{1i} + E_{2i}\exp(-j\phi))\sqrt{k}, \tag{2.58}$$

Fig. 2.33 Optical hybrid

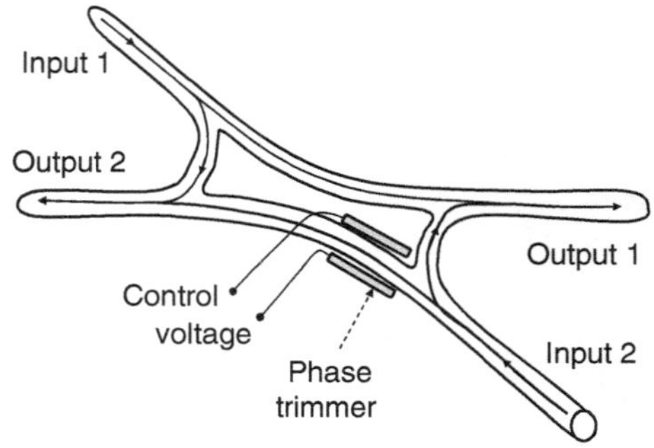

where k is the power splitting ratio and ϕ is the phase shift introduced by the phase trimmer (see Fig. 2.33). Equation (2.58) can also be written in terms of scattering (S-) matrix as follows:

$$\boldsymbol{E}_{\mathrm{o}} = \begin{bmatrix} E_{\mathrm{o}1} \\ E_{\mathrm{o}2} \end{bmatrix} = \boldsymbol{S} \begin{bmatrix} E_{\mathrm{o}1} \\ E_{\mathrm{o}2} \end{bmatrix} = \boldsymbol{S}\boldsymbol{E}_{i}, \quad \boldsymbol{S} = \begin{bmatrix} s_{11} & s_{12} \\ s_{21} & s_{22} \end{bmatrix} = \begin{bmatrix} \sqrt{1-k} & \sqrt{1-k} \\ \sqrt{k} & \mathrm{e}^{-\mathrm{j}\phi}\sqrt{k} \end{bmatrix}. \tag{2.59}$$

In expressions (2.58) and (2.59), we assumed that hybrid is lossless device, which leads to: $s_{11} = |s_{11}|$, $s_{12} = |s_{12}|$, $s_{21} = |s_{21}|$, and $s_{22} = |s_{22}|\exp(\mathrm{j}\theta_{22})$. Popular hybrids are π hybrid, which S-matrix can be written as (by setting $\phi = \pi$ in (2.59))

$$\boldsymbol{S} = \begin{bmatrix} \sqrt{1-k} & \sqrt{1-k} \\ \sqrt{k} & -\sqrt{k} \end{bmatrix},$$

and $\pi/2$ hybrid, which S-matrix can be written as

$$\boldsymbol{S} = \begin{bmatrix} s_{11} & s_{12} \\ s_{21} & s_{22}\mathrm{e}^{-\mathrm{j}\pi/2} \end{bmatrix}. \tag{2.60}$$

Well-known π hybrid is 3-dB coupler ($k = 1/2$). If $\pi/2$ hybrid is symmetric $|s_{ij}| = 1/L\,(\forall i, j)$, the phase difference between the input electrical fields $E_{1\mathrm{i}} = |E_{1\mathrm{i}}|$, $E_{2\mathrm{i}} = |E_{2\mathrm{i}}|\mathrm{e}^{\mathrm{j}\theta_i}$ can be chosen in such a way that total output ports power

$$\boldsymbol{E}_0^{\dagger}\boldsymbol{E}_0 = \frac{2}{L}\left[P_{1\mathrm{i}} + P_{2\mathrm{i}} + \sqrt{P_{1\mathrm{i}}P_{2\mathrm{i}}}(\cos\theta_i + \sin\theta_i)\right] \tag{2.61}$$

is maximized. (We use $\dagger$ to denote Hermitian transposition–simultaneous transposition and complex conjugation.) For equal input powers, the maximum of (2.61) is obtained when $\theta_i = \pi/4$, leading to $L \geq 2 + \sqrt{2}$. The corresponding loss is $10\log_{10}(L/2) = 2.32\,\mathrm{dB}$. For Costas loop- and DDL-based homodyne systems, there exists optimum k in corresponding S-matrix:

$$\boldsymbol{S} = \frac{1}{\sqrt{L}}\begin{bmatrix} \sqrt{1-k} & \sqrt{1-k} \\ \sqrt{k} & -\mathrm{j}\sqrt{k} \end{bmatrix}.$$

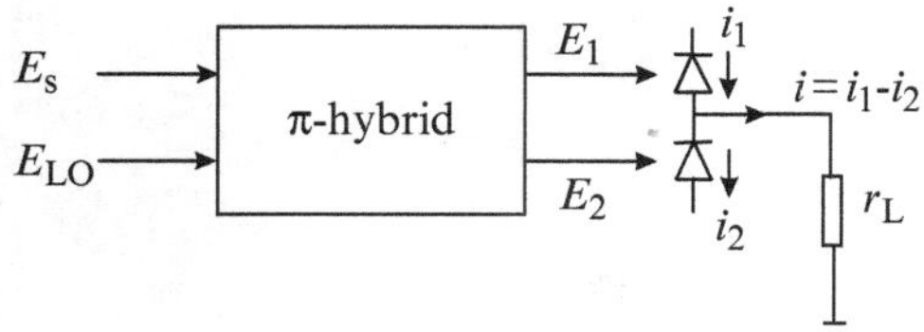

Fig. 2.34 Balanced detector

To reduce the relative intensity noise (RIN) of transmitting laser and to eliminate the direct-detection and signal-cross-signal interferences, the most deleterious sources for multichannel applications, the balanced receiver, shown in Fig. 2.34, is commonly used. The upper and lower photodetector's output currents can, respectively, be written as

$$i_1(t) = R|E_1|^2 = \frac{1}{2}R\left(P_s + P_{LO} + 2\sqrt{P_s P_{LO}}\cos\theta_s\right) + n_1(t),$$
$$i_2(t) = R|E_2|^2 = \frac{1}{2}R\left(P_s + P_{LO} - 2\sqrt{P_s P_{LO}}\cos\theta_s\right) + n_2(t), \quad (2.62)$$

where θ_s is the phase of incoming optical signal and $n_i(t)$ $(i = 1, 2)$ is the ith photodetector shot noise process of PSD $S_{n_i} = qR|E_i|^2$. The balanced receiver output current (see Fig. 2.34) can be written as

$$i(t) = i_1(t) - i_2(t) = 2R\sqrt{P_s P_{LO}}\cos\theta_s + n(t), \quad n(t) = n_1(t) - n_2(t), \quad (2.63)$$

where $n(t)$ is a zero-mean Gaussian process of PSD $S_n = S_{n_1} + S_{n_2} = qR(P_s + P_{LO}) \cong qRP_{LO}$. For binary PSK signaling, $\theta_s = \pm\pi$ and expression (2.63) is consistent with (2.56).

2.4.2 Dominant Coherent Detector Noise Sources

In this section, we describe the basic coherent detector noise processes, and a more detailed description of different channel impairments and receiver processes is provided in Chap. 3. The dominant sources of performance degradation in coherent detection are the laser-phase noise, photodiode shot noise, polarization noise, data-to-phase lock crosstalk for receiver with residual carrier, and adjacent channel interference for multichannel applications.

The PIN photodiode shot noise is zero-mean white Gaussian with PSD $\text{PSD}_n(f) = qR(P_s + P_{LO}) \cong qRP_{LO}$ if the following is valid

$$\text{Bandwidth of interest} \ll 1/T_d \ll hP_s/h\nu.$$

Noise, especially spontaneous emission, causes phase fluctuations in lasers, leading to a nonzero spectral linewidth $\Delta\nu$, which is illustrated in Fig. 2.35. In gas or solid-state lasers, the linewidth $\Delta\nu$ typically ranges from the subhertz to the kilohertz

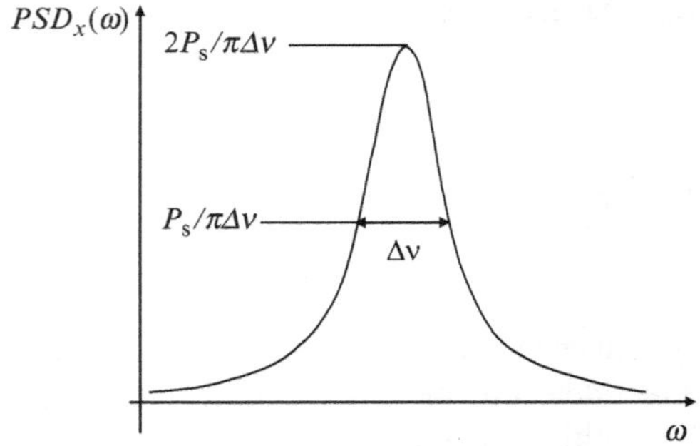

Fig. 2.35 Lorentzian spectrum of laser diode

range. In semiconductor lasers, the linewidth $\Delta\nu$ is often much larger, up the megahertz range, because a large number of photons are stored in the small cavity and due to the nonnegligible value of the linewidth enhancement factor.

The laser lightwave signal can be written as

$$x(t) = \sqrt{P_S}\mathrm{e}^{\mathrm{j}[\omega_0 t + \phi(t) + \theta]}, \tag{2.64}$$

where $\phi(t)$ is the laser phase noise process, which is commonly modeled as Wiener–Lévy process [7], that is a zero-mean Gaussian process with variance $\sigma_\phi = 2\pi\Delta\nu|t|$. The autocorrelation function of $x(t)$ is given by

$$R_x(\tau) = P_S \mathrm{e}^{\mathrm{j}\omega_0\tau}\,\mathrm{e}^{-\pi\Delta\nu|\tau|}. \tag{2.65}$$

The corresponding PSD of $x(t)$ can be found as Fourier transform of (2.65)

$$\mathrm{PSD}_x(\omega) = \frac{2P_S}{\pi\,\Delta\nu}\left[1 + \left(\frac{\omega - \omega_0}{\pi\,\Delta\nu}\right)^2\right]^{-1} \tag{2.66}$$

and has therefore the Lorentzian shape.

The influence of laser phase noise for BPSK signaling is illustrated in Fig. 2.36. The effect on BER curves is twofold (1) the BER curves are shifted to the right and (2) BER floor appears. For state-of-the art optical communication systems, the laser phase noise does impose a serious limitation. However, for large medium and large multilevel modulations, such as M-ary PSK and M-ary QAM, the laser phase noise is very important factor of performance degradation.

The polarization noise comes from discrepancies of the state of polarization (SOP) of incoming optical signal and local laser signal. Different polarization noise avoidance techniques can be classified as follows (1) polarization control, (2) polarization maintenance fibers, (3) polarization scrambling, (4) polarization diversity, and (5) polarization multiplexing. The polarized electromagnetic field launched into the fiber can be represented as

$$\boldsymbol{E}(t) = \begin{bmatrix} e_x(t) \\ e_y(t) \end{bmatrix} \mathrm{e}^{\mathrm{j}\omega_c t}, \tag{2.67}$$

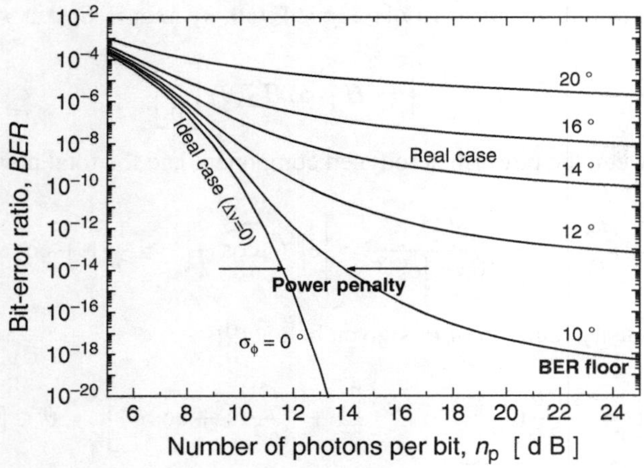

Fig. 2.36 The laser phase noise influence on BER performance of homodyne binary PSK signaling

where e_x and e_y represent two orthogonal SOP components and ω_c is the carrier frequency. The received field can be represented by

$$\boldsymbol{E}_s(t) = \boldsymbol{H}' \begin{bmatrix} e_x(t) \\ e_y(t) \end{bmatrix} e^{j\omega_c t}, \tag{2.68}$$

where $\boldsymbol{H}'$ is the Jones matrix of birefringence. The additional transformation is needed to match the SOPs of local laser with that of incoming optical signal:

$$\boldsymbol{E}'_s(t) = \boldsymbol{H}''\boldsymbol{H}' \begin{bmatrix} e_x(t) \\ e_y(t) \end{bmatrix} e^{j\omega_c t} = \boldsymbol{H} \begin{bmatrix} e_x(t) \\ e_y(t) \end{bmatrix} e^{j\omega_c t}, \quad \boldsymbol{H} = \boldsymbol{H}''\boldsymbol{H}'. \tag{2.69}$$

The SOP of local laser in Stokes coordinates can be represented by $\boldsymbol{S}_{LO} = (S_{1,LO}\ S_{2,LO}\ S_{3,LO})$. The heterodyning is possible only if $\boldsymbol{S}_{LO} = \boldsymbol{S}_R$, where $\boldsymbol{S}_R$ is the SOP of received signal. The action of birefringence corresponds to rotating the point, which represents the launched SOP, on the surface of the Poincaré sphere. This rotation can be represented as product of three matrices, each of them corresponding to the rotation of reference axes in the Stokes space around axes s_1, s_2, and s_3 for α, β, and γ, respectively [9]:

$$\boldsymbol{H}(\alpha,\beta,\gamma) = \boldsymbol{H}_1(\alpha)\boldsymbol{H}_2(\beta)\boldsymbol{H}_3(\gamma),$$

$$\boldsymbol{H}_1(\alpha) = \begin{bmatrix} e^{j\alpha/2} & 0 \\ 0 & e^{-j\alpha/2} \end{bmatrix}, \quad \boldsymbol{H}_2(\beta) = \begin{bmatrix} \cos(\beta/2) & j\sin(\beta/2) \\ j\sin(\beta/2) & \cos(\beta/2) \end{bmatrix},$$

$$\boldsymbol{H}_3(\gamma) = \begin{bmatrix} \cos(\lambda/2) & \sin(\gamma/2) \\ -\sin(\lambda/2) & \cos(\gamma/2) \end{bmatrix}. \tag{2.70}$$

If we assume that the SOP of LO is aligned with s_1-axis in Stokes space we can write

$$\begin{bmatrix} e'_x(t) \\ e'_y(t) \end{bmatrix} = \boldsymbol{H}_1(\phi)\boldsymbol{H}_2(\theta)\begin{bmatrix} e \\ 0 \end{bmatrix}. \tag{2.71}$$

The ratio between the power heterodyned component and the total power is

$$p(\theta,\phi) = \frac{P_{\text{het}}}{P_{\text{tot}}} = \frac{|e''_x|^2}{|e''_x|^2 + |e''_y|^2} = \left|e^{j\phi/2}\cos\frac{\theta}{2}\right|^2 = \frac{1}{2}(1+\cos\theta). \tag{2.72}$$

Probability density function of θ is given below [9]:

$$\text{PDF}(\theta) = \frac{\sin\theta}{2}e^{-\frac{A^2}{4\sigma^2}(1-\cos\theta)}\left[1+\frac{A^2}{4\sigma^2}(1+\cos\theta)\right], \quad \theta\in[0,\pi];$$
$$A = 2R\sqrt{P_s P_{\text{LO}}}, \quad \sigma^2 = qRP_{\text{LO}}. \tag{2.73}$$

The typical optical receiver is based on transimpedance FET stage, as illustrated in Fig. 2.37. The PSD of receiver noise can be written as [28]

$$\text{PSD}_{\text{rec}}(f) = \frac{4k_B T_a}{R_f} + 4k_B T_a g_m \Gamma\left(\frac{f}{f_{\text{T,eff}}}\right)^2, \quad f_{\text{T,eff}} = g_m/2\pi C_T, \quad C_T = C_i + C_{\text{PIN}}, \tag{2.74}$$

where g_m is transconductance, Γ is the FET channel-noise factor, and C_T is total capacitance (FET input capacitance C_i plus PIN photodetector capacitance C_{PIN}). (R_L is the load resistor, R_f is feedback resistor, T_a is absolute temperature, and k_B is the Boltzmann constant.)

This amplifier stage is popular because it has large gain–bandwidth product (GB_i) defined below as [28]:

$$\text{GB}_i = \left|-R_f\frac{g_m R_L}{1+g_m R_L}\right|\frac{1+g_m R_L}{2\pi C_T(R_f+R_i)} = f_{\text{T,eff}}(R_f||R_L). \tag{2.75}$$

The intensity noise comes from the variation of optical power of transmitting laser $P = \langle P\rangle + \Delta P$ (<·> denotes the statistical averaging). The RIN is defined

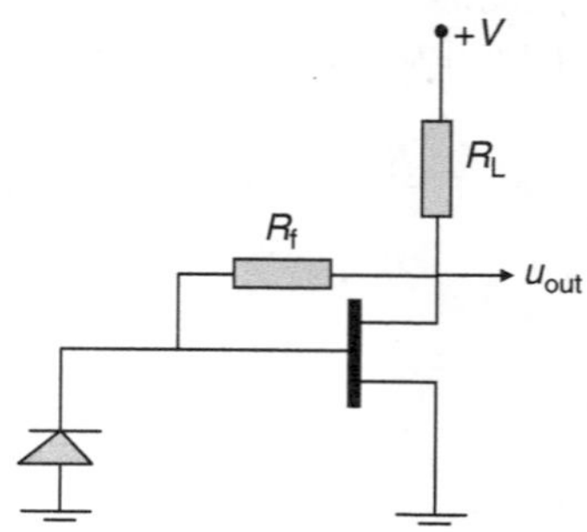

Fig. 2.37 Transimpedance FET receiver

as $\mathrm{RIN} = \langle \Delta P^2 \rangle / \langle P \rangle^2$. Because the power of transmitting laser fluctuates, the photocurrent fluctuates as well, $I = \langle I \rangle + \Delta I$. The corresponding shot noise PSD can be determined as $2q\langle I \rangle = 2qR\langle P \rangle$. The intensity noise PSD is simply $\langle \Delta I^2 \rangle = R^2 \mathrm{RIN} \langle P \rangle^2$. The SNR in the presence of shot, receiver, and RIN can be determined as signal power over total noise power:

$$\mathrm{SNR} = \frac{2R^2 P_\mathrm{s} P_\mathrm{LO}}{2qRP_\mathrm{LO}B + \mathrm{RIN} \cdot R^2 P_\mathrm{LO}^2 B + \mathrm{PSD_{rec}} B}, \tag{2.76}$$

where B is the receiver bandwidth, and $\mathrm{PSD_{rec}}$ is introduced by (2.74). The SNR for balanced reception can be written as

$$\mathrm{SNR} = \frac{2R^2 P_\mathrm{s} P_\mathrm{LO}}{2qRP_\mathrm{LO}B + \frac{\mathrm{RIN}}{\mathrm{CMRR}} R^2 P_\mathrm{LO}^2 B + 2\mathrm{PSD_{rec}} B}, \tag{2.77}$$

where CMRR is the common mode rejection ratio of FET. Therefore, the RIN of balanced receiver is significantly reduced by balanced detection.

2.4.3 Homodyne Coherent Detection

Homodyne coherent detection receiver can be classified as either residual carrier receivers or suppressed carrier receivers [29]. In systems with residual carrier, shown in Fig. 2.38, the phase deviation between the mark- and space-state bits is less than $\pi/2$ rad, so that the part of transmitted signal power is used for the nonmodulated carrier transmission and as a consequence some power penalty occurs.

Costas loop and DDL [29] based receivers, shown in Fig. 2.39, are two alternatives to the receivers with residual carrier. Both these alternatives employ a fully suppressed carrier transmission, in which the entire transmitted power is used for data transmission. However, at the receiver side a part of the power is used for the carrier extraction, so some power penalty is incurred with this approach too.

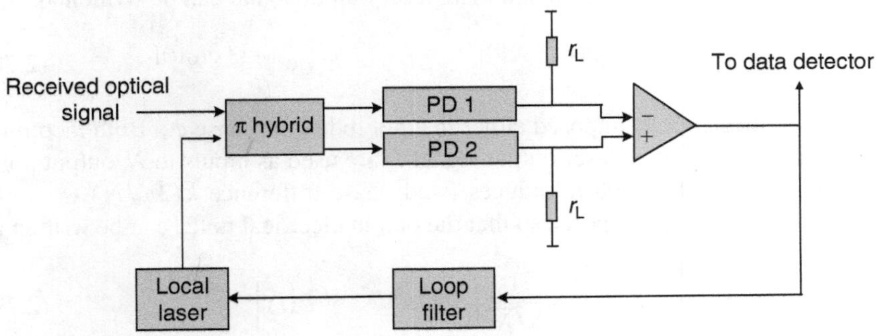

Fig. 2.38 Balanced-loop-based receiver. *PD* Photodetector

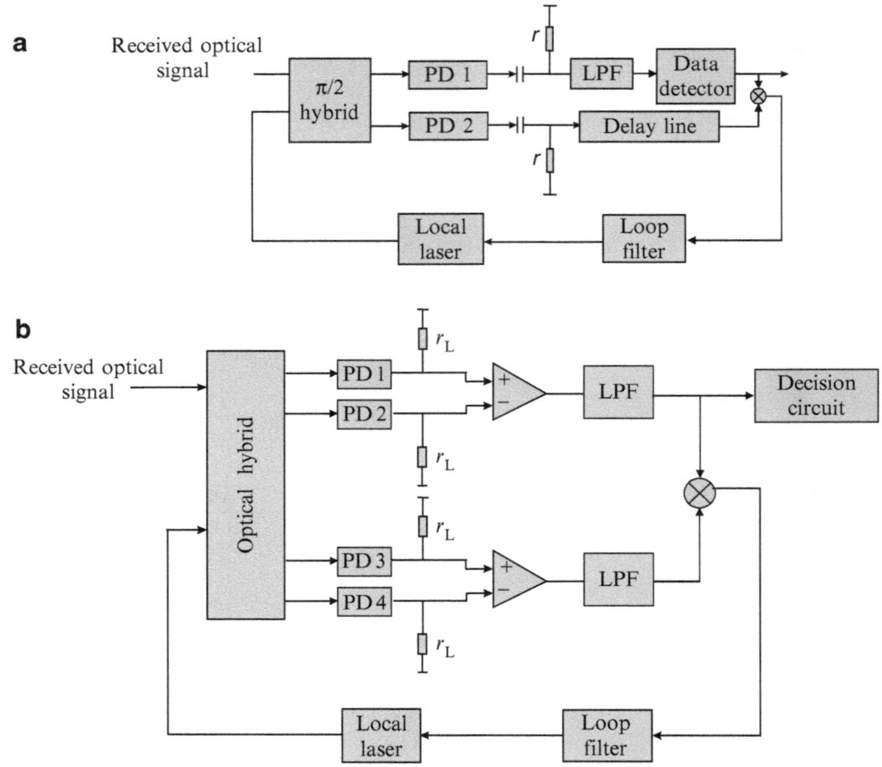

Fig. 2.39 (**a**) Decision-driven loop based receiver and (**b**) Costas loop based receiver. *PD* Photodetector, *LPF* low-pass filter

2.4.4 Phase Diversity Receivers

The general architecture of a multiport homodyne receiver is shown in Fig. 2.40a [9]. The incoming optical signal and local laser output signal can be written as

$$S(t) = aE_{\mathrm{s}}\mathrm{e}^{\mathrm{j}[\omega_{\mathrm{c}}t+\phi_{\mathrm{s}}(t)]}, \quad L(t) = E_{\mathrm{LO}}\mathrm{e}^{\mathrm{j}[\omega_{\mathrm{c}}t+\phi_{\mathrm{LO}}(t)]}, \tag{2.78}$$

where the information is imposed either in amplitude a or phase ϕ_{s}. Both incoming optical signal S and local laser output signal L are used as inputs to N output ports of an optical hybrid, which introduces fixed phase difference $k(2\pi/N)$ ($k = 0, 1, \ldots, N-1$) between the ports, so that the output electrical fields can be written as

$$E_k(t) = \frac{1}{\sqrt{N}}\left[S(t)\mathrm{e}^{\mathrm{j}k\frac{2\pi}{N}} + L(t)\right]. \tag{2.79}$$

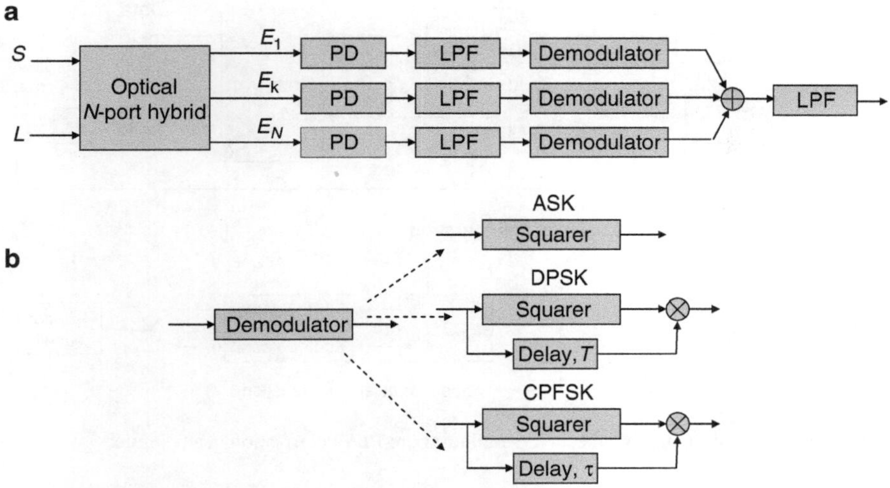

Fig. 2.40 Phase diversity receivers (**a**) and demodulator configurations (**b**)

The corresponding photodetector outputs are as follows:

$$\begin{aligned} i_k(t) &= R|E_k(t)| + i_{nk}(t) \\ &= \frac{R}{N}\left\{P_{\mathrm{LO}} + aP_{\mathrm{s}} + 2a\sqrt{P_{\mathrm{s}}P_{\mathrm{LO}}}\cos\left[\phi_{\mathrm{s}}(t) - \phi_{\mathrm{LO}}(t) + k\frac{2\pi}{N}\right]\right\} + i_{nk}(t), \end{aligned} \tag{2.80}$$

where $i_{nk}(t)$ is the kth photodetector shot noise. Different versions of demodulators for ASK, DPSK, and DPFSK are shown in Fig. 2.40b. For ASK, we simply have to square photodetector outputs and add them together:

$$y = \sum_{k=1}^{N} i_k^2. \tag{2.81}$$

2.4.5 Polarization Control and Polarization Diversity

The coherent receivers require matching the SOP of the local laser with that of the received optical signal. In practice, only the SOP of local laser can be controlled, and one possible polarization control receiver configuration is shown in Fig. 2.41. Polarization controller is commonly implemented by using four squeezers as shown in [9].

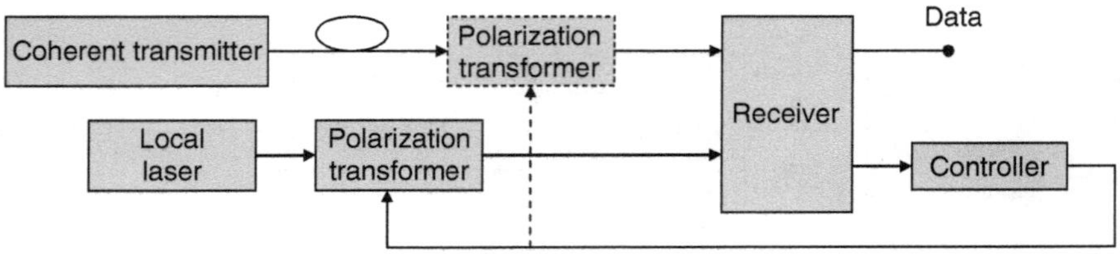

Fig. 2.41 Polarization control receiver configuration

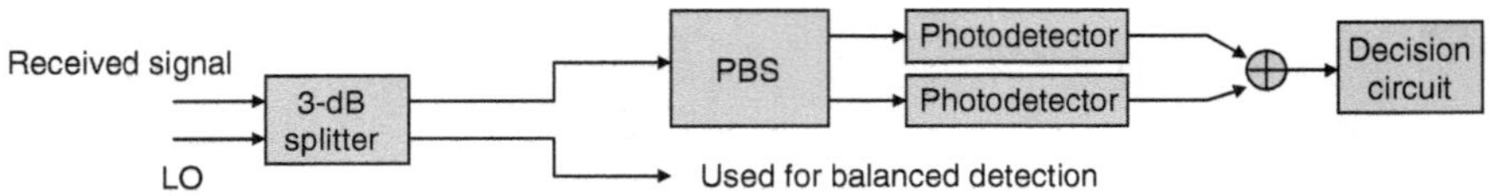

Fig. 2.42 Polarization diversity receiver configuration. *PBS* Polarization beam splitter

Insensitivity with respect to polarization fluctuations is possible if the receiver derives two demodulated signals from two orthogonal polarizations of the received signal, which is illustrated in Fig. 2.42. This scheme is known as polarization diversity receiver. In polarization diversity receivers, however, only one polarization is used and spectral efficiency is reduced. To double the spectral efficiency of polarization diversity schemes, the polarization multiplexing is advocated in [30–33].

2.4.6 Polarization Multiplexing and Coded Modulation

In polarization multiplexing [30–35], both polarizations carry independent multilevel modulated streams, which is illustrated in Fig. 2.43. *M*-ary PSK, *M*-ary QAM, and *M*-ary DPSK achieve the transmission of $\log_2 M(=m)$ bits per symbol, providing bandwidth-efficient communication. In coherent detection, the data phasor $\phi_l \in \{0, 2\pi/M, \ldots, 2\pi(M-1)/M\}$ is sent at each lth transmission interval. In direct detection, the modulation is differential, the data phasor $\phi_l = \phi_{l-1} + \Delta\phi_l$ is sent instead, where $\Delta\phi_l \in \{0, 2\pi/M, \ldots, 2\pi(M-1)/M\}$ is determined by the sequence of $\log_2 M$ input bits using an appropriate mapping rule. Let us now introduce the transmitter architecture employing forward error correction (FEC) codes. More details about coded modulation can be found in Chap. 6. If the component FEC codes are of different code rates but of the same length, the corresponding scheme is commonly referred to as multilevel coding (MLC) [34]. If all component codes are of the same code rate, corresponding scheme is referred to as the bit-interleaved coded modulation (BICM) [35]. The use of MLC allows us to adapt the code rates to the constellation mapper and channel. In MLC, the bit streams originating from m different information sources are encoded using different (n, k_i) FEC codes of code rate $r_i = k_i/n$. k_i denotes the number of information bits of the ith $(i = 1, 2, \ldots, m)$ component FEC code and n denotes the codeword length, which

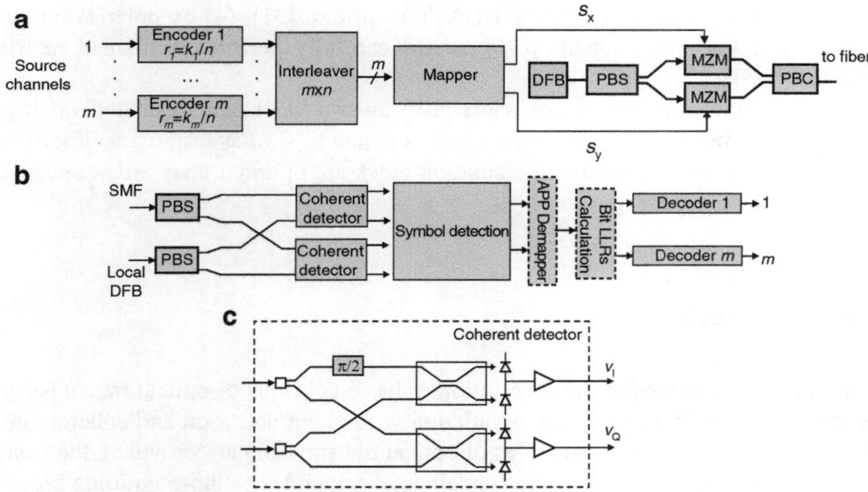

Fig. 2.43 The polarization-multiplexed multilevel-coded modulation scheme (**a**) transmitter configuration, (**b**) receiver configuration, and (**c**) coherent balanced detector architecture. *DFB* Distributed feedback laser, *PBS/C* polarization beam splitter/combiner, *MZM* dual-drive Mach–Zehnder modulator

is the same for all FEC codes. The mapper accepts m bits, $\boldsymbol{c} = (c_1, c_2, \ldots, c_m)$, at time instance i from the $(m \times n)$ interleaver column-wise and determines the corresponding M-ary ($M = 2^m$) constellation point $\boldsymbol{s}_i = (I_i, Q_i) = |\boldsymbol{s}_i| \exp(\mathrm{j}\phi_i)$ (see Fig. 2.43a). Two dual-drive MZMs are needed, one for each polarization. The outputs of the MZMs are combined using the polarization beam combiner (PBC). The same DFB laser is used as CW source, with x- and y-polarization being separated by a polarization beam splitter (PBS).

The coherent detector receiver architecture is shown in Fig. 2.43b. The balanced outputs of I- and Q-channel branches for x-polarization at time instance l can be written as

$$v_{I,l}^{(x)} = R\left|S_l^{(x)}\right|\left|L^{(x)}\right| \cos\left(\varphi_l^{(x)} + \varphi_{S,\mathrm{PN}}^{(x)} - \varphi_{L,\mathrm{PN}}^{(x)}\right),$$

$$v_{Q,l}^{(x)} = R\left|S_l^{(x)}\right|\left|L^{(x)}\right| \sin\left(\varphi_l^{(x)} + \varphi_{S,\mathrm{PN}}^{(x)} - \varphi_{L,\mathrm{PN}}^{(x)}\right), \tag{2.82}$$

where R is photodiode responsivity while $\varphi_{S,\mathrm{PN}}$ and $\varphi_{L,\mathrm{PN}}$ represent the laser phase noise of transmitting and receiving (local) laser, respectively. $S_l^{(x)}$ and $L^{(x)}$ represent the receiver incoming signal in x-polarization and local laser x-polarization signal, respectively. Similar expressions hold for y-polarization. In symbol detection block, the PMD is compensated for using one of the following possible approaches (1) blind equalization [30], (2) polarization-time coding [32] similar to space–time coding proposed for use in multi-input multi-output (MIMO) wireless communication systems, (3) using BLAST algorithm [31], (4) by polarization interference cancellation scheme [31], or (5) carefully performed channel matrix inversion [35].

For soft decoding, the a posteriori probability (APP) demapper and bit log-likelihood ratios (LLRs) calculation blocks operate in similar fashion as described in [31–35]. The APP and LLRs calculation block are optional; they are not needed if hard decision decoding is used.

2.5 Summary

This chapter is devoted to the description of basic concepts of optical transmission systems based on both intensity modulation with direct detection and coherent detection. In Sect. 2.2, basic principles of optical transmission are provided, the basic building blocks are identified, and fundamental principles of those building blocks are described. Section 2.3 is devoted to the description of basic direct detection modulations schemes such as NRZ, RZ, AMI, duobinary modulation, carrier-suppressed RZ, NRZ-DPSK, and RZ-DSPK.

In Sect. 2.4, the basic concepts of coherent detection are introduced; including description and comparison of different coherent detection schemes for shot-noise-dominated scenario, description of dominant receiver noises, description of homodyne detection principle, phase diversity and polarization diversity principles, and polarization multiplexing.

References

1. Cvijetic M (2004) Optical transmission systems engineering. Artech House, Boston, MA
2. Ramaswami R, Sivarajan K (2002) Optical networks: a practical perspective, 2nd edn. Morgan Kaufman, San Fransisco, CA
3. Agrawal GP (2002) Fiber-optic communication systems, 3rd edn. Wiley, New York
4. Agrawal GP (2004) Lightwave technology: components and devices. Wiley, New York
5. Agrawal GP (2005) Lightwave technology: telecommunication systems. Wiley, New York
6. Agrawal GP (2007) Nonlinear fiber optics, 4th edn. Academic, New York
7. Cvijetic M (1996) Coherent and nonlinear lightwave communications. Artech House, Boston, MA
8. Keiser G (2000) Optical fiber communications, 3rd edn. McGraw-Hill, New York
9. Kazovsky L, Benedetto S, Willner A (1996) Optical fiber communication systems. Artech House, Boston, MA
10. Palais JC (2005) Fiber optic communications, 5th edn. Pearson Prentice Hall, Upper Saddle River, NJ
11. Ramamurthy B (2000) Switches, wavelength routers, and wavelength converters. In: Sivalingam KM, Subramaniam S (eds) Optical WDM networks: principles and practice. Kluwer Academic, New York
12. Djordjevic IB (2006/2007) ECE 430/530: optical communication systems (lecture notes). University of Arizona, Tuczon, AZ

13. Kostuk R (2002/2004) ECE 430/530: optical communication systems (lecture notes). University of Arizona, Tuczon, AZ
14. Djordjevic IB (2007) ECE 632: advanced optical communication systems (lecture notes). University of Arizona, Tuczon, AZ
15. Yariv A (1997) Optical electronics in modern communications. Oxford University Press, Oxford
16. Amersfoort M (1998) Arrayed waveguide grating, Application note A1998003. Available at http://www.c2v.nl
17. Staif M, Mecozzi A, Nagel J (2000) Mean square magnitude of all orders PMD and the relation with the bandwidth of principal states. IEEE Photon Technol Lett 12:53–55
18. Kogelnik H, Nelson LE, Jobson RM (2002) Polarization mode disperison. In: Kaminow IP, Li T (eds) Optical fiber telecommunications. Academic, San Diego, CA
19. Foschini GJ, Pole CD (1991) Statistical theory of polarization mode dispersion in single-mode fibers. IEEE/OSA J Lightwave Technol LT-9:1439–1456
20. Bendelli G et al (2000) Optical performance monitoring techniques. In: Proceedings of ECOC 2000, Munich, vol 4, pp 213–216
21. Djordjevic IB, Vasic B (2004) An advanced direct detection receiver model. J Opt Commun 25(1):6–9
22. Winzer PJ, Pfennigbauer M, Strasser MM, Leeb WR (2001) Optimum filter bandwidths for optically preamplified NRZ receivers. J Lightwave Technol. 19:1263–1272
23. Hui R et al (2004) Technical report: advanced optical modulation formats and their comparison in fiber-optic systems. University of Kansas, Lawrence, KS
24. Wongpaibool V (2003) Improvement of fiber optic system performance by synchronous phase modulation and filtering at the transmitter. PhD dissertation, Virginia Polytechnic Institute and State University
25. Sano A, Miyamoto Y (2001) Performance evaluation of prechirped RZ and CS-RZ formats in high-speed transmission systems with dispersion management. IEEE/OSA J Lightwave Technol 19:1864–1871
26. Proakis JG (2001) Digital communications. McGraw-Hill, Boston, MA
27. Jacobsen G (1994) Noise in digital optical transmission systems. Artech House, Boston, MA
28. Hooijmans PW (1994) Coherent optical system design. Wiley, New York
29. Djordjevic IB (1999) Analysis and optimization of coherent optical systems with the phase-locked loop. PhD dissertation, University of Nis, Serbia
30. Savory SJ (2008) Digital filters for coherent optical receivers. Opt Express 16:804–817
31. Djordjevic IB, Xu L, Wang T (2008) PMD compensation in multilevel coded-modulation schemes with coherent detection using BLAST algorithm and iterative polarization cancellation. Opt Express 16(19):14845–14852
32. Djordjevic IB, Xu L, Wang T (2008) PMD compensation in coded-modulation schemes with coherent detection using Alamouti-type polarization-time coding. Opt Express 16(18):14163–14172
33. Minkov LL, Djordjevic IB, Xu L, Wang T (2009) IEEE Photon. Technol. Lett. 21(23): 1773–1775
34. Djordjevic IB, Arabaci M, Minkov L (2009), Next generation FEC for high-capacity communication in optical transport networks. IEEE/OSA J Lightwave Technol 27(16):3518–3530 (Invited Paper)
35. Djordjevic IB, Xu L, Wang T (2009) Beyond 100 Gb/s optical transmission based on polarization multiplexed coded-OFDM with coherent detection. IEEE/OSA J Opt Commun Netw 1(1):50–56

Chapter 3
Channel Impairments and Optical Communication Systems Engineering

In this chapter, we describe basic signal and noise parameters; major channel impairments including chromatic dispersion, multimode dispersion, polarization mode dispersion (PMD), and fiber nonlinearities; and system design process. This chapter is based on [1–23]. This chapter is organized as follows. In Sect. 3.1, different noise sources are identified and explained. Section 3.2 is devoted to different channel impairments. Section 3.3 deals with different figures of merit to describe system performance and basic guidelines for system design.

3.1 Noise Sources

The total noise is a stochastic process that has both multiplicative (introduced only when the signal is present) and additive (always present) components [1]. *Multiplicative noise components* are mode partition noise (MPN), relative intensity noise (RIN), modal noise, quantum shot noise, and avalanche shot noise. *Additive noise components* are dark current noise, thermal noise, amplified-spontaneous emission (ASE) noise, and crosstalk noise.

Originators of the noise in an optical transmission system can be identified as follows. In semiconductor lasers, the laser intensity noise, laser phase noise, and MPN are present. In optical cable, in fiber and splicing, the modal noise and reflection-induced noise are present. In optical amplifiers, the spontaneous emission and ASE noises are present. At the receiver side, during the photodetection process, the thermal noise and quantum noise are generated. The noise components at the receiver side are cumulative. The most relevant optical noise components at the input of optical preamplifier are intensity noise and spontaneous emission noise, which contain the spontaneous emission noise from the preceding in-line amplifier and accumulated ASE noise from other in-line amplifiers. The optical preamplifier will enhance all optical inputs in proportion to the gain, and it will also generate additional spontaneous emission noise. During the photodetection process, as already indicated above the quantum noise and thermal noise will be generated, in addition, different beating components, such as signal–noise beating and noise–noise

I. Djordjevic et al., *Coding for Optical Channels*,
DOI 10.1007/978-1-4419-5569-2_3, © Springer Science+Business Media, LLC 2010

beating components, will be created. The most important noise source coming to the front-end amplifiers are ASE noise and signal–noise beating component. The front-end amplifier will generate additional thermal noise.

3.1.1 Mode Partition Noise

The MPN is the relative fluctuation in power between the main and side modes. The total power can remain unchanged but the distribution of power among modes changes. MPN is present (up to certain amount) even in DFBs. MPN can affect the RIN significantly by enhancing it. It will occur even when the mode suppression ratio (MSR) is around 20 dB. Different modes will travel at different velocities due to dispersion, so that MPN can lead to the ISI.

3.1.2 Reflection-Induced Noise

The reflection-induced noise is related to the appearance of the back-reflected light due to refractive index discontinuities at optical splices, connectors, and optical fiber ends. The amount of reflected light can be estimated by the *reflection coefficient* [1]:

$$r_{\text{ref}} = \left(\frac{n_a - n_b}{n_a + n_b}\right)^2, \tag{3.1}$$

where n_a and n_b denote the refractive index coefficients of materials facing each other.

The strongest reflection occurs at the glass–air interface, $r_{ref} \approx [(1.46–1)/(1.46+1)]^2 \approx 3.5\%$ (-14.56 dB). It can be reduced below 0.1% if index-matching oils or gels are used. The considerable amount of back-reflected light can come back and enter the laser cavity, negatively affecting the laser operation and leading to excess intensity noise. The RIN can be increased by up to 20 dB if the back-reflected light exceeds -30 dBm. The multiple back and forth reflections between splices and connectors can be the source of additional intensity noise.

3.1.3 Relative Intensity Noise (RIN) and Laser Phase Noise

The operating characteristics of semiconductor lasers are well described by the set of ordinary differential equations – *the rate equations,* which describe the interaction between photons and electrons inside the active region [1, 3, 4]:

$$\frac{\mathrm{d}P}{\mathrm{d}t} = \mathrm{GP} + R_{\mathrm{sp}} - \frac{P}{\tau_{\mathrm{p}}} + F_{\mathrm{P}}(t), \tag{3.2a}$$

$$\frac{dN}{dt} = \frac{I}{q} - \frac{N}{\tau_c} - GP + F_N(t), \text{ and} \tag{3.2b}$$

$$\frac{d\phi}{dt} = \frac{1}{2}\alpha_{\text{chirp}}\left[G - \frac{1}{\tau_p}\right] + F_\phi(t). \tag{3.2c}$$

In (3.2a) the term GP denotes the increase in number of photons (P) due to stimulated emission, the term R_{sp} denotes the increase in P due to spontaneous emission and can be calculated by $R_{sp} = n_{sp}G$ (n_{sp} – the spontaneous emission factor, typically $n_{sp} \approx 2$), the term $-P/\tau_p$ denotes the decrease in P due to emission through the mirrors and scattering/absorption by free carriers, and $F_P(t)$ corresponds to the noise process. In (3.2c) the term I/q denotes the increase in electron numbers (N) due to injection current I, the term $-N/\tau_c$ denotes the decrease in N due to spontaneous emission and nonradiative recombination, the term –GP the decrease in N due to stimulated emission, and $F_N(t)$ corresponds to the noise process. Finally, in (3.2b) the first term corresponds to dynamic chirp, the second term to the adiabatic chirp, and $F_\phi(t)$ corresponds to the noise process. The noise terms $F_P(t)$, $F_N(t)$, and $F_\phi(t)$ introduced above are known as *Langevin forces* that are commonly modeled as zero-mean Gaussian random processes. The photon lifetime τ_p is related to the internal losses (α_{int}) and mirror losses (α_{mir}) by $\tau_p = \left[v_g(\alpha_{int} + \alpha_{mir})\right]^{-1}$, with v_g being the group velocity. The carrier lifetime τ_c is related to the spontaneous recombination time (τ_{spon}) and nonradiation recombination time (τ_{nr}) by $\tau_c = \tau_{spon}\tau_{nr}/(\tau_{spon} + \tau_{nr})$. The net rate of stimulate emission G is related to the material gain g_m by $G = \Gamma v_g g_m$, where Γ is the confinement factor.

Noise in semiconductor lasers originates from two sources (1) spontaneous emission (the dominant noise source) and (2) electron–hole recombination shot noise. As mentioned above, the noise effects can be modeled by random driving terms in laser rate equations known as *Langevin forces*, which can be described as zero-mean Gaussian random processes with autocorrelation function (*Markoffian approximation*) [3, 4]:

$$\langle F_i(t)F_j(t')\rangle = 2D_{ij}\delta(t - t'); \quad i, j \in \{P, N, \phi\}, \tag{3.3}$$

with dominating factor being $D_{PP} = R_{sp}P$ and $D_{\phi\phi} = R_{sp}/4P$. Fluctuation in intensity can be described by *intensity autocorrelation function* [3, 4]:

$$C_{PP}(\tau) = \langle \delta P(t)\delta P(t+\tau)\rangle / \overline{P}^2, \quad \overline{P} = \langle P\rangle, \quad \delta P = P - \overline{P}. \tag{3.4}$$

The RIN spectrum can be found by FT of the autocorrelation function [3, 4]:

$$\text{RIN}(\omega) = \int_{-\infty}^{\infty} C_{PP}(\tau)e^{-j\omega t}d\tau \sim \begin{cases} 1/\overline{P}^3 \text{ at low } \overline{P} \\ 1/\overline{P} \text{ at high } \overline{P} \end{cases} . \tag{3.5}$$

Finally, the SNR can be estimated by [3, 4]

$$\mathrm{SNR} = [C_{\mathrm{PP}}(0)]^{-1/2} \sim \left(\frac{\varepsilon_{\mathrm{NL}}}{R_{\mathrm{sp}}\tau_{\mathrm{p}}}\right)^{1/2} \overline{P}, \tag{3.6}$$

where the approximation is valid for SNR above 20 dB.

Noise, especially spontaneous emission, causes phase fluctuations in lasers, leading to a nonzero spectral linewidth $\Delta\nu$. In gas or solid-state lasers, the linewidth $\Delta\nu$ typically ranges from the subhertz to the kilohertz range. In semiconductor lasers, the linewidth $\Delta\nu$ is often much larger, up the megahertz range, because of (1) the small number of photons stored in the small cavity and (2) the nonnegligible value of the linewidth enhancement factor α_{chirp}. The spectrum of emitted light electric field $E(t) = \sqrt{(P)}\exp(\mathrm{j}\phi)$ is related to the field-autocorrelation function $\Gamma_{\mathrm{EE}}(\tau)$ by

$$S(\omega) = \int_{-\infty}^{\infty} \Gamma_{\mathrm{EE}}(t)\mathrm{e}^{-\mathrm{j}t(\omega-\omega_0)\tau}\mathrm{d}\tau,$$
$$\Gamma_{\mathrm{EE}}(t) = \langle E^*(t)E(t+\tau)\rangle = \langle \exp(\mathrm{j}\Delta\phi(t))\rangle = \exp\left(-\left\langle \Delta\phi^2(\tau)\right\rangle/2\right), \tag{3.7}$$

where $\Delta\phi(\tau) = \phi(t+\tau) - \phi(t)$. By describing the laser phase noise process as Wiener–Lévy process [7], that is a zero-mean Gaussian process with variance $2\pi\Delta\nu|t|$ (where $\Delta\nu$ is the laser linewidth), the spectrum of emitted light is found to be Lorentzian:

$$S_{\mathrm{EE}}(\nu) = \frac{\overline{P}}{2\pi\Delta\nu}\left[\frac{1}{1+\left(\frac{2(\nu+\nu_0)}{\Delta\nu}\right)^2} + \frac{1}{1+\left(\frac{2(\nu-\nu_0)}{\Delta\nu}\right)^2}\right], \tag{3.8}$$

where ν_0 is the central frequency and other parameters are already introduced above. The phase fluctuation variance (neglecting the relaxation oscillation) is determined by [3, 4]

$$\left\langle \Delta\phi^2(\tau)\right\rangle = \frac{R_{\mathrm{sp}}}{2\overline{P}}\left(1+\alpha_{\mathrm{chirp}}^2\right)|\tau| = 2\pi\Delta\nu\,|\tau|. \tag{3.9a}$$

The laser linewidth can therefore be evaluated by

$$\Delta\nu = \frac{R_{\mathrm{sp}}}{4\pi\overline{P}}\left(1+\alpha_{\mathrm{chirp}}^2\right). \tag{3.9b}$$

From (3.9b), we see that the chirp effect enhances the laser linewidth by factor $\left(1+\alpha_{\mathrm{chirp}}^2\right)$.

3.1.4 Modal Noise

Modal noise is related to multimode optical fibers. The optical power is nonuniformly distributed among a number of modes in multimode fibers (MMFs), causing the so-called speckle pattern at the receiver end containing brighter and darker sports in accordance to the mode distribution. If the speckle pattern is stable, the photodiode effectively eliminates it by registering the total power over the photodiode area. If the speckle pattern changes with time, it will induce the fluctuation in the received optical power, known as the modal noise. The modal noise is inversely proportional to the laser linewidth, so it is a good idea to use LEDs in combination with MMFs.

3.1.5 *Quantum Shot Noise*

The optical signal arriving at the photodetector contains a number of photons generating the electron–hole pairs through the photoelastic effect. The electron–hole pairs are effectively separated by the reversed-bias voltage. However, not every photon is going to generate the electron–hole pair contributing to the total photocurrent. The probability of having n electron–hole pairs during the time interval Δt is governed by the *Poisson probability density function* (PDF) as given by

$$p(n) = \frac{N^n \mathrm{e}^{-N}}{n!}, \tag{3.10}$$

where N is the average number of generated electron–hole pairs. The Poisson PDF approaches Gaussian for large N. The *mean* of *photocurrent intensity* can be determined by

$$I = \langle i(t) \rangle = \frac{qN}{\Delta t} = \frac{qN}{T}, \quad \Delta t = T. \tag{3.11}$$

For Poisson distribution, the variance equals the mean $\langle (n-N)^2 \rangle = N$ so the photocurrent can be calculated by $i(t) = qn/T$. The *shot-noise mean-square value* can be determined by

$$\begin{aligned} \langle i^2 \rangle_{\mathrm{sn}} &= \langle [i(t) - I]^2 \rangle = \frac{q^2 \langle [n-N]^2 \rangle}{T^2} = \frac{q^2 N}{T^2} = \frac{qI}{T} = 2qI\Delta f, \\ \Delta f &= \frac{R_{\mathrm{b}}}{2} \quad \left(R_{\mathrm{b}} = \frac{1}{T} \right), \end{aligned} \tag{3.12}$$

where R_{b} is the bit rate. Equation (3.12) is derived for NRZ modulation format, but validity is more general. Therefore, the power-spectral density of shot noise is determined by $S_{\mathrm{sn}}(f) = 2qI$. To determine the avalanche photodiode (APD) *shot-noise mean-square value*, the p–i–n shot-noise mean-square value has to be multiplied by the excess noise factor $F(M)$ to account for the randomness of impact ionization effect [1–3]:

$$\langle i^2 \rangle_{\mathrm{sn}}^{\mathrm{APD}} = S_{\mathrm{sn}}^{\mathrm{APD}}(f)\Delta f = 2qM^2 F(M) I \Delta f; \quad F(M) = k_N M + (1-k_N)\left[2 - \frac{1}{M}\right], \tag{3.13}$$

where M is the average APD multiplication factor and k_N is the ratio of impact ionization factors of holes and electrons, respectively. The often used excess noise factor approximation is the following one [1]:

$$F(M) \cong M^x, \quad x \in [0, 1]. \tag{3.14}$$

3.1.6 Dark Current Noise

The dark current I_d flows through the reversed-biased photodiode even in the absence of incoming light. The dark current consists of electron–hole pairs created due to thermal effects in p–n junction. The dark current noise power can be calculated by modifying 3.13 as follows:

$$\left\langle i^2 \right\rangle_{\mathrm{dcn}}^{\mathrm{APD}} = S_{\mathrm{dcn}}^{\mathrm{APD}}(f)\Delta f = 2q \left\langle M \right\rangle^2 F(M) I_\mathrm{d} \Delta f. \tag{3.15}$$

3.1.7 Thermal Noise

The photocurrent generated during photodetection process is converted to the voltage through the load resistance. The load voltage is further amplified by the front-end amplifier stage. Due to random thermal motion of electrons in load resistance, the already generated photocurrent exhibits additional noise component, also known as *Johnson noise*. The front-end amplifier enhances the thermal noise generated in the load resistance, which is described by the *amplifier noise figure* $F_{\mathrm{n,el}}$. The *thermal noise PSD* in the load resistance R_L, and corresponding *thermal noise power* are defined as [1–4]

$$S_{\mathrm{thermal}} = \frac{4k_\mathrm{B}\Theta}{R_\mathrm{L}}; \quad \left\langle i^2 \right\rangle_{\mathrm{thermal}} = \frac{4k_\mathrm{B}\Theta}{R_\mathrm{L}}\Delta f, \tag{3.16}$$

where Θ is the absolute temperature, k_B the Boltzmann's constant, and Δf is the receiver bandwidth.

3.1.8 Spontaneous Emission Noise

The spontaneous emission of light appears during the optical signal amplification, and it is not correlated with signal (it is additive in nature). The noise introduced by spontaneous emission has a flat frequency characterized by Gaussian PDF, and the (double-sided) PSD in one state of polarization (SOP) is given by [1–10]

$$S_{sp}(\nu) = (G-1)F_{no}h\nu/2, \tag{3.17}$$

where G is the amplifier gain and $h\nu$ is the photon energy. The *optical amplifier noise figure*, F_{no}, defined as the ratio of SNRs at the input and output of the optical amplifier, is related to the *spontaneous emission* factor $n_{sp} = N_2/(N_2 - N_1)$ (for the two-level system) by [3]

$$F_{no} = 2n_{sp}\left(1-\frac{1}{G}\right) \cong 2n_{sp} \geq 2. \tag{3.18}$$

In the most of practical cases, the noise figure is between 3 and 7 dB. The effective noise figure of the cascade of K amplifiers with corresponding gains G_i and noise figures $F_{no,i}\,(i = 1, 2, \ldots, K)$ [1]:

$$F_{no} = F_{no,1} + \frac{F_{no,2}}{G_1} + \frac{F_{no,3}}{G_1G_2} + \cdots + \frac{F_{no,K}}{G_1G_2\cdots G_{K-1}}. \tag{3.19}$$

The total power of the spontaneous emission noise, for an amplifier followed by an optical filter of bandwidth B_{op} is determined by

$$P_{sp} = 2\left|E_{sp}\right|^2 = 2S_{sp}(\nu)B_{op} = (G-1)F_{no}h\nu B_{op}, \tag{3.20}$$

where the factor 2 is used to account for both polarizations.

3.1.9 Noise Beat Components

Optical preamplifiers are commonly used *to improve the receiver sensitivity* by preamplifying the signal before it reaches the photodetector, which is illustrated in Fig. 3.1. The gain of the optical amplifier is denoted with G, the optical filter bandwidth is denoted with B_{op}, and electrical filter bandwidth is denoted with B_{el}.

An optical amplifier introduces the spontaneous emission noise in addition to the signal amplification:

$$E(t) = \sqrt{2P}\cos(2\pi f_c t + \theta) + n(t), \tag{3.21}$$

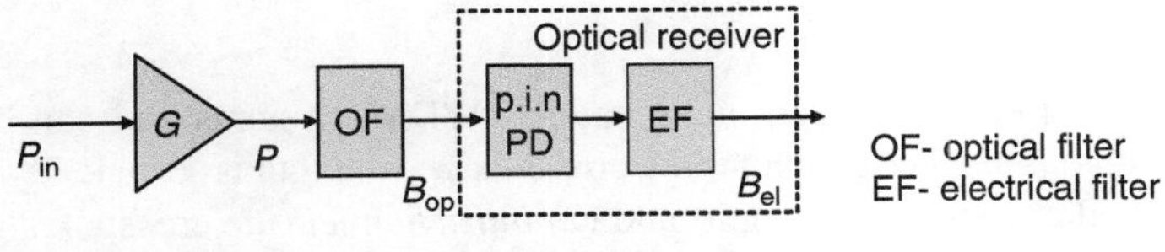

Fig. 3.1 The optical receiver with the preamplifier

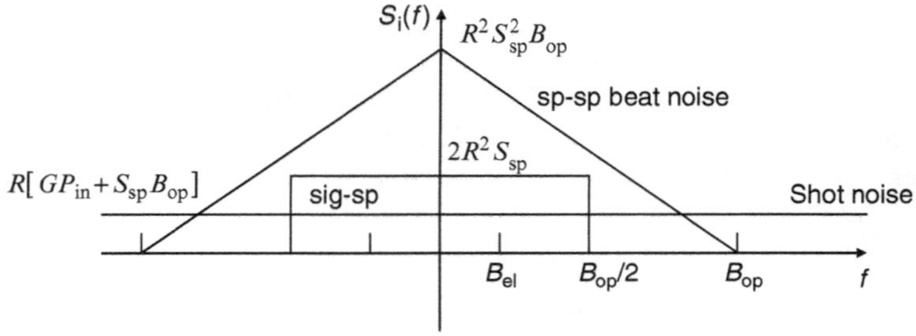

Fig. 3.2 The components of PSD of photocurrent

where $E(t)$ denotes the electrical field at the receiver input and $n(t)$ denotes the amplifier spontaneous emission noise, which is the Gaussian process with PSD given by (3.20). The photocurrent has the following form:

$$i(t) = RP(t) = RE^2(t) = 2RP\cos^2(2\pi f_c t+\theta)+ 2R\sqrt{2P}n(t)\cos(2\pi f_c t+\theta)+Rn^2(t). \tag{3.22}$$

Following the procedure described in [2], we can determine the variance of different noise and beating components as follows:

$$\begin{aligned}
\sigma^2 &= \int_{-B_{el}}^{B_{el}} S_i(f)\mathrm{d}f = \sigma^2_{sh} + \sigma^2_{sig-sp} + \sigma^2_{sp-sp},\\
\sigma^2_{sh} &= 2qR\left[GP_{in} + S_{sp}B_{op}\right]B_{el}, \quad \sigma^2_{sig-sp} = 4R^2GP_{in}S_{sp}B_{el},\\
\sigma^2_{sp-sp} &= R^2S^2_{sp}(2B_{op} - B_{el})B_{el},
\end{aligned} \tag{3.23}$$

where we used subscripts "sh" to denote the variance of shot noise, "sp–sp" to denote the variance of spontaneous–spontaneous beat noise, and "sig–sp" to denote the variance of signal–spontaneous beating noise. S_{sp} is the PSD of spontaneous emission noise (3.20), P_{in} is the average power of incoming signal (see Fig. 3.1), and the components of the PSD of photocurrent $i(t)$, denoted by $S_i(f)$, are shown in Fig. 3.2.

3.1.10 Crosstalk Components

The crosstalk effects occurs in multichannel (WDM) systems and can be classified as (1) *interchannel* component (the crosstalk wavelength is sufficiently different from observed channel wavelength) and (2) *intrachannel* (the crosstalk signal is at the same wavelength as observed channel or sufficiently close) component.

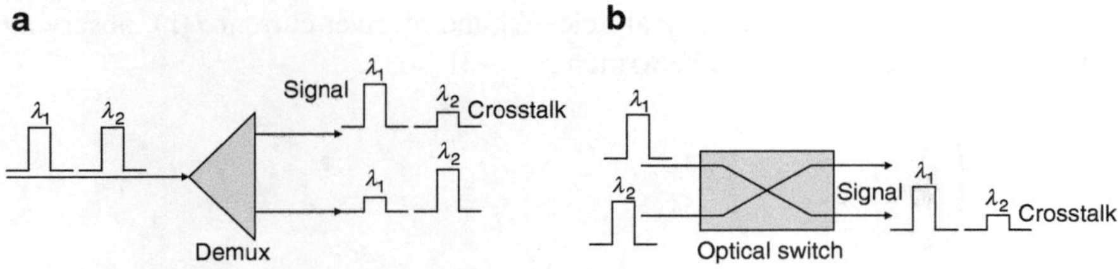

Fig. 3.3 Sources of interchannel crosstalk (**a**) demux and (**b**) the optical switch

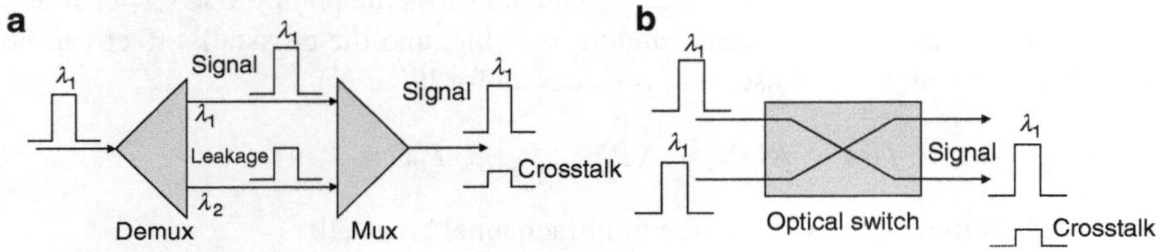

Fig. 3.4 Sources of intrachannel crosstalk (**a**) a demux–mux pair and (**b**) the optical switch

Interchannel crosstalk (also known as out-of-band or heterowavelength crosstalk) is introduced by either an *optical filter* or *demultiplexer* that selects desired channel and imperfectly rejects the neighboring channels, which is illustrated in Fig. 3.3a. Another source of interchannel crosstalk is an *optical switch*, shown in Fig. 3.3b, in which the crosstalk arises because of imperfect isolation among different wavelength ports. The crosstalk signal behaves as noise and it is a source of linear (noncoherent) crosstalk, so that the photocurrent at the receiver can be written as [1, 3]

$$I = R_m P_m + \sum_{n \neq m} R_n X_{mn} P_n \equiv I_{\mathrm{ch}} + I_X, \quad R_m = \eta_m q / h\nu_m, \quad I_X = \sum_{n \neq m} R_n X_{mn} P_n, \tag{3.24}$$

where R_m is the photodiode responsivity of mth WDM channel with average power P_m, $I_{\mathrm{ch}} = R_m P_m$ is the photocurrent corresponding to the selected channel, I_X is the crosstalk photocurrent, and X_{mn} is the portion of nth channel power captured by the optical receiver of the mth optical channel.

The intrachannel crosstalk in transmission systems may occur due to multiple reflections. However, it is much more important in optical networks. In optical networks, it arises from cascading optical (WDM) demux and mux or from an optical switch, which is illustrated in Fig. 3.4. Receiver current $I(t) = R|E_m(t)|^2$ contains interference or beat terms, in addition to the desired signal (1) signal–crosstalk beating terms (like $E_m E_n$) and (2) crosstalk–crosstalk beating terms (like $E_n E_k$,

$k \neq m, n \neq m$). The total electrical field E_m and receiver current $I(t)$, observing only the in-band crosstalk, can be written as [1–3]

$$E_m(t) = \left(E_m + \sum_{n \neq m} E_n \right) e^{-j\omega_m t},$$

$$I(t) = R\,|E_m(t)|^2 \approx RP_m(t) + 2R \sum_{n \neq m}^{N} \sqrt{P_m P_n} \cos[\phi_m(t) - \phi_n(t)], \quad (3.25)$$

where $\phi_n(t)$ denotes the phase of nth channel and R is the photodiode responsivity. Each term acts as an independent random variable, and the crosstalk effect can be observed as the intensity noise (that is Gaussian for large N):

$$I(t) = R(P_m + \Delta P) \quad (P_n \ll P_m, n \neq m),$$

where ΔP is the intensity noise due to intrachannel crosstalk.

3.2 Channel Impairments

In this section, we describe different channel impairments, including the fiber attenuation (Sect. 3.2.1), insertion losses (Sect. 3.2.2), chromatic dispersion (Sect. 3.2.3), multimode dispersion (Sect. 3.2.4), PMD (Sect. 3.2.5), and fiber nonlinearities (Sect. 3.2.6). The frequency chirp effect is already introduced in Chap. 2.

3.2.1 Fiber Attenuation

Fiber attenuation can be described by the general relation $\mathrm{d}P/\mathrm{d}z = -\alpha P$, where α is the *power attenuation coefficient per unit length*. If P_{in} is the power launched into the fiber, the power remaining after propagating a length L within the fiber P_{out} is $P_{\text{out}} = P_{\text{in}} \exp(-\alpha L)$. The absorption coefficient varies with wavelength as many of the absorption and scattering mechanisms vary with λ. For instance *Rayleigh scattering* in fiber is due to microscopic variations in the density of glass (density fluctuation sites $<\lambda$) and varies as [3]

$$\alpha_{\mathrm{R}} = C/\lambda^4, \quad C = 0.7 - 0.9 \quad (\mathrm{dB/km})\,\mu\mathrm{m}^4.$$

Mie scattering is caused by imperfections (scattering) in the core–cladding interface that are larger than λ. *Intrinsic absorption* can be identified as (1) *infrared absorption*: in SiO_2 glass, vibrational modes of Si–O bonds cause an absorption peak at $\lambda > 7\,\mu$m, which has a tail extending to the $\lambda = 1.55 - \mu$m range; and (2) *ultraviolet absorption* is due to the electronic resonances occurring in ultraviolet region ($\lambda < 0.4\,\mu$m).

Extrinsic absorption results from the presence of impurities, such as (1) *transition-metal impurities* (Fe, Cu, Co, Ni, Mn, and Cr) absorb strongly in 0.6–1.6 μm, and the loss level is reduced below 1 dB/km if their amount should be kept below 1 part per billion and (2) *residual OH–* ions cause peaks near $\lambda = 0.95\,\mu\text{m}$, $1.24\,\mu\text{m}$, $1.39\,\mu\text{m}$.

3.2.2 Insertion Losses

Fiber loss is not the only source of optical signal attenuation along the transmission lines. The *fiber splices* and *fiber connectors* also cause the signal attenuation. The fiber splices can be fused or joined together by some mechanical means, with typical attenuation being 0.01–0.1 dB per fused splice and slightly above 0.1 dB per mechanical splice. Optical connectors are removable and allow many repeated connections and disconnections, with typical insertion loss for high-quality single mode fiber (SMF) not above 0.25 dB. To minimize the *connector return loss* (a fraction of the optical power reflected back into the fiber at the connector point) the angled fiber-end surfaces are commonly used. The number of optical splices and connectors depends on transmission length, and must be taken into account, unless the total attenuation due to fiber joints is distributed and added to the optical fiber attenuation.

3.2.3 Chromatic Dispersion and Single Mode Fibers

Dispersion problem can be described as follows. Short optical pulses entering a dispersive channel such as an optical fiber are spread out into a much broader temporal distribution. Both the *intermodal dispersion* and *chromatic dispersion* cause the distortion in multimode optical fibers, while chromatic dispersion is the only cause of the signal distortion in SMF. The intermodal dispersion can be specified through the *optical fiber bandwidth* B_{fib} that is related to a 1-km long optical fiber, and this parameter is specified by the manufacturers and is commonly measured at wavelength around 1,310 nm (chromatic dispersion is negligible in this region compared to intermodal dispersion) [1]:

$$B_{\text{fib},L} = \frac{B_{\text{fib}}}{L^{\mu}}, \quad \mu = 0.5 - 1. \tag{3.26}$$

Single-mode optical fibers do, however, introduce another signal impairment known as the chromatic dispersion. *Chromatic dispersion* is caused by difference in velocities among different spectral components within the same mode and has two components (1) *material dispersion* and (2) *waveguide dispersion* component. The material dispersion is caused by the fact that the refractive index is a function of wavelength, defined by Sellmeier equation [1,3]:

$$n(\lambda) = \left[1 + \sum_{i=1}^{M} \frac{B_i \lambda^2}{\lambda^2 - \lambda_i^2}\right], \tag{3.27}$$

with typical B_i and λ_i parameters for pure silica being:

$$B_1 = 0.6961663 \quad \text{at} \quad \lambda_1 = 0.0684043\,\mu\text{m},$$
$$B_2 = 0.4079426 \quad \text{at} \quad \lambda_2 = 0.1162414\,\mu\text{m},$$
$$B_3 = 0.8974794 \quad \text{at} \quad \lambda_3 = 9.896161\,\mu\text{m}.$$

The group index n_g can be determined by using the Sellmeier equation and the following definition expression:

$$n_g = n + \omega\, dn/d\omega = n - \lambda\, dn/d\lambda.$$

Waveguide dispersion is related to the physical design of the optical fiber. Since the value of Δ is typically small, the refractive indices of the core–cladding are nearly equal, the light is not strictly confined in the fiber core, and the fiber modes are said to be *weakly guided.* For a given mode, say fundamental mode, the portion of light energy that propagates in the core depends on wavelength, giving rise to the pulse spreading phenomenon known as waveguide dispersion. By changing the power distribution across the cross-sectional area, the overall picture related to the chromatic dispersion can be changed. The distribution of the mode power and the total value of waveguide dispersion can be manipulated by *multiple cladding layers.* This is commonly done in optical fibers for special application purposes, such as *dispersion compensation.* Different refractive index profiles of single-mode fibers used today are shown in Fig. 3.5 [1]. The conventional SMF index profile is shown in Fig. 3.5a. The nonzero dispersion shifted fiber (NZDSF), with refractive index profile shown in Fig. 3.5b, is suitable for use in WDM systems. Large effective area NZDSFs, with index profile shown in Fig. 3.5c, are suitable to reduce the effect of fiber nonlinearities. Dispersion compensating fiber (DCF), with index profile shown in Fig. 3.5d, is suitable to compensate for positive chromatic dispersion accumulated along the transmission line.

Because the source is nonmonochromatic, different spectral components within a pulse will travel at different velocities, inducing pulse broadening. When the neighboring pulses cross their allocated time slots, the ISI occurs, and the signal bit rates that can be effectively transmitted can be severely limited. A specific spectral component, characterized by the angular optical frequency ω, will arrive at the output end of the fiber of length L after some *delay* τ_g, known as the group delay:

$$\tau_g = \frac{L}{v_g} = L\frac{d\beta}{d\omega} = \frac{L}{c}\frac{d\beta}{dk} = -L\frac{\lambda^2}{2\pi c}\frac{d\beta}{d\lambda}, \tag{3.28}$$

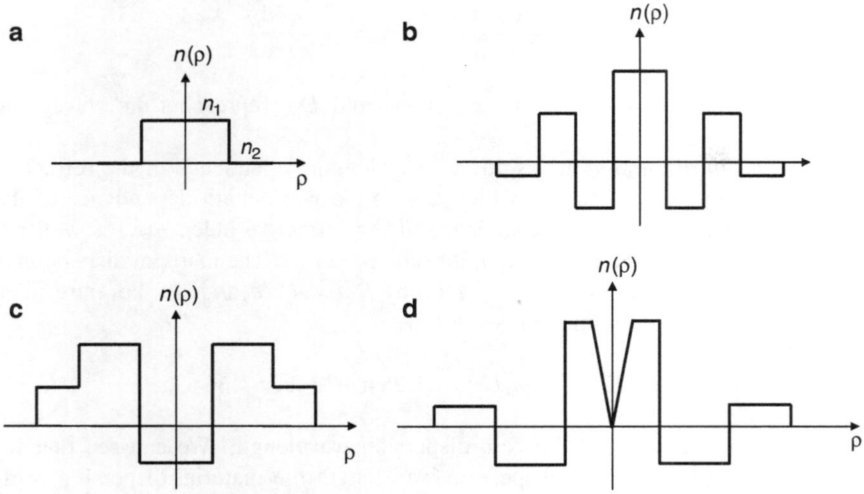

Fig. 3.5 Refractive index profiles of different SMFs (**a**) standard SMF, (**b**) NZDSF with reduced dispersion slope, (**c**) NZDSF with large effective area, and (**d**) DCF

where β is the propagation constant introduced earlier and $v_g = [d\beta/d\omega]^{-1}$ is the group velocity, the speed at which the energy of an optical pulse travels. As a result of the difference in time delays, the optical pulse disperses after traveling a certain distance, and the *pulse broadening* ($\Delta\tau_g$) can be characterized by

$$\Delta\tau_g = \frac{d\tau_g}{d\omega}\Delta\omega = \frac{d\tau_g}{d\lambda}\Delta\lambda, \tag{3.29}$$

where $\Delta\omega$ represents the frequency bandwidth of the source and $\Delta\lambda$ represents the wavelength bandwidth of the source. By substituting (3.28) into (3.29) we get

$$\Delta\tau_g = \frac{d\tau_g}{d\omega}\Delta\omega = \frac{d\left(L\frac{d\beta}{d\omega}\right)}{d\omega}\Delta\omega = L\frac{d^2\beta}{d\omega^2}\Delta\omega = L\frac{d^2\beta}{d\omega^2}\left(-\frac{2\pi c}{\lambda^2}\Delta\lambda\right) = LD\Delta\lambda, \tag{3.30}$$

where D [ps/nm km] represents the chromatic dispersion parameter defined by

$$D = -\frac{2\pi c}{\lambda^2}\frac{d^2\beta}{d\omega^2} = -\frac{2\pi c}{\lambda^2}\beta_2, \quad \beta_2 = \frac{d^2\beta}{d\omega^2}, \tag{3.31}$$

where β_2 denotes the group-velocity dispersion (GVD) parameter. The chromatic dispersion can be expressed as the sum of two contributing factors:

$$D = \frac{d\left(-\frac{\lambda^2}{2\pi c}\frac{d\beta}{d\lambda}\right)}{d\lambda} = -\frac{1}{2\pi c}\left(2\lambda\frac{d\beta}{d\lambda} + \lambda^2\frac{d^2\beta}{d\lambda^2}\right) = D_M + D_W, \tag{3.32a}$$

$$D_{\mathrm{M}} = -\frac{\lambda^2}{2\pi c}\frac{\mathrm{d}^2\beta}{\mathrm{d}\lambda^2} \quad D_{\mathrm{W}} = -\frac{\lambda}{\pi c}\frac{\mathrm{d}\beta}{\mathrm{d}\lambda}, \tag{3.32b}$$

where D_{M} represents the material dispersion and D_{W} represents the waveguide dispersion.

The *material dispersion* arises due to wavelength dependence of the refractive index on the fiber-core material, which causes the wavelength dependence of the group delay. The wavelength dependence of the refractive index $n(\lambda)$ is well approximated by the *Sellmeier equation*, introduced earlier. The material dispersion is related to the slope of $n_{\mathrm{g}} = n - \lambda \mathrm{d}n(\lambda)/\mathrm{d}\lambda$ by $D_{\mathrm{M}} = (1/c)\mathrm{d}n_{\mathrm{g}}/\mathrm{d}\lambda$. For pure silica fiber the following approximation is valid [3]:

$$D_{\mathrm{M}} \approx 122(1 - \lambda_{\mathrm{ZD}}/\lambda), \quad 1.25\,\mu\mathrm{m} < \lambda < 1.66\,\mu\mathrm{m}, \tag{3.33}$$

where $\lambda_{\mathrm{ZD}} = 1.276\,\mu\mathrm{m}$ is the zero-dispersion wavelength. We can see that for wavelengths larger than zero-dispersion wavelength the material dispersion coefficient is positive, while the GVD is negative, and this regime is known as the *anomalous* dispersion region.

The *waveguide dispersion* occurs because the propagation constant is a function of the fiber parameters (core radius and difference between refractive indices in fiber core and cladding) and at the same time is a function of wavelength. Therefore, the propagation paths for a mode due to different boundary matching conditions is slightly different at different wavelengths. Waveguide dispersion is related to the physical design of the optical fiber. Since the value of Δ is typically small, the refractive indices of the core–cladding are nearly equal, and the light is not strictly confined in the fiber core. For a given mode, say fundamental mode, the portion of light energy that propagates in the core depends on wavelength: the longer the wavelength, the more the power in the cladding. The effect of waveguide dispersion on pulse spreading can be approximated by assuming that refractive index of the material is independent of wavelength. To make the result independent of fiber configuration, we should express the group delay in terms of the normalized propagation constant b, introduced in Chap. 2 [see (2.18)], so that the propagation constant can be expressed as

$$\beta \approx n_2 k(b\Delta + 1), \quad \Delta \ll 1. \tag{3.34}$$

The group delay due to waveguide dispersion can be found as [8]

$$\tau_{\mathrm{W}} = \frac{L}{c}\frac{\mathrm{d}\beta}{\mathrm{d}k} \ll \frac{L}{c}\left(n_2 + n_2\Delta\frac{\mathrm{d}(Vb)}{\mathrm{d}V}\right), \quad \frac{\mathrm{d}(Vb)}{\mathrm{d}V} = b\left[1 - \frac{2J_m^2(pa)}{J_{m+1}(pa)J_{m-1}(pa)}\right], \tag{3.35}$$

where p is the core parameter introduced earlier ($p^2 = n_1^2k^2 - \beta^2$). The pulse broadening due to waveguide dispersion can be determined using (3.29) by

$$\Delta\tau_{\mathrm{W}} = \frac{\mathrm{d}\tau_{\mathrm{W}}}{\mathrm{d}\lambda}\Delta\lambda = \frac{\mathrm{d}\tau_{\mathrm{W}}}{\mathrm{d}V}\frac{\mathrm{d}V}{\mathrm{d}k}\frac{\mathrm{d}k}{\mathrm{d}\lambda}\Delta\lambda = L\frac{n_2\Delta}{c}\frac{\mathrm{d}^2(Vb)}{\mathrm{d}V^2}\frac{V}{k}\left(-\frac{k}{\lambda}\right)\Delta\lambda = LD_{\mathrm{W}}\Delta\lambda, \tag{3.36}$$

where the waveguide dispersion parameter D_W is defined by

$$D_W = -\frac{n_2\Delta}{c}\frac{1}{\lambda}\left[V\frac{d^2(Vb)}{dV^2}\right], \tag{3.37}$$

and it is negative for the normalized frequencies between 0 and 3.

The total dispersion can be written as sum of two contributions, $D = D_M + D_W$. The waveguide dispersion (that is negative) shifts the zero-dispersion wavelength to approximately 1.308 μm. $D \sim$ 15–18 ps/(km nm) near 1.55 μm. This is the low loss region for fused silica optical fibers. This approach for mutual cancellation was used to produce several types of single-mode optical fibers that are different in design compared to standard SMF, standardized by ITU-T. In addition to standard SMF, there are two major fiber types [1, 2] (1) dispersion shifted fibers (DSF) (described in ITU-T recommendation G.653) with dispersion minimum shifted from the 1,310-nm wavelength region to the 1,550-nm region and (ii) nonzero-dispersion shifted fibers (NZDSF) (described in G.655 recommendation), with dispersion minimum shifted from 1,310-nm window to anywhere within C or L bands (commercial examples: TrueWave fiber or LEAF).

Dispersion effects do not disappear completely at zero-dispersion wavelength. Residual dispersion due to higher order dispersive effects still exists. Higher order dispersive effects are governed by the *dispersion slope parameter* $S = dD/d\lambda$. Parameter S is also known as differential dispersion parameter or second-order dispersion parameter, and it is related to the GVD parameter and second-order GVD parameter $\beta_3 = d\beta_2/d\omega$ by

$$S = \frac{dD}{d\lambda} = \frac{d}{d\lambda}\left(-\frac{2\pi c}{\lambda^2}\beta_2\right) = \frac{4\pi c}{\lambda^3}\beta_2 + \left(\frac{2\pi c}{\lambda^2}\right)^2\beta_3. \tag{3.38}$$

The *geometrical optics* (or *ray theory*) approach, used to describe the light confinement in step-index fibers through the total internal reflection, is valid when the fiber has a core radius a much larger than the operating wavelength λ. Once the core radius becomes comparable to the operating wavelength, the propagation of light in step-index fiber is governed by Maxwell's equations describing the change of electric ($\boldsymbol{E}$) and magnetic ($\boldsymbol{H}$) in space and time [1, 4, 6, 8–10]:

$$\nabla \times \boldsymbol{E} = -\partial \boldsymbol{B}/\partial t, \tag{3.39a}$$

$$\nabla \times \boldsymbol{H} = \boldsymbol{J} + \partial \boldsymbol{D}/\partial t, \tag{3.39b}$$

$$\nabla \cdot \boldsymbol{D} = \rho, \tag{3.39c}$$

$$\nabla \cdot \boldsymbol{B} = 0, \tag{3.39d}$$

where $\boldsymbol{B}$ denotes the magnetic flux density, $\boldsymbol{D}$ is the electric flux density, and $\boldsymbol{J} = \sigma\boldsymbol{E}$ is the current density. Given the fact that there is no free charge in fiber, the charge density $\rho = 0$, and that the conductivity of silica is extremely low $\sigma \approx 0$, the equations above can be simplified. The flux densities are related to the field vectors by constitutive relations:

$$\boldsymbol{D} = \varepsilon_0 \boldsymbol{E} + \boldsymbol{P}, \tag{3.40a}$$

$$\boldsymbol{B} = \mu_0 \boldsymbol{H} + \boldsymbol{M}, \tag{3.40b}$$

where $\boldsymbol{P}$ and $\boldsymbol{M}$ denoted the induced electric and magnetic densities, ε_0 is the permeability in the vacuum, and μ_0 is the permittivity in the vacuum. Given the fact that silica is nonmagnetic material $\boldsymbol{M} = \boldsymbol{0}$. $\boldsymbol{P}$ and $\boldsymbol{E}$ are mutually connected by

$$\boldsymbol{P}(\boldsymbol{r},t) = \varepsilon_0 \int_{-\infty}^{\infty} \chi(\boldsymbol{r},t-t')\boldsymbol{E}(\boldsymbol{r},t')\mathrm{d}t', \tag{3.41}$$

where the linear susceptibility χ is a second-rank tensor that becomes scalar for an isotropic medium. After certain manipulations the *wave equation* is obtained [1–4, 6]:

$$\nabla^2 \tilde{\boldsymbol{E}} + n^2(\omega)k_0^2 \tilde{\boldsymbol{E}} = 0, \tag{3.42}$$

where $k_0 = 2\pi/\lambda$ is the free space wave number, and with $\tilde{\boldsymbol{E}}$ we denoted the Fourier transform of electric field vector.

An *optical mode* refers to a specific solution of the wave equation, subject to appropriate boundary conditions. The spatial field distributions, $\boldsymbol{E}(\boldsymbol{r},\omega) = \boldsymbol{E}_0(\rho,\phi)\exp[\mathrm{j}\beta(\omega)z - \mathrm{j}\omega t]$ and $\boldsymbol{H}(\boldsymbol{r},\omega) = \boldsymbol{H}_0(\rho,\phi)\exp[\mathrm{j}\beta(\omega)z - \mathrm{j}\omega t]$, of a mode do not change as the mode propagates along z-axis except for a multiplicative factor $\exp[\mathrm{j}\beta(\omega)z]$, with $\beta(\omega)$ being the *propagation constant* of a mode. The fiber modes may be classified as *guided modes*, *leaky modes*, and *radiation modes*. Different modes in fiber propagate with different values of propagation constant, that is with different speeds, causing the pulse spreading. Therefore, it is desirable to design the fiber which supports only one mode-*fundamental mode*. Such a fiber is called SMF, as we mentioned earlier. If the *weak guidance condition* ($\Delta << 1$) is not satisfied the conventional modes TE, TM, EH, and HE modes can be found. The cylindrical symmetry of fiber suggests the use of *cylindrical coordinate system* (ρ, ϕ, z). The wave equation is to be solved for six components: E_ρ, E_ϕ, E_z and H_ρ, H_ϕ, H_z. We can solve the wave equation for axial components only (E_z and H_z), and use the system of Maxwell's equations to express the other components as functions of axial ones. The wave equation for E_z component is as follows [1–4, 6]:

$$\frac{\partial^2 E_z}{\partial \rho^2} + \frac{1}{\rho}\frac{\partial E_z}{\partial \rho} + \frac{1}{\rho^2}\frac{\partial^2 E_z}{\partial z^2} + n^2 k_0^2 E_z = 0, \quad n = \begin{cases} n_1, & \rho \leq a \\ n_2, & \rho > a \end{cases}. \tag{3.43}$$

The wave equation for E_z component can easily be solved using the *method of separation of variables*, leading to the following overall solution [3, 4, 6]:

$$E_z(\rho,\phi,z) = \begin{cases} AJ_m(p\rho)\mathrm{e}^{\mathrm{j}m\phi}\mathrm{e}^{\mathrm{j}\beta z}, & \rho \leq a \\ BK_m(q\rho)\mathrm{e}^{\mathrm{j}m\phi}\mathrm{e}^{\mathrm{j}\beta z}, & \rho > a \end{cases}, \tag{3.44}$$

where $J_m(x)$ and $K_m(x)$ are corresponding Bessel functions of mth order (A and B are constants to be determined from boundary conditions). The similar equation is valid for H_z.

The other four components can be expressed in terms of axial ones by using the Maxwell's equations, and in the core region we obtain [3, 4, 6]

$$E_\rho = \frac{\mathrm{j}}{p^2}\left(\beta\frac{\partial E_z}{\partial\rho} + \mu_0\frac{\omega}{\rho}\frac{\partial H_z}{\partial\phi}\right); \quad E_\phi = \frac{\mathrm{j}}{p^2}\left(\frac{\beta}{\rho}\frac{\partial E_z}{\partial\phi} - \mu_0\omega\frac{\partial H_z}{\partial\rho}\right),$$
$$H_\rho = \frac{\mathrm{j}}{p^2}\left(\beta\frac{\partial H_z}{\partial\rho} - \varepsilon_0 n^2\frac{\omega}{\rho}\frac{\partial E_z}{\partial\phi}\right); \quad H_\phi = \frac{\mathrm{j}}{p^2}\left(\frac{\beta}{\rho}\frac{\partial H_z}{\partial\phi} + \varepsilon_0 n^2\omega\frac{\partial H_z}{\partial\rho}\right), \quad (3.45)$$

where $p^2 = n_1^2k_0^2 - \beta^2$. The similar equations can be obtained for cladding by replacing p^2 with $-q^2$, where $q^2 = \beta^2 - n_2^2k_0^2$. By satisfying the boundary conditions, the homogeneous system of equation is obtained, which has the nontrivial solution only if corresponding determinant is zero, which leads to the following characteristic (eigenvalue) equation [3, 4, 6]:

$$\left[\frac{J'_m(pa)}{paJ_m(pa)} + \frac{K'_m(qa)}{qaK_m(qa)}\right]\cdot\left[\frac{J'_m(pa)}{paJ_m(pa)} + \frac{n_2^2}{n_1^2}\frac{K'_m(qa)}{qaK_m(qa)}\right]$$
$$= m^2\left[\frac{1}{(pa)^2} + \frac{1}{(qa)^2}\right]\cdot\left[\frac{1}{(pa)^2} + \frac{n_2^2}{n_1^2}\frac{1}{(qa)^2}\right], \quad (3.46)$$

where pa and qa are related by the following equation:

$$(pa)^2 + (pq)^2 = V^2. \quad (3.47)$$

For given fiber parameters n_1, n_2, a, and the operating wavelength, we can determine the normalized frequency V. For fixed V and a given m, there exist multiple solutions $n = 1,2,3,\ldots$ that lead to different modes – *propagating modes*. The case $m = 0$ corresponds to meridional rays (the rays that periodically intersect the center axis of the fiber) has the electric/magnetic components independent of ϕ, and the corresponding modes are classified as: $\mathrm{TE}_{0n}(E_z = 0)$, and $\mathrm{TM}_{0n}(H_z = 0)$ modes. The case $m \neq 0$ corresponds to skew rays, the electric and magnetic fields components are functions of ϕ, and the corresponding modes are classified as: HE_{mn} (H_z dominates over E_z) and EH_{mn} (E_z dominates over H_z) modes. Once the mode is identified (m and n are fixed) from (3.46) and by using the definition expressions for p and q above, we can easily determine the propagation constant of the mode β_{mn}.

The propagating light pulse is a composite optical signal containing a number of monochromatic spectral components. Each spectral component behaves differently in a dispersion medium, such as optical fiber, leading to the light pulse distortion. Each axial component of the monochromatic electromagnetic wave can be represented by its complex electric field function [1, 3, 6]:

$$E(z,t) = E_a(z,t)\exp[\mathrm{j}\beta(\omega)z]\exp[-\mathrm{j}\omega' t]. \quad (3.48)$$

The light pulse distortion, which is observed through the pulse broadening along the fiber, can be evaluated by knowing the frequency dependence of the propagation constant $\beta = \beta(\omega)$ at a specific distance z along the fiber. Each spectral component will experience a phase shift proportional to $\beta(\omega)z$. The amplitude spectrum observed at point z along the fiber length, in the frequency domain is given by

$$\tilde{E}_a(z,\omega) = \tilde{E}_a(0,\omega)\exp[\mathrm{j}\beta(\omega)z]. \tag{3.49}$$

The behavior of the pulse envelope during the propagation process can be evaluated through the inverse Fourier transform of previous equation, which is very complicated to calculate, unless additional simplifications are made, for example by expressing the propagation constant in terms of a Taylor series:

$$\beta(\omega)\approx\beta(\omega_c)+\frac{\mathrm{d}\beta}{\mathrm{d}\omega}|_{\omega=\omega_c}(\omega-\omega_c)+\frac{1}{2}\frac{\mathrm{d}^2\beta}{\mathrm{d}\omega^2}|_{\omega=\omega_c}(\omega-\omega_c)^2+\frac{1}{6}\frac{\mathrm{d}^3\beta}{\mathrm{d}\omega^3}|_{\omega=\omega_c}(\omega-\omega_c)^3+\cdots, \tag{3.50}$$

where $\beta_1 = (\mathrm{d}\beta/\mathrm{d}\omega)|_{\omega=\omega_c}$ is related to the group velocity v_g by $\beta_1 = 1/v_g$, $\beta_2 = (\mathrm{d}^2\beta/\mathrm{d}\omega^2)\big|_{\omega=\omega_c}$ is the GVD parameter, and $\beta_3 = (\mathrm{d}^3\beta/\mathrm{d}\omega^3)\big|_{\omega=\omega_c}$ is the second-order GVD parameter. Introducing the concept of slow varying amplitude $A(z,t)$ [1,3,6]:

$$E(z,t) = E_a(z,t)\exp[\mathrm{j}\beta(\omega)z]\exp(-j\omega_c t) = A(z,t)\exp(\mathrm{j}\beta_c z - \mathrm{j}\omega_c t), \tag{3.51}$$

we get

$$A(z,t)=\frac{1}{2\pi}\int_{-\infty}^{\infty}\tilde{A}(0,\omega)\exp\left[\mathrm{j}\left(\beta_1\Delta\omega+\frac{\beta_2}{2}\Delta\omega^2+\frac{\beta_3}{6}\Delta\omega^3\right)z\right]\exp(-\mathrm{j}(\Delta\omega)t)\mathrm{d}(\Delta\omega). \tag{3.52}$$

By taking partial derivative of $A(z,t)$ with respect to propagation distance z we obtain

$$\begin{aligned}\frac{\partial A(z,t)}{\partial z} &= \frac{1}{2\pi}\int_{-\infty}^{\infty}\left[\mathrm{j}\left(\beta_1\Delta\omega+\frac{\beta_2}{2}\Delta\omega^2+\frac{\beta_3}{6}\Delta\omega^3\right)\right]\tilde{A}(0,\Delta\omega)\\ &\quad\times\exp\left[\mathrm{j}\left(\beta_1\Delta\omega+\frac{\beta_2}{2}\Delta\omega^2+\frac{\beta_3}{6}\Delta\omega^3\right)z\right]\exp(-\mathrm{j}(\Delta\omega)t)\mathrm{d}(\Delta\omega).\end{aligned} \tag{3.53}$$

Finally, by taking the inverse Fourier transform of (3.53) we derive basic propagation equation describing the *pulse propagation* in single-mode optical fibers:

$$\frac{\partial A(z,t)}{\partial z} = -\beta_1\frac{\partial A(z,t)}{\partial t} - \mathrm{j}\frac{\beta_2}{2}\frac{\partial^2 A(z,t)}{\partial t^2} + \frac{\beta_3}{6}\frac{\partial^3 A(z,t)}{\partial t^3}. \tag{3.54}$$

This equation can be generalized by including the fiber attenuation and the Kerr nonlinear index:

$$\frac{\partial A(z,t)}{\partial z} = -\frac{\alpha}{2}A(z,t) - \beta_1\frac{\partial A(z,t)}{\partial t} - \mathrm{j}\frac{\beta_2}{2}\frac{\partial^2 A(z,t)}{\partial t^2} + \frac{\beta_3}{6}\frac{\partial^3 A(z,t)}{\partial t^3} + \mathrm{j}\gamma\left|A(z,t)\right|^2 A(z,t), \tag{3.55}$$

where α is the attenuation coefficient and $\gamma = 2\pi n_2/(\lambda A_{\mathrm{eff}})$ is the nonlinear coefficient with n_2 being the nonlinear Kerr coefficient and A_{eff} being the effective cross-sectional area of the fiber core.

After the introduction of a new coordinate system with $T = t - \beta_1 z$, *the nonlinear Schrödinger equation* (NLSE) is obtained:

$$\frac{\partial A(z,T)}{\partial z} = -\frac{\alpha}{2}A(z,T) - \mathrm{j}\frac{\beta_2}{2}\frac{\partial^2 A(z,t)}{\partial T^2} + \frac{\beta_3}{6}\frac{\partial^3 A(z,T)}{\partial T^3} + \mathrm{j}\gamma\left|A(z,T)\right|^2 A(z,T). \tag{3.56}$$

The NLSE can be solved using the *Fourier split-step algorithm* [6]. The length of the fiber is divided into small steps of size h, as shown in Fig. 3.6. The algorithm proceeds by first applying the linear operator of the equation (L) for one-half step length ($h/2$). The nonlinear operator (NL) is applied for one step length h. The linear operator is again applied for the full step size. Therefore, the linear operator is applied at every half-step ($h/2$) and the nonlinear one at every full-step (h). The linear operator D in the time domain can be described as follows:

$$D = -\frac{\alpha}{2} - \mathrm{j}\frac{\beta_2}{2}\frac{\partial^2}{\partial T^2} + \frac{\beta_3}{6}\frac{\partial^3}{\partial t^3}, \tag{3.57}$$

and the linear operator in the frequency domain is obtained by applying the Fourier transform (FT) on (3.57):

$$\tilde{D} = -\frac{\alpha}{2} + \mathrm{j}\frac{\beta_2}{2}\omega^2 + \mathrm{j}\frac{\beta_3}{6}\omega^3. \tag{3.58}$$

The nonlinear operator is defined as

$$N = \mathrm{j}\gamma\left|A(z,T)\right|^2, \tag{3.59}$$

so that the Fourier split-step algorithm can be symbolically described by

$$A(z_0 + h, T) = \exp\left(h \cdot \frac{D}{2}\right)\left[\exp(h \cdot N)A(z_0, T)\right]\exp\left(h \cdot \frac{D}{2}\right). \tag{3.60}$$

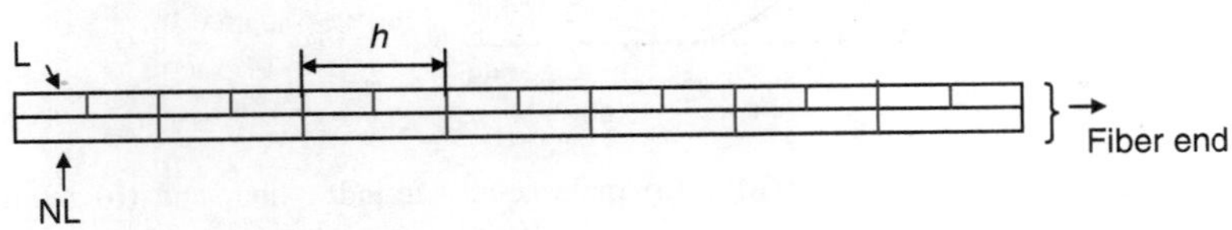

Fig. 3.6 Illustrating the Fourier split-step algorithm

The linear operator is applied in frequency domain, while the nonlinear operator is applied in time domain. If fiber nonlinearities can be neglected, the study of propagation is greatly simplified. We have to evaluate the FT of pulse $A(0,t)$ to get $\tilde{A}(0,\omega) = \mathrm{FT}[A(0,t)]$. We have to multiply the corresponding FT by dispersion factor to get

$$\tilde{A}(z,\omega) = \tilde{A}(0,\omega)\mathrm{e}^{\mathrm{j}\left[\frac{\beta_2}{2}\omega^2+\frac{\beta_3}{6}\omega^3\right]z}\mathrm{e}^{-\frac{\alpha}{2}z}. \tag{3.61}$$

Finally, to get the pulse shape at distance z we have to evaluate the inverse Fourier transform (FT^{-1}):

$$A(z,t) = \mathrm{FT}^{-1}\left[\tilde{A}(z,\omega)\right].$$

3.2.4 Multimode Dispersion and Multimode Fibers

MMFs transfer the light through a collection of spatial transversal modes, with each mode being defined through a specified combination of electrical and magnetic components, which occupies a different cross section of the optical fiber core and takes a slightly distinguished path along the optical fiber. The difference in mode path lengths in MMFs causes a difference in arrival times at the receiver side, and this effect is known as multimode (intermodal) dispersion and causes signal distortion and imposes the limitations in signal bandwidth. The effect of intermodal dispersion is illustrated in Fig. 3.7 for meridional rays, for both step index MMF (Fig. 3.7a) and graded-index MMF (Fig. 3.7b). The steeper the angle of the propagation of ray of congruence, the higher is the mode number, and slower the axial group velocity [8]. This variation in group velocities of different modes results in group delay spread, which is the root cause of intermodal dispersion. The maximum pulse broadening can be obtained as the difference of travel time of the highest-order mode $T_{\max}$ and that of fundamental mode $T_{\min}$ as follows:

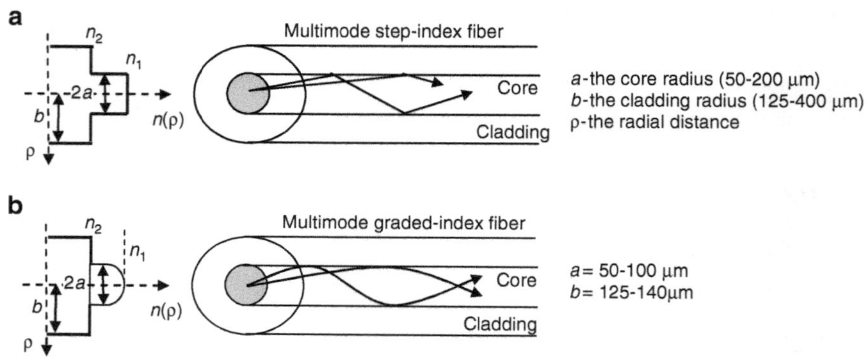

Fig. 3.7 Intermodal dispersion in MMFs (**a**) multimode step-index fiber and (**b**) multimode graded-index fiber

$$\Delta T_{\text{mod}} = T_{\max} - T_{\min} = \frac{Ln_1^2}{cn_2} - \frac{Ln_1}{c} = \frac{Ln_1^2}{cn_2}\Delta \approx \frac{L}{2cn_2}(\text{NA})^2, \quad \Delta \ll 1, \quad (3.62)$$

where L is the MMF length, n_1 and n_2 are refractive indices related to the fiber core and cladding respectively, and NA is numerical aperture defined as the light gathering capacity of MMF:

$$\text{NA} = \sqrt{n_1^2 - n_2^2} \approx n_1\sqrt{2\Delta} \quad (\Delta \ll 1),$$

$\Delta = (n_1 - n_2)/n_1$ is the normalized index difference. The delay spread is larger for fibers with larger numerical aperture or larger normalized index difference. A rough measure of the delay spread that can be tolerated is half of the bit duration, which results in the following bit maximum bit rate × distance product:

$$BL < \frac{n_2 c}{2n_1^2\Delta} \approx \frac{n_2 c}{(\text{NA})^2}. \quad (3.63)$$

For example, for typical MMF parameters, $n_1 \approx n_2 \approx 1.5$ and $\Delta = 0.01$, we obtain $BL < 10$ (Mb/s) km, for non-return-to zero (NRZ) systems. To improve BL product, graded-index MMFs are used, whose index profile can be described by [8]

$$n(\rho) = \begin{cases} n_1[1 - 2\Delta(\rho/a)^\alpha]^{1/2}, & \rho \le a \\ n_1(1 - 2\Delta)^{1/2} \simeq n_1(1 - \Delta) = n_2, & \rho > a \end{cases} \quad (\alpha - \text{profile}). \quad (3.64)$$

A step-index fiber is approached in the limit for large α and for $\alpha = 2$ the parabolic-index profile is obtained. Because the index of refraction is lower at the outer edges of the core, the light rays will travel faster in that region than in the center where the refractive index is higher. Therefore, the ray congruence characterizing the higher order mode will travel further than the fundamental ray congruence, but at faster rate, resulting in similar arrival times at receiver side. Multimode dispersion in graded-index fibers is reduced compared to that in step-index fibers.

The trajectory of a light ray is obtained by solving the following differential equation:

$$\frac{\mathrm{d}}{\mathrm{d}s}\left[n(\boldsymbol{r})\frac{\mathrm{d}\boldsymbol{r}}{\mathrm{d}s}\right] = \nabla n(\boldsymbol{r}), \quad (3.65)$$

where $\boldsymbol{r}$ is the point along trajectory, s is the path along trajectory, and $n(\boldsymbol{r})$ is index of refraction. By observing meridional paraxial rays $\boldsymbol{r} = (\rho, z)$ and noticing that $n(\boldsymbol{r}) = n(\rho)$, the differential equation (3.65) becomes

$$\frac{\mathrm{d}^2\rho}{\mathrm{d}z^2} = \frac{1}{n}\frac{\mathrm{d}n}{\mathrm{d}\rho}, \quad (3.65\text{a})$$

which can easily be solve for parabolic profile to get

$$\rho(z) = \rho_0 \cos(pz) + \frac{\rho_0'}{p} \sin(pz), \quad p = \sqrt{\frac{2\Delta}{a^2}}, \tag{3.66}$$

where ρ_0 and ρ_0' are the slope and slope radius at $z = 0$, respectively. Therefore, the trajectory of meridional rays is periodic with a period $2\pi/p$.

The root-mean square (rms) pulse broadening σ in a graded-index fiber can be obtained by

$$\sigma = \sqrt{\sigma_{\text{intermodal}}^2 + \sigma_{\text{intramodal}}^2}, \tag{3.67}$$

where $\sigma_{\text{intermodal}}$ is the rms pulse spread resulting from intermodal dispersion, while $\sigma_{\text{intramodal}}$ is the rms pulse spread resulting from intramodal dispersion. The intermodal delay distortion can be found by

$$\sigma_{\text{intermodal}} = \sqrt{\left\langle \tau_g^2 \right\rangle - \left\langle \tau_g \right\rangle^2}, \tag{3.68}$$

where τ_g is the group delay of the mode, which is function of the order (v,m) of the mode:

$$\tau_g = \frac{L}{c} \frac{d\beta_{vm}}{dk}, \tag{3.69}$$

with L being the fiber length, β_{vm} being the propagation constant, and $k = 2\pi/\lambda$ being the wave number. The quantities $\left\langle \tau_g \right\rangle$ and $\left\langle \tau_g^2 \right\rangle$ can be obtained from mode distribution P_{vm} by [8]

$$\left\langle \tau_g \right\rangle = \sum_{v,m} \frac{P_{vm}\tau_g(v,m)}{M}; \quad \left\langle \tau_g^2 \right\rangle = \sum_{v,m} \frac{P_{vm}\tau_g^2(v,m)}{M}, \tag{3.70}$$

where M is the total number of modes. To determine the group delay, we use the following expression for propagation constant [8]:

$$\beta = kn_1 \left[1 - 2\Delta \left(\frac{m}{M} \right)^{\alpha/(\alpha+2)} \right]^{1/2}, \tag{3.71}$$

where m is the number of modes with propagation constant between kn_1 and β. By substituting (3.71) into (3.69), and assuming $\Delta \ll 1$, we obtain

$$\tau_g = \frac{N_1 L}{c} \left[1 + \frac{\alpha - 2 - \varepsilon}{\alpha + 2} \Delta \left(\frac{m}{M} \right)^{\alpha/(\alpha+1)} + \frac{3\alpha - 2 - 2\varepsilon}{2(\alpha + 2)} \Delta^2 \left(\frac{m}{M} \right)^{\alpha/(\alpha+1)} + O(\Delta^3) \right], \tag{3.72}$$

where $N_1 = n_1 + k\partial n_1/\partial k$, and $\varepsilon = [2n_1 k/(N_1 \Delta)]\partial\Delta/\partial k$. The equation above indicates that to the first order in Δ, the group delay difference between the modes is zero if $\alpha = 2 + \varepsilon$. If all modes are equally excited $P_{vm} = P$ and the number of modes are large, the summations in (3.70) become integrals, and upon substituting (3.72) in (3.70) we obtain the following expression for intermodal pulse spread [8]:

$$\sigma_{\text{intermodal}} = \frac{LN_1}{2c}\frac{\alpha}{\alpha+1}\left(\frac{\alpha+2}{3\alpha+2}\right)^{1/2}\left[c_1^2 + \frac{4c_1c_2(\alpha+1)\Delta}{2\alpha+1} + \frac{16\Delta^2c_2^2(\alpha+1)^2}{(5\alpha+2)(3\alpha+2)}\right]^{1/2}, \tag{3.73}$$

where $c_1 = (\alpha - 2 - \varepsilon)/(\alpha + 2)$ and $c_2 = (3\alpha - 2 - 2\varepsilon)/(\alpha + 2)$. The optimum index profile can be found by minimizing the (3.73) to get

$$\alpha_{\text{opt}} = 2 + \varepsilon - \Delta\frac{(4+\varepsilon)(3+\varepsilon)}{5+2\varepsilon}. \tag{3.74}$$

By setting $\varepsilon = 0$ and substituting (3.74) into (3.73) we obtain

$$\sigma_{\text{opt}} = \frac{n_1\Delta^2 L}{20\sqrt{3}c}. \tag{3.75}$$

The corresponding expression for step-index fiber is obtained by setting $\alpha \to \infty$ (and also assuming $\varepsilon = 0$) as follows:

$$\sigma_{\text{step}} = \frac{n_1\Delta L}{c}\frac{1}{2\sqrt{3}}\left(1 + 3\Delta + \frac{12\Delta^2}{5}\right)^{1/2} \approx \frac{n_1\Delta L}{2\sqrt{3}c}. \tag{3.76}$$

The ratio of (3.76) and (3.75) gives $\sigma_{\text{step}}/\sigma_{\text{opt}} = 10/\Delta$. Since typically $\Delta = 0.01$ turns up that a graded-index fiber capacity is three orders of magnitude larger than that of step index. The corresponding bandwidth × length product is then $BL < 10$ (Gb/s) km.

The impulse response of graded-index MMFs can be written as follows [23]:

$$h(t) = \begin{cases} \dfrac{\alpha+2}{\alpha}\left|\dfrac{\alpha+2}{\Delta(\alpha-2)}\right|^{(2/\alpha)+1} |t|^{2/\alpha}, & \alpha \neq 2 \\ \dfrac{2}{\Delta^2}, & \alpha \approx 2 \end{cases}, \quad 0 \le t < T; \quad T = \begin{cases} \dfrac{\alpha-2}{\alpha+2}\Delta, & \alpha \neq 2 \\ \dfrac{\Delta^2}{2}, & \alpha \approx 2 \end{cases}, \tag{3.77}$$

where the total delay of channel is normalized with delay of the central mode. Notice that for $\alpha < 2$, the delay spread is negative meaning that higher modes arrive earlier than the central mode. The impulse responses for different α-profile MMFs are shown in Fig. 3.8. For $\alpha \to \infty$, the fiber becomes a step-index.

3.2.5 *Polarization-Mode Dispersion*

The polarization unit vector, representing the SOP of the electric field vector, does not remain constant in practical optical fibers; rather, it changes in a random fashion

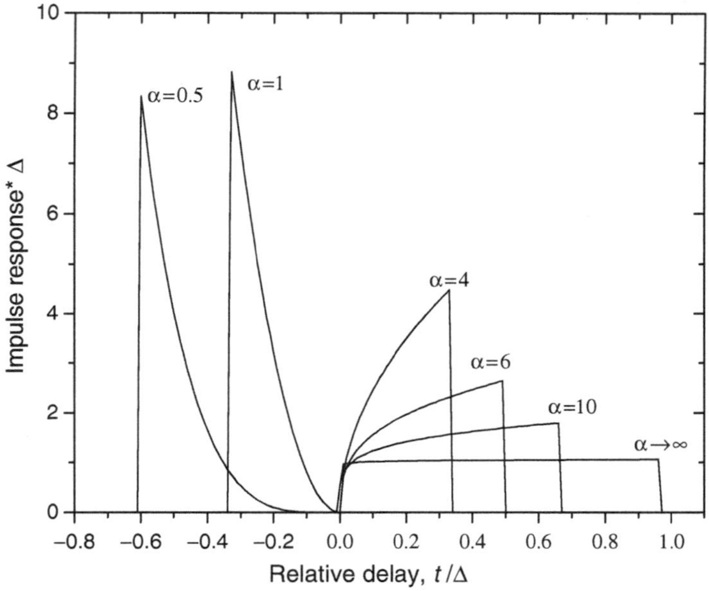

Fig. 3.8 Impulse responses of MMFs with $\alpha = 0.5, 1, 2, 4, 6, 10$, and ∞ (step-index) for index profile given by (3.64)

along the fiber because of its fluctuating birefringence [1–3]. Two common birefringence sources are (1) *geometric birefringence* (related to small departures from perfect cylindrical symmetry) and (2) *anisotropic stress* (produced on the fiber core during manufacturing or cabling of the fiber). The degree of birefringence is described by the difference in refractive indices of orthogonally polarized modes $B_m = |n_x - n_y| = \Delta n$. The corresponding difference in propagation constants of two orthogonally polarized modes is $\Delta\beta = |\beta_x - \beta_y| = (\omega/c)\Delta n$. Birefringence leads to a periodic power exchange between the two polarization components, described by *beat length* $L_B = 2\pi/\Delta\beta = \lambda/\Delta n$. Typically $B_m \sim 10^{-7}$, and therefore $L_B \sim 10\,\text{m}$ for $\lambda \sim 1\,\mu\text{m}$. Linearly polarized light remains linearly polarized only when it is polarized along one of the principal axes; otherwise, its SOP changes along the fiber length from linear to elliptical, and then back to linear, in a periodic manner over the length L_B.

In certain applications, it is needed to transmit a signal through a fiber that maintains its SOP. Such a fiber is called a *polarization-maintaining fiber* (PMF). (PANDA fiber is well-known PMF.) Approaches to design PMF having high but constant birefringence are (1) the *shape birefringence* (the fiber having an elliptical core) and (2) *stress birefringence* (stress-inducing mechanism is incorporated in fiber). Typical PMF has a birefringence $B_m \sim 10^{-3}$ and a beat length $B_m \sim 1\,\text{mm}$, resulting in a polarization crosstalk smaller than -30 dB/km. Unfortunately, the loss of PMFs is high ($\alpha \sim 2\,\text{dB/km}$).

The *modal group indices* and *modal group velocities* are related by

$$\bar{n}_{gx,y} = \frac{c}{v_{gx,y}} = \frac{c}{1/\beta_{1x,y}}, \quad \beta_{1x,y} = \frac{\mathrm{d}\beta x, y}{\mathrm{d}\omega}, \tag{3.78}$$

so that the difference in time arrivals (at the end of fiber of length L) for two orthogonal polarization modes, known as the *differential group delay* (DGD) can be calculated by [3]

$$\Delta\tau = \left| \frac{L}{v_{gx}} - \frac{L}{v_{gy}} \right| = L \left| \beta_{1x} - \beta_{1y} \right| = L\Delta\beta_1. \tag{3.79}$$

DGD can be quite large in PMF (~1 ns/km) due to their large birefringence. Conventional fibers exhibit much smaller birefringence, but its magnitude and polarization orientation change randomly at scale known as the *correlation length* l_c (with typical values in the range 10–100 m). The analytical treatment of PMD is quite complex due to its statistical nature. A simple model divides the fiber into a large number of segments (tens to thousands of segments), with both the degree of birefringence and the orientation of the principal states being constant in each segment but change randomly from section to section, as shown in Fig. 3.9. Even though the fiber is composed of many segments having different birefringence, there exist *principal states of polarization* (PSP). If we launch a pulse light into one PSP, it will propagate to the corresponding output PSP without spreading due to PMD. Output PSPs are different from input PSPs. Output PSPs are frequency independent up to the first-order approximation. A pulse launched into the fiber with arbitrary polarization can be represented as a superposition of two input PSPs. The fiber is subject to time-varying perturbations (such as temperature and vibrations) so that PSPs and DGD vary randomly over time.

DGD is *Maxwellian distributed* random variable with the mean-square DGD value [3]:

$$\left\langle (\Delta T)^2 \right\rangle = 2(\Delta\beta_1)^2 l_c^2 \left[\exp\left(-\frac{L}{l_c}\right) + \frac{L}{l_c} - 1 \right], \tag{3.80}$$

where the parameters in (3.80) are already introduced above. For long fibers, $L \ll l_c$, the RMS DGD follows:

$$\left\langle (\Delta T)^2 \right\rangle^{1/2} = \Delta\beta_1 \sqrt{2 l_c L} = D_p \sqrt{L}, \tag{3.81}$$

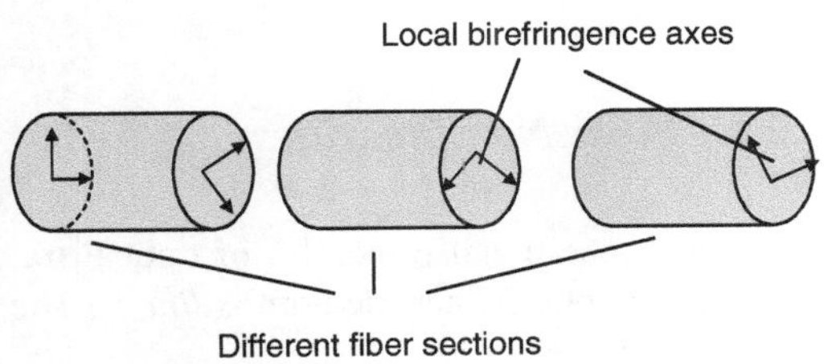

Fig. 3.9 A simple PMD model

where D_p [ps/$\sqrt{\text{km}}$] is the PMD parameter. Typical values for PMD parameter are in the range 0.01–10 ps/(km)$^{1/2}$, in new fibers $D_p < 0.1\,\text{ps/(km)}^{1/2}$. The first-order PMD coefficient, D_p, can be characterized again by Maxwellian PDF [1]:

$$p(D_\text{p}) = \sqrt{\frac{2}{\pi}}\frac{D_\text{p}^2}{\alpha^3}\exp\left(-\frac{D_\text{p}^2}{2\alpha^2}\right), \tag{3.82}$$

where α is the coefficient with a typical value around 30 ps. The mean value $\langle D_\text{p}\rangle$ determined from Maxwellian PDF: $\langle D_\text{p}\rangle = (8\pi)^{1/2}\alpha$. The overall probability $P(D_\text{p})$ that coefficient D_p will be larger than prespecified value can be determined by [1]

$$P(D_\text{p}) = \int_0^{D_\text{p}} p(D_\text{p})\text{d}D_\text{p}, \tag{3.83}$$

$P\left(3\langle D_\text{p}\rangle\right)$ is about 4×10^{-5}, so that the practical expression to characterize the first-order PMD is [1]:

$$\Delta T = 3\langle D_\text{p}\rangle\sqrt{L}. \tag{3.84}$$

The *second-order PMD* occurs due to frequency dependence of both DGD and PSP and can be characterized by coefficient D_{P2} [1, 17]:

$$D_{\text{P2}} = \left[\left(\frac{1}{2}\frac{\partial D_{\text{P1}}}{\partial\omega}\right)^2 + \left(\frac{D_{\text{P1}}}{2}\frac{\partial\,|s|}{\partial\omega}\right)^2\right]^{1/2}, \tag{3.85}$$

where the first term describes the frequency dependence of DGD and the second term describes the frequency dependence of Stokes vector s (describing the position of PSP) [18]. The statistical nature of second-order PMD coefficient in real fiber is characterized by PDF of D_{P2} [1, 19]:

$$p(D_{\text{P2}}) = \frac{2\sigma^2 D_{\text{P2}}}{\pi}\frac{\tanh(\sigma D_{\text{P2}})}{\cosh(\sigma D_{\text{P2}})}. \tag{3.86}$$

The probability that the second-order PMD coefficient is larger than 3 $\langle D_{\text{P2}}\rangle$ is still not negligible. However, it becomes negligible for larger values of 5 $\langle D_{\text{P2}}\rangle$, so that the total pulse spreading due to the second-order PMD effect can be expressed as [1]

$$\Delta\tau_{\text{P2}} = 5\,\langle D_{\text{P2}}\rangle\,L. \tag{3.87}$$

3.2.6 Fiber Nonlinearities

The basic operational principles of optical transmission can be explained assuming that the optical fiber medium is *linear*. The linearity assumption is valid if the

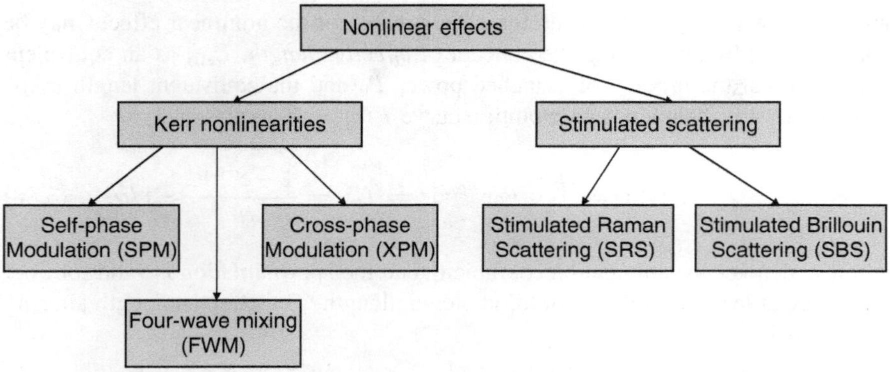

Fig. 3.10 Classification of fiber nonlinearities

launched power does not exceed several mW in a single channel system. In modern WDM technology, high-power semiconductor lasers and optical amplifiers are employed, and the influence of fiber nonlinearities becomes important. Moreover, in some special cases, the fiber nonlinearities may be used to enhance the fiber transmission capabilities (e.g., soliton transmission). There are two major groups of fiber nonlinearities related either to nonlinear refractive index (*Kerr effect*) or to *nonlinear optical scattering*, which is illustrated in Fig. 3.10.

The Kerr effect occurs due to dependence of index of refraction on light intensity, and fiber nonlinearities belonging to this category are (1) self-phase modulation (SPM), (2) cross-phase modulation (XPM), and (3) four-wave mixing (FWM). The SPM is related to the single optical channel. The variation of power within the channel causes the changes in refractive index, which leads to the pulse distortion. In XPM, the refractive index changes not only due to variations in power in observed channel but also due to variation in powers of other wavelength channels leading to the pulse distortion. In FWM, several wavelength channels interact to create new channels. This effect is dependent on both the powers of interacting channel and chromatic dispersion.

The stimulated scattering effects are caused by parametric interaction between the light and materials. There are two types of stimulated scattering effects (1) stimulated Raman scattering (SRS) and (2) stimulated Brillouin scattering (SBS). In the SRS, the light interacts with material through the vibrations and causes the energy transfer from short wavelength channels to longer wavelength channels, leading to the interchannel crosstalk. In the SBS, the light interacts with the matter through acoustic waves, leading to the coupling with backward propagating waves, thus limiting the available power per channel.

The nonlinear effects introduced above are functions of the transmission length and cross-sectional area of the optical fiber. The nonlinear interaction is stronger for longer fibers and smaller cross-sectional area. On the other hand, the nonlinear interaction decreases along the transmission line because of the signal getting attenuated as it propagates. Therefore, the strength of the nonlinear effects may be characterized by introducing the concept of *effective length*, L_{eff}, as an equivalent length so that the product of launched power P_0 and the equivalent length is the same as the area below power evolution curve $P(z) = P_0 \exp(-\alpha z)$:

$$P_0 L_{\mathrm{eff}} = \int_{z=0}^{L} P(z)\mathrm{d}z = \int_{z=0}^{L} P_0 \mathrm{e}^{-\alpha z}\mathrm{d}z \Rightarrow L_{\mathrm{eff}} = \frac{1-\mathrm{e}^{-\alpha L}}{\alpha} \approx 1/\alpha, \quad (3.88)$$

where α denotes the attenuation coefficient, and the approximation is valid for fiber which are at least several tens of kilometers in length. The affective length after M amplifiers is

$$L_{\mathrm{eff,\,total}} = \frac{1-\mathrm{e}^{-\alpha L}}{\alpha} M = \frac{1-\mathrm{e}^{-\alpha L}}{\alpha} \frac{L}{l}, \quad (3.89)$$

where L is the total transmission length, l is the amplifier spacing, and M is the number of amplifiers. If the amplifier spacing increases, the optical amplifier gain increases in proportion to $\exp(\alpha l)$, to compensate for the fiber losses. On the other hand, the increase in power enhances the nonlinear effects. Therefore, what matters is the product of launched power P_0 and the total effective length $L_{\mathrm{eff,total}}$ as follows: $P_0 L_{\mathrm{eff,total}}$.

The nonlinear effects are inversely proportional to the area of the fiber core because the concentration of the optical power per unit cross-sectional area (power density) is higher for smaller cross-sectional area and vice versa. Optical power distribution across the cross-sectional area is closely related to the overall refractive index, and to characterize it the concept of *effective cross-sectional area* (A_{eff}) is introduced [1,2]:

$$A_{\mathrm{eff}} = \frac{\left[\int_r \int_\theta r\,\mathrm{d}r\,\mathrm{d}\theta\,|E(r,\theta)|^2\right]^2}{\int_r \int_\theta r\,\mathrm{d}r\,\mathrm{d}\theta\,|E(r,\theta)|^4}, \quad (3.90)$$

where $E(r,\theta)$ is the distribution of electrical field in cross section and (r,θ) denote the polar coordinates. For the Gaussian approximation $E(r,\theta) = E_0 \exp(-r^2/w^2)$, the effective cross-sectional area is simply $A_{\mathrm{eff}} = \pi w^2$, where w is the mode radius. The Gaussian approximation is not applicable to the optical fibers having more complicated index profile, such as DSFs, NZDFs, and DCFs. In these cases, the effective cross-sectional area can be found by [1]

$$A_{\mathrm{eff}} = k\pi w^2, \quad (3.91)$$

where k is smaller than 1 for DCFs and some NZDSFs and larger than 1 for NZDSF with larger effective cross-sectional area.

3.2.6.1 Self-Phase Modulation

For relatively high power density, the index of refraction acts as the function density, the effect is known as *Kerr effect* [1–3]:

$$n(P) = n_0 + n_2 \frac{P}{A_{\text{eff}}}, \tag{3.92}$$

where n_2 is the Kerr coefficient (second order refractive index coefficient), with typical values being 2.2–3.4 × 10^{-20} m^2/W. The variation in refractive index due to Kerr effect yield to the propagation constant β variations, because $\beta = 2\pi n/\lambda$:

$$\beta(P) = \beta_0 + \gamma P, \quad \beta_0 = 2\pi n_0/\lambda, \quad \gamma = \frac{2\pi n_2}{\lambda A_{\text{eff}}}, \tag{3.93}$$

where γ is the nonlinear coefficient, with typical values being 0.9–2.75 W^{-1} km^{-1} at 1,550 nm [1, 2]. The propagation constant β will vary along the duration of optical pulse, because the different points along the pulse will "see" different optical powers, and the *frequency chirping* is introduced. The propagation constant associated with the leading edge of the pulse will be lower than that related to the central part of the pulse. The difference in propagation constants will cause the difference in phases associated with different portions of the pulse. The central part of the pulse will acquire phase more rapidly than the leading and trailing edges. The total nonlinear phase shift after some length L can be found by

$$\begin{aligned}\Delta\Phi[P(t)] &= \int_0^L [\beta(P_0) - \beta]\mathrm{d}z = \int_0^L \gamma P(z)\mathrm{d}z \\ &= \frac{\gamma P_0(t)[1 - \mathrm{e}^{-\alpha L}]}{\alpha} = \gamma P_0(t) L_{\text{eff}} = \frac{L_{\text{eff}}}{L_{\text{NL}}}, \quad L_{\text{NL}} = \frac{1}{\gamma P_0},\end{aligned} \tag{3.94}$$

where L_{NL} is the nonlinear length.

The frequency variation (chirp) can be found as the first derivative of the nonlinear phase shift with respect to time:

$$\delta\omega_{\text{SPM}}(t) = \frac{\mathrm{d}[\Delta\Phi(t)]}{\mathrm{d}t} = \frac{2\pi}{\lambda} \frac{L_{\text{eff}} n_2}{A_{\text{eff}}} \frac{\mathrm{d}P_0(t)}{\mathrm{d}t}. \tag{3.95}$$

In general case, SPM causes the pulse broadening. However, in some situations, the frequency chirp due to SPM could be opposite to that due to chromatic dispersion (such as *the anomalous dispersion region*). In such a case, SPM helps to reduce the impact of chromatic dispersion. In order to understand the relative effects of chromatic dispersion and SPM, let us observe the NLSE (3.56), assuming that the fiber loss and the second-order GVD can be neglected:

$$\frac{\partial A(z,t)}{\partial z} = -\beta_1 \frac{\partial A(z,t)}{\partial t} - \mathrm{j}\frac{\beta_2}{2}\frac{\partial^2 A(z,t)}{\partial t^2} + \mathrm{j}\gamma\, |A(z,t)|^2\, A(z,t). \tag{3.96}$$

Substituting: $\tau = (t - \beta_1 z)/\tau_0$, $\xi = z/L_D$ $\left(L_D = \tau_0^2/|\beta_2|\right)$, and $U = A/\sqrt{P_0}$ $\left(P_0 = A_0^2\right)$ we get

$$\mathrm{j}\frac{\partial U}{\partial \xi} - \frac{\mathrm{sgn}(\beta_2)}{2}\frac{\partial^2 U}{\partial \tau^2} + N^2 |U|^2 U = 0, \quad N^2 = L_D/L_{NL}. \tag{3.97}$$

When $N \ll 1$, the nonlinear length is much larger than the dispersion length so that the nonlinear effects can be neglected compared with those of chromatic dispersion. *If chromatic dispersion can be neglected* ($L_D \to \infty$), the NLSE can be solved analytically [2]:

$$U(z, \tau) = U(0, \tau)\mathrm{e}^{\mathrm{j}z|U(0,\tau)|^2/L_{NL}}. \tag{3.98}$$

Interestingly, the SPM modulation causes a phase change but no change in the envelope of the pulse. Thus, SPM by itself leads only to chirping, regardless of the pulse shape. The SPM-induced chirp, however, modifies the pulse-broadening effects of chromatic dispersion.

3.2.6.2 Cross-Phase Modulation

XPM is another effect caused by the intensity dependence of the refractive index, but it is related to the multichannel transmission and occurs during propagation of a composite signal. The nonlinear phase shift of a specific optical channel is affected not only by the power of that channel but also by the optical power of the other channels:

$$\Delta\Phi_m(t) = \gamma L_{\mathrm{eff}} P_{0m}(t) + \sum_{i=1,\, i\neq m}^{M} P_{0i}(t) = \frac{2\pi}{\lambda}\frac{L_{\mathrm{eff}} n_2}{A_{\mathrm{eff}}}\left[P_{0m}(t) + \sum_{i=1,\, i\neq m}^{M} P_{0i}(t)\right], \tag{3.99}$$

where $P_{0m}(t)$ denotes the pulse shape in mth channel, other parameters are the same as before. The nonlinear phase shift is bit-pattern dependent. In the *worst-case scenario* (assuming that all channels are transmitting ones and are loaded with the same power P_{0m}):

$$\Delta\Phi_m(t) = \frac{2\pi}{\lambda}\frac{L_{\mathrm{eff}} n_2}{A_{\mathrm{eff}}}(2M - 1)P_{0m}(t). \tag{3.100}$$

In practice, the optical pulses from different optical channels propagate at different speeds (they have different group velocities). The phase shift given above can occur only during the overlapping time. The overlapping among neighboring channels is longer than the overlapping of channels spaced apart, and it will produce the most significant impact on the phase shift. If pulses walk through one another quickly (due to significant chromatic dispersion or when the channels are widely separated), the described effects on both pulses are diminished because the distortion done by the trailing edge is undone by the leading edge.

3.2.6.3 Four-Wave Mixing

FWM is another effect that occurs in optical fibers during the propagation of a composite optical signal, such as the WDM signal. It gives rise to the new optical signals. The three optical signals with different carrier frequencies f_i, f_j, and f_k $(i, j, k = 1, \ldots, M)$ interact to generate the new optical signal at frequency $f_{ijk} = f_i + f_j - f_k$, providing that the *phase-matching condition* is satisfied:

$$\beta_{ijk} = \beta_i + \beta_j - \beta_k, \tag{3.101}$$

where βs are corresponding propagation constants. The measure of the phase-matching condition of wavelength channels involved in the interaction is defined by

$$\Delta\beta = \beta_i + \beta_j - \beta_k - \beta_{ijk}, \tag{3.102}$$

The FWM is effective only if $\Delta\beta$ approaches zero. Therefore, the phase-matching condition is a requirement for the *momentum conservation*. The FWM process can also be considered from quantum-mechanic point of view as the *annihilation* of two photons with energies hf_i and hf_j and *generation* of two new photons with energies hf_k and hf_{ijk}. In FWM process, the indices i and j do not need to be necessarily distinct, meaning that only two channels may interact to create the new one, this case is known as *degenerate* case.

The power of the newly generated optical frequency is a function of the powers of optical signals involved in the process, the Kerr coefficient, and the degree of satisfaction of the phase-matching condition, so that we can write [1,3,5]

$$P_{ijk} \sim \left(\frac{2\pi f_{ijk} n_2 d_{ijk}}{3cA_{\mathrm{eff}}}\right)^2 P_i P_j P_k L_{\mathrm{eff}}^2, \tag{3.103}$$

where d_{ijk} is the measure of degeneracy, which takes value 3 in degenerate case or value 6 in nondegenerate case. FWM can produce significant signal degradation in WDM systems since several newly generated frequencies can coincide with any specific channel. Namely, the total number of newly generated frequencies N through the FWM process is $N = M^2(M-1)/2$. Fortunately, some of the newly generated frequencies have negligible impact. However, some of them will have significant impact especially those coinciding with already existing WDM channels.

The FWM can be reduced either by reducing the power per channel or by preventing the perfect phase matching by increasing the chromatic dispersion or increasing the channel spacing. One option would be to use nonuniform channel spacing. On the other hand, the FWM process can be used to perform some useful functions, such as the *wavelength conversion* through the process of *phase conjugation* [3].

In the rest of this subsection, we provide another interpretation of Kerr nonlinearities [2]. Under isotropy assumption, in the presence of nonlinearities, induced polarization $\boldsymbol{P}(\boldsymbol{r}, t)$ is along the same direction as the electric field $\boldsymbol{E}(\boldsymbol{r}, t)$, so that corresponding vector notation can be substituted by the scalar functions, and the induced polarization be written in the form [2]:

$$P(\boldsymbol{r},t) = \varepsilon_0 \int_{-\infty}^{t} \chi^{(1)}(t - t'E(\boldsymbol{r},t'\mathrm{d}t'\varepsilon_0\chi^{(3)}E^3(\boldsymbol{r},t), \tag{3.104}$$

where the first term corresponds to the linear part and the second term, $P_{\mathrm{NL}}(\boldsymbol{r},t)$, to the nonlinear part. $\chi^{(1)}$ denotes the linear susceptibility, $\chi^{(3)}$ denotes the third-order susceptibility, and ε_0 is the permittivity of vacuum. (Because of the symmetry of silica molecules only odd terms are present in Taylor expansion of $P(\boldsymbol{r},t)$.) A WDM signal can be represented as the sum of n monochromatic plane waves of angular frequency ω_i (the modulation process is ignored in this simplified analysis):

$$E(\boldsymbol{r},t) = \sum_{i=1}^{M} E_i \cos(\omega_i t - \beta_i z). \tag{3.105}$$

The nonlinear polarization, after substitution of (3.105) into the second term of (3.104), can be written as follows:

$$\begin{aligned}
P_{\mathrm{NL}}(\boldsymbol{r},t) &= \varepsilon_0\chi^{(3)} \sum_{i=1}^{M}\sum_{j=1}^{M}\sum_{k=1}^{M} E_i \cos(\omega_i t - \beta_i z) E_j \cos(\omega_j t - \beta_j z) E_k \cos(\omega_k t - \beta_k z) \\
&= \frac{3}{4}\varepsilon_0\chi^{(3)} \sum_{i=1}^{M} E_i^2 E_i \cos(\omega_i t - \beta_i z) + \frac{3}{4}\varepsilon_0\chi^{(3)} \sum_{i=1}^{M}\sum_{j\neq i} 2E_i E_j E_i \cos(\omega_i t - \beta_i z) \\
&+ \frac{3}{4}\varepsilon_0\chi^{(3)} \sum_{i=1}^{M}\sum_{j\neq i} E_i^2 E_j \cos\left[(2\omega_i - \omega_j)t - (2\beta_i - \beta_j)z\right] \\
&+ \frac{6}{4}\varepsilon_0\chi^{(3)} \sum_{i=1}^{M}\sum_{j>i}\sum_{k>j} E_i E_j E_k \cos\left[(\omega_i + \omega_j - \omega_k)t - (\beta_i + \beta_j - \beta_k)z\right] + \cdots
\end{aligned} \tag{3.106}$$

The first term of the second line [of (3.106)] corresponds to the SPM, the second term in the same line to the XPM, the first term of the third line corresponds to the degenerate FWM case, and the last term corresponds to the nondegenerate FWM.

3.2.6.4 Stimulated Raman Scattering

Raman scattering is a nonlinear effect that occurs when a propagating optical signal interacts with glass molecules in the fiber undergoing a wavelength shift. The result of interaction is a transfer of energy from some photons of the input optical signal into vibrating silica molecules and a creation of new photons with lower energy than the energy of incident photons. The incident optical signal is commonly referred to as a pump, and the generated signal is called the *Stokes signal.* The difference in frequencies between the pump (ω_{p}) and the Stokes signal (ω_{S}) is known as the *Raman frequency shift* $\omega_{\mathrm{R}} = \omega_{\mathrm{p}} - \omega_{\mathrm{S}}$. Scattered photons are not in phase with each other and do not follow the same scattering pattern, meaning that energy transfer from pump to Stokes photons *is not a uniform process.* As a result, there exists the frequency band that includes frequencies of all scattered Stokes photons. Scattered Stokes photons can take any direction that can be either forward or backward with respect to the direction of the pump, meaning that *Raman scattering is isotropic process*. If the pump power is lower than a certain *threshold value*, the Raman scattering process will have a spontaneous character, characterized by the small number of pump photons that will be scattered and converted to the Stokes photons. However, when the pump power exceeds the threshold value, Raman scattering becomes a *stimulated process.* SRS can be explained as *a positive feedback process* in the pump signal that interacts with Stokes signal and creates the beat frequency $\omega_R = \omega_p - \omega_S$, and the beat frequency then serves as a stimulator of molecular oscillations, and the process is enhanced or amplified. Assuming that the Stokes signal propagates in the same direction as the pump signal, the intensity of the pump signal $I_p = P_p/A_{eff}$ and the intensity of the Stokes signal $I_S = P_S/A_{eff}$ are related by the following systems of coupled differential equations [1, 3]:

$$\frac{dI_p}{dz} = -g_R\left(\frac{\omega_p}{\omega_S}\right) I_p I_S - \alpha_p I_p; \quad \frac{dI_S}{dz} = g_R I_p I_S - \alpha_S I_S, \tag{3.107}$$

where α_p (α_S) is the fiber attenuation coefficient of the pump (Stokes) signal, and g_R is the Raman gain coefficient. Because the SRS process is not the uniform process, the scattered Stokes photons will occupy a certain frequency band, and the Raman gain is not constant, but a function of frequency. It can be roughly approximated by the *Lorentzian spectral profile* [1]:

$$g_R(\omega_R) = \frac{g_R(\Omega_R)}{1 + (\omega_R - \Omega_R)^2 T_R^2}, \tag{3.108}$$

where T_R is vibration state's decay time (in the order 0.1 ps for silica-based materials), and Ω_R is the Raman frequency shift of a corresponding Raman gain peak. The actual gain profile extends over 40 THz (~320 nm), with a peak around 13.2 THz. The Raman gain peak $g_R(\Omega_R) = g_{R,max}$ is between 10^{-12} and 10^{-13} m/W for wavelengths above 1,300 nm. The gain profile can also be approximated by a *triangle* function, which is commonly used in analytical studies of SRS [1, 3]:

$$g_R(\omega_R) = \frac{g_R(\Omega_R)\omega_R}{\Omega_R}. \tag{3.109}$$

The *Raman threshold*, defined as the incident power at which half of the pump power is converted to the Stokes signal, can be estimated using the system of differential equations by replacing g_R by $g_{R,max}$. The amplification of Stokes power along distance L can be determined by

$$P_S(L) = P_{S,0} \exp\left(\frac{g_{R,max} P_{S,0} L}{2A_{eff}}\right), \tag{3.110}$$

where $P_{S,0} = P_S(0)$. The Raman threshold can be determined by [1–3]

$$P_{\mathrm{R,threshold}} = P_{\mathrm{S},0} \approx \frac{16A_{\mathrm{eff}}}{g_{\mathrm{R,max}}} L_{\mathrm{eff}}, \tag{3.111}$$

and for typical SMF parameters ($A_{\mathrm{eff}} = 50\,\mu\mathrm{m}^2$, $L_{\mathrm{eff}} = 20\,\mathrm{km}$, and $g_{\mathrm{R,max}} = 7 \times 10^{-13}\,\mathrm{m/W}$), it is around 500 mW.

The SRS can effectively be used for optical signal amplification (*Raman amplifiers*), as we explained earlier. In WDM systems, however, the SRS effect can be quite detrimental, because the Raman gain spectrum is very broad, which enables the energy transfer from the lower to higher wavelength channels. The shortest wavelength channel within the WDM system acts as a pump for several long wavelength channels, while undergoing the most intense depletion. The fraction of power coupled out of the channel 0 to all the other channels (1, 2, ..., M - 1; M is the number of wavelengths), when the Raman gain shape is approximated as a triangle, can be found as [2]:

$$P_0 = \sum_{i=1}^{M-1} P_0(i) = \frac{g_{\mathrm{R,max}} \Delta\lambda P L_{\mathrm{eff}}}{2\Delta\lambda_{\mathrm{C}} A_{\mathrm{eff}}} \frac{M(M-1)}{2}, \tag{3.112}$$

where $\Delta\lambda$ is the channel spacing and $\Delta\lambda_{\mathrm{C}} \cong 125\,\mathrm{nm}$. In order to keep power penalty, $-10\log(1 - P_0)$, below 0.5 dB, the following inequality must be satisfied [2]:

$$PM(M-1)\Delta\lambda L_{\mathrm{eff}} < 40{,}000\,\mathrm{mW\,nm\,km}. \tag{3.113}$$

With chromatic dispersion present, the previous inequality can be relaxed to 80,000 mW nm km. For example, in a 32-wavelength channels spaced apart 0.8 nm and $L_{\mathrm{eff}} = 20\,\mathrm{km}$, $P \leq 2.5\,\mathrm{mW}$. The energy transfer between two channels is bit-pattern dependent and occurs only if both wavelengths are synchronously loaded with 1 bit, meaning that the energy transfer will be reduced if dispersion is higher due to the *walk-off effect* (difference in velocities of different wavelength channels will reduce the time of overlapping).

3.2.6.5 Stimulated Brillouin Scattering

Brillouin scattering is a physical process that occurs when an optical signal interacts with acoustical phonons, rather than the glass molecules. During this process an incident optical signal reflects backward from the grating formed by acoustic vibrations and downshifts in frequency. The acoustic vibrations originate from the thermal effect if the power of an incident optical signal is relatively small. If the power of incident light goes up, it increases the material density through the electrostrictive effect. The change in density enhances acoustic vibrations and forces Brillouin scattering to become *stimulated*. The SBS process can also be explained as a positive feedback mechanism, in which the incident light (the pump) interacts with Stokes signal and creates the beat frequency $\omega_{\mathrm{B}} = \omega_{\mathrm{p}} - \omega_{\mathrm{S}}$. The parametric interaction between the pump, Stokes signal, and acoustical waves requires both the *energy and the momentum conservation*: the energy is effectively preserved through the downshift in frequency, while the momentum conservation occurs through the backward direction of the Stokes signal. The *Brillouin shift* (the frequency downshift) can be determined by [1]

$$f_{\mathrm{B}} = \frac{2nv_{\mathrm{A}}}{\lambda_{\mathrm{p}}}, \tag{3.114}$$

where v_{A} is the velocity of the acoustic wave. The frequency shift is fiber material dependent and can vary from 10.5 GHz to 12 GHz. The SBS is governed by the following set of coupled equations, similarly as in SRS [1, 3]:

$$\frac{\mathrm{d}I_{\mathrm{p}}}{\mathrm{d}z} = -g_{\mathrm{B}}I_{\mathrm{p}}I_{\mathrm{S}} - \alpha_{\mathrm{p}}I_{\mathrm{p}}; \quad -\frac{\mathrm{d}I_{\mathrm{S}}}{\mathrm{d}z} = g_{\mathrm{B}}I_{\mathrm{p}}I_{\mathrm{S}} - \alpha_{\mathrm{S}}I_{\mathrm{S}}, \tag{3.115}$$

where I_{p} and I_{S} are the intensities of the pump and Stokes waves, respectively; g_{B} is the Brillouin gain coefficient; and α_{p} (α_{S}) is the attenuation coefficient of the pump (Stokes) signal. The scattered Stokes photons will not have equal frequencies, but will be dispersed within certain frequency band. The spectrum of Brillouin gain is related to the acoustic phonons lifetime (characterized by the time constant T_{B}) and can be approximated by a Lorentzian spectral profile [1]:

$$g_{\mathrm{B}}(\omega_{\mathrm{B}}) = \frac{g_{\mathrm{B}}(\Omega_{\mathrm{B}})}{1 + (\omega_{\mathrm{B}} - \Omega_{\mathrm{B}})^2 T_{\mathrm{B}}^2}, \tag{3.116}$$

where $\Omega_{\mathrm{B}} = 2\pi f_{\mathrm{B}}$ is the Brillouin shift. The Brillouin gain is also dependent on the fiber waveguide characteristics. The SBS gain bandwidth Δf_{B} is about 17 MHz at $\lambda_{\mathrm{p}} = 1{,}520$ nm for pure silica, but it is almost 100 MHz in doped silica fibers, and the typical bandwidth is about 50 MHz. The maximum value of the gain $g_{\mathrm{B,max}}$ is also dependent on the type of fiber material, and it is between 10^{-10} and 10^{-11} m/W for silica-based optical fiber above 1 μm. The threshold pump power at which the Brillouin scattering becomes stimulated can be calculated in similar fashion as done for SRS [1]:

$$P_{\mathrm{B,\,threshold}} \approx \frac{21A_{\mathrm{eff}}}{g_{\mathrm{B,max}}L_{\mathrm{eff}}}, \tag{3.117}$$

and the estimated Brillouin threshold is about 1.05 mW for typical values of fiber parameters ($A_{\mathrm{eff}} = 50\,\mu\mathrm{m}^2$, $L_{\mathrm{eff}} = 20$ km, and $g_{\mathrm{B,max}} = 5 \times 10^{-11}$ m/W). The preceding expression assumes that the pump signal has a negligible line width and lies within the gain bandwidth of SBS. The threshold power is considerably increased if the signal has a broad linewidth, and thus much of the pump power lies outside of the gain bandwidth of SBS [2]:

$$P_{\mathrm{B,threshold}} \approx \frac{21A_{\mathrm{eff}}}{g_{\mathrm{B,max}}L_{\mathrm{eff}}}\left(1 + \frac{\Delta f_{\mathrm{s}}}{\Delta f_{\mathrm{B}}}\right), \tag{3.118}$$

where Δf_s is the spectral width of the source. For example, with source bandwidth is ten times larger than the SBS gain bandwidth, the SBS threshold increases to 11.55 mW. The SBS suppression can be achieved by *dithering the source* (directly modulating the laser with a sinusoid at a frequency much lower than the receiver low-frequency cutoff).

3.3 Transmission System Performance Assessment and System Design

The transmission quality can be estimated by using different figures of merit such as SNR, optical SNR (OSNR), Q-factor, eye opening penalty, and the BER. The SNR at decision point is defined as the ratio between the signal power to the noise power. The optical amplification process is accompanied with ASE noise that accumulates along the transmission line and degrades the OSNR. Therefore, the OSNR can be defined as the ratio of the received optical power and accumulated noise power. It is a common practice in optical communications to use the Q-factor as a figure of merit instead of SNR. It is defined as

$$Q = \frac{I_1 - I_0}{\sigma_1 + \sigma_0}, \tag{3.119}$$

where I_1 and I_0 represent the average photocurrents corresponding to one- and zero-symbol levels, and σ_1 and σ_0 are corresponding standard deviations. The total variance in the photocurrents corresponding to symbol 1/0 is:

$$\sigma_{1,0}^2 = \left\langle i_{\mathrm{sn},1,0}^2 \right\rangle + \left\langle i_{\mathrm{thermal},1,0}^2 \right\rangle, \tag{3.120}$$

where $\left\langle i_{\mathrm{sn},1,0}^2 \right\rangle$ is the variance corresponding to the shot noise and $\left\langle i_{\mathrm{thermal},1,0}^2 \right\rangle$ is the variance corresponding to the thermal noise.

The BER is an important parameter in digital transmission quality assessment. It defines the probability that the bit being transmitted will be mistaken by the decision circuit. The fluctuating signal levels corresponding to 0 and 1 bits can be characterized by corresponding PDFs. If at decision point the received sample is larger than the threshold ($I > I_{\mathrm{tsh}}$) we decide in favor of bit 1, otherwise we decide in favor of bit 0. An error occurs if a bit 1 was transmitted but we decided in favor of 0 or a bit 0 was transmitted but we decided in favor of 1.

The bit-error ratio is defined by

$$\mathrm{BER} = \Pr(0|1)\Pr(1) + \Pr(1|0)\Pr(0), \tag{3.121}$$

where Pr(0|1) is the conditional probability that symbol 1 was transmitted by the decision circuit decided in favor of 1, Pr(1|0) is the conditional probability that symbol 0 was transmitted by the decision circuit decided in favor of 1, and Pr(0) and Pr(1) are a priori probabilities of symbols 0 and 1, respectively. Assuming an equiprobable transmission, Pr(0) = Pr(1) = 1/2, we obtain BER = 0.5(Pr(0|1) + Pr(1|0)). The PDFs for symbols 0 and 1 are commonly considered as Gaussian [1, 3, 5, 7–9], so that conditional error probabilities can be determined by

$$\Pr(0|1) = \frac{1}{\sigma_1\sqrt{2\pi}}\int_{-\infty}^{I_{\mathrm{tsh}}}\exp\left[-\frac{(I-I_1)^2}{2\sigma_1^2}\right]\mathrm{d}I = \frac{1}{2}\mathrm{erfc}\left(\frac{I_1-I_{\mathrm{tsh}}}{\sigma_1\sqrt{2}}\right),$$

$$\Pr(1|0) = \frac{1}{\sigma_0\sqrt{2\pi}}\int_{I_{\mathrm{tsh}}}^{\infty}\exp\left[-\frac{(I-I_0)^2}{2\sigma_0^2}\right]\mathrm{d}I = \frac{1}{2}\mathrm{erfc}\left(\frac{I_{\mathrm{tsh}}-I_0}{\sigma_0\sqrt{2}}\right), \quad (3.122)$$

where the complementary error function erfc(x) is defined by

$$\mathrm{erfc}(x) = \frac{2}{\sqrt{\pi}}\int_x^{+\infty}\mathrm{e}^{-z^2}\mathrm{d}z.$$

The optimum decision, minimizing the BER, is the *maximum a posteriori probability* (MAP) decision rule, which can be formulated as follows:

$$p(I|0)\Pr(0) \underset{H_1}{\overset{H_0}{\gtrless}} p(I|1)\Pr(1), \quad p(I|i) = \frac{1}{\sigma_i\sqrt{2\pi}}\exp\left[-\frac{(I-I_i)^2}{2\sigma_i^2}\right], \quad i\in\{0,1\}, \quad (3.123)$$

where we used H_i ($i = 0, 1$) to denote the hypothesis the symbol i was transmitted, and $p(I|i)$ ($i = 0, 1$) is the corresponding PDF. For the equiprobable transmission [Pr(0) = Pr(1) = 1/2], the MAP rule becomes the *maximum-likelihood* (ML) decision rule [Pr(0) = Pr(1) = 1/2]:

$$p(I|0) \underset{H_1}{\overset{H_0}{\gtrless}} p(I|1). \quad (3.124)$$

The resulting BER can be obtained by substituting (3.123) into (3.121):

$$\mathrm{BER} = \frac{1}{4}\left[\mathrm{erfc}\left(\frac{I_1-I_{\mathrm{tsh}}}{\sigma_1\sqrt{2}}\right) + \mathrm{erfc}\left(\frac{I_{\mathrm{tsh}}-I_0}{\sigma_0\sqrt{2}}\right)\right]. \quad (3.125)$$

The optimum threshold can be obtained by minimizing the BER, which yields to the following equation to be solved numerically:

$$\ln\left[\frac{\Pr(0)}{\Pr(1)}\right] + \ln\left(\frac{\sigma_1}{\sigma_0}\right) = \frac{(I_1-I_{\mathrm{tsh}})^2}{2\sigma_1^2} - \frac{(I_{\mathrm{tsh}}-I_0)^2}{2\sigma_0^2}. \quad (3.126)$$

When the standard deviations are close to each other ($\sigma_1 \approx \sigma_0$), the following approximation is valid:

$$I_{\text{tsh}} \approx \frac{\sigma_0 I_1 + \sigma_1 I_0}{\sigma_0 + \sigma_1}. \tag{3.127}$$

By substituting (3.127) into (3.125) we obtain the following relation:

$$\text{BER} = \frac{1}{2}\text{erfc}\left(\frac{Q}{\sqrt{2}}\right), \quad Q = \frac{I_1 - I_0}{\sigma_1 + \sigma_0} \tag{3.128}$$

establishing the connection between the BER and Q-factor introduced in (3.119). For example, $Q = 7$ (16.9 dB) corresponds to BER of 10^{-12}, and $Q = 8$ (18.06 dB) corresponds to BER of 10^{-15}.

Another important receiver parameter is the receiver sensitivity. *Receiver sensitivity* is defined as minimum average received optical power $P_R = (P_0 + P_1)/2$ (P_i is the average power corresponding to symbol i) required to operate at a given BER (such as 10^{-12} or 10^{-15}). By including the impact of the basic noise components (*shot, thermal, and beat noise components*) into three basic real case detection scenarios, we can establish the new reference point. All other impairments degrade the receiver sensitivity by increasing the value of the optical power P_R required to achieve a desired BER. Accordingly, each individual impairment causes the receiver sensitivity degradation, which is evaluated by the *power penalty*. The power penalty, denoted as ΔP, is defined as follows:

$$\Delta P\ [\text{dB}] = 10\log_{10}\frac{P_R(\text{in the presence of impairment, at given BER})}{P_R(\text{in the absence of impairment, at given BER})}. \tag{3.129}$$

3.3.1 Quantum Limit for Photodetection

The quantum limit is defined for an ideal photodiode (no thermal noise, no dark current, no shot noise, and 100% of quantum efficiency) and can be used as the reference point to compare different receiver architectures. In the quantum limit case the Gaussian assumption is no longer valid, and Poisson statistics should be used instead. Assuming that N_p is the average number of photons in each 1 bit (the number of photons in a 0 bit is zero), the probability for the formation of m e–h pairs is given by *Poisson distribution*:

$$P_m = \exp(-N_p)N_p^m/m!. \tag{3.130}$$

Using our earlier definitions, $\Pr(1|0) = 0$ because no e–h pairs are formed when $N_p = 0$, while $\Pr(0|1) = P_0 = \exp(-N_p)$. Therefore, the BER for quantum limit is as follows:

$$\text{BER} = \frac{1}{2}\exp(-N_p). \tag{3.131}$$

The receiver sensitivity is related to the average number of photons extended over the stream of 0s and 1s $\langle N_p \rangle$ by

$$P_R = \frac{P_0 + P_1}{2} = \frac{N_p h\nu}{2T} = \langle N_p \rangle R_b h\nu, \tag{3.132}$$

where $h\nu$ is the photon energy, T is the bit duration, and R_b is the bit rate. For example, $N_p = 34(\langle N_p \rangle = 17)$ for BER of 10^{-15}, while corresponding receiver sensitivity is $P_R = -46.61\,\text{dBm}$ at $R_b = 10\,\text{Gb/s}$ and $P_R = -40.59\,\text{dBm}$ at $R_b = 40\,\text{Gb/s}$.

3.3.2 Shot Noise and Thermal Noise Limit

The impact of shot noise and thermal noise is unavoidable in any optical receiver and as such can be used as the referent case. Let us assume that $I_0 = 0(P_0 = 0$, infinite extinction ratio), then $P_1 = 2P_R$ and $I_1 = 2\langle M \rangle RP_R$ for avalanche photodiodes (APDs). For the bit one the variance of noise includes both thermal and shot noise terms, while for bit 0 includes only the thermal noise term, and in case of APDs the corresponding variances can be written as [1–3]

$$\sigma_1^2 = \sigma_{sn}^2 + \sigma_{thermal}^2, \quad \sigma_0^2 = \sigma_{thermal}^2,$$
$$\sigma_{sn}^2 = \langle i_{sn}^2 \rangle = 4qM^2F(M)RP_R B_{el} \quad \sigma_{thermal}^2 = \langle i_{thermal}^2 \rangle = \frac{4k\Theta F_{ne}}{R_L} B_{el}, \tag{3.133}$$

where $F(M)$ is the excess noise factor introduced earlier, M the APD multiplication factor, R the photodiode responsivity, k_B the Boltzmann's constant, Θ the absolute temperature, F_{ne} the noise figure of postamplifier (see Fig. 3.2), and B_{el} is the electrical filter bandwidth. The corresponding expressions for PIN photodetectors can be obtained by setting $M = 1$ and $F(M) = 1$. The receiver sensitivity can be obtained from the Q-factor

$$Q = \frac{I_1}{\sigma_1 + \sigma_0} = \frac{MR(2P_R)}{\left(2qM^2F(M)R(2P_R)B_{el} + \sigma_{thermal}^2\right)^{1/2} + \sigma_{thermal}}, \tag{3.134}$$

as follows

$$P_R = \frac{Q_{req}}{R}\left(qF(M)Q_{req}B_{el} + \frac{\sigma_{thermal}}{M}\right). \tag{3.135}$$

The optimum multiplication factor is obtained by solving for $M = M_{opt}$ from $dP_R/dM = 0$ by

$$M_{opt} = \sqrt{\frac{\frac{\sigma_{thermal}}{qQB_{el}} + k_N - 1}{k_N}} \approx \sqrt{\frac{\sigma_{thermal}}{k_N qQB_{el}}}, \tag{3.136}$$

where k_N is the ratio of impact ionization factors of holes and electrons, respectively. By substituting (3.136) into (3.135) we obtain

$$P_{\mathrm{R,APD}} = \frac{2qB_{\mathrm{el}}}{R} Q^2 (k_N M_{\mathrm{opt}} + 1 - k_N). \tag{3.137}$$

The receiver sensitivity for the thermal noise limit is obtained by setting the shot noise contributor in (3.135) to zero, and this case corresponds to the PIN photodiode:

$$P_{\mathrm{R,PIN}} \approx \frac{\sigma_{\mathrm{thermal}} Q}{R} = \frac{Q\sqrt{4k\Theta F_{\mathrm{ne}} B_{\mathrm{el}}}}{R\sqrt{R_{\mathrm{L}}}}. \tag{3.138}$$

3.3.3 Receiver Sensitivity for Receivers with Optical Preamplifier

The receiver sensitivity can be greatly improved by using the optical preamplifier in front of PIN photodiode. By following the similar procedure to that from previous section, following expression for the receiver sensitivity is obtained [1,3] (assuming again that $I_0 = 0$):

$$\begin{aligned} P_{\mathrm{R}} &= \frac{2S_{\mathrm{sp}} B_{\mathrm{el}} \left[Q^2 - Q\sqrt{B_{\mathrm{op}}/B_{\mathrm{el}}}\right]}{G-1} = F_{\mathrm{no}} hf B_{\mathrm{el}} \left[Q^2 - Q\sqrt{B_{\mathrm{op}}/B_{\mathrm{el}}}\right], \\ S_{\mathrm{sp}} &= F_{\mathrm{no}} hf (G-1)/2, \end{aligned} \tag{3.139}$$

where S_{sp} is the PSD of spontaneous emission noise, G the gain of optical preamplifier, F_{no} the corresponding noise figure, hf a photon energy, B_{op} the optical filter bandwidth, and B_{el} is the electrical filter bandwidth.

3.3.4 Optical Signal-to-Noise Ratio

The optical amplification process is accompanied with ASE noise that accumulates along the transmission line degrading the OSNR, defined by

$$\mathrm{OSNR} = \frac{P_{\mathrm{s}}}{P_{\mathrm{ASE}}} = \frac{P_{\mathrm{s}}}{S_{\mathrm{sp}} B_{\mathrm{op}}} = \frac{P_{\mathrm{s}}}{2n_{\mathrm{sp}} hf (G-1) B_{\mathrm{op}}}, \tag{3.140}$$

where P_{s} is the signal power and P_{ASE} is the noise power, other parameters are already defined in Sect. 3.3.3. The OSNR can be measured at the receiver entrance point (just before photodetector) and as such can be related to the Q-factor as follows [1,2]:

$$Q = \frac{2\mathrm{OSNR}\sqrt{B_{\mathrm{op}}/B_{\mathrm{el}}}}{1 + \sqrt{1 + 4\mathrm{OSNR}}}, \tag{3.141}$$

where B_{op} and B_{el} are optical and electrical filters bandwidths, respectively. Notice that expression above is valid only for NRZ modulation format.

3.3.5 Power Penalty Due to Extinction Ratio

In previous subsections, we considered different detection scenarios by including the basic receiver noise components. All other factors degrading the system performance, both noise components and signal impairments, can be characterized by introducing the power penalty, (3.129), that is the increase in receiver sensitivity required to keep the same quality of transmission.

The power penalty (PP) due to extinction ratio (the extinction ratio is defined as $r_{ex} = P_0/P_1$), denoted as ΔP_{ex}, for the thermal-noise-dominated scenario can be obtained by using the PP definition expression (3.130) and receiver sensitivity expression as follows:

$$\Delta P_{ex}\,[\mathrm{dB}] = 10\log_{10}\frac{P_R(r_{ex}\neq 0)}{P_R(r_{ex}=0)} = 10\log_{10}\frac{1+r_{ex}}{1-r_{ex}}. \tag{3.142}$$

The power penalty due to extinction ratio for the shot-noise-dominated scenario can be obtained by repeating the similar procedure as above to obtain

$$\Delta P_{ex}\,[\mathrm{dB}] = 10\log_{10}\frac{1+r_{ex}}{\left(1-\sqrt{r_{ex}}\right)^2}. \tag{3.143}$$

3.3.6 Power Penalty Due to Intensity Noise

The intensity noise is related to the fluctuations of the incoming optical power, mostly due to the light source. The power penalty due to intensity noise can be obtained by calculating first the Q-factors in the presence and absence of intensity noise, then to calculate the corresponding receiver sensitivities, and finally to calculate the power penalty as the ratio of receiver sensitivities in the presence and absence of intensity noise to get

$$\Delta P_{int}\,[\mathrm{dB}] = 10\log_{10}\frac{P_R(r_{int}\neq 0)}{P_R(r_{int}=0)} = -10\log_{10}\left(1-r_{int}^2Q^2\right), \tag{3.144}$$

where r_{int} is the intensity noise parameter defined as $\sqrt{(2\mathrm{RIN}_{laser}\Delta f)}$, where RIN_{laser} is the average value of RIN spectrum [see (3.3.6)], which is typically below $-160\,\mathrm{dBm/Hz}$ ($r_{int} \sim 0.0004$) for high-quality lasers, and Δf is the receiver bandwidth. The other sources of intensity noise include reflections (characterized by intensity reflection noise parameter r_{ref}), MPN (characterized by r_{MPN}), and phase noise to intensity noise conversion (characterized by r_{phase}), so that the effective intensity noise parameter can be determined by [1]

$$r_{\text{eff}}^2 = r_{\text{int}}^2 + r_{\text{ref}}^2 + r_{\text{MPN}}^2 + r_{\text{phase}}^2, \quad r_{\text{ref}} \approx \frac{(r_1 r_2)^{1/2}}{\alpha_{\text{iso}}},$$
$$r_{\text{MPN}} \approx \left(k/\sqrt{2}\right)\left\{1 - \exp\left[-(\pi R_{\text{b}} D L \sigma_\lambda)^2\right]\right\}, \tag{3.145}$$

where r_1 and r_2 are reflection coefficients of two disjoints, α_{iso} the attenuation of optical isolator, R_{b} the bit rate, D the chromatic dispersion coefficient, L the transmission distance, σ_λ the light source spectral linewidth, and k is the MPN coefficient ($k = 0.6$–0.8).

3.3.7 Power Penalty Due to Timing Jitter

Timing is typically determined by the clock-recovery circuit. Because the input to this circuit is noisy, the sampling time fluctuates from bit to bit, and this fluctuation is commonly referred to as *timing skew* or *timing jitter*. Timing skew occurs when the sampling time error is fixed (e.g., offsets in circuits or insufficient dc gain in PLL). Timing jitter occurs when the sampling time error is dynamic (e.g., interplay of optical amplifier noise, GVD, and fiber nonlinearities; or due to clock-recovery circuit). By assuming that PIN photodiode is used, the thermal-noise-dominated scenario is being observed, and the extinction ratio is ideal, the following power penalty expression is obtained:

$$\Delta P_{\text{jitter}}\,[\text{dB}] = 10\log_{10}\frac{P_{\text{R}}(b \neq 0)}{P_{\text{R}}(b = 0)} = -5\log_{10}\left[(1-b)^2 - 2b^2 Q^2\right],$$
$$b = \left(\frac{\pi R_{\text{b}} \left\langle \Delta t^2 \right\rangle^{1/2}}{2}\right)^2, \tag{3.146}$$

where R_{b} is the bit rate and Δt is the timing error.

3.3.8 Power Penalty Due to GVD

The impact of GVD is difficult to evaluate because of the nonlinear relationship between transmitted and received intensities and because of the presence of fiber nonlinearities. Under assumptions that the noise and variation in intensity due to ISI can be modeled using Gaussian approximation (and neglecting the extinction ratio), the power penalty due to GVD for thermal-noise-dominated scenario can be evaluated as follows [12, 13]:

$$\Delta P_{\mathrm{GVD}}\,[\mathrm{dB}] = 10\log_{10}\frac{1+\langle\Delta I_1\rangle-\langle\Delta I_0\rangle}{\sqrt{\left[(1+\langle\Delta I_1\rangle-\langle\Delta I_0\rangle)^2-Q^2\left(\sigma_{I_0}^2+\sigma_{I_1}^2\right)\right]^2-4Q^2\sigma_{I_0}^2\sigma_{I_1}^2}}, \tag{3.147}$$

where $\langle\Delta I_1\rangle$ and $\langle\Delta I_0\rangle$ are the average values of variations in intensities for bit 1 and bit 0 (normalized with I_1 – the intensity of bit 1 in the absence of GVD) respectively, while the σ_{I_1} and σ_{I_0} denote corresponding standard deviations.

3.3.9 Power Penalty Due to Signal Crosstalk

The crosstalk noise is related to multichannel systems (WDM systems) and can be either out-of-band or in-band in nature. Following the similar procedure as we did before the power penalty expression due to signal crosstalk can be found as follows [1]:

$$\Delta P_{\mathrm{cross}} = -10\log_{10}\left(1-r_{\mathrm{cross}}^2 Q^2\right), \quad r_{\mathrm{cross}} = \sqrt{r_{\mathrm{cross,out}}^2 + r_{\mathrm{cross,in}}^2}, \tag{3.148}$$

where $r_{\mathrm{cross,out}}^2 = \left(\sum_{n=1,n\neq m}^{M} X_{n,\mathrm{out}}\right)^2$ and $r_{\mathrm{cross,in}}^2 = 2\left(\sum_{n=1,n\neq m}^{M}\sqrt{X_{n,\mathrm{in}}}\right)^2$. $X_{n,\mathrm{out}}$ denotes the out-of-band crosstalk captured by observed channel, $X_{n,\mathrm{in}}$ denotes the in-band crosstalk originating from nth channel, and M denotes the number of wavelength channels.

3.3.10 Accumulation Effects

A typical optical transmission system contains the amplifiers that are spaced l km apart, and the span loss between two amplifiers is $\exp(-\alpha l)$, where α is the fiber attenuation coefficient. Each amplifier generates spontaneous emission noise in addition to providing the signal amplification. Both the signal and noise propagate together toward the next amplifier stage to be amplified. The buildup of amplifier noise is the most critical factor in achieving prespecified performance, when dealing with long distances. The total *ASE noise* will be *accumulated* and increased after each amplifier stage, thus contributing to SNR degradation. Moreover, the ASE noise power accumulation will contribute to the *saturation of optical amplifiers* reducing therefore the amplifier gain. The optical amplifier gain is usually adjusted just to compensate for the span loss. *Spatial steady-state condition* is achieved when the amplifier output power and gain remain the same from stage to stage [2]:

$$P_{\mathrm{o,sat}}e^{-\alpha l}G_{\mathrm{sat}} + P_{\mathrm{sp}} = P_{\mathrm{out}}, \quad P_{\mathrm{sp}} = 2S_{\mathrm{sp}}B_{\mathrm{op}} = F_{\mathrm{no}}h\nu(G_{\mathrm{sat}}-1)B_{\mathrm{op}}, \tag{3.149}$$

where $P_{\mathrm{o,sat}}$ is the optical amplifier saturation power, P_{sp} is the spontaneous emission factor, B_{op} is the optical filter bandwidth, F_{no} is the optical amplifier noise figure, $h\nu$ is the photon energy, G_{sat} is the saturation gain determined by $G_{\text{sat}} = 1 + (P_{\text{sat}}/P_{\text{in}})\ln(G_{\max}/G_{\text{sat}})$ ($G_{\max}$ is the maximum gain and P_{in} is the amplifier input). The *total noise power* at the end of transmission line, related to the steady-state condition [2]:

$$NP_{\text{sp}} = NF_{\text{no}}h\nu(G_{\text{sat}} - 1)B_{\text{op}} = (L/l)F_{\text{no}}h\nu(e^{\alpha l} - 1)B_{\text{op}}, \tag{3.150}$$

where N is the number of optical amplifies and L is the total transmission length.

The *OSNR* calculated per channel basis, at the end of amplifier chain, can be obtained as follows [1]:

$$\text{OSNR} = \frac{P_{\text{s,sat}}/M - F_{\text{no}}h\nu(e^{\alpha l} - 1)B_{\text{op}}L/l}{F_{\text{no}}h\nu(e^{\alpha l} - 1)B_{\text{op}}L/l} = \frac{P_{\text{ch}} - F_{\text{no}}h\nu(e^{\alpha l} - 1)B_{\text{op}}N}{F_{\text{no}}h\nu(e^{\alpha l} - 1)B_{\text{op}}N}, \tag{3.151}$$

where P_{ch} is the required launch power to obtain a given *OSNR*. Equation (3.137) can be used to design a system satisfying a given BER (or OSNR). The minimum launched power to achieve the prespecified *OSNR* is as follows:

$$P_{\text{ch}} \geq (1 + \text{OSNR})\left[F_{\text{no}}h\nu(e^{\alpha l} - 1)B_{\text{op}}N\right] \approx \text{OSNR}\left[F_{\text{no}}h\nu(e^{\alpha l} - 1)B_{\text{op}}N\right]. \tag{3.152}$$

The minimum launched power expressed in dB scale obtained from (3.152) is:

$$P_{\text{ch}}\,[\text{dB}] \geq \text{OSNR}\,[\text{dB}] + F_{\text{no}}\,[\text{dB}] + 10\log_{10}N + 10\log_{10}(h\nu B_{\text{op}}). \tag{3.153}$$

Using (3.141), (3.151) can be written in the following form [1]:

$$\text{OSNR}\,[\text{dB}] \approx 10\log_{10}\left(\frac{Q^2 B_{\text{el}}}{B_{\text{op}}}\right) \geq P_{\text{ch}}\,[\text{dB}] - F_{\text{no}}\,[\text{dB}] - \alpha l - 10\log_{10}N - 10\log_{10}(h\nu B_{\text{op}}). \tag{3.154}$$

The OSNR can be increased by increasing the output optical power, decreasing the amplifier noise figure, and decreasing the optical loss per fiber. However, the number of amplifiers can be decreased only when the span length is increased (meaning the total signal attenuation will be increased). The transmission system design is based on prespecified transmission system quality. The basic steps involved in design process include [1] (1) identification of the systems requirements in terms of transmission distance, bit rate, and BER; (2) identification system elements; (3) set up major system parameters (output signal power, amplifier spacing, ...) that satisfy initial requirements; (4) allocation of the margins to account for penalties; and finally (5) repeat the design process again in order to perform fine tuning of parameters and identify possible trade-offs. In Sect. 3.3.11, the design process is described for several systems of interest (1) power-budget limited point-to-point lightwave system, (2) bandwidth-limited system, and (3) high-speed optical transmission system.

3.3.11 Systems Design

When the optical amplifiers are not employed, the power budget is expressed as follows:

$$P_{\mathrm{out}} - \alpha L - \alpha_{\mathrm{c}} - \Delta P_{\mathrm{M}} \geq P_{\mathrm{R}}(Q, B_{\mathrm{el}}), \tag{3.155}$$

where P_{out} is the output power from the light source pigtail, α the fiber attenuation coefficient, ΔP_{M} the system margin to be allocated for different system imperfections, and P_{R} is the receiver sensitivity that is the function of required Q-factor and electrical filter bandwidth. All parameters from (3.155) are expressed in dB-scale.

The performance of an optical transmission system can be limited due to limited frequency bandwidth of the some of the key components that are used, such as light source, photodetector, or optical fiber. Available fiber bandwidth at length L, for different fiber types and bit rate R_{b}, can be summarized as follows [1]:

$$\begin{aligned} R_{\mathrm{b}} L^{\mu} &\leq B_{\mathrm{fiber}}, \quad \text{for MMFs} \\ R_{\mathrm{b}} L &\leq (4 D \sigma_{\lambda})^{-1}, \quad \text{for SMFs and large source linewidth} \\ R_{\mathrm{b}}^{2} L &\leq (16 |\beta_{2}|)^{-1}, \quad \text{for SMFs and narrow source linewidth.} \end{aligned} \tag{3.156}$$

The maximum transmission system length to the power budget can be expressed as follows [1]:

$$L(\lambda, B) \leq \frac{P_{\mathrm{out}}(\lambda, B) - [P_{\mathrm{R}}(\lambda, B) + \alpha_{\mathrm{c}} + \Delta P_{\mathrm{M}}]}{\alpha(\lambda)}. \tag{3.157}$$

The bandwidth limitation is related to the pulse rise times occurring in individual modules that can be expressed using the following expression [1, 3]:

$$T_{\mathrm{r}}^{2} \geq T_{\mathrm{Tx}}^{2} + T_{\mathrm{fiber}}^{2} + T_{\mathrm{Rx}}^{2}, \tag{3.158}$$

where T_{r} is the overall response time of the system, T_{Tx} the rise time of the transmitter, T_{fiber} the rise time of the fiber, and T_{Rx} is the rise time of receiver. The response time and 3-dB system bandwidth are related by [3] $T_{\mathrm{r}} = 0.35/B_{\mathrm{el}}$, where $B_{\mathrm{el}} = R_{\mathrm{b}}$ for RZ and $B_{\mathrm{el}} = 0.5 R_{\mathrm{b}}$ for NRZ, with R_{b} being the bit rate. Based on this relationship and previous expression, we can impose the following design criteria [1]:

$$\frac{a^{2}}{R_{\mathrm{b}}^{2}} \geq \frac{1}{B_{\mathrm{el,Tx}}^{2}} + \frac{1}{B_{\mathrm{fiber},L}^{2}} + \frac{1}{B_{\mathrm{el,Rx}}^{2}}, \tag{3.159}$$

where $a = 1$ for RZ and $a = 2$ for NRZ. In (3.159), the optical fiber bandwidth is determined from (3.156), $B_{\mathrm{el,Tx}}$ denotes the transmitter bandwidth, and $B_{\mathrm{el,Rx}}$ denotes the receiver bandwidth.

The results related to point-to-point transmission can be generalized for an optically amplified system, with certain modifications. For example, the power-budget equation should be replaced by the OSNR equation (3.154):

$$\text{OSNR}\,[\text{dB}] \approx 10\log_{10}\left(\frac{Q^2 B_{\text{el}}}{B_{\text{op}}}\right) \geq P_{\text{ch}}\,[\text{dB}] - F_{\text{no}}\,[\text{dB}] - \alpha l - 10\log_{10} N - 10\log_{10}(h\nu B_{\text{op}}).$$

The in-line optical amplifiers are employed to compensate not only for attenuation loss, but also for chromatic dispersion. Such compensation is not perfect in multichannel systems, and some residual penalty remains, thus causing the power penalty, so that required OSNR expression should be modified to account for this imperfection as follows:

$$\begin{aligned}\text{OSNR}_{\text{req}}\,[\text{dB}] &= \text{OSNR} + \Delta P \approx 10\log_{10}\left(\frac{Q^2 B_{\text{el}}}{B_{\text{op}}}\right) + \Delta P \geq P_{\text{ch}}\,[\text{dB}] \\ &\quad - F_{\text{no}}\,[\text{dB}] - \alpha l - 10\log_{10} N - 10\log_{10}(h\nu B_{\text{op}}), \qquad (3.160)\end{aligned}$$

where OSNR_{req} is the OSNR needed to account for different system imperfections and ΔP is the system margin.

The transmission system design process presented above represents a conservative scenario and has the purpose to provide understanding of design parameters and design processes [1]. This conservative scenario can be used as a reference case and feasibility check study before we continue with further considerations. The most complex case from the system design point of view is the high-speed long haul transmission with many WDM channels and possible optical routing. In this case, the conservative scenario can be used only as the guidance; while the more precise design should be performed by using the computer-aided deign (CAD) approach. The commercially available software packages include Virtual Photonics-VPI, OptiWave, and BroadNeD-BNeD. Simulation software can be written by using different tools, including C/C + +, Matlab, Mathcad, Fortran, etc.

3.3.12 Optical Performance Monitoring

Performance monitoring is very important for the network operator to control overall status of transmission lines, and the status of network, and to deliver desired quality of service to the end user. The performance monitoring techniques can be either analog or digital. Commonly used analog techniques for performance monitoring include [1] (1) optical spectrum analysis (OSNR, optical power, and optical frequency monitoring), (2) detection of a special pilot tone (the amplitude of pilot tone is related to the signal power and can be used to extract the OSNR), and (3) the histogram method (suitable to identify dispersion and nonlinear distortions). The OSNR measured by OSA or pilot method is related to the Q-factor by [1]

$$\begin{aligned}Q &= \frac{2\text{OSNR}_{\text{reference band}}\sqrt{B_{\text{op}}/B_{\text{el}}}}{1+\sqrt{1+4\text{OSNR}_{\text{reference band}}}}, \\ \text{OSNR}_{\text{reference band}} &= (B_{\text{reference band}}/B_{\text{op}})\text{OSNR}, \qquad (3.161)\end{aligned}$$

where $OSNR_{\text{reference band}}$ is the OSNR defined in the referent bandwidth (0.1 nm is commonly used) and B_{op} and B_{el} are the optical and electrical filter bandwidths, respectively. The Q-factor penalty can be expressed by

$$\Delta Q = 10\log_{10}\frac{Q}{Q_{ref}}, \tag{3.162}$$

where the Q_{ref} is the referent Q-factor. The histogram method provides more details about signal degradation [1,20–22]:

$$\text{BER} = \frac{1}{2p(1)}\sum_i H(I_{1,i})\text{erfc}\left(\frac{I_{1,i}-I_{tsh}}{\sigma_{1,i}\sqrt{2}}\right) + \frac{1}{2p(0)}\sum_i H(I_{0,i})\text{erfc}\left(\frac{I_{tsh}-I_{0,i}}{\sigma_{1,i}\sqrt{2}}\right), \tag{3.163}$$

where $H(I_{0,i})$ and $H(I_{1,i})$ represent the occurrences associated with bits 0 and 1, while $p(i)$ is the a priori probability of bit i ($i = 0, 1$).

The digital methods of performance monitoring are based on error detection codes, for example CRC codes and bit-interleaved parity (BIP) codes. The information is processed by receiving data in blocks and performing the parity checks. Those methods are suitable for in-service monitoring, but not suitable for fault localization.

3.4 Summary

This chapter is devoted to the description of channel impairments, noise sources, and transmission system engineering. In Sect. 3.1 different noise sources, both additive and multiplicative, are described. Section 3.2 is devoted to the description of basic channel impairments including fiber attenuation, chromatic dispersion, multimode dispersion, PMD, and fiber nonlinearities. In Sect. 3.3, the basic figures of merit of an optical transmission system are introduced, including SNR, optical SNR, receiver sensitivity, and BER. Further, the receiver sensitivity is determined in the presence of shot, thermal, and beat noise; all other noise sources and channel impairments are described through the power penalty, the increase in receiver sensitivity needed to preserve the desired BER. This approach leads to the worst-case scenario, but helps to verify the feasibility of particular system design before further considerations. The guidelines for the system design are provided as well in the same section.

References

1. Cvijetic M (2004) Optical transmission systems engineering. Artech House, Boston, MA
2. Ramaswami R, Sivarajan K (2002) Optical networks: a practical perspective, 2nd edn. Morgan Kaufman, San Fransisco, CA
3. Agrawal GP (2002) Fiber-optic communication systems, 3rd edn. Wiley, New York

4. Agrawal GP (2004) Lightwave technology: components and devices. Wiley, New York
5. Agrawal GP (2005) Lightwave technology: telecommunication systems. Wiley, New York
6. Agrawal GP (2007) Nonlinear fiber optics, 4th edn. Academic, New York
7. Cvijetic M (1996) Coherent and nonlinear lightwave communications. Artech House, Boston, MA
8. Keiser G (2000) Optical fiber communications, 3rd edn. McGraw-Hill, New York
9. Kazovsky L, Benedetto S, Willner A (1996) Optical fiber communication systems. Artech House, Boston, MA
10. Palais JC (2005) Fiber optic communications, 5th edn. Pearson Prentice Hall, Upper Saddle River, NJ
11. Ramamurthy B (2000) Switches, wavelength routers, and wavelength converters. In: Sivalingam KM, Subramaniam S (eds) Optical WDM networks: principles and practice. Kluwer Academic, New York
12. Djordjevic IB (2006/2007), ECE 430/530: Optical communication systems (lecture notes). University of Arizona, Tucson, AZ
13. Kostuk R (2002/2004) ECE 430/530: optical communication systems (lecture notes). University of Arizona, Tucson, AZ
14. Djordjevic IB (2007) ECE 632: advanced optical communication systems (lecture notes). University of Arizona, Tucson, AZ
15. Yariv A (1997) Optical electronics in modern communications. Oxford University Press, Oxford
16. Amersfoort M (1998) Arrayed waveguide grating, Application note A1998003. Available at http://www.c2v.nl
17. Staif M, Mecozzi A, Nagel J (2000) Mean square magnitude of all orders PMD and the relation with the bandwidth of principal states. IEEE Photon Technol Lett 12:53–55
18. Kogelnik H, Nelson LE, Jobson RM (2002) Polarization mode dispersion. In: Kaminow IP, Li T (eds) Optical fiber telecommunications. Academic, San Diego, CA
19. Foschini GJ, Pole CD (1991) Statistical theory of polarization mode dispersion in single-mode fibers. IEEE/OSA J Lightwave Technol LT-9:1439–1456
20. Bendelli G et al (2000) Optical performance monitoring techniques. In Proceedings of ECOC 2000, Munich , vol 4, pp 213–216
21. Djordjevic IB, Vasic B (2004) An advanced direct detection receiver model. J Opt Commun 25(1):6–9
22. Winzer PJ, Pfennigbauer M, Strasser MM, Leeb WR (2001) Optimum filter bandwidths for optically preamplified NRZ receivers. J Lightwave Technol 19:1263–1272
23. Gloge D Marcatili EAJ (1973) Multimode theory of graded-core fibers Bell Syst Tech J 52:1563–1578

Chapter 4
Channel Coding for Optical Channels

In this chapter, we describe different forward error correction (FEC) schemes currently in use or suitable for use in optical communication systems. We start with the description of standard block codes. The state-of-the-art in optical communication systems standardized by the ITU employ concatenated Bose–Chaudhuri–Hocquenghem (BCH)/RS codes [1, 2]. The RS(255,239), in particular, has been used in a broad range of long-haul communication systems [1, 2], and it is commonly considered as the first-generation of FEC [3,4]. The elementary FEC schemes (BCH, RS, or convolutional codes) may be combined to design more powerful FEC schemes, e.g., RS(255,239) + RS(255,233). Several classes of concatenation codes are listed in ITU-T G975.1. Different concatenation schemes, such as the concatenation of two RS codes or the concatenation of RS and convolutional codes, are commonly considered as second generation of FEC [3, 4].

In recent years, iteratively decodable codes, such as turbo codes [3–10] and low-density parity-check (LDPC) codes [11–20], have generated significant research attention. In ref. [9] Sab and Lemarie proposed a FEC scheme based on block turbo code for long-haul DWDM optical transmission systems. In several recent papers [12–18], we have shown that iteratively decodable LDPC codes outperform turbo product codes in bit-error rate (BER) performance. The decoder complexity of these codes is comparable (or lower) to that of turbo product codes, and significantly lower than that of serial/parallel concatenated turbo codes. For reasons mentioned above, LDPC code is a viable and attractive choice for the FEC scheme of 40 Gb/s and 100 Gb/s optical transmission systems. The soft iteratively decodable codes, turbo and LDPC codes, are commonly referred to as the third generation of FEC [3, 4]. This chapter is devoted to the first and the second generation of FEC for optical communications. The third generation of FEC is described in next chapter.

This chapter is organized as follows. In Sect. 4.1 we introduce channel coding preliminaries, namely, basic definitions, channel models, concept of channel capacity, and statement of channel coding theorem. The Sect. 4.2 is devoted to basics of linear block codes (LBCs), such as definition of generator and parity-check matrices, syndrome decoding, distance properties of LBCs, and some important coding bounds. In Sect. 4.3 cyclic codes are introduced. The Sect. 4.4 is devoted to description of BCH codes. The special subclass of BCH codes, the RS codes is described in Sect. 4.5. To deal with simultaneous burst and random errors we describe the

I. Djordjevic et al., *Coding for Optical Channels*,
DOI 10.1007/978-1-4419-5569-2_4,

concatenated and interleaved codes in the same section. In Sect. 4.6 we introduce the trellis description of LBCs and describe the Viterbi algorithm. Finally, in Sect. 4.7 we describe convolutional codes.

4.1 Channel Coding Preliminaries

Two key system parameters are transmitted power and channel bandwidth, which together with additive noise sources determine the signal-to-noise ratio (SNR) and, correspondingly, BER. In practice, we very often come into situation when the target BER cannot be achieved with a given modulation format. For the fixed SNR, the only practical option to change the data quality transmission from unacceptable to acceptable is through use of *channel coding*. Another practical motivation of introducing the channel coding is to reduce required SNR for a given target BER. The amount of energy that can be saved by coding is commonly described by coding gain. *Coding gain* refers to the savings attainable in the energy per information bit to noise spectral density ratio (E_b/N_0) required to achieve a given bit error probability when coding is used compared to that with no coding. A typical digital optical communication system employing channel coding is shown in Fig. 4.1. The discrete source generates the information in the form of sequence of symbols. The channel encoder accepts the message symbols and adds redundant symbols according to a corresponding prescribed rule. The channel coding is the act of transforming of a length-k sequence into a length-n codeword. The set of rules specifying

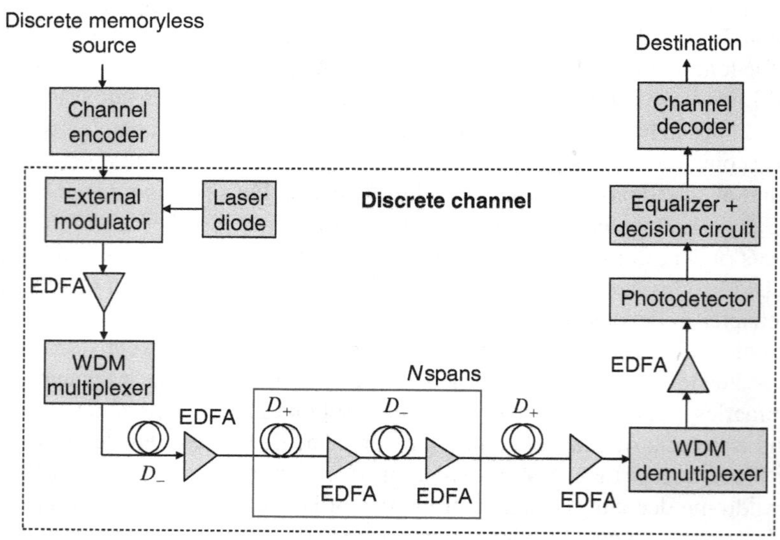

Fig. 4.1 Block diagram of a point-to-point digital optical communication system

this transformation are called the channel code, which can be represented as the following mapping:

$$C : M \to X,$$

where C is the channel code, M is the set of information sequences of length k, and X is the set of codewords of length n. The decoder exploits these redundant symbols to determine which message symbol was actually transmitted. Encoder and decoder consider whole digital transmission system as a discrete channel. Other blocks, shown in Fig. 4.1, are already explained in Chap. 3, here we are concerned with channel encoders and decoders. Different classes of channel codes can be categorized into three broad categories: (1) *error detection* in which we are concerned only with detecting the errors occurred during transmission (examples include automatic request for transmission-ARQ), (2) FEC, where we are interested in correcting the errors occurred during transmission, and (3) hybrid channel codes that combine the previous two approaches. In this chapter, we are concerned only with FEC.

The key idea behind the forward error correcting codes is to add extra redundant symbols to the message to be transmitted, and use those redundant symbols in decoding procedure to correct the errors introduced by the channel. The redundancy can be introduced in time, frequency or space domain. For example, the redundancy in time domain is introduced if the same message is transmitted at least twice, the technique is known as the *repetition code*. The space redundancy is used as a means to achieve high spectrally efficient transmission, in which the modulation is combined with error control.

The codes commonly considered in fiber-optics communications belong either to the class of *block codes* or to the class of *convolutional codes*. In an (n, k) *block code* the channel encoder accepts information in successive k-symbol blocks, adds $n - k$ redundant symbols that are algebraically related to the k message symbols; thereby producing an overall encoded block of n symbols $(n > k)$, known as a *codeword.* If the block code is *systematic*, the information symbols stay unchanged during the encoding operation, and the encoding operation may be considered as adding the $n - k$ generalized parity checks to k information symbols. Since the information symbols are statistically independent (a consequence of source coding or scrambling), the next codeword is independent of the content of the current codeword. The *code rate* of an (n, k) block code is defined as $R = k/n$, and *overhead* by $OH = (1/R - 1) \cdot 100\%$. In *convolutional code*, however, the encoding operation may be considered as the discrete-time convolution of the input sequence with the impulse response of the encoder. Therefore, the $n - k$ generalized parity checks are functions of not only k information symbols but also the functions of m previous k-tuples, with $m + 1$ being the encoder impulse response length. The statistical dependence is introduced to the window of length $n(m + 1)$, the parameter known as *constraint length* of convolutional codes.

As mentioned above, the channel code considers whole transmission system as a discrete channel, in which the sizes of input and output alphabets are finite. Two examples of such channel are shown in Fig. 4.2. In Fig. 4.2a we show an example of a discrete memoryless channel (DMC), which is characterized by channel (transition) probabilities. Let $X = \{x_0, x_1, \ldots, x_{I-1}\}$ denote the channel input alphabet, and

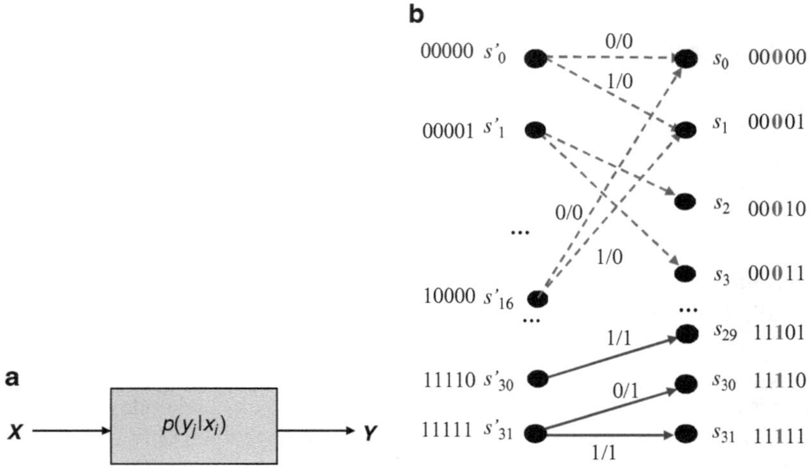

Fig. 4.2 Two examples of discrete channels: (**a**) discrete memoryless channel (DMC), and (**b**) discrete channel with memory described as dynamic trellis [20]

$Y = \{y_0, y_1, \ldots, y_{J-1}\}$ denote the channel output alphabet. This channel is completely characterized by the following set of transition probabilities:

$$p(y_j|x_i) = P(Y = y_j|X = x_i), \quad 0 \leq p(y_j|x_i) \leq 1, \quad i \in \{0, 1, \ldots, I-1\}, \\ j \in \{0, 1, \ldots, J-1\}, \tag{4.1}$$

where I and J denote the sizes of input and output alphabets, respectively. The transition probability $p(y_j|x_i)$ represents the conditional probability that channel output $Y = y_j$ given the channel input $X = x_i$. The channel introduces the errors, and if $j \neq i$ the corresponding $p(y_j|x_i)$ represents the conditional probability of error, while for $j = i$ it represents the conditional probability of correct reception. For $I = J$, the *average symbol error probability* is defined as the probability that output random variable Y_j is different from input random variable X_i, with averaging being performed for all $j \neq i$:

$$P_e = \sum_{i=0}^{I-1} p(x_i) \sum_{j=0, j\neq i}^{J-1} p(y_j|x_i), \tag{4.2}$$

where the inputs are selected from the following distribution $\{p(x_i) = P(X = x_i),\ i = 0, 1, \ldots, I-1\}$, with $p(x_i)$ being known as *a priori* probability of input symbol x_i. The corresponding probabilities of output symbols can be calculated by:

$$p(y_j) = \sum_{i=0}^{I-1} P(Y = y_j|X = x_i)P(X = x_i) = \sum_{i=0}^{I-1} p(y_j|x_i)p(x_i), \quad j = 0, 1, \ldots, J-1. \tag{4.3}$$

The decision rule that minimizes average symbol error probability (4.2), denoted as $D(y_j) = x^*$, is known as maximum a posteriori (MAP) rule, and can be formulated as follows:

$$D(y_j) = x^* : \quad P(x^*|y_j) \geq P(x_i|y_j), \quad i = 0,\ 1, \ldots, I-1. \tag{4.4}$$

Therefore, the symbol error probability P_e will be minimal when to every output symbol y_j the input symbol x^* is assigned having largest a posteriori probability $P(x^*|y_j)$. By using the Bayes' rule (4.4) can be rewritten as

$$D(y_j) = x^* : \quad \frac{P(y_j|x^*)P(x^*)}{P(y_j)} \geq \frac{P(y_j|x_i)P(x_i)}{P(y_j)}, \quad i = 0, 1, \ldots, I-1. \tag{4.5}$$

If all input symbols are equally likely $P(x_i) = 1/I(i = 0, \ldots, I-1)$, the corresponding decision rule is known as maximum-likelihood (ML) decision rule:

$$D(y_j) = x^* : \quad P(y_j|x^*) \geq P(y_j|x_i), \quad i = 0, 1, \ldots, I-1. \tag{4.6}$$

In Fig. 4.2b we show a discrete channel model with memory [20], which is more suitable for optical communications, because the optical channel is essentially the channel with memory. We assume that the optical channel has the memory equal to $2m+1$, with $2m$ being the number of bits that influence the observed bit from both sides. This dynamical trellis is uniquely defined by the set of previous state, the next state, in addition to the channel output. The state (the bit-pattern configuration) in the trellis is defined as $\boldsymbol{s}_j = (x_{j-m}, x_{j-m+1}, \ldots, x_j, x_{j+1}, \ldots, x_{j+m}) = \boldsymbol{x}[j-m, j+m]$, where $x_k \in \boldsymbol{X} = \{0, 1\}$. An example trellis of memory $2m+1 = 5$ is shown in Fig. 4.2b. The trellis has $2^5 = 32$ states $(\boldsymbol{s}_0, \boldsymbol{s}_1, \ldots, \boldsymbol{s}_{31})$, each of which corresponds to a different five-bit pattern. For the complete description of the trellis, the transition probability density functions (PDFs) $p(y_j|x_j) = p(y_j|\boldsymbol{s}), \boldsymbol{s} \in \boldsymbol{S}$ can be determined from *collected histograms*, where y_j represents the sample that corresponds to the transmitted bit x_j, and $\boldsymbol{S}$ is the set of states in the trellis.

One important figure of merit for DMCs is the amount of information conveyed by the channel, which is known as the *mutual information* and it is defined as

$$\begin{aligned} I(\boldsymbol{X};\boldsymbol{Y}) = H(\boldsymbol{X}) - H(\boldsymbol{X}|\boldsymbol{Y}) &= \sum_{i=0}^{I-1} p(x_i)\log_2\left[\frac{1}{p(x_i)}\right] \\ &- \sum_{j=0}^{J-1} p(y_j)\sum_{i=0}^{I-1} p(x_i|y_j)\log_2\left[\frac{1}{p(x_i|y_j)}\right], \end{aligned} \tag{4.7}$$

where $H(\boldsymbol{X})$ denotes the uncertainty about the channel input before observing the channel output, also known as entropy; while $H(\boldsymbol{X}|\boldsymbol{Y})$ denotes the conditional entropy or the amount of uncertainty remaining about the channel input after the channel output has been received. Therefore, the mutual information represents the

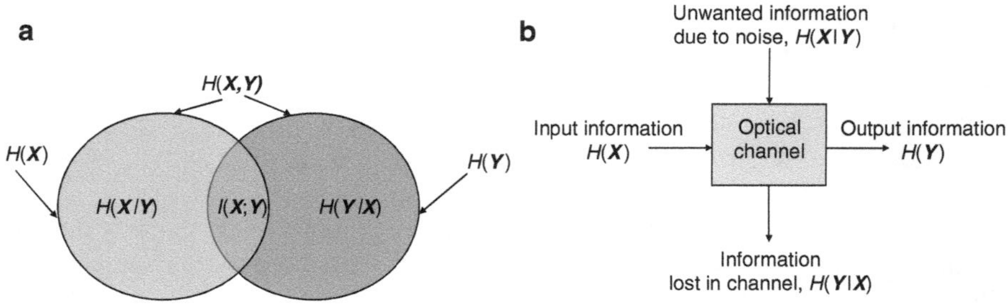

Fig. 4.3 Interpretation of the mutual information: (**a**) using Venn diagrams, and (**b**) using the approach due to Ingels

amount of information (per symbol) that is conveyed by the channel, i.e., the uncertainty about the channel input that is resolved by observing the channel output. The mutual information can be interpreted by means of Venn diagram shown in Fig. 4.3a. The left circle represents the entropy of channel input, the right circle represents the entropy of channel output, and the mutual information is obtained in intersection of these two circles. Another interpretation due to Ingels [21] is shown in Fig. 4.3b. The mutual information, i.e., the information conveyed by the channel, is obtained as output information minus information lost in the channel.

It is clear from (4.7) that mutual information is independent on the channel, and someone may try to maximize the information conveyed by the channel, to obtain the so-called *channel capacity*:

$$C = \max_{\{p(x_i)\}} I(\boldsymbol{X};\boldsymbol{Y}); \quad \text{subject to}: p(x_i) \geq 0 \quad \sum_{i=0}^{I-1} p(x_i) = 1. \tag{4.8}$$

Now we have built enough knowledge to formulate the *channel coding theorem* [22,23]:

Let a discrete memoryless source with an alphabet $\boldsymbol{S}$ have entropy $H(\boldsymbol{S})$ and emit the symbols every T_s seconds. Let a DMC have capacity C and be used once in T_c seconds. Then, if

$$H(\boldsymbol{S})/T_s \leq C/T_c, \tag{4.9}$$

there exists a coding scheme for which the source output can be transmitted over the channel and reconstructed with an arbitrary small probability of error. The parameter $H(\boldsymbol{S})/T_s$ is related to the average information rate, while the parameter C/T_c is related to the channel capacity per unit time. For binary symmetric channel (BSC) $(I = J = 2)$ the inequality (4.9) simply becomes

$$R \leq C, \tag{4.10}$$

where R is the code rate introduced above.

Another very much important theorem is the Shannon's third theorem, also known as the *information capacity theorem*, and can be formulated as follows [22, 23]:

The information capacity of a continuous channel of bandwidth B Hz, perturbed by AWGN of PSD $N_0/2$ and limited in bandwidth B, is given by

$$C = B \log_2 \left(1 + \frac{P}{N_0 B} \right) \text{ [bits/s]}, \tag{4.11}$$

where P is the average transmitted power. This theorem represents remarkable result of information theory, because it connects all important system parameters (transmitted power, channel bandwidth, and noise power spectral density) in only one formula. What is also interesting is that LDPC codes can approach the Shannon's limit within 0.0045dB [19]. By using (4.11) and Fano's inequality [23]

$$H(X|Y) \leq H(P_e) + P_e \log_2(I-1), \quad H(P_e) = -P_e \log_2 P_e - (1-P_e)\log_2(1-P_e) \tag{4.12}$$

for amplified spontaneous emission (ASE) noise-dominated scenario and binary phase-shift keying (BPSK) at 40 Gb/s in Fig. 4.4 we report the minimum BERs against optical SNR for different code rates.

In the rest of this section an elementary introduction to LBCs, BCH codes, RS codes, concatenated codes, and product codes is given. These classes of codes are already employed in fiber-optics communication systems. For a detailed treatment of different error-control coding schemes an interested reader is referred to [24–30]. Since the convolutional codes are used in turbo codes for deep-space optical communications we also provide short description of convolutional codes.

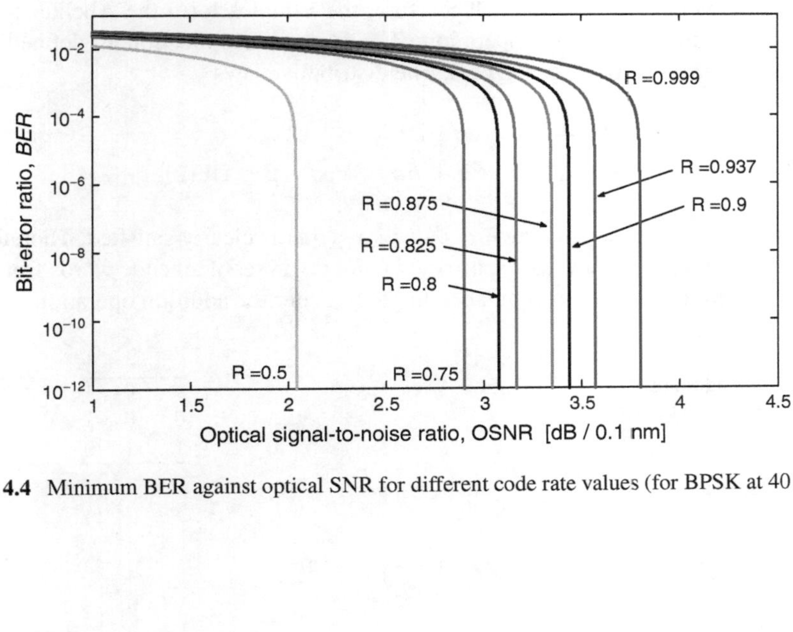

Fig. 4.4 Minimum BER against optical SNR for different code rate values (for BPSK at 40 Gb/s)

4.2 Linear Block Codes

The *linear block code* (n, k), using the language of vector spaces, can be defined as a subspace of a vector space over finite field GF(q), with q being the prime power. Every space is described by its *basis* – a set of linearly independent vectors. The number of vectors in the basis determines the dimension of the space. Therefore, for an (n, k) linear block code the dimension of the space is n, and the dimension of the code subspace is k.

Example: $(n, 1)$ repetition code. The repetition code has two code words $\boldsymbol{x}_0 = (00\ldots0)$ and $\boldsymbol{x}_1 = (11\ldots1)$. Any linear combination of these two code words is another code word as shown below:

$$\begin{aligned}\boldsymbol{x}_0 + \boldsymbol{x}_0 &= \boldsymbol{x}_0 \\ \boldsymbol{x}_0 + \boldsymbol{x}_1 &= \boldsymbol{x}_1 + \boldsymbol{x}_0 = \boldsymbol{x}_1. \\ \boldsymbol{x}_1 + \boldsymbol{x}_1 &= \boldsymbol{x}_0\end{aligned}$$

The set of code words from an LBC forms a group under the addition operation, because all-zero code word serves as the identity element, and the code word itself serves as the inverse element. This is the reason why the LBCs are also called the group codes. The linear block code (n,k) can be observed as a k-dimensional subspace of the vector space of all n-tuples over the binary filed GF(2) $= \{0, 1\}$, with addition and multiplication rules given in Table 4.1. All n-tuples over GF(2) form the vector space. The sum of two n-tuples $\boldsymbol{a} = (a_1\ a_2 \cdots a_n)$ and $\boldsymbol{b} = (b_1\ b_2\ \cdots\ b_n)$ is clearly an n-tuple and commutative rule is valid because $\boldsymbol{c} = \boldsymbol{a}+\boldsymbol{b} = (a_1+b_1\ a_2+b_2 \cdots a_n+b_\mathrm{n}) = (b_1+a_1\ b_2+a_2 \cdots b_n+a_n) = \boldsymbol{b}+\boldsymbol{a}$. The all-zero vector $\boldsymbol{0} = (00\cdots0)$ is the identity element, while n-tuple $\boldsymbol{a}$ itself is the inverse element $\boldsymbol{a} + \boldsymbol{a} = 0$. Therefore, the n-tuples form the Abelian group with respect to the addition operation. The scalar multiplication is defined by: $\alpha\boldsymbol{a} = (\alpha a_1\ \alpha a_2\ \cdots\ \alpha a_n), \alpha \in \mathrm{GF(2)}$. The distributive laws

$$\begin{aligned}\alpha(\boldsymbol{a} + \boldsymbol{b}) &= \alpha\boldsymbol{a} + \alpha\boldsymbol{b} \\ (\alpha + \beta)\boldsymbol{a} &= \alpha\boldsymbol{a} + \beta\boldsymbol{a}, \quad \forall\alpha, \quad \beta \in \mathrm{GF(2)}\end{aligned}$$

are also valid. The associate law $(\alpha \cdot \beta)\boldsymbol{a} = \alpha \cdot (\beta\boldsymbol{a})$ is clearly satisfied. Therefore, the set of all n-tuples is a vector space over GF(2). The set of all code words from an (n, k) linear block code forms an abelian group under the addition operation. It can

Table 4.1 Addition (+) and multiplication (·) rules

+	0	1	·	0	1
0	0	1	0	0	0
1	1	0	1	0	1

be shown, in a fashion similar to that above, that all code words of an (n, k) linear block codes form the vector space of dimensionality k. There exists k basis vectors (code words) such that every code word is a linear combination of these code words.

Example: $(n, 1)$ repetition code: $C = \{(00\ldots0), (11\ldots1)\}$. Two code words in C can be represented as linear combination of all-ones basis vector: $(11\ldots1) = 1 \cdot (11\ldots1), (00\ldots0) = 1 \cdot (11\ldots1) + 1 \cdot (11\ldots1)$.

4.2.1 Generator Matrix for Linear Block Code

Any code word $\boldsymbol{x}$ from the (n, k) linear block code can be represented as a linear combination of k basis vectors $\boldsymbol{g}_i (i = 0, 1, \ldots, k-1)$ as given below:

$$\boldsymbol{x} = m_0\boldsymbol{g}_0 + m_1\boldsymbol{g}_1 + \cdots + m_{k-1}\boldsymbol{g}_{k-1} = \boldsymbol{m}\begin{bmatrix} \boldsymbol{g}_0 \\ \boldsymbol{g}_1 \\ \cdots \\ \boldsymbol{g}_{k-1} \end{bmatrix} = \boldsymbol{mG}, \quad \boldsymbol{G} = \begin{bmatrix} \boldsymbol{g}_o \\ \boldsymbol{g}_1 \\ \cdots \\ \boldsymbol{g}_{k-1} \end{bmatrix},$$

$$\boldsymbol{m} = \begin{pmatrix} m_0 \; m_1 \cdots m_{k-1} \end{pmatrix}, \tag{4.13}$$

where $\boldsymbol{m}$ is the message vector, and $\boldsymbol{G}$ is the generator matrix (of dimensions $k \times n$), in which every row represents a vector from the coding subspace. Therefore, in order to encode, the message vector $\boldsymbol{m}(m_0, m_1, \ldots, m_{k-1})$ has to be multiplied with a generator matrix $\boldsymbol{G}$ to get $\boldsymbol{x} = \boldsymbol{mG}$, where $\boldsymbol{x}(x_0, x_1, \ldots, x_{n-1})$ is a codeword.

Example: Generator matrices for repetition $(n, 1)$ code $\boldsymbol{G}_{\text{rep}}$ and $(n, n-1)$ single-parity-check code $\boldsymbol{G}_{\text{par}}$ are given, respectively, as

$$\boldsymbol{G}_{\text{rep}} = [11\ldots1] \quad \boldsymbol{G}_{\text{par}} = \begin{bmatrix} 100\ldots01 \\ 010\ldots01 \\ \cdots \\ 000\ldots11 \end{bmatrix}.$$

By elementary operations on rows in the generator matrix, the code may be transformed into systematic form

$$\boldsymbol{G}_{\text{s}} = [\boldsymbol{I}_k | \boldsymbol{P}], \tag{4.14}$$

where $\boldsymbol{I}_k$ is unity matrix of dimensions $k \times k$, and $\boldsymbol{P}$ is the matrix of dimensions $k \times (n-k)$ with columns denoting the positions of parity checks

$$\boldsymbol{P} = \begin{bmatrix} p_{00} & p_{01} & \cdots & p_{0,n-k-1} \\ p_{10} & p_{11} & \cdots & p_{1,n-k-1} \\ \cdots & & \cdots & \cdots \\ p_{k-1,0} & p_{k-1,1} & \cdots & p_{k-1,n-k-1} \end{bmatrix}.$$

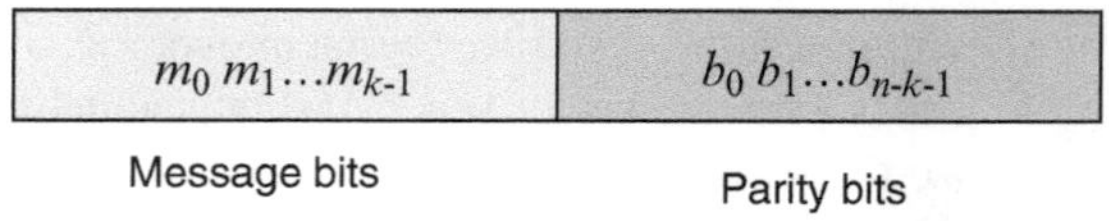

Fig. 4.5 Structure of systematic code word

The codeword of a systematic code is obtained by

$$\boldsymbol{x} = [\boldsymbol{m}|\boldsymbol{b}] = \boldsymbol{m}[\boldsymbol{I}_k|\boldsymbol{P}] = \boldsymbol{m}\boldsymbol{G}, \quad \boldsymbol{G} = [\boldsymbol{I}_k|\boldsymbol{P}], \tag{4.15}$$

and the structure of systematic codeword is shown in Fig. 4.5.

Therefore, during encoding the message vector stays unchanged and the elements of vector of parity checks $\boldsymbol{b}$ are obtained by

$$b_i = p_{0i}m_0 + p_{1i}m_1 + \cdots + p_{k-1,i}m_{k-1}, \tag{4.16}$$

where

$$p_{ij} = \begin{cases} 1, & \text{if } b_i \text{ depends on } \mathrm{m}_j, \\ 0, & \text{otherwise.} \end{cases}$$

During transmission an optical channel introduces the errors so that the received vector $\boldsymbol{r}$ can be written as $\boldsymbol{r} = \boldsymbol{x} + \boldsymbol{e}$, where $\boldsymbol{e}$ is the error vector (pattern) with elements components determined by

$$e_i = \begin{cases} 1, & \text{if an error occured in the } i\text{th location,} \\ 0, & \text{otherwise.} \end{cases}$$

To determine whether the received vector $\boldsymbol{r}$ is a codeword vector, we are introducing the concept of a *parity-check matrix.*

4.2.2 Parity-Check Matrix for Linear Block Code

Another useful matrix associated with the LBCs is the parity-check matrix. Let us expand the matrix equation $\boldsymbol{x} = \boldsymbol{m}\boldsymbol{G}$ in scalar form as follows:

$$\begin{aligned}
x_0 &= m_0 \\
x_1 &= m_1 \\
&\ldots \\
x_{k-1} &= m_{k-1} \\
x_k &= m_0 p_{00} + m_1 p_{10} + \cdots + m_{k-1} p_{k-1,0} \\
x_{k+1} &= m_0 p_{01} + m_1 p_{11} + \cdots + m_{k-1} p_{k-1,1} \\
&\ldots \\
x_{n-1} &= m_0 p_{0,n-k-1} + m_1 p_{1,n-k-1} + \cdots + m_{k-1} p_{k-1,n-k-1}
\end{aligned} \quad . \tag{4.17}$$

By using the first k equalities, the last $n-k$ equations can be rewritten as follows:

$$\begin{array}{l} x_0 p_{00} + x_1 p_{10} + \cdots + x_{k-1} p_{k-1,0} + x_k = 0 \\ x_0 p_{01} + x_1 p_{11} + \cdots + x_{k-1} p_{k-1,0} + x_{k+1} = 0 \\ \cdots \\ x_0 p_{0,n-k+1} + x_1 p_{1,n-k-1} + \ldots + x_{k-1} p_{k-1,n-k+1} + x_{n-1} = 0 \end{array}. \quad (4.18)$$

The matrix representation of (4.18) is given below:

$$\begin{aligned} &\left[x_0\ x_1 \ldots x_{n-1} \right] \begin{bmatrix} p_{00} & p_{10} & \ldots & p_{k-1,0} & 1 & 0 & \ldots & 0 \\ p_{01} & p_{11} & \ldots & p_{k-1,1} & 0 & 1 & \ldots & 0 \\ \ldots & & \ldots & & \ldots & & & \ldots \\ p_{0,n-k-1} & p_{1,n-k-1} & \ldots & p_{k-1,n-k-1} & 0 & 0 & \ldots & 1 \end{bmatrix}^{\mathrm{T}} \\ &= \boldsymbol{x} \left[\boldsymbol{P}^{\mathrm{T}}\ \boldsymbol{I}_{n-k} \right] = \boldsymbol{x}\boldsymbol{H}^{\mathrm{T}} = \boldsymbol{0}, \quad \boldsymbol{H} = \left[\boldsymbol{P}^{\mathrm{T}}\ \boldsymbol{I}_{n-k} \right]_{(n-k)kn}. \end{aligned} \quad (4.19)$$

The $\boldsymbol{H}$-matrix in (4.19) is known as the parity-check matrix. We can easily verify that:

$$\boldsymbol{G}\boldsymbol{H}^{\mathrm{T}} = \left[\boldsymbol{I}_k\ \boldsymbol{P} \right] \begin{bmatrix} \boldsymbol{P} \\ \boldsymbol{I}_{n-k} \end{bmatrix} = \boldsymbol{P} + \boldsymbol{P} = \boldsymbol{0}, \quad (4.20)$$

meaning that the parity-check matrix of an (n,k) linear block code $\boldsymbol{H}$ is a matrix of rank $n-k$ and dimensions $(n-k) \times n$ whose null-space is k-dimensional vector with basis being the generator matrix $\boldsymbol{G}$.

Example: Parity-Check Matrices for $(n,1)$ repetition code $\boldsymbol{H}_{\text{rep}}$ and $(n,n-1)$ single parity-check code $\boldsymbol{H}_{\text{par}}$ are given, respectively, as

$$\boldsymbol{H}_{\text{rep}} = \begin{bmatrix} 100\ldots01 \\ 010\ldots01 \\ \ldots \\ 000\ldots11 \end{bmatrix}, \quad \boldsymbol{H}_{\text{par}} = [11\ldots1].$$

Example: For Hamming (7,4) code the generator $\boldsymbol{G}$ and parity check $\boldsymbol{H}$ matrices are given, respectively, as

$$\boldsymbol{G} = \begin{bmatrix} 1000|110 \\ 0100|011 \\ 0010|111 \\ 0001|101 \end{bmatrix} \quad \boldsymbol{H} = \begin{bmatrix} 1011|100 \\ 1110|010 \\ 0111|001 \end{bmatrix}.$$

Every (n,k) linear block code with generator matrix $\boldsymbol{G}$ and parity-check matrix $\boldsymbol{H}$ has a dual code with generator matrix $\boldsymbol{H}$ and parity-check matrix $\boldsymbol{G}$. For example, $(n,1)$ repetition code and $(n,n-1)$ single parity check code are dual.

4.2.3 Distance Properties of Linear Block Codes

To determine the *error correction capability* of the code we have to introduce the concept of Hamming distance and Hamming weight. *Hamming distance* between two codewords $\boldsymbol{x}_1$ and $\boldsymbol{x}_2$, $d(\boldsymbol{x}_1, \boldsymbol{x}_2)$, is defined as the number of locations in which their respective elements differ. *Hamming weight*, $w(\boldsymbol{x})$, of a codeword vector $\boldsymbol{x}$ is defined as the number of nonzero elements in the vectors. The *minimum distance*, d_{min}, of an LBC is defined as the smallest Hamming distance between any pair of code vectors in the code. Since the zero-vector is a codeword, the minimum distance of an LBC can be determined simply as the smallest Hamming weight of the nonzero code vectors in the code. Let the parity-check matrix be written as $\boldsymbol{H} = [\boldsymbol{h}_1\ \boldsymbol{h}_2\ \cdots\ \boldsymbol{h}_n]$, where $\boldsymbol{h}_i$ is the ith column in $\boldsymbol{H}$. Since every codeword $\boldsymbol{x}$ must satisfy the syndrome equation, $\boldsymbol{x}\boldsymbol{H}^{\mathrm{T}} = \boldsymbol{0}$ (see (4.19)) the minimum distance of an LBC is determined by the minimum number of columns of the $\boldsymbol{H}$-matrix whose sum is equal to the zero vector. For example, (7,4) Hamming code in example above has the minimum distance $d_{\mathrm{min}} = 3$ since the addition of first, fifth, and sixth columns leads to zero vector. The codewords can be represented as points in n-dimensional space, as shown in Fig. 4.6. Decoding process can be visualized by creating the spheres of radius t around codeword points. The received word vector $\boldsymbol{r}$ in Fig. 4.6a will be decoded as a codeword $\boldsymbol{x}_i$ because its Hamming distance $d(\boldsymbol{x}_i, \boldsymbol{r}) \leq t$ is closest to the codeword $\boldsymbol{x}_i$. On the other hand, in example shown in Fig. 4.6b the Hamming distance $d(\boldsymbol{x}_i, \boldsymbol{x}_j) \leq 2t$ and the received vector $\boldsymbol{r}$ that falls in intersection area of two spheres cannot be uniquely decoded.

Therefore, an (n, k) linear block code of minimum distance d_{min} can correct up to t errors if, and only if, $t \leq \lfloor 1/2(d_{\mathrm{min}} - 1) \rfloor$ (where $\lfloor\rfloor$ denotes the largest integer less than or equal to the enclosed quantity) or equivalently $d_{\mathrm{min}} \geq 2t + 1$. If we are only interested in detecting e_{d} errors then $d_{\mathrm{min}} \geq e_{\mathrm{d}} + 1$. Finally, if we are interested in detecting e_{d} errors and correcting e_{c} errors then $d_{\mathrm{min}} \geq e_{\mathrm{d}} + e_{\mathrm{c}} + 1$. The Hamming (7,4) code is, therefore, a single-error-correcting and double-error-detecting code. More generally, a family of (n, k) linear block codes with following parameters:

- Block length: $n = 2^m - 1$
- Number of message bits: $k = 2^m - m - 1$
- Number of parity bits: $n - k = m$
- $d_{\mathrm{min}} = 3$

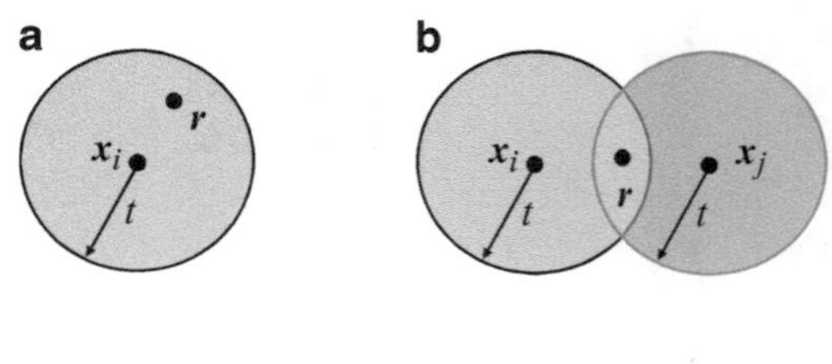

Fig. 4.6 Illustration of Hamming distance: (**a**) $d(\boldsymbol{x}_i, \boldsymbol{x}_j) \geq 2t + 1$ and (**b**) $d(\boldsymbol{x}_i, \boldsymbol{x}_j) < 2t + 1$

where $m \geq 3$, are known as *Hamming* codes. Hamming codes belong to the class of perfect codes, the codes that satisfy the Hamming inequality below with equality sign [22, 23]:

$$2^{n-k} \geq \sum_{i=0}^{t} \binom{n}{i}. \tag{4.21}$$

This bound gives how many errors t can be corrected with an (n, k) linear block code by using the syndrome decoding (described in Sect. 4.2.6).

4.2.5 Coding Gain

A very important characteristics of an (n, k) linear block code is the so-called coding gain, which was introduced in introductory section of this chapter as being the savings attainable in the energy per information bit to noise spectral density ratio (E_b/N_0) required to achieve a given bit error probability when coding is used compared to that with no coding. Let E_c denote the transmitted bit energy, and E_b denote the information bit energy. Since the total information word energy kE_b must be the same as the total codeword energy nE_c, we obtain the following relationship between E_c and E_b:

$$E_c = (k/n)E_b = RE_b. \tag{4.22}$$

The probability of error for BPSK on an AWGN channel, when coherent hard decision (bit-by-bit) demodulator is used, can be obtained as follows:

$$p = \frac{1}{2}\text{erfc}\left(\sqrt{\frac{E_c}{N_0}}\right) = \frac{1}{2}\text{erfc}\left(\sqrt{\frac{RE_b}{N_0}}\right), \tag{4.23}$$

where erfc(x) function is defined by

$$\text{erfc}(x) = \frac{2}{\sqrt{\pi}} \int_{x}^{+\infty} e^{-z^2} dz.$$

For high SNRs the word error probability (remained upon decoding) of a t-error-correcting code is dominated by a $t + 1$ error event:

$$P_w(e) \approx \binom{n}{t+1} p^{t+1}(1-p)^{n-t+1} \approx \binom{n}{t+1} p^{t+1}. \tag{4.24}$$

The bit error probability P_b is related to the word error probability by

$$P_b \approx \frac{2t+1}{n} P_w(e) \approx c(n,t) p^{t+1}, \tag{4.25}$$

because $2t + 1$ and more errors per codeword cannot be corrected and they can be located anywhere on n codeword locations, and $c(n,t)$ is a parameter dependent on error correcting capability t and codeword length n. By using the upper bound on erfc(x) we obtain

$$P_b \approx \frac{c(n,t)}{2}\left[\exp\left(\frac{-RE_b}{N_0}\right)\right]^{t+1}. \tag{4.26}$$

The corresponding approximation for uncoded case is

$$P_{b,\mathrm{uncoded}} \approx \frac{1}{2}\exp\left(-\frac{E_b}{N_0}\right). \tag{4.27}$$

By equating (4.26) and (4.27) and ignoring the parameter $c(n,t)$ we obtain the following expression for hard decision decoding coding gain

$$\frac{(E_b/N_0)_{\mathrm{uncoded}}}{(E_b/N_0)_{\mathrm{coded}}} \approx R(t+1). \tag{4.28}$$

The corresponding soft decision asymptotic coding gain of convolutional codes is [24, 26–28, 30]

$$\frac{(E_b/N_0)_{\mathrm{uncoded}}}{(E_b/N_0)_{\mathrm{coded}}} \approx Rd_{\min}, \tag{4.29}$$

and it is about 3 dB better than hard decision decoding (because $d_{\min} \geq 2t + 1$).

In optical communications it is very common to use the Q-factor as the figure of merit instead of SNR, which is related to the BER on an AWGN, as shown in Chap. 3, as follows

$$\mathrm{BER} = \frac{1}{2}\mathrm{erfc}\left(\frac{Q}{\sqrt{2}}\right). \tag{4.30}$$

Let $\mathrm{BER_{in}}$ denote the BER at the input of FEC decoder, let $\mathrm{BER_{out}}$ denote the BER at the output of FEC decoder, and let BER_{ref} denote target BER (such as either 10^{-12} or 10^{-15}). The corresponding coding gain GC and net coding gain NCG are, respectively, defined as [10]

$$\mathrm{CG} = 20\log_{10}\left[\mathrm{erfc}^{-1}(2\mathrm{BER_{ref}})\right] - 20\log_{10}\left[\mathrm{erfc}^{-1}(2\mathrm{BER_{in}})\right] \quad [\mathrm{dB}], \tag{4.31}$$

$$\mathrm{NCG} = 20\log_{10}\left[\mathrm{erfc}^{-1}(2\mathrm{BER_{ref}})\right] - 20\log_{10}\left[\mathrm{erfc}^{-1}(2\mathrm{BER_{in}})\right] + 10\log_{10} R \quad [\mathrm{dB}]. \tag{4.32}$$

All coding gains reported in this chapter are in fact NCG, although they are sometimes called the coding gains only, because this is a common practice in coding theory literature [24, 26–28, 30].

4.2.6 Syndrome Decoding and Standard Array

The received vector $\boldsymbol{r} = \boldsymbol{x} + \boldsymbol{e}$ ($\boldsymbol{x}$ is the codeword and $\boldsymbol{e}$ is the error patter introduced above) is a codeword if the following *syndrome equation* is satisfied $\boldsymbol{s} = \boldsymbol{r}\boldsymbol{H}^{\mathrm{T}} = \boldsymbol{0}$. The syndrome has the following important properties:

1. The syndrome is only function of the error pattern. This property can easily be proved from definition of syndrome as follows: $\boldsymbol{s} = \boldsymbol{r}\boldsymbol{H}^{\mathrm{T}} = (\boldsymbol{x} + \boldsymbol{e})\boldsymbol{H}^{\mathrm{T}} = \boldsymbol{x}\boldsymbol{H}^{\mathrm{T}} + \boldsymbol{e}\boldsymbol{H}^{\mathrm{T}} = \boldsymbol{e}\boldsymbol{H}^{\mathrm{T}}$.
2. All error patterns that differ by a codeword have the same syndrome. This property can also be proved from syndrome definition. Let $\boldsymbol{x}_i$ be the ith ($i = 0, 1, \ldots, 2^{k-1}$) codeword. The set of error patterns that differ by a codeword is known as cosset: $\{\boldsymbol{e}_i = \boldsymbol{e} + \boldsymbol{x}_i, i = 0, 1, \ldots, 2^{k-1}\}$. The syndrome corresponding to ith error pattern from this set $\boldsymbol{s}_i = \boldsymbol{r}_i\boldsymbol{H}^{\mathrm{T}} = (\boldsymbol{x}_i + \boldsymbol{e})\boldsymbol{H}^{\mathrm{T}} = \boldsymbol{x}_i\boldsymbol{H}^{\mathrm{T}} + \boldsymbol{e}\boldsymbol{H}^{\mathrm{T}} = \boldsymbol{e}\boldsymbol{H}^{\mathrm{T}}$ is only a function of the error pattern, and therefore all error patterns from the cosset have the same syndrome.
3. The syndrome is function of only those columns of a parity-check matrix corresponding to the error locations. The parity-check matrix can be written into following form: $\boldsymbol{H} = [\boldsymbol{h}_1 \cdots \boldsymbol{h}_n]$, where the ith element $\boldsymbol{h}_i$ denotes the ith column of $\boldsymbol{H}$. Based on syndrome definition for an error patter $\boldsymbol{e} = [e_1 \cdots e_n]$ the following is valid:

$$\boldsymbol{s} = \boldsymbol{e}\boldsymbol{H}^T = \begin{bmatrix} e_1 \ e_2 \cdots e_n \end{bmatrix} \begin{bmatrix} \boldsymbol{h}_1^T \\ \boldsymbol{h}_2^T \\ \cdots \\ \boldsymbol{h}_n^T \end{bmatrix} = \sum_{i=1}^{n} e_i \boldsymbol{h}_i^{\mathrm{T}}, \tag{4.33}$$

   which proves the claim of property 3.
4. With syndrome decoding an (n, k) linear block code can correct up to t errors, providing that Hamming bound (4.21) is satisfied. (This property will be proved in next subsection.)

By using the property 2, 2^k codewords partition the space of all received words into 2^k disjoint subsets. Any received word within subset will be decoded as the unique codeword. A *standard array* is a technique by which this partition can be achieved, and can be constructed using the following two steps [22, 24, 26–28, 30–32]:

1. Write down 2^k code words as elements of the first row, with the all-zero codeword as the leading element.
2. Repeat the steps 2(a) and 2(b) until all 2^n words are exhausted.
   (a) Out of the remaining unused n-tuples, select one with the least weight for the leading element of the next row (the *current row*).
   (b) Complete the current row by adding the leading element to each nonzero code word appearing in the first row and writing down the resulting sum in the corresponding column.

Fig. 4.7 The standard array architecture

$x_1 = 0$	x_2	x_3	...	x_i	...	x_{2^k}
e_2	$x_2 + e_2$	$x_3 + e_2$	...	$x_i + e_2$	...	$x_{2^k} + e_2$
e_3	$x_2 + e_3$	$x_3 + e_3$	...	$x_i + e_2$	...	$x_{2^k} + e_3$
...	...	...	...	...	...	...
e_j	$x_2 + e_j$	$x_3 + e_j$	...	$x_i + e_j$	...	$x_{2^k} + e_j$
...	...		...			...
$e_{2^{n-k}}$	$x_2 + e_{2^{n-k}}$	$x_3 + e_{2^{n-k}}$	...	$x_i + e_{2^{n-k}}$	...	$x_{2^k} + e_{2^{n-k}}$

Table 4.2 Standard array of (5,2) code and corresponding decoding table

	Codewords				Syndrome s	Error pattern
	00000	11010	10101	01111	000	00000
	00001	11011	10100	01110	101	00001
Coset leader	00010	11000	10111	01101	110	00010
	00100	11110	10001	01011	001	00100
	01000	10010	11101	00111	010	01000
	10000	01010	00101	11111	100	10000
	00011	11001	10110	01100	011	00011
	00110	11100	10011	01001	111	00110

The standard array for an (n, k) block code obtained by this algorithm is illustrated in Fig. 4.7. The columns represent 2^k disjoint sets, and every row represents the coset of the code with leading elements being called the coset leaders.

Example: Standard array of (5,2) code $C = \{(00000), (11010), (10101), (01111)\}$ is given in Table 4.2. The parity-check matrix of this code is given by

$$H = \begin{bmatrix} 1\,0\,0\,1\,1 \\ 0\,1\,0\,1\,0 \\ 0\,0\,1\,0\,1 \end{bmatrix}.$$

Because the minimum distance of this code is 3 (first, second, and fourth columns add to zero), this code is able to correct all single errors. For example, if the word 01010 is received it will be decoded to the top-most codeword 11010 of column in which it lies. In the same Table corresponding syndromes are provided as well.

The syndrome decoding procedure is a three-step procedure [22, 24, 26–28, 30–32]:

1. For the received vector $\boldsymbol{r}$, compute the syndrome $\boldsymbol{s} = \boldsymbol{r}\boldsymbol{H}^{\mathrm{T}}$. From property 3 we can establish one-to-one correspondence between the syndromes and error patterns (see Table 4.2), leading to the lookup table (LUT) containing the syndrome and corresponding error pattern (the coset leader).
2. Within the coset characterized by the syndrome $\boldsymbol{s}$, identify the coset leader, say $\boldsymbol{e}_0$. The coset leader corresponds to the error pattern with the largest probability of occurrence.
3. Decode the received vector as $\boldsymbol{x} = \boldsymbol{r} + \boldsymbol{e}_0$.

Example: Let the received vector for (5,2) code example above be $\boldsymbol{r} = (01010)$. The syndrome can be computed as $\boldsymbol{s} = \boldsymbol{r}\boldsymbol{H}^{\mathrm{T}} = (100)$, and corresponding error pattern from LUT is found to be $\boldsymbol{e}_0 = (10000)$. The decoded word is obtained by adding the error pattern to received word $\boldsymbol{x} = \boldsymbol{r} + \boldsymbol{e}_0 = (11010)$, and the error on the first bit position is corrected.

The standard array can be used to determine probability of word error as follows:

$$P_w(e) = 1 - \sum_{i=0}^{n} \alpha_i \; p^i (1-p)^{n-i}, \tag{4.34}$$

where α_i is the number of coset leaders of weight i (distribution of weights is also known as *weight distribution* of cosset leaders) and p is the crossover probability of BSC. Any error pattern that is not a coset leader will result in decoding error. For example, the weight distribution of coset leaders in (5,2) code are $\alpha_0 = 1$, $\alpha_1 = 5$, $\alpha_2 = 2$, $\alpha_i = 0$, $i = 3, 4, 5$, which leads to the following word error probability:

$$P_w(e) = 1 - (1-p)^5 - 5p(1-p)^4 - 2p^2(1-p)^3|_{p=10^{-3}} = 7.986 \cdot 10^{-6}.$$

We can use (4.34) to estimate the coding gain of a given LBC. For example, the word error probability for Hamming (7,4) code is

$$P_w(e) = 1 - (1-p)^7 - 7p(1-p)^6 = \sum_{i=2}^{7} \binom{7}{i} p^i (1-p)^{7-i} \approx 21p^2.$$

In the previous section, we established the following relationship between bit- and word-error probabilities: $P_b \approx P_w(e)(2t+1)/n = (3/7)P_w(e) \approx (3/7)21p^2 = 9p^2$. Therefore, the crossover probability can be evaluated as

$$p = \sqrt{P_b}/3 = (1/2)\mathrm{erfc}\left(\sqrt{\frac{RE_b}{N_0}}\right).$$

From this expression we can easily calculate the required SNR to achieve target P_{b}. By comparing such obtained SNR with corresponding SNR for uncoded BPSK we can evaluate the corresponding coding gain.

In order to evaluate the probability of undetected error, we have to determine the other codeword weights as well. Because the undetected errors are caused by error patterns being identical to the nonzero codewords, the undetected error probability can be evaluated by:

$$P_u(e) = \sum_{i=1}^{n} A_i \; p^i (1-p)^{n-i} = (1-p)^n \sum_{i=1}^{n} A_i \left(\frac{p}{1-p}\right), \tag{4.35}$$

where p is the crossover probability and A_i denotes the number of codewords of weight i. The codeword weight can be determined by *McWilliams identity* that establishes the connection between codeword weights A_i and the codeword weights of corresponding dual code B_i by [24]:

$$A(z) = 2^{-(n-k)}(1+z)^n B\left(\frac{1-z}{1+z}\right), \quad A(z) = \sum_{i=0}^{n} A_i z^i, \quad B(z) = \sum_{i=0}^{n} B_i z^i \tag{4.36}$$

where $A(z)(B(z))$ represents the polynomial representation of codeword weights (dual-codeword weights). By substituting $z = p/(1-p)$ in (4.36) and knowing that $A_0 = 1$ we obtain

$$A\left(\frac{p}{1-p}\right) - 1 = \sum_{i=1}^{n} A\left(\frac{p}{1-p}\right)^i. \tag{4.37}$$

Substituting (4.37) into (4.35) we obtain

$$P_u(e) = (1-p)^n \left[A\left(\frac{p}{1-p}\right) - 1\right]. \tag{4.38}$$

An alternative expression for $P_{\mathrm{u}}(e)$ in terms of $B(z)$ can be obtained from (4.36), which is more suitable for use when $n-k < k$, as follows

$$P_u(e) = 2^{-(n-k)} B(1-2p) - (1-p)^n. \tag{4.39}$$

For large n, k, $n-k$ the use of McWilliams identity is impractical, instead an upper bound on average probability of undetected error of an (n,k) systematic codes should be used instead:

$$\overline{P_u}(e) \leq 2^{-(n-k)}\left[1-(1-p)^n\right]. \tag{4.40}$$

For a q-ary maximum-distance separable code, which satisfies the Singleton bound introduced in the next section with equality, we can determine a closed formula for weight distribution [24]:

$$A_i = \binom{n}{i}(q-1)\sum_{j=0}^{i-d_{\min}}(-1)^j\binom{i-1}{j}q^{i-d_{\min}-j}, \tag{4.41}$$

where $d_{\min}$ is the minimum distance of the code, $A_0 = 1$, and $A_i = 0$ for $i \in [1, d_{\min}-1]$.

4.2.7 Important Coding Bounds

In this section, we describe several important coding bounds including Hamming, Plotkin, Gilbert-Varshamov, and Singleton bounds [22,24,26–28,30–32]. The *Hamming* bound has already been introduced for binary LBCs by (4.21). The Hamming bound for q-ary (n, k) LBC is given by

$$\left[1 + (q-1)\binom{n}{1} + (q-1)^2\binom{n}{2} + \cdots + (q-1)^i\binom{n}{i} + \cdots + (q-1)^t\binom{n}{t}\right] q^k \leq q^n, \tag{4.42}$$

where t is the error correction capability and $(q-1)^i\binom{n}{i}$ is the number of received words that differ from a given code word in i symbols. Namely there are n choices i ways in which symbols can be chosen out of n and there are $(q-1)^i$ possible choices for symbols. The codes satisfying the Hamming bound with equality sign are known as *perfect codes*. Hamming codes are perfect codes because $n = 2^{n-k} - 1$, which is equivalent to $(1+n)2^k = 2^n$ so that inequality above is satisfied with equality. $(n, 1)$ repetition code is also a perfect code. The 3-error correcting (23, 12) Golay code is another example of perfect code because:

$$\left[1 + \binom{23}{1} + \binom{23}{2} + \binom{23}{3}\right] 2^{12} = 2^{23}.$$

The *Plotkin* bound is the bound on the minimum distance of a code:

$$d_{\min} \leq \frac{n2^{k-1}}{2^k - 1}. \tag{4.43}$$

Namely, if all code word are written as the rows of a $2^k \times n$ matrix, each column will contain 2^{k-1} 0s and 2^{k-1} 1s, with the total weight of all code words being $n2^{k-1}$.

Gilbert-Varshamov bound is based on the property that the minimum distance $d_{\min}$ of a linear (n, k) block code, can be determined as the minimum number of columns in $\boldsymbol{H}$-matrix that sum to zero:

$$\binom{n-1}{1} + \binom{n-1}{2} + \cdots + \binom{n-1}{d_{\min}-2} < 2^{n-k} - 1. \tag{4.44}$$

Another important bound is *Singleton bound*:

$$d_{\min} \leq n - k + 1. \tag{4.45}$$

This bound is straightforward to prove. Let only one 1-bit be present in information vector. If it is involved in $n - k$ parity checks, then the total number of ones in code word cannot be larger than $n-k+1$. The codes satisfying the Singleton bound with equality sign are known as the maximum-distance separable (MDS) codes (e.g., RS codes are MDS codes).

4.3 Cyclic Codes

The most commonly used class of LBCs is the class of cyclic codes. Examples of cyclic codes include BCH codes, Hamming codes, and Golay codes. RS codes are also cyclic but nonbinary codes. Even LDPC codes can be designed in cyclic or quasi-cyclic fashion.

Let us observe the vector space of dimension n. The subspace of this space is *cyclic code* if for any codeword $\boldsymbol{c}(c_0, c_1, \ldots, c_{n-1})$ arbitrary cyclic shift $\boldsymbol{c}_j(c_{n-j}, c_{n-j+1}, \ldots, c_{n-1}, c_0, c_1, \ldots, c_{n-j-1})$ is another codeword. With every codeword $\boldsymbol{c}(c_0, c_1, \ldots, c_{n-1})$ from a cyclic code, we associate the *codeword polynomial*

$$c(x) = c_0 + c_1 x + c_2 x^2 + \cdots + c_{n-1} x^{n-1}. \tag{4.46}$$

The jth cyclic shift, observed mod $(x^n - 1)$, is also a codeword polynomial

$$c^{(j)}(x) = x^j c(x) \mathrm{mod}(x^n - 1). \tag{4.47}$$

It is straightforward to show that observed subspace is cyclic if composed from polynomials divisible by a polynomial $g(x) = g_0 + g_1 x + \cdots + g_{n-k} x^{n-k}$ that divides $x^n - 1$ at the same time. The polynomial $g(x)$, of degree $n-k$, is called the *generator polynomial* of the code. If $x^n - 1 = g(x)h(x)$ then the polynomial of degree k is called the *parity-check polynomial*. The generator polynomial has the following three important properties [22, 24, 26–28, 30–32]:

1. The generator polynomial of an (n,k) cyclic code is unique (usually proved by contradiction)
2. Any multiple of generator polynomial is a codeword polynomial
3. The generator polynomial and parity-check polynomial are factors of $x^n - 1$

The generator polynomial $g(x)$ and the parity-check polynomial $h(x)$ serve the same role as the generator matrix $\boldsymbol{G}$ and parity-check matrix $\boldsymbol{H}$ of an LBC. n-Tuples related to the k polynomials $g(x), xg(x), \ldots, x^{k-1}g(x)$ may be used in rows of the $k \times n$ generator matrix $\boldsymbol{G}$, while n-tuples related to the $(n-k)$ polynomials $x^k h(x^{-1}), x^{k+1} h(x^{-1}), \ldots, x^{n-1} h(x^{-1})$ may be used in rows of the $(n-k) \times n$ parity-check matrix $\boldsymbol{H}$.

To encode we have simply to multiply the message polynomial $m(x) = m_0 + m_1 x + \cdots + m_{k-1} x^{k-1}$ with the generator polynomial $g(x)$, i.e., $c(x) = m(x)g(x) \mathrm{mod}(x^n - 1)$, where $c(x)$ is the codeword polynomial. To encode in *systematic* form we have to find the reminder of $x^{n-k} m(x)/g(x)$ and add it to the shifted version of message polynomial $x^{n-k} m(x)$, i.e., $c(x) = x^{n-k} m(x) + \mathrm{rem}[x^{n-k} m(x)/g(x)]$, where with rem[] is denoted the remainder of a given entity. The general circuit for generating the codeword polynomial in systematic form is given in Fig. 4.8. The encoder operates as follows. When the switch S is in position 1 and Gate is closed (on), the information bits are shifted into the shift register and at the same time transmitted onto the channel. Once all information bits are shifted into register in k shifts, with Gate being open (off), the switch S is moved in position 2, and the content of $(n-k)$-shift register is transmitted onto the channel.

To check if the received word polynomial is the codeword polynomial $r(x) = r_0 + r_1x + \cdots + r_{n-1}x^{n-1}$ we have simply to determine the *syndrome polynomial* $s(x) = \mathrm{rem}[r(x)/g(x)]$. If $s(x)$ is zero then there is no error introduced during transmission. Corresponding circuit is given in Fig. 4.9.

For example, the encoder and syndrome calculator for [4, 7] Hamming code are given in Fig. 4.10a, b, respectively. The generating polynomial is given by $g(x) = 1 + x + x^3$. The polynomial $x^7 + 1$ can be factorized as follows: $\mathrm{x}^7 + 1 = (1+x)(1+x^2+x^3)(1+x+x^3)$. If we select $g(x) = 1 + x + x^3$ as the generator polynomial, based on property 3 of generator polynomial, the corresponding parity-check polynomial will be $h(x) = (1+x)(1+x^2+x^3) = 1 + x + x^2 + x^4$. The message sequence 1001 can be represented in polynomial form by $m(x) = 1 + x^3$. For the representation in systematic form, we have to multiply $m(x)$ by x^{n-k} to obtain: $x^{n-k}m(x) = x^3m(x) = x^3 + x^6$. The codeword polynomial is obtained by $c(x) = x^{n-k}m(x) + \mathrm{rem}[x^{n-k}m(x)/g(x)] = x + x^3 + \mathrm{rem}[(x + x^3)/(1 + x + x^3)] = x + x^2 + x^3 + x^6$. The corresponding codeword is 0111001. To obtain the generator matrix of this code we can use the following polynomials $g(x) = 1 + x + x^3$, $xg(x) = x + x^2 + x^4$, $x^2g(x) = x^2 + x^3 + x^5$, and

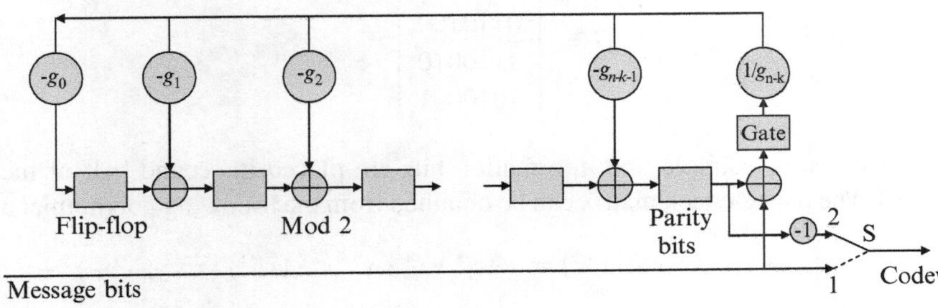

Fig. 4.8 Systematic cyclic encoder

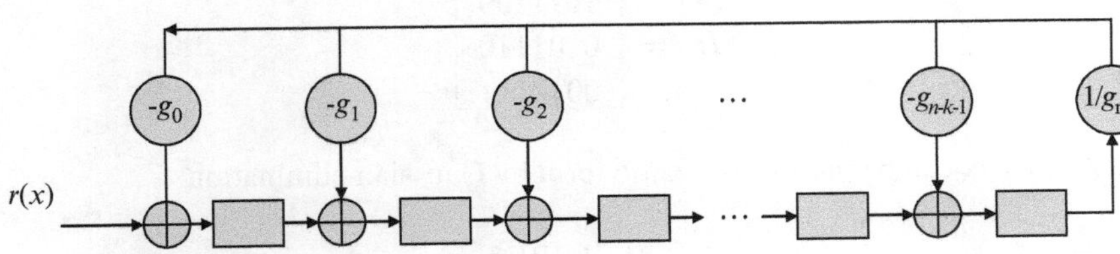

Fig. 4.9 Syndrome calculator

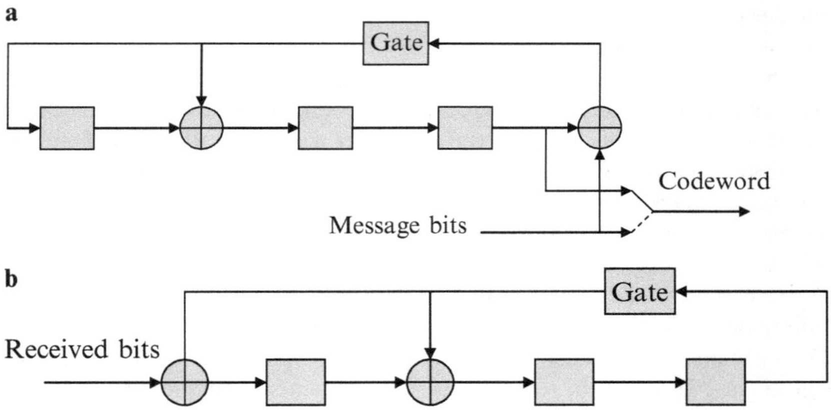

Fig. 4.10 Hamming (7,4) encoder (**a**) and syndrome calculator (**b**)

$x^3g(x) = x^3 + x^4 + x^6$; and write the corresponding n-tuples in the form of matrix as follows:

$$\boldsymbol{G}' = \begin{bmatrix} 1101000 \\ 0110100 \\ 0011010 \\ 0001101 \end{bmatrix}.$$

By Gaussian elimination we can put the generator matrix in a systematic form:

$$\boldsymbol{G} = \begin{bmatrix} 1101000 \\ 0110100 \\ 1110010 \\ 1010001 \end{bmatrix}.$$

Notice that in this example, the information bits are placed in second half of the codeword. The parity-check matrix can be obtained from the following polynomials:

$$x^4h\left(x^{-1}\right) = 1+x^2+x^3+x^4, x^5h\left(x^{-1}\right) = x+x^3+x^4+x^5,\ x^6h\left(x^{-1}\right) = x^2+x^3+x^5+x^6,$$

by writing down the corresponding n-tuples in the form of a matrix:

$$\boldsymbol{H}' = \begin{bmatrix} 1011100 \\ 0101110 \\ 0010111 \end{bmatrix}.$$

The $\boldsymbol{H}$-matrix can be put in systematic form by Gaussian elimination:

$$\boldsymbol{H} = \begin{bmatrix} 1001011 \\ 0101110 \\ 0010111 \end{bmatrix}.$$

The syndrome polynomial $s(x)$ has the following three important properties, that can be used to simplify the implementation of decoders [22, 24, 26–28, 30–32]:

1. The syndrome of received word polynomial $r(x)$ is also the syndrome of corresponding error polynomial $e(x)$.
2. The syndrome of a cyclic shift of $r(x)$, $xr(x)$ is determined by $xs(x)$.
3. The syndrome polynomial $s(x)$ is identical to the error polynomial $e(x)$, if the errors are confined to the $(n-k)$ parity-check bits of the received word polynomial $r(x)$.

Maximal-length codes $(n = 2^m - 1, m)(m \geq 3)$ are dual of Hamming codes and have minimum distance $d_{\min} = 2^m - 1$. The parity-check polynomial for $(7,3)$ maximal-length codes is therefore $h(x) = 1 + x + x^3$. The encoder for $(7,3)$ maximum-length code is given in Fig. 4.11. The generator polynomial gives one period of maximum-length code, providing that encoder is initialized to $0\ldots01$. For example, the generator polynomial for (7,3) maximum-length code above is $g(x) = 1 + x + x^2 + x^4$, and the output sequence is given by:

$$\underbrace{100}_{\text{initial state}} \quad \underbrace{1110100}_{g(x)=1+x+x^2+x^4} .$$

Cyclic redundancy check (CRC) codes are very popular codes for error detection. An (n,k) CRC codes are capable of detecting [22, 24, 26–28, 30–32]:

- All error bursts of length $n-k$, with an error burst of length $n-k$ being defined as a contiguous sequence of $n-k$ bits in which the first and the last bits or any other intermediate bits are received in error.
- A fraction of error bursts of length equal to $n-k+1$; the fraction equals $1-2^{-(n-k-1)}$.
- A fraction of error of length greater than $n-k+1$; the fraction equals $1-2^{-(n-k-1)}$.
- All combinations of $d_{\min}-1$ (or fewer) errors.
- All error patterns with an odd number of errors if the generator polynomial $g(x)$ for the code has an even number of nonzero coefficients.

In Table 4.3 we listed the generator polynomials of several CRC codes, currently in use in various communication systems.

Decoding of cyclic codes is composed of the same three steps used in decoding of LBCs, namely, the syndrome computation, the error pattern identification, and the error correction [22, 24, 26–28, 30–32]. The Meggit decoder configuration, which

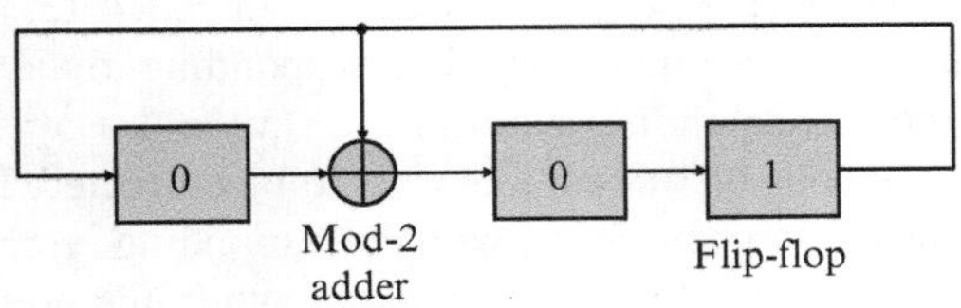

Fig. 4.11 Encoder for the (7, 3) maximal-length code

Table 4.3 Generator polynomials of several CRC codes

CRC Codes	Generator polynomial	$n-k$
CRC-8 code (IEEE 802.16, WiMax)	$1+x^2+x^8$	8
CRC-16 code (IBM CRC-16, ANSI, USB, SDLC)	$1+x^2+x^{15}+x^{16}$	16
CRC-ITU (X25, V41, CDMA, Bluetooth, HDLC, PPP)	$1+x^5+x^{12}+x^{16}$	16
CRC-24 (WLAN, UMTS)	$1+x+x^5+x^6+x^{23}+x^{24}$	24
CRC-32 (Ethernet)	$1+x+x^2+x^4+x^5+x^7+x^8+x^{10}+x^{11}+x^{12}+x^{16}+x^{22}+x^{23}+x^{26}+x^{32}$	32

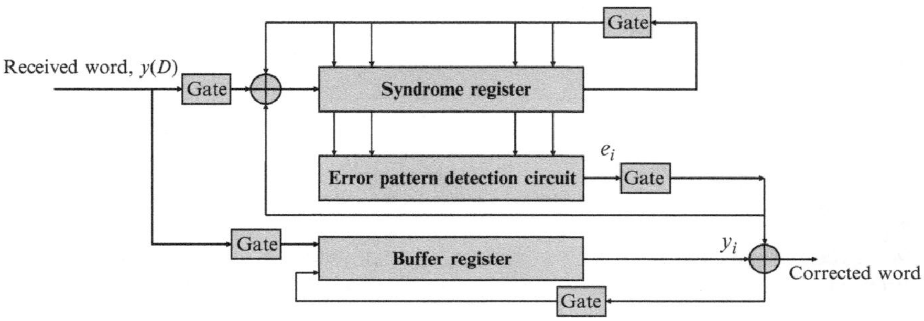

Fig. 4.12 A Meggit decoder configuration

is implemented based on syndrome property 2 (also known as Meggit theorem), is shown in Fig. 4.12. The syndrome is calculated by dividing the received word by generating polynomial $g(x)$, and at the same time the received word is shifted into the buffer register. Once the last bit of received word enters the decoder, the gate is turned off. The syndrome is further read into the error pattern detection circuit, implemented as the combinational logic circuit, which generates 1 if and only if the content of the syndrome register corresponds to a correctible error pattern at the highest-order position x^{n-1}. By adding the output of the error pattern detector to the bit in error, the error can be corrected, and at the same time the syndrome has to be modified. If an error occurred on position x^l, by cyclically shifting the received word $n-l-1$ times, the erroneous bit will appear in position x^{n-1}, and can be corrected. Decoder, therefore, corrects the errors in a bit-by-bit fashion until entire received word is read out from the buffer register.

The Hamming (7,4) cyclic decoder configuration, for generating polynomial $g(x) = 1+x+x^3$, is shown in Fig. 4.13. Because we expect only single errors, the error polynomial corresponding to the highest order position is $e(x) = x^6$ and corresponding syndrome is $s(x) = 1+x^2$. Once this syndrome is detected, the erroneous bit at position x^6 is to be corrected. Let us now assume that an error occurred on position x^i, with corresponding error pattern $e(x) = x^i$. Once the entire received word is shifted into the syndrome register, the error syndrome is not 101.

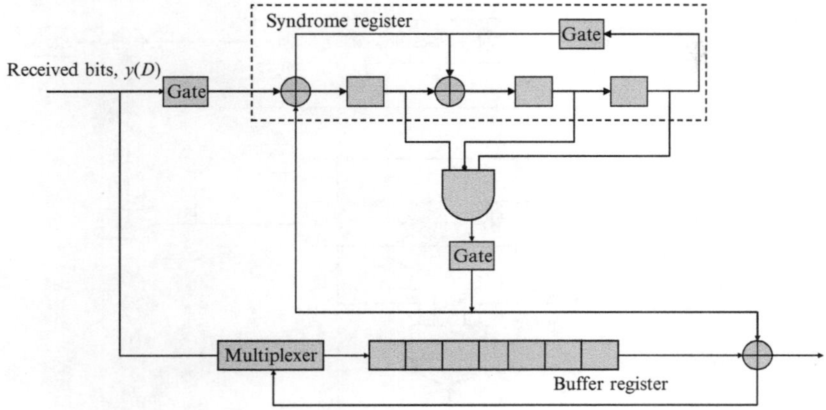

Fig. 4.13 Hamming (7,4) decoder configuration

But after 6-i additional shifts, the content of syndrome register will become 101, and error at position x^6 (initially at position x^i) will be corrected.

There exists several versions of Meggit decoders, however, the basic idea is essentially similar to that described above. The complexity of this decoder increases very fast as the number of errors to be corrected increases, and it is rarely used for correction of more than three single errors or one burst of errors. The Meggit decoder can be simplified under certain assumptions. Let us assume that errors are confined to only highest order information bits of the received polynomial $r(x) : x^k, x^{k+1}, \ldots, x^{n-1}$, so that syndrome polynomial $s(x)$ is given by

$$s(x) = r(x) \bmod g(x) = [c(x) + e(x)] \bmod g(x) = e(x) \bmod g(x), \tag{4.48}$$

where $r(x)$ is the received polynomial, $c(x)$ is the codeword polynomial and $g(x)$ is the generator polynomial. The corresponding error polynomial can be estimated by

$$\begin{aligned} e'(x) &= e'(x) \bmod g(x) = s'(x) = x^{n-k} s(x) \bmod g(x) \\ &= e_k + e_{k+1}x + \cdots + e_{n-2}x^{n-k-2} + e_{n-1}x^{n-k-1}. \end{aligned} \tag{4.49}$$

The error polynomial will be at most of degree $n-k-1$ because $\deg[g(x)] = n-k$. Therefore, the syndrome register content is identical to the error pattern, and we say that the error pattern is *trapped* in the syndrome register, and corresponding decoder, known as *error-trapping decoder*, is shown in Fig. 4.14 [24, 26]. If t or fewer errors occur in $n-k$ consecutive locations, it can be shown that the error pattern is trapped in the syndrome register only when the weight of the syndrome $w(s)$ is less than or equal to t [24, 26]. Therefore, the test for the error-trapping condition is to check if the weight of the syndrome is t or less. With gates 1, 2, and 3 closed (4 and 5 open), the received word is shifted into the syndrome register from the right end, which is equivalent to premultiplying the received polynomial $r(x)$ with x^{n-k}. Once the highest order k bits (corresponding to information bits) are

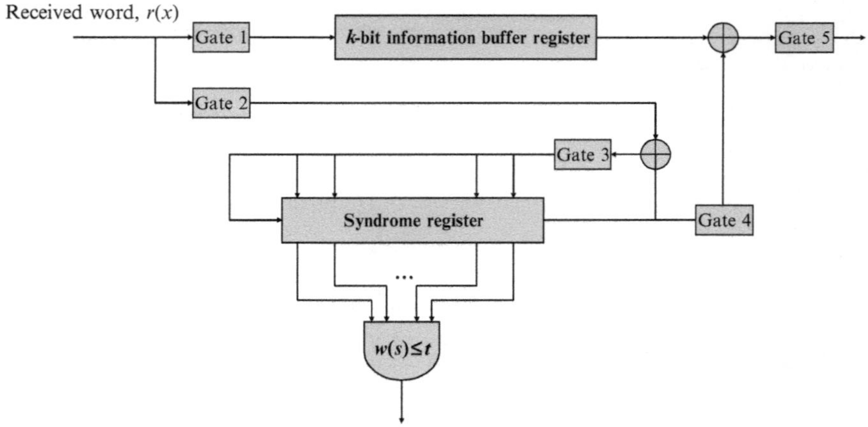

Fig. 4.14 Error-trapping decoding architecture

shifted into information buffer, the gate 1 is opened. The gate 2 is opened once all n bits from received word are shifted into syndrome register. The syndrome register at this point contains the syndrome corresponding to $x^{n-k}r(x)$. If its weight is t or less, the gates 4 and 5 are closed (all others are opened), and the corrected information is shifted out. If $w(s) > t$, the errors are not confined to the $n-k$ higher-order positions of $s(x)$, we keep shifting with gate 3 being on (other gates are switched off), until $w(s) \leq t$. If $w(s)$ never goes $\leq t$ and the syndrome register is shifted k times, either an error pattern with errors confined to $n-k$ consecutive end-around locations has occurred or uncorrectable error pattern has occurred. For additional explanation on this decoder, and other types of cyclic decoders an interested reader is referred to [24].

4.4 Bose–Chaudhuri–Hocquenghem Codes

The Bose–Chaudhuri–Hocquenghem (BCH) codes, the most famous cyclic codes, were discovered by Hocquenghem in 1959 and by Bose and Chaudhuri in 1960 [24, 26–28]. Among many different decoding algorithms, the most important are Massey–Berlekamp algorithm and Chien's search algorithm. An important subclass of BCH is a class of Reed–Solomon codes proposed in 1960. Before we continue further with study of BCH codes, we have to introduce some properties of finite fields [24, 28, 33, 34].

4.4.1 Galois Fields

Before we can proceed further with finite fields, we introduce the concepts of ring, field and congruencies from abstract algebra. A *ring* is a set of elements R with two operations, addition "+" and multiplication "·", satisfying the following three properties: (a) R is an Abelian (commutative) group under addition, (b) multiplication operation is associative, and (c) multiplication is associative over addition. A *field* is a set of elements F with two operations, addition "+" and multiplication "· ", satisfying the following properties: (a) F is an Abelian group under addition operation, with 0 being the identity element, (b) the nonzero elements of F form an Abelian group under the multiplication, with 1 being the identity element, and (c) the multiplication operation is distributive over the addition operation.

The quantity a is said to be *congruent* to quantity b observed per modulus n, denoted as $a \equiv b(\text{mod}\, n)$, if $a - b$ is divisible by n. If $x \equiv a(\text{mod}\, n)$, then a is called a *residue* to x to modulus n. A *class of residues* to modulus n is the class of all integers congruent to a given residue (mod n), and every member of the class is called a representative of the class. There are n classes, represented by (0), (1), (2), ..., $(n-1)$, and the representative of these classes are called a *complete system* of incongruent residues to modulus n. If i and j are two members of a complete system of incongruent residues to modulus n, then addition and multiplication between i and j can be introduced by

1. $i + j = (i + j)(\text{mod}\, n)$
2. $i \cdot j = (i \cdot j)(\text{mod}\, n)$

A complete system of residues (mod n) forms a *commutative ring* with unity element. Let s be a nonzero element of these residues. Then s possesses an inverse element if and only if n is prime, p. When p is a prime, a complete system of residues (mod p) forms a *Galois* (finite) *field*, and is commonly denoted by GF(p).

Let $P(x)$ be any given polynomial in x of degree m with coefficients belonging to GF(p), and let $F(x)$ be any polynomial in x with integral coefficients. Then $F(x)$ may be expressed as [34]

$$F(x) = f(x) + pq(x) + P(x)Q(x),$$

where $f(x) = a_0 + a_1x + a_2x^2 + \cdots + a_{m-1}x^{m-1}$, $a_i \in \text{GF}(p)$. This relationship may be written as $F(x) \equiv f(x)\text{mod}\{p, P(x)\}$, and we say that $f(x)$ is the *residue* of $F(x)$ modulus p and $P(x)$. If p and $P(x)$ are kept fixed but $f(x)$ varied, p^m classes can be formed (because each coefficient of $f(x)$ may take p values of GF(p)). The classes defined by $f(x)$ form a commutative (Abelian) ring, which will be a field if an only if $P(x)$ is *irreducible* over GF(p) (not divisible with any other polynomial of degree $m-1$ or less) [24,26,28]. The finite field formed by p^m classes of residues is called a *Galois field of order* p^m and is denoted by GF(p^m). Two important *properties of* GF(q), $q = p^m$ are given below [24,26,28]:

1. The roots of polynomial $x^{q-1} - 1$ are all nonzero elements of GF(q).
2. Let $P(x)$ be an irreducible polynomial of degree m with coefficients from GF(p) and β be a root from the extended field GF($q = p^m$). Then all m roots of $P(x)$ are $\beta, \beta^p, \beta^{p^2}, \ldots, \beta^{p^{m-1}}$.

The nonzero elements of GF(p^m) can be represented as polynomials of degree at most $m - 1$ or as powers of a *primitive root* α such that [24, 26, 28]

$$\alpha^{p^m-1} = 1, \quad \alpha^d \neq 1 \text{ (for } d \text{ dividing } p^m - 1).$$

Therefore, the primitive element (root) is a field element that generates all nonzero filed elements as its successive powers. An irreducible polynomial that has a primitive element as its root is called a *primitive polynomial.*

The function $P(x)$ is said to be *minimum polynomial* for generating the elements of GF(p^m) and represents the smallest degree polynomial over GF(p) having a field element $\beta \in$ GF(p^m) as a root. To obtain a minimum polynomial we have divide $x^q - 1(q = p^m)$ by the least common multiple (LCM) of all factors of the form $x^d - 1$, where d is a divisor of $p^m - 1$; and obtain so-called *cyclotomic equation* (that is the equation having for its roots all primitive roots of equation $x^{q-1} - 1 = 0$). The order of this equation is $O(p^m - 1)$, where $O(k)$ is the number of all positive integers less than k and relatively prime to it. By substituting each coefficient in this equation by least nonzero residue to modulus p, we get the cyclotomic polynomial of order $O(p^m - 1)$. Let $P(x)$ be an irreducible factor of this polynomial, then $P(x)$ is a minimum polynomial, which is in general not the unique one.

Example: Let us determine the minimum polynomial for generating the elements of GF(2^3). The cyclotomic polynomial is $(x^7 - 1)/(x - 1) = x^6 + x^5 + x^4 + x^3 + x^2 + x + 1 = (x^3 + x^2 + 1)(x^3 + x + 1)$. Hence, $P(x)$ can be either $x^3 + x^2 + 1$ or $x^3 + x + 1$. Let us choose $P(x) = x^3 + x^2 + 1$. The degree of this polynomial is $\deg[P(x)] = 3$. Now we explain how we can construct GF(2^3) using the $P(x)$. The construction always starts with elements from basic field (in this case GF(2) $= \{0, 1\}$). All nonzero elements can be obtained as successive powers of primitive root α, until no new element is generated, which is given in the first column of Table 4.4. The second column is obtained exploiting the primitive polynomial $P(x) = x^3 + x^2 + 1$. α is the root of $P(x)$ and, therefore, $P(\alpha) = \alpha^3 + \alpha^2 + 1 = 0$ and α^3 can be expressed as $\alpha^2 + 1$. α^4 can be expressed as $\alpha\alpha^3 = \alpha\alpha^2 + 1) = \alpha^3 + \alpha = \alpha^2 + 1 + \alpha$. The third column in Table 4.4 is obtained by reading off coefficients in second column, with leading coefficient multiplying the α^2.

Table 4.4 Three different representations of GF(2^3) generated by $x^3 + x^2 + 1$

Power of α	Polynomial	3-tuple
0	0	000
α^0	1	001
α^1	α	010
α^2	α^2	100
α^3	$\alpha^2 + 1$	101
α^4	$\alpha^2 + \alpha + 1$	111
α^5	$\alpha + 1$	011
α^6	$\alpha^2 + \alpha$	110
α^7	1	001

4.4.2 The Structure and Decoding of BCH Codes

Equipped we this knowledge of Galois fields we can continue our description of structure of BCH codes. Let the finite field GF(q) (*symbol field*) and extension field GF(q^m) (*locator field*), $m \geq 1$, be given. For every $m_0 (m_0 \geq 1)$ and Hamming distance d there exists a BCH code with the generating polynomial $g(x)$, if and only if it is of smallest degree with coefficients from GF(q) and the roots from the extension field GF(q^m) as follows [24, 31]:

$$\alpha^{m_0}, \alpha^{m_0,+1}, \ldots, \alpha^{m_0+d-2}, \tag{4.50}$$

where α is from GF(q^m). The codeword length is determined as the least common multiple of orders of roots. (The order of an element β from finite field is the smallest positive integer j such that $\beta^j = 1$.)

It can be shown that for any positive integer m ($m \geq 3$) and t ($t < 2^{m-1}$) there exists a *binary* BCH code having the following properties [24, 26, 31]:

- Codeword length: $n = 2^m - 1$
- Number of parity bits: $n - k \leq mt$
- Minimum Hamming distance: $d \geq 2t + 1$

This code is able to correct up to t errors. The generator polynomial can be found as the LCM of the minimal polynomials of α^i [24, 26, 31]

$$g(x) = \text{LCM}\left[P_{\alpha^1}(x), P_{\alpha^3}(x), \ldots, P_{\alpha^{2t-1}}(x)\right], \tag{4.51}$$

where α is a primitive element in GF(2^m), and $P_{\alpha^i}(x)$ is the minimal polynomial of α^i.

Let $c(x) = c_0 + c_1 x + c_2 x^2 + \cdots + c_{n-1} x^{n-1}$ be the codeword polynomial, and let the roots of generator polynomial be $\alpha, \alpha^2, \ldots, \alpha^{2t}$, where t is the error correction capability of BCH code. Because the generator polynomial $g(x)$ is the factor of codeword polynomial $c(x)$, the rots of $g(x)$ must also be the roots of $c(x)$:

$$c(\alpha^i) = c_0 + c_1\alpha^i + \cdots + c_{n-1}\alpha^{(n-1)i} = 0, \quad 1 \leq i \leq 2t \tag{4.52}$$

This equation can also be written as inner (scalar) product of codeword vector $\boldsymbol{c} = [c_0 c_1 \ldots c_{n-1}]$ and the following vector $[1\, \alpha^i \alpha^{2i} \cdots \alpha^{2(n-1)i}]$:

$$[c_0\, c_1 \cdots c_{n-1}] \begin{bmatrix} 1 \\ \alpha^i \\ \cdots \\ \alpha^{(n-1)i} \end{bmatrix} = 0, \quad 1 \leq i \leq 2t \tag{4.53}$$

Equation (4.53) can also be written as the following matrix

$$[c_{n-1}\, c_{n-2} \cdots c_0] \begin{bmatrix} \alpha^{n-1} & \alpha^{n-2} & \cdots & \alpha & 1 \\ (\alpha^2)^{n-1} & (\alpha^2)^{n-2} & \cdots & \alpha^2 & 1 \\ (\alpha^3)^{n-1} & (\alpha^3)^{n-2} & \cdots & \alpha^3 & 1 \\ \cdots & \cdots & \cdots & \cdots & \cdots \\ (\alpha^{2t})^{n-1} & (\alpha^{2t})^{n-2} & \cdots & \alpha^{2t} & 1 \end{bmatrix}^T = \boldsymbol{c}\boldsymbol{H}^T = \boldsymbol{0},$$

$$\boldsymbol{H} = \begin{bmatrix} \alpha^{n-1} & \alpha^{n-2} & \cdots & \alpha & 1 \\ (\alpha^2)^{n-1} & (\alpha^2)^{n-2} & \cdots & \alpha^2 & 1 \\ (\alpha^3)^{n-1} & (\alpha^3)^{n-2} & \cdots & \alpha^3 & 1 \\ \cdots & \cdots & \cdots & \cdots & \cdots \\ (\alpha^{2t})^{n-1} & (\alpha^{2t})^{n-2} & \cdots & \alpha^{2t} & 1 \end{bmatrix}, \tag{4.54}$$

where $\boldsymbol{H}$ is the parity-check matrix of BCH code. Using the property 2 of GF(q) from the previous section, we conclude that α^i and α^{2i} are the roots of the same minimum polynomial, so that the even rows in $\boldsymbol{H}$ can be omitted to get the final version of the parity-check matrix of BCH codes:

$$H = \begin{bmatrix} \alpha^{n-1} & \alpha^{n-2} & \cdots \alpha & 1 \\ \alpha^{3(n-1)} & \alpha^{3(n-2)} & \cdots \alpha^3 & 1 \\ \alpha^{5(n-1)} & \alpha^{5(n-2)} & \cdots \alpha^5 & 1 \\ \cdots & \cdots & \cdots\cdots & \cdots \\ \alpha^{(2t-1)(n-1)} & \alpha^{(2t-1)(n-2)} & \cdots \alpha^{2t-1} & 1 \end{bmatrix} \tag{4.55}$$

For example, (15,7) 2-error-correcting BCH code has the generator polynomial [35]

$$g(x) = \text{LCM}\left[\phi_\alpha(\text{x}),\ \text{phi}_{\alpha^3}(\text{x})\right] = \text{LCM}\left[\text{x}^4 + \text{x} + 1, \left(\text{x} + \alpha^3\right)\left(\text{x} + \alpha^6\right) \right.$$
$$\left. \times \left(x + \alpha^9\right)\left(x + \alpha^{12}\right)\right] = x^8 + x^7 + x^6 + x^4 + 1$$

and the parity-check matrix [35]

$$H = \begin{bmatrix} \alpha^{14} & \alpha^{13} & \alpha^{12} & \alpha^{14} & \cdots & \alpha & 1 \\ \alpha^{42} & \alpha^{39} & \alpha^{36} & \alpha^{33} & \cdots & \alpha^3 & 1 \end{bmatrix}$$
$$= \begin{bmatrix} \alpha^{14} & \alpha^{13} & \alpha^{12} & \alpha^{11} & \alpha^{10} & \alpha^9 & \alpha^8 & \alpha^7 & \alpha^6 & \alpha^5 & \alpha^4 & \alpha^3 & \alpha^2 & \alpha & 1 \\ \alpha^{12} & \alpha^9 & \alpha^6 & \alpha^3 & 1 & \alpha^{12} & \alpha^9 & \alpha^6 & \alpha^3 & 1 & \alpha^{12} & \alpha^9 & \alpha^6 & \alpha^3 & 1 \end{bmatrix}$$

In the previous expression, we have used the fact that in GF(2^4) $\alpha^{15} = 1$. The primitive polynomial used to design this code was $p(x) = x^4 + x + 1$. Every element in GF(2^4) can be represented as 4-tuple, as shown in Table 4.5.

To create the second column we have used the relation $\alpha^4 = \alpha + 1$, and the 4-tuples are obtained reading off coefficients in second column. By replacing the powers of α in parity-check matrix above by corresponding 4-tuples, the parity-check matrix can be written in following binary form:

Table 4.5 GF(2^4) generated by $x^4 + x + 1$

Power of α	Polynomial of α	4-tuple
0	0	0000
α^0	1	0001
α^1	α	0010
α^2	α^2	0100
α^3	α^3	1000
α^4	$\alpha + 1$	0011
α^5	$\alpha^2 + \alpha$	0110
α^6	$\alpha^3 + \alpha^2$	1100
α^7	$\alpha^3 + \alpha + 1$	1011
α^8	$\alpha^2 + 1$	0101
α^9	$\alpha^3 + \alpha$	1010
α^{10}	$\alpha^2 + \alpha + 1$	0111
α^{11}	$\alpha^3 + \alpha^2 + \alpha$	1110
α^{12}	$\alpha^3 + \alpha^2 + \alpha + 1$	1111
α^{13}	$\alpha^3 + \alpha^2 + 1$	1101
α^{14}	$\alpha^3 + 1$	1001

$$H = \begin{bmatrix} 1 & 1 & 1 & 1 & 0 & 1 & 0 & 1 & 1 & 0 & 0 & 1 & 0 & 0 & 0 \\ 0 & 1 & 1 & 1 & 1 & 0 & 1 & 0 & 1 & 1 & 0 & 0 & 1 & 0 & 0 \\ 0 & 0 & 1 & 1 & 1 & 1 & 0 & 1 & 0 & 1 & 1 & 0 & 0 & 1 & 0 \\ 1 & 1 & 1 & 0 & 1 & 0 & 1 & 1 & 0 & 0 & 1 & 0 & 0 & 0 & 1 \\ 1 & 1 & 1 & 1 & 0 & 1 & 1 & 1 & 1 & 0 & 1 & 1 & 1 & 1 & 0 \\ 1 & 0 & 1 & 0 & 0 & 1 & 0 & 1 & 0 & 0 & 1 & 0 & 1 & 0 & 0 \\ 1 & 1 & 0 & 0 & 0 & 1 & 1 & 0 & 0 & 0 & 1 & 1 & 0 & 0 & 0 \\ 1 & 0 & 0 & 0 & 1 & 1 & 0 & 0 & 0 & 1 & 1 & 0 & 0 & 0 & 1 \end{bmatrix}.$$

In Table 4.6 we listed the parameters of several BCH codes generated by primitive elements of order less than $2^5 - 1$ that are of interest for optical communications. The complete list can be found in Appendix C of [24].

Generally speaking there is no need for q to be a prime, it could be a prime power. However, the symbols must be taken from GF(q), and the roots from GF(q^m). From nonbinary BCH codes, the Reed–Solomon codes are the most famous and these codes are briefly explained in the next section.

The BCH codes can be decoded as any other cyclic codes class. For example, in Fig. 4.15 we provide the error-trapping decoder for BCH (15,7) double-error correcting code. The operation principle of this circuit is already explained in the previous section. Here we explain the decoding process by employing the algorithms especially developed for decoding of BCH codes. Let $g(x)$ be the generator polynomial with corresponding roots $\alpha, \alpha^2, \ldots, \alpha^{2t}$. Let $c(x) = c_0 + c_1x + c_2x^2 + \cdots + c_{n-1}x^{n-1}$ be the codeword polynomial, $r(x) = r_0 + r_1x + r_2x^2 + \cdots + r_{n-1}x^{n-1}$ be the received word polynomial, and $e(x) = e_0 + e_1x + e_2x^2 + \cdots + e_{n-1}x^{n-1}$ be the error polynomial. The roots of generator polynomial are also roots of the codeword polynomial, that is,

$$c(\alpha^i) = 0, \quad i = 0, 1, \ldots, 2t. \tag{4.56}$$

Table 4.6 Primitive binary BCH codes of interest for optical communications

n	k	t	Generator polynomial (in octal form)
15	11	1	23
63	57	1	103
63	51	2	12471
63	45	3	1701317
127	120	1	211
127	113	2	41567
127	106	3	11554743
127	99	4	3447023271
255	247	1	435
255	239	2	267543
255	231	3	156720665
255	223	4	75626641375
255	215	5	23157564726421
255	207	6	16176560567636227
255	199	7	7633031270420722341
255	191	8	2663470176115333714567
255	187	9	52755313540001322236351
255	179	10	22624710717340432416300455

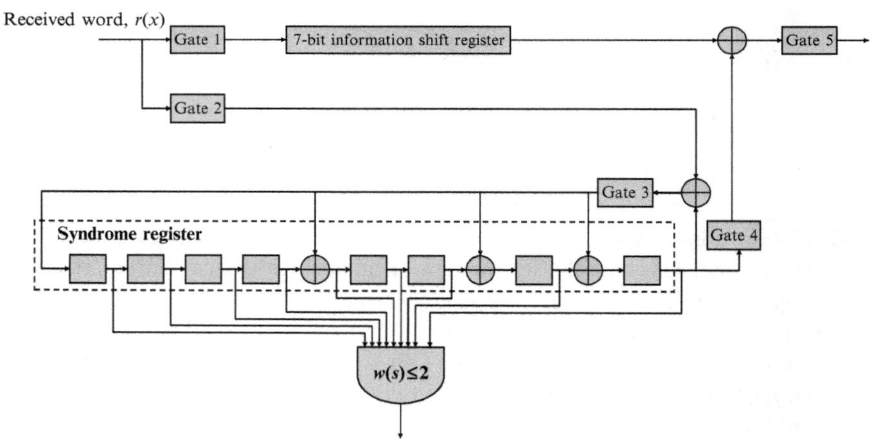

Fig. 4.15 The Error-trapping decoder for (15,7) BCH code generated by $g(x) = 1 + x^4 + x^6 + x^7 + x^8$

For binary BCH codes the only nonzero element is 1, therefore, the indices i of coefficients $e_i \neq 0$ (or $e_i = 1$) determine the error locations. For nonbinary BCH codes the error magnitudes are also important in addition to error locations. By evaluating the received word polynomial $r(x)$ for α^i we obtain

$$r(\alpha^i) = c(\alpha^i) + e(\alpha^i) = e(\alpha^i) = S_i, \tag{4.57}$$

where S_i is the ith component of syndrome vector, defined by

$$\boldsymbol{S} = [S_1\, S_2 \cdots S_{2t}] = \boldsymbol{r}\boldsymbol{H}^{\mathrm{T}}. \tag{4.58}$$

The BCH code is able to correct up to t errors. Let assume that error polynomial $e(x)$ does not have more than t errors, which can then be written as

$$e(x) = e_{j_1}x^{j_1} + e_{j_2}x^{j_2} + \cdots + e_{j_l}x^{j_l} + \cdots + e_{j_v}x^{j_v}, \quad 0 \le v \le t \tag{4.59}$$

e_{j_1} is the error magnitude, α^{j_l} is the error-location number, while j_l is the error location. Notice that the error magnitudes are from symbol field, while the error location numbers are from extension field. The corresponding syndrome components can be obtained from (4.56) and (4.59) as follows:

$$S_i = e_{j_1}(\alpha^i)^{j_1} + e_{j_2}(\alpha^i)^{j_2} + \cdots + e_{j_l}(\alpha^i)^{j_l} + \cdots + e_{j_v}(\alpha^i)^{j_v}, \quad 0 \le v \le t \tag{4.60}$$

In order to avoid the double-indexing let us introduce the following notation $X_l = \alpha^{j_l}, Y_l = e_{j_l}$. The pairs (X_l, Y_l) completely identifies the errors ($l \in [1, v]$). We have to solve the following set of equations:

$$\begin{aligned} S_1 &= Y_1X_1 + Y_2X_2 + \cdots + Y_vX_v \\ S_2 &= Y_1X_1^2 + Y_2X_2^2 + \cdots + Y_vX_v^2 \\ &\ldots \\ S_{2t} &= Y_1X_1^{2t} + Y_2X_2^{2t} + \cdots + Y_vX_v^{2t} \end{aligned} \tag{4.61}$$

The procedure to solve this system of equations represents the corresponding decoding algorithm. Direct solution of this system of equations is impractical. There exists many different algorithms to solve the system of equations (4.61), ranging from iterative to Euclidean algorithms [24, 26–28]. The very popular decoding algorithm of BCH codes is *Massey–Berlekamp algorithm* [24, 26–28]. In this algorithm the BCH decoding is observed as shift register synthesis problem: given the syndromes S_i we have to find the minimal length shift register that generates the syndromes. Once we determine the coefficients of this shift register, we construct the *error locator* polynomial [26, 28]:

$$\sigma(x) = \prod_{i=1}^{v}(1 + X_i x) = \sigma_v x^v + \sigma_{v-1}x^{v-1} + \cdots + \sigma_1 x + 1, \tag{4.62a}$$

where the σ_i's, also known as elementary symmetric functions, are given by Viète's formulas:

$$\begin{aligned} \sigma_1 &= X_1 + X_2 + \cdots + X_v \\ \sigma_2 &= \sum_{i<j} X_i X_j \\ \sigma_3 &= \sum_{i<j<k} X_i X_j X_k \\ &\ldots \\ \sigma_v &= X_1\, X_2 \cdots X_v \end{aligned} \tag{4.62b}$$

Because $\{X_l\}$ are the inverses of the roots of $\sigma(x), \sigma(1/X_l) = 0 \forall\ l$, and we can write [26, 28]:

$$X_l^v \sigma(X_l^{-1}) = X_l^v + \sigma_1 X_l^{v-1} + \cdots + \sigma_v. \quad (4.63)$$

By multiplying the previous equation by X_l^j and performing summation over l for fixed j we obtain

$$S_{v+j} + \sigma_1 S_{v+j-1} + \cdots + \sigma_v S_j = 0, \quad j = 1, 2, \cdots, v \quad (4.64)$$

This equation can be rewritten as follows:

$$S_{v+j} = -\sum_{i=1}^{v} \sigma_i S_{v+j-i}, \quad j = 1, 2, \ldots, v \quad (4.65)$$

S_{v+j} represents the output of shift register shown in Fig. 4.16. The Massey–Berlekamp algorithm is summarized by flowchart shown in Fig. 4.17, which is self-explanatory. Once the locator polynomial is determined, we have to find the roots and invert them to obtain the error locators.

To determine the error magnitudes, we have to define another polynomial, known as the *error-evaluator* polynomial, defined as follows [26]:

$$\varsigma(x) = 1 + (S_1 + \sigma_1)x + (S_2 + \sigma_1 S_1 + \sigma_2)x^2 + \cdots + (S_v + \sigma_1 S_{v-1} + \cdots + \sigma_v)x^v. \quad (4.66)$$

The error magnitudes are then obtained from [26]:

$$Y_l = \frac{\varsigma(X_l^{-1})}{\prod_{i=1, i \neq l}^{v} (1 + X_i X_l^{-1})}. \quad (4.67)$$

Fir binary BCH codes $Y_l = 1$ so that we do not need to evaluate the error magnitudes.

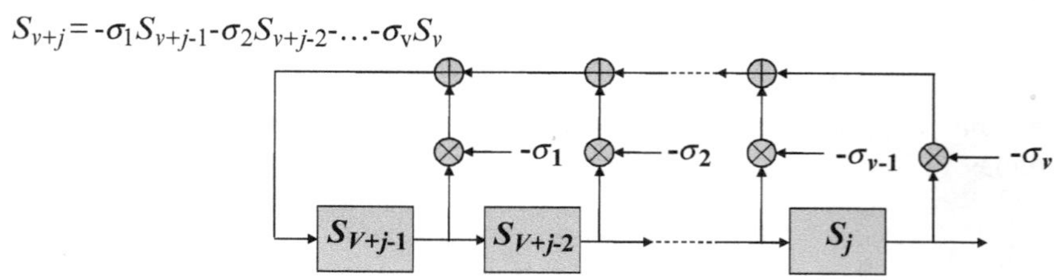

Fig. 4.16 A shift register that generates the syndromes S_j

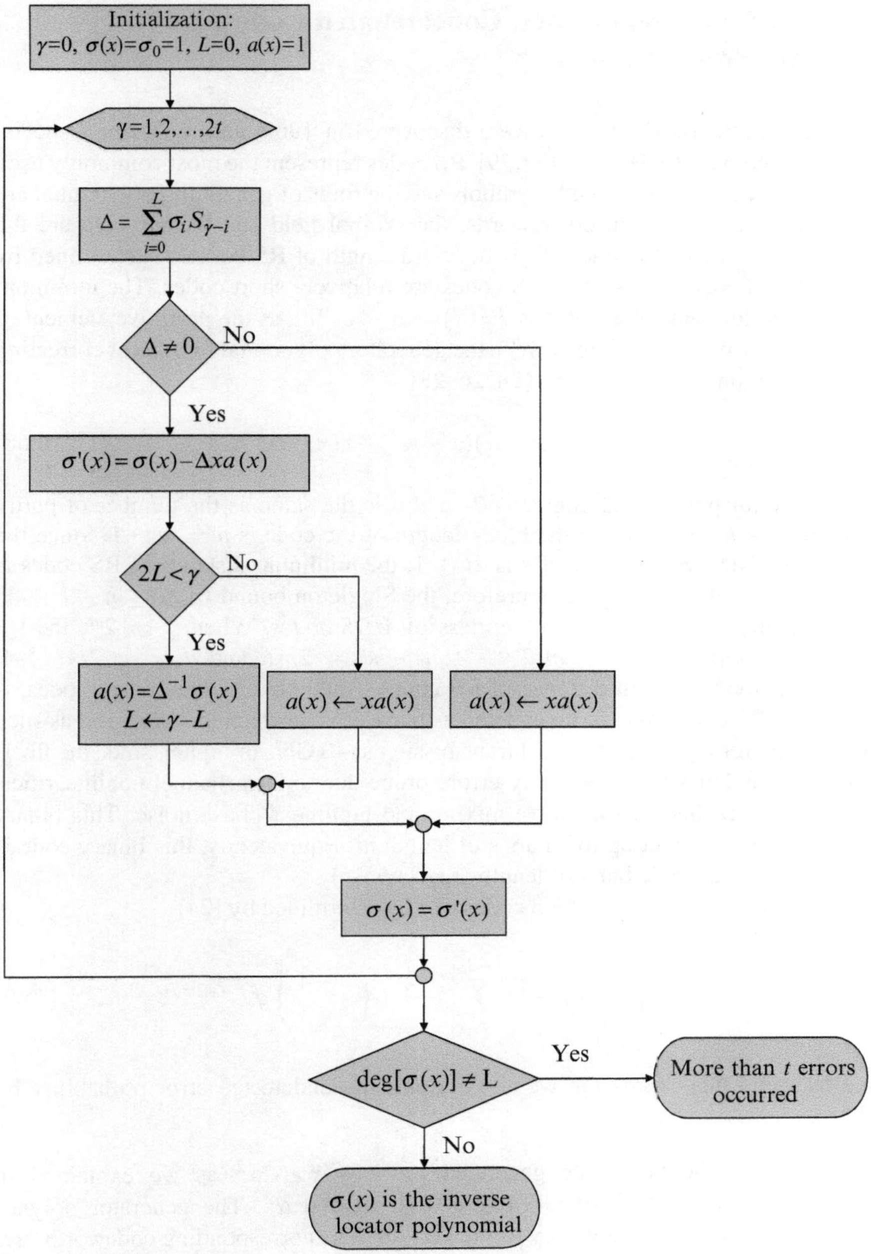

Fig. 4.17 Flowchart of Massey–Berlekamp algorithm. Δ denotes the error (discrepancy) between the syndrome and the shift register output, $a(x)$ stores the content of shift register (normalized by Δ^{-1}) prior to the lengthening

4.5 Reed–Solomon Codes, Concatenated Codes, and Product Codes

The Reed-Solomon (RS) codes were discovered in 1960, and represent a special class of nonbinary BCH codes [38,39]. RS codes represent the most commonly used nonbinary codes. Both the code symbols and the roots of generating polynomial are from the locator field. In other words, the symbol field and locator field are the same ($m = 1$) for RS codes. The codeword length of RS codes is determined by $n = q^m - 1 = q - 1$, so that RS codes are relatively short codes. The minimum polynomial for some element β is $P_\beta(x) = x - \beta$. If α is the primitive element of GF (q) (q is a prime or prime power), the generator polynomial for t-error correcting Reed–Solomon code is given by [24, 26–28]

$$g(x) = (x - \alpha)(x - \alpha^2)\cdots(x - \alpha^{2t}). \tag{4.68}$$

The generator polynomial degree is $2t$ and it is the same as the number of parity symbols $\mathrm{n} - k = 2t$, while the block length of the code is $n = q - 1$. Since the minimum distance of BCH codes is $2t + 1$, the minimum distance of RS codes is $d_{\min} = n - k + 1$, satisfying, therefore, the Singleton bound ($d_{\min} \leq n - k + 1$) with equality and belonging to the class of *MDS codes*. When $q = 2^m$, the RS codes parameters are: $n = m(2^m - 1), n - k = 2mt$, and $d_{\min} = 2mt + 1$. Therefore, the minimum distance of RS codes, when observed as binary codes, is large. The RS codes may be considered as burst-error-correcting codes, and as such are suitable for high-speed optical transmission at 40 Gb/s or higher, since the fiber-optics channel at 40 Gb/s is bursty-errors-prone due to intrachannel nonlinearities, especially intrachannel four-wave mixing and nonlinear phase noise. This binary code is able to correct up to t bursts of length m. Equivalently, this binary code is able to correct a single burst of length $(t - 1)m + 1$.

The weight distribution of RS codes can be determined by [24]

$$A_i = \binom{n}{i}(q - 1)\sum_{j=0}^{i-d_{\min}}(-1)^j\binom{i-1}{j}q^{i-d_{\min}-j}, \tag{4.69}$$

and by using this expression we can evaluate the undetected error probability by (4.38).

Example: Let the GF(4) be generated by $1 + x + x^2$ as we explained in Sect. 4.4.1. The symbols of GF(4) are 0, 1, α, and α^2. The generator polynomial for RS(3,2) code is given by $g(x) = x - \alpha$. The corresponding codewords are: $000, 101, \alpha 0\alpha, \alpha^2 0\alpha^2, 011, 110, \alpha 1\alpha^2, \alpha^2 1\alpha, 0\alpha\alpha, 1\alpha\alpha^2, \alpha\alpha 0, \alpha^2\alpha 1, 0\alpha^2\alpha^2, 1\alpha^2\alpha,$ $\alpha\alpha^2 1$, and $\alpha^2\alpha^2 0$. This code is essentially the even parity-check code ($\alpha^2 + \alpha + 1 = 0$). The generator polynomial for RS(3,1) is $g(x) = (x - \alpha)(x - \alpha^2) = x^2 + x + 1$, while the corresponding codewords are: $000, 111, \alpha\alpha\alpha$, and $\alpha^2\alpha^2\alpha^2$. Therefore, this code is in fact the repetition code.

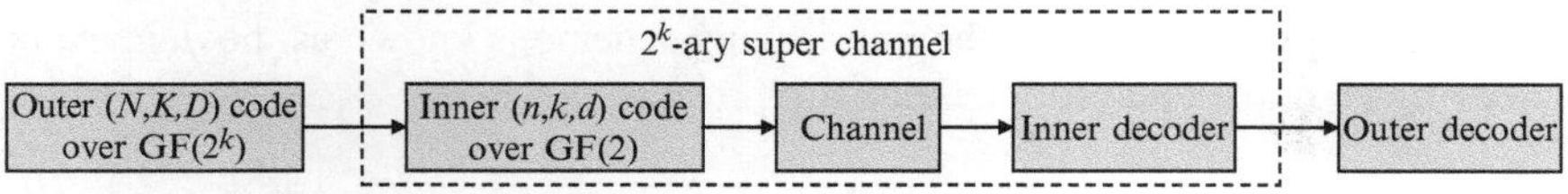

Fig. 4.18 The concatenated $(Nn, Kk, \geq Dd)$code

Since RS codes are special class of nonbinary BCH codes, they can be decoded using the same decoding algorithm already explained in the previous section.

To improve the bust error correction capability of RS codes, RS code can be combined with an inner binary block code in a *concatenation* scheme as shown in Fig. 4.18. The key idea behind the concatenation scheme can be explained as follows [26]. Consider the codeword generated by inner (n, k, d) code (with d being the minimum distance of the code), and transmitted over the bursty channel. The decoder processes the erroneously received codeword and decodes it correctly. However, occasionally the received codeword is decoded incorrectly. Therefore, the inner encoder, the channel and the inner decoder may be considered as a super channel whose input and output alphabets belong to GF(2^k). The outer encoder (N, K, D) (D-the minimum distance of outer code) encodes input K symbols and generates N output symbols transmitted over the super channel. The length of each symbol is k information digits. The resulting scheme, known as concatenated code and proposed initially by Forney [29], is an $(Nn, Kk, \geq Dd)$ code with the minimum distance of at least Dd. For example, RS(255, 239, 8) code can be combined with the (12, 8, 3) single parity-check code in the concatenation scheme $(12 \cdot 255, 239 \cdot 8, \geq 24)$. The outer RS decoder can be implemented using the Massey–Berlekamp algorithm as described above, while the inner decoder can be implemented using the MAP decoding based on BCJR algorithm [36]. The concatenated scheme from Fig. 4.18 can be generalized to q-ary channels, the inner code operating over GF(q), and outer over GF(q^k).

Two RS codes can be combined in a concatenated scheme by interleaving. An *interleaved code* is obtained by taking L codewords (of length N) of a given code $\boldsymbol{x}_j = (x_{j1}, x_{j2}, \ldots, x_{jN}) (j = 1, 2, \ldots, L)$, and forming the new codeword by interleaving the L codewords as follows $\boldsymbol{y}_i = (x_{11}, x_{21}, \ldots, x_{L1}, x_{12}, x_{22}, \ldots, x_{L2}, \ldots, x_{1N}, x_{2N}, \ldots, x_{LN})$. The process of interleaving can be visualized as the process of forming an $L \times N$ matrix of L codewords written row by row and transmitting the matrix column by column, as given below:

$$\begin{matrix} x_{11} & x_{12} & \cdots & x_{1N} \\ x_{21} & x_{22} & \cdots & x_{2N} \\ \cdots & \cdots & \cdots & \\ x_{L1} & x_{L2} & \cdots & x_{LN} \end{matrix}.$$

The parameter L is known as the interleaving *degree*. The transmission must be postponed until L codewords are collected. To be able to transmit a column whenever a new codeword becomes available, the codewords should be arranged down diagonals as given below, and the interleaving scheme is known as the *delayed interleaving* (1-frame delayed interleaving):

$$\begin{array}{llllll}
x_{i-(N-1),1} \cdots & x_{i-2,1} & x_{i-1,1} & x_{i,1} & & \\
 & x_{i-(N-1),2} \cdots & x_{i-2,2} & x_{i-1,2} & x_{i,2} & \\
 & \cdots & \cdots & \cdots & \cdots & \\
 & & x_{i-(N-1),(N-1)} & x_{i-(N-2),(N-1)} & & \\
 & & & x_{i-(N-1),N} & x_{i-(N-2),N} &
\end{array}.$$

Each new codeword completes one column of this array. In the example above the codeword $\boldsymbol{x}_i$ completes the column (frame) $x_{i,1}, x_{i-1,2}, \ldots, x_{i-(N-1),N}$. A generalization of this scheme, in which the components of i-th codeword $\boldsymbol{x}_i$ say $x_{i,j}$ and $x_{i,j+1}$ are spaced λ frames apart, is known as λ-frame delayed interleaved.

Another way to deal with burst errors is to arrange two RS codes in a *product* manner as shown in Fig. 4.19. A product code [3–6] is an $(n_1 n_2, k_1 k_2, d_1 d_2)$ code in which codewords form an $n_1 \times n_2$ array such that each row is a codeword from an (n_1, k_1, d_1) code C_1, and each column is a codeword from an (n_2, k_2, d_2) code C_2; with n_i, k_i, and $d_i\,(i = 1, 2)$ being the codeword length, dimension, and minimum distance, respectively, of ith component code. The product codes were proposed by Elias [25]. Both binary (such as binary BCH codes) and nonbinary codes (such as RS codes) may be arranged in the turbo product manner. It is possible to show [26] that the minimum distance of a product codes is the product of minimum distances of component codes. It is straightforwardly to show that the product code is able to correct the burst error of length $b = \max(n_1 b_2, n_2 b_1)$, where b_i is the burst error capability of component code $i = 1, 2$.

The results of Monte Carlo simulations for different RS concatenation schemes and an AWGN channel are shown in Fig. 4.20. Interestingly, the concatenation scheme RS(255,239) + RS(255,223) of code rate $R = 0.82$ outperforms the concatenation scheme RS(255,223) + RS(255, 223) of lower code rate $R = 0.76$, as well as the concatenation scheme RS(255, 223) + RS(255, 239) of the same code rate.

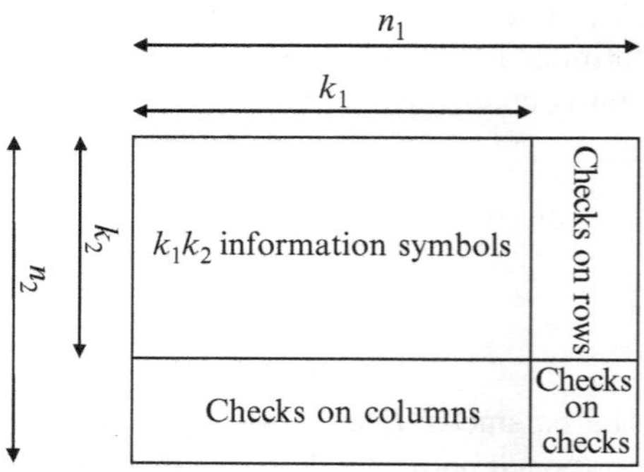

Fig. 4.19 The structure of a codeword of a product code

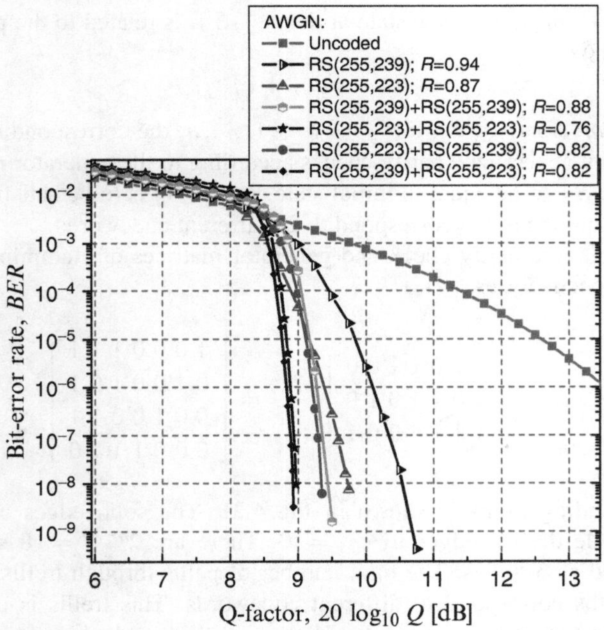

Fig. 4.20 BER performance of concatenated RS codes

4.6 Trellis Description of Linear Block Codes and Viterbi Algorithm

Since the bits in codeword are statistically dependent, the encoding process can be described by the finite state machine, with precisely determined initial and terminal states. We can further describe the encoding by means of trellis [36], and perform maximum likelihood decoding using the Viterbi algorithm [40, 41].

Let linear block code (n, k) be defined over GF(2). The corresponding parity-check matrix of dimension $(n - k) \times n$ can be written as

$$\boldsymbol{H} = [\boldsymbol{h}_1 \, \boldsymbol{h}_2 \cdots \boldsymbol{h}_n], \tag{4.70}$$

where $\boldsymbol{h}_i$ denotes the ith column of parity-check matrix. The codeword vector $c = (c_1 \, c_2 \cdots c_n)$ satisfies the following equation

$$\boldsymbol{c}\boldsymbol{H}^{\mathrm{T}} = 0. \tag{4.71}$$

Let the code be systematic with the first k bits representing the information bits and last $n - k$ bits representing the parity bits. The state at depth j is denoted by $\boldsymbol{S}_j$. The number of states in trellis is determined by 2^{n-k}, each represented in binary form as a vector of length $n - k$. Let the information vector be denoted by

$m = (m_1 m_2 \cdots m_k)$. The new state at depth j, $\boldsymbol{S}_j$, is related to the previous state $\boldsymbol{S}_{j-1}$ by [36]:

$$S_j = S_{j-1} + c_j h_j, \tag{4.72}$$

where $c_j = m_j$(for $j \leq k$). For $j = k + 1, \ldots, n$, the corresponding parity bits are obtained by generalized parity-checks according to the generator matrix $\boldsymbol{G}$. For terminal state $\boldsymbol{S}_n$ to be equal to initial state $\boldsymbol{S}_0$, (4.71) is to be satisfied. Different paths throughout the trellis correspond to 2^k different codewords.

For example, the parity-check and generator matrices of Hamming (7, 4) code are given, respectively, as

$$H = \begin{bmatrix} 1\,1\,0\,1\,1\,0\,0 \\ 1\,0\,1\,1\,0\,1\,0 \\ 1\,1\,1\,0\,0\,0\,1 \end{bmatrix} \quad G = \begin{bmatrix} 1\,0\,0\,0\,1\,1\,1 \\ 0\,1\,0\,0\,1\,0\,1 \\ 0\,0\,1\,0\,0\,1\,1 \\ 0\,0\,0\,1\,1\,1\,0 \end{bmatrix}.$$

The corresponding trellis is shown in Fig. 4.21. The solid edges correspond to $c_j = 1$, while dashed edges to $c_j = 0$. There are $2^{n-k} = 8$ states in trellis represented as 3-tuples. The total number of paths through trellis is $2^k = 16$, and these paths correspond to different codewords. This trellis is created based on (4.72). For example, for $j = 1$ and $c_1 = 0$ the terminal state is obtained by $\boldsymbol{S}_1 = \boldsymbol{S}_0 + 0 \cdot \boldsymbol{h}_1 = (000) + 0 \cdot (111) = (000)$; while for $c_1 = 1$ the terminal state is obtained by $\boldsymbol{S}_1 = \boldsymbol{S}_0 + 1 \cdot \boldsymbol{h}_1 = (000) + 0 \cdot (111) = (111)$. For $j = 2$, initial state $\boldsymbol{S}_1 = (111)$ and $c_2 = 1$ we arrive at terminal state $\boldsymbol{S}_2 = \boldsymbol{S}_1 + 1 \cdot \boldsymbol{h}_2 = (111) + 1 \cdot (101) = (010)$, and so on. After $j = 3$ we can see that trellis is fully developed, there exist 16 paths leading to 8 nodes (with two edges reaching the terminal state). These 16 paths correspond to 16 possible information sequences of length 4. The number of possible transitions after $j = k$ reduces dramatically because the remaining bits c_5, c_6, and c_7 are parity bits, which are algebraically related to the previous information bits. For q-ary linear block

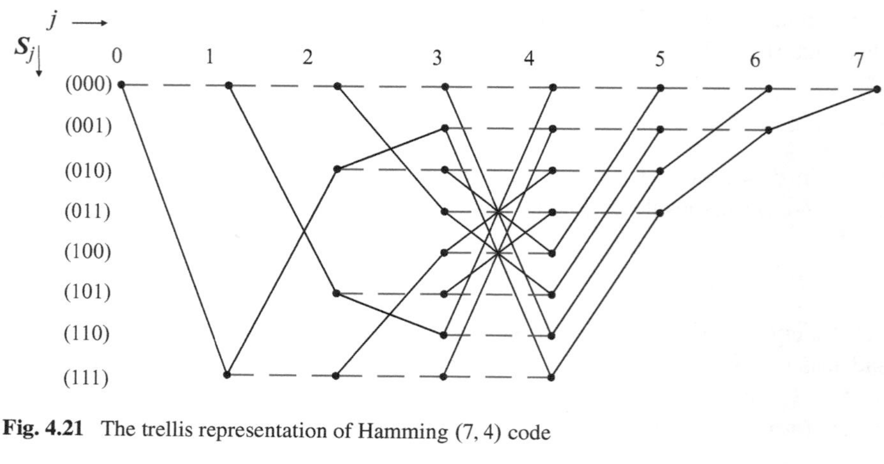

Fig. 4.21 The trellis representation of Hamming (7, 4) code

codes, once the trellis is fully developed, the every node will have q outgoing and q incoming edges. It is interesting to notice that the trellis for liner-block codes is *time-variant*.

The decoding rule for choosing an estimate of the code vector $\boldsymbol{C}$, given the received vector $\boldsymbol{r}$, is optimum when the probability of decoding error is minimized. The maximum likelihood (ML) decoder decision rule can be formulated as follows:

Choose the estimate of $\boldsymbol{C}$ if the log-likelihood function $\log p_{\boldsymbol{R}|\boldsymbol{C}}(\boldsymbol{r}\,|\boldsymbol{c})$ is maximum.

The log-likelihood function, given the transmitted codeword $\boldsymbol{c} = (c_1\, c_2\, \cdots\, c_n)$, assuming that components of receiver word $\boldsymbol{r} = (r_1 r_2 \cdots r_n)$ are statistically independent and noise is additive can be written as follows:

$$\log p_{\boldsymbol{R}|\boldsymbol{C}}(\boldsymbol{r}\,|\boldsymbol{c}) = \sum_{i=1}^{n} \log[p_{N_i}(r_i\,|c_i)] = -\sum_{i=1}^{n} M_i(r_i, c_i) \triangleq -M(\boldsymbol{C}), \tag{4.73}$$

where $p_{\boldsymbol{R}|\boldsymbol{C}}(\boldsymbol{r}\,|\boldsymbol{c})$ is the joint PDF, $p_{N_i}(r_i\,|c_i)$ is the noise PDF, and M_i is the branch metric. The ML decoder provides the codeword $\boldsymbol{C}$ that minimizes the path metric $M(\boldsymbol{C})$. Consider BPSK coherent optical communication system with homodyne balanced detection and ASE noise-dominated scenario. The corresponding PDF function will be:

$$p_{N_i}(r_i\,|c_i) = \frac{1}{\sigma\sqrt{2\pi}} \exp\left\{-\frac{(r_i - c_i 2R_{\mathrm{PD}}\sqrt{P_{\mathrm{S}} P_{\mathrm{LO}}})^2}{2\sigma^2}\right\}, \tag{4.74}$$

where R_{PD} is the photodiode responsivity, P_{s} and P_{LO} are the average powers of incoming and local laser, respectively, and σ^2 is the variance of equivalent noise process dominated by ASE noise. The corresponding log-likelihood function can be obtained by

$$\begin{aligned}\log p_{\boldsymbol{R}|\boldsymbol{C}}(\boldsymbol{r}\,|\boldsymbol{c}) &= -\sum_{i=1}^{n} \frac{(r_i - c_i 2R_{\mathrm{PD}}\sqrt{P_{\mathrm{S}} P_{\mathrm{LO}}})^2}{2\sigma^2} - \frac{n}{2}\log(\pi\sigma^2) \\ &= -A\sum_{i=1}^{n} r_i c_i + B = -A(r \cdot c) + B,\end{aligned} \tag{4.75}$$

where A and B are constants independent on transmitted codeword. The path metric is therefore the inner (scalar) product of the received word and transmitted codeword.

The brute force decoding method would mean to try all possible codewords 2^k and select one that minimizes the path metric $M(\boldsymbol{C}) = \boldsymbol{r} \cdot \boldsymbol{c}$. The better option would be to use *Viterbi algorithm* [24, 26–28, 31, 40, 41]; which is a recursive algorithm, whereby many codewords can be discarded from consideration in determination of $\boldsymbol{c}$ that minimizes the path metric. Alternatively, we can use the square Euclidean metric, which is equivalent to the correlation metric as shown by (4.75). The square Euclidean metric can also be used for ASK transmission, wherein $c_i \in \{0, 1\}$. Below we provide an illustrative example that employs the squared

Euclidean metric and ASK. To keep the exposition simple we perform the following normalization: $2\sigma^2 = 1$, and $2R_{\mathrm{PD}}\sqrt{(P_{\mathrm{s}}P_{\mathrm{LO}})} = 1$; and assume that Hamming (7, 4) code (with corresponding trellis shown in Fig. 4.21) is used. Let the received word be as follows $\boldsymbol{r} = (0.4, 0.2, 0.5, 0.1, 0.2, 0.1, 0.2)$, while the all-zero codeword was transmitted. The first four samples correspond to information bits, and last three to the parity bits. The decoding starts when all branches merge together for the first time, which occurs at trellis depth $j = 4$. The cumulative metric for $(000) \rightarrow (000)$ path at depth $j = 4$, corresponding to the information sequence 0000, is $(0.4-0)^2 + (0.2-0)^2 + (0.5-0)^2 + (0.1-0)^2 = 0.46$; while the cumulative metric for $(000) \rightarrow (000)$ path at depth $j = 4$, corresponding to the information sequence 0111, is $(0.4-0)^2 + (0.2-1)^2 + (0.5-1)^2 + (0.1-1)^2 = 1.86$. Out of two possible paths, the path with lower accumulated metric (0.46) is preserved, and is commonly referred to as the *survivor* path. The other path is discarded from further consideration. The similar procedure is repeated for all state nodes at level $j = 4$, and the survivor paths are denoted as bolded letters in Table 4.7. In Step 2 ($j = 5$), we repeat the similar procedure using the accumulated metric from survived paths as the starting point. When two paths have the same accumulated metric we flip the coin to determine which one to chose without affecting the final result. We see that

Table 4.7 Illustrating the soft Viterbi decoding of Hamming (7,4) code (adopted from [31])

	Step 1	Step 2	Step 3	Step 4
Received samples →	0.4, 0.2, 0.5, 0.1	0.2	0.1	0.2
Paths ↓				
(000)→(000)	0000: **0.46**	0: 0.46+(0.2–0)^2 = **0.5**	0: 0.5+(0.1–0)^2 = **0.51**	0: 0.51+(0.2–0)^2 = **0.55**
	0111: 1.86	1: 0.66+(0.2–1)^2 = 1.3	1: 1.3+(0.1–1)^2 = 2.11	1: 1.31 + (0.2–1)^2 = 1.95
(000)→(001)	1110: **1.26**	0: 1.26+(0.2–0)^2 = **1.3**	0: 1.3+(0.1–0)^2 = 1.31	
	1001: 1.46	1: 1.06+(0.2–1)^2 = 1.7	1: 0.5+(0.1–1)^2 = **1.31**	
(000)→(010)	1100: **1.26**	0: 1.26+(0.2–0)^2 = **1.3**		
	1011: 1.46	1: 1.06+(0.2–1)^2 = 1.7		
(000)→(011)	0010: **0.46**	0: 0.46+(0.2–0)^2 = **0.5**		
	0101: 1.86	1: 0.66+(0.2–1)^2 = 1.3		
(000)→(100)	1101: 2.06			
	1010: **0.66**			
(000)→(101)	0011: 1.26			
	0100: **1.06**			
(000)→(110)	0001: 1.26			
	0110: **1.06**			
(000)→(111)	1111: 2.06			
	1000: **0.66**			

in Step 5 the only survived path is all-zeros path, and the transmitted codeword is properly decoded. We can also introduce a single error, for example, by setting 0.5 sample to 0.6, and repeat the decoding procedure. The all-zero codeword will be detected again. However, when two errors are introduced the Viterbi decoder will fail to determine the correct codeword. On the other hand, by checking the syndrome equation we can detect the double error.

If the probability of error is independent of the transmitted bit value, the corresponding channel is called the binary symmetric channel (BSC). This channel is completely described by transition (crossover) probability: $P\,(r_i = 1|c_i = 0) = P\,(r_i = 0|c_i = 1) = p$. Because

$$p(r_i|c_i) = \begin{cases} p, & r_i \neq c_i, \\ 1-p, & r_i = c_i, \end{cases}$$

the corresponding log-likelihood function can be written as follows:

$$\log p(\boldsymbol{r}|\boldsymbol{c}) = d_H \log p + (n - d_H)\log(1-p) = d_H \log\left(\frac{p}{1-p}\right) + n\log(1-p), \tag{4.76}$$

where d_H is the Hamming distance between the received word $\boldsymbol{r}$ and transmitted codeword $\boldsymbol{c}$. Therefore, for $p < 1/2$ the ML rule for BSC can be formulated as

Choose the estimate C that minimizes the Hamming distance between the received vector r and codeword c.

For intensity modulation with direct detection (IM/DD) we have found in [42] that distribution of PDF tails is exponential rather than Gaussian, so that square root metric is more suitable than Euclidean metric. Some other metrics suitable for IM/DD systems are discussed in [35]. For high-speed long-haul optical communication system, such as those operating at 40 Gb/s and above, the influence of self-phase modulation (intrachannel four-wave mixing in systems with direct detection, and nonlinear phase noise in systems with coherent detection) is important and we have to use the metric given by (4.73), providing that interleaving is used so that neighboring symbols are independent of each other. The PDF has to be estimated by histogram methods. An alternative would be to use the edgeworth expansion method as we described in [44].

So far we were concerned with different classes of linear block-codes. Another important class of codes is based on convolutional codes. Convolutional codes of practical importance are commonly $(n, 1)$ codes, whose code rate is low $1/n$, and as such are not suitable for fiber-optics communication, because the penalty due to chromatic dispersion is highly severe at high-data rates. The code rate can be increased by puncturing, at the expense of performance degradation. On the other hand, the RSC codes are usually employed as component codes for turbo codes that are used in deep-space communications [45]. This is the reason why we devoted the next section to the fundamentals of convolutional codes, so that the potential reader will be able easier to follow the section on turbo codes, introduced in next chapter.

4.7 Convolutional Codes

In a systematic (n, k) linear block code, information bits get unchanged during the encoding process, only $n - k$ parity-check bits are added that are algebraically related to the information bits. If the information bits are statistically independent, the parity (redundant) bits will only be dependent on the information block of current codeword, so that the codewords are statistically independent. On the other hand in an $(n, k\ M)$ convolutional code, the parity bits are not only the function of current information k-tuple, but also the function of previous m information k-tuples. The statistical dependence is therefore introduced in a window of $K = n(M + 1)$ symbols, which is known as the *constraint length.* The constrained length represents the number of coded symbols influenced by a single message symbol. The general architecture of a convolutional code is shown in Fig. 4.22. The k information bits are taken during encoding process from a serial-to-parallel converter (S/P) that follows an information buffer as the input of the encoder memory of size k-by-M. The same bits are also used as the input to the combinational logic, which determines the n-outputs based on current k-tuple and previous Mk bits. The logic outputs are written in parallel to the output shift register, and the codeword is transmitted in a serial fashion over the channel.

An example of nonsystematic convolutional (2,1,2) code is shown in Fig. 4.23. This encoder has the code rate $R = 1/2$, and has $M = 2$ flip-flops as memory elements. Let the input sequence to the encoder be denoted by $\boldsymbol{m} = (m_0\ m_1 \cdots m_l \cdots)$. The corresponding output sequences are denoted with $\boldsymbol{c}^{(1)} = (c_0^{(1)}\ c_1^{(1)} \cdots c_l^{(1)} \cdots)$ and $\boldsymbol{c}^{(2)} = (c_0^{(2)}\ c_1^{(2)} \cdots c_l^{(2)} \cdots)$, respectively. The encoder sequence is obtained by multiplexing two output sequences as follows: $\boldsymbol{c} = (c_0^{(1)}\ c_0^{(2)}\ c_1^{(1)}\ c_1^{(2)} \cdots c_l^{(1)}\ c_l^{(2)} \cdots)$. For convolutional (2,1,2) code shown in Fig. 4.23, the output sequences are related to the input sequence by:

$$c_l^{(1)} = m_l + m_{l-2}, \quad c_l^{(2)} = m_l + m_{l-1} + m_{l-2}.$$

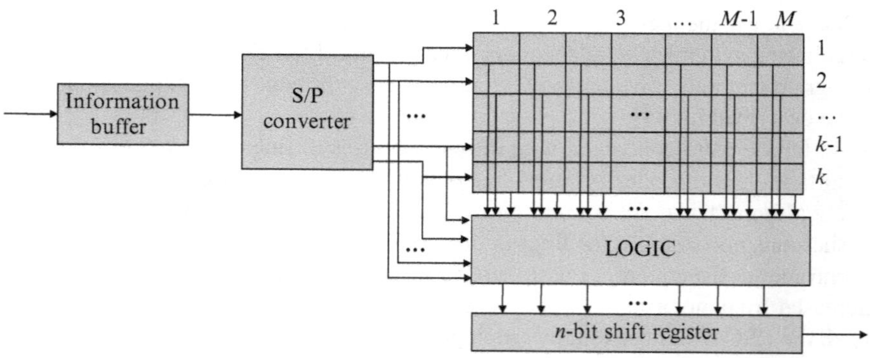

Fig. 4.22 Illustrating the operational principle of convolutional codes

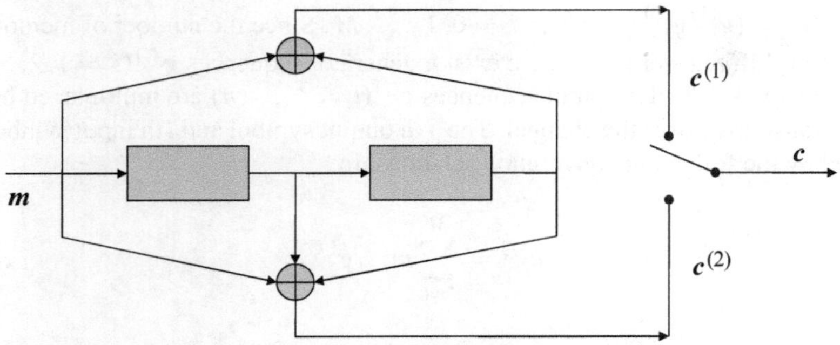

Fig. 4.23 Nonsystematic convolutional (2,1,2) encoder

Another approach to describe the encoding operation is by convolution of the input sequence and corresponding *generator sequences* $\boldsymbol{g}^{(1)} = (101)$ and $\boldsymbol{g}^{(2)} = (111)$ as follows:

$$\boldsymbol{c}^{(1)} = \boldsymbol{m} * \boldsymbol{g}^{(1)}, \quad \boldsymbol{c}^{(2)} = \boldsymbol{m} * \boldsymbol{g}^{(2)},$$

where the generator sequences are obtained as the response to the unit step sequence (100...). The generator sequence, representing the impulse response of corresponding input–output path can also be determined as the sequence of connections from the shift register to the pertinent adder, with a 1 representing the existence of connection and a 0 representing the absence of connection.

The encoding process in convolutional encoder can also be represented in similar fashion to that of the block codes, as matrix product of the input message vector $\boldsymbol{m}$ and generator matrix $\boldsymbol{G}$:

$$\boldsymbol{c} = \boldsymbol{m}\boldsymbol{G}, \quad \boldsymbol{G} = \begin{bmatrix} g_0^{(1)} & g_0^{(2)} & g_1^{(1)} & g_1^{(2)} & g_2^{(1)} & g_2^{(2)} & 0 & 0 & 0 & 0 & \cdots \\ 0 & 0 & g_0^{(1)} & g_0^{(2)} & g_1^{(1)} & g_1^{(2)} & g_2^{(1)} & g_2^{(2)} & 0 & 0 & \cdots \\ 0 & 0 & 0 & 0 & g_0^{(1)} & g_0^{(2)} & g_1^{(1)} & g_1^{(2)} & g_2^{(1)} & g_2^{(2)} & \cdots \\ \vdots & \vdots & & & \ddots & & \ddots & & \vdots & & \vdots \\ 0 & 0 & 0 & 0 & \cdots & g_0^{(1)} & g_0^{(2)} & g_1^{(1)} & g_1^{(2)} & g_2^{(1)} & g_2^{(2)} \end{bmatrix}.$$

The generator matrix of an $(n, 1, m)$ convolutional code, of rate $R = 1/n$, can be represented by [24, 26, 31]:

$$\boldsymbol{G} = \begin{bmatrix} \boldsymbol{G}_0 \, \boldsymbol{G}_1 \cdots \boldsymbol{G}_M & & \\ & \boldsymbol{G}_0 \, \boldsymbol{G}_1 \cdots \boldsymbol{G}_M & \\ & & \boldsymbol{G}_0 \, \boldsymbol{G}_1 \cdots \boldsymbol{G}_M \\ & & \ddots \end{bmatrix}, \tag{4.77}$$

where $\boldsymbol{G}_l = (g_l^{(1)} g_l^{(2)} \ldots g_l^{(n)})$; $l = 0, 1, \ldots, M$. Since the number of memory elements in shift register is M, there exist n generator sequences $\boldsymbol{g}^{(i)} (i = 1, 2, \ldots, n)$ of length $M + 1$, and n output sequences $\boldsymbol{c}^{(l)} (l = 1, \ldots, n)$ are multiplexed before being transmitted over the channel. The jth output symbol and lth input symbol are related by the following convolution relationship:

$$c_l^{(j)} = \sum_{i=0}^{M} m_{l-i} g_i^{(j)}. \tag{4.78}$$

(The summation in (4.78) is to be performed by mod 2.)

By employing the delay operator D (or z^{-1} in digital signal processing books), we can simplify the encoding process and perform the multiplication in transform-domain instead of convolution in time-domain. The transform-domain representations of message sequence $\boldsymbol{m}$ and jth output sequence can, respectively, be written as

$$m(D) = m_0 + m_1 D + m_2 D^2 + \cdots \quad c^{(j)}(D) = c_0^{(j)} + c_1^{(j)} D + c_2^{(j)} D^2 + \cdots \tag{4.79}$$

For (2,1,2) convolutional encoder shown in Fig. 4.23, the corresponding transform-domain representations of generator polynomials can be written as

$$g_1(D) = 1 + D^2 \quad g_2(D) = 1 + D + D^2.$$

The corresponding transform-domain generator matrix $\boldsymbol{G}(D)$ will be

$$\boldsymbol{G}(D) = [g_1(D) \quad g_2(D)] = [1 + D^2 \quad 1 + D + D^2].$$

The output sequences in transform-domain can be written as

$$c^{(1)}(D) = m(D) g_1(D) \qquad c^{(2)}(D) = m(D) g_2(D).$$

The multiplexed sequence in transform-domain is as follows:

$$c(D) = c^{(1)}(D) + D c^{(2)}(D).$$

For an arbitrary $(n, 1, M)$ convolutional code the corresponding generator matrix in transform-domain can be represented as

$$\boldsymbol{G}(D) = [g_1(D) \quad g_2(D) \quad \cdots \quad g_n(D)], \tag{4.80}$$

While the transform-domain of output sequence can be represented by

$$c(D) = m(D)\boldsymbol{G}(D), \quad c(D) = \left[c^{(1)}(D) \quad c^{(2)}(D) \quad \cdots \quad c^{(n)}(D)\right]. \tag{4.81}$$

For an arbitrary (n, k, M) convolutional code, we have to split the input sequence into k sub-sequences $\boldsymbol{m} = [\boldsymbol{m}_1 \boldsymbol{m}_2 \ldots \boldsymbol{m}_k]$, and define the generator matrix as

$$\boldsymbol{G} = \begin{bmatrix} \boldsymbol{G}_0 & \boldsymbol{G}_1 & \cdots & \boldsymbol{G}_M & & & \\ & \boldsymbol{G}_0 & \boldsymbol{G}_1 & \cdots & \boldsymbol{G}_M & & \\ & & \boldsymbol{G}_0 & \boldsymbol{G}_1 & \cdots & \boldsymbol{G}_M & \\ & & & \ddots & & & \end{bmatrix}, \tag{4.82}$$

where

$$\boldsymbol{G}_l = \begin{bmatrix} g_l^{(11)} & g_l^{(12)} & \cdots & g_l^{(1n)} \\ g_l^{(21)} & g_l^{(22)} & \cdots & g_l^{(2n)} \\ \vdots & \vdots & \ddots & \vdots \\ g_l^{(k1)} & g_l^{(k2)} & \cdots & g_l^{(kn)} \end{bmatrix},$$

and proceed in similar fashion as above.

The convolutional codes can also be put in systematic form, by transforming the generator matrix and represent it into the following form:

$$\boldsymbol{G}(D) = [\boldsymbol{I} \quad \boldsymbol{P}(D)], \tag{4.83}$$

where $\boldsymbol{I}$ is $k \times k$ identity matrix, and $\boldsymbol{P}(D)$ is $k \times (n-k)$ matrix of polynomials. The nonsystematic generator matrix of LBCs can easily be transformed in systematic form by elementary operations per rows, while this operation is not always possible to do for the convolutional codes. Namely, the two generator matrices are equivalent if they generate the same convolutional code. Let us again observe the (2,1,2) convolutional code shown in Fig. 4.23, whose generator matrix is $\boldsymbol{G}(D) = [1 + D^2 1 + D + D^2]$. We can perform the following transformation:

$$\begin{aligned} c(D) &= m(D)\boldsymbol{G}(D) = m(D)\left[1 + D^2 \quad 1 + D + D^2\right] \\ &= m(D)(1 + D^2)\left[1 \quad \frac{1 + D + D^2}{1 + D^2}\right] = m'(D)\boldsymbol{G}'(D), \end{aligned}$$

where

$$m'(D) = m(D)\boldsymbol{T}(D), \quad \boldsymbol{T}[D] = [1 + D^2], \quad \boldsymbol{G}'(D) = \left[1 \quad \frac{1 + D + D^2}{1 + D^2}\right].$$

Because both generator matrices, $\boldsymbol{G}$(D) and $\boldsymbol{G}'(D)$, will generate the same output sequence for the same input sequence we can say that that these two matrices are equivalent, which represents a necessary condition. For two generator matrices to be equivalent, the $\boldsymbol{T}$(D) matrix should be invertible, which represents the sufficient condition. In previous example, m(D) can be obtained from $m'(D)$ by

$$m(D) = m'(D)\boldsymbol{T}^{-1}(D), \quad \boldsymbol{T}^{-1}[D] = \left[\frac{1}{1 + D^2}\right].$$

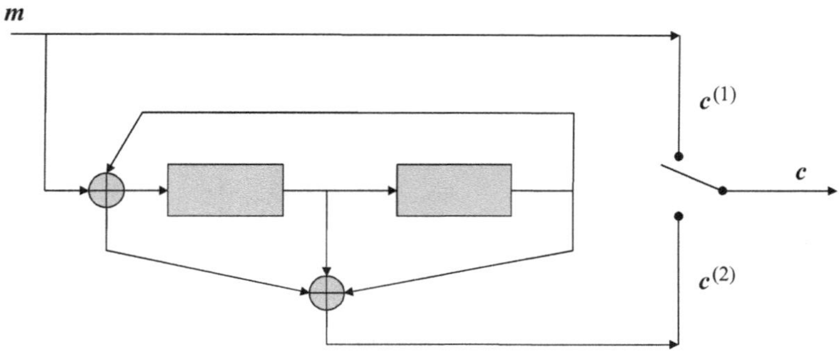

Fig. 4.24 The systematic encoder equivalent to the nonsystematic encoder shown in Fig. 4.23

Therefore, two generator matrices $\boldsymbol{G}(\boldsymbol{D})$ and $\boldsymbol{G}'(\boldsymbol{D}) = \boldsymbol{T}(D)\boldsymbol{G}(D)$ are equivalent if the matrix $\boldsymbol{T}(D)$ is invertible. In the example above, $\boldsymbol{G}'(\boldsymbol{D})$ is in systematic form; unfortunately, its elements are not polynomials but rather rational functions. The parity sequence can be generated by first multiplying the input sequence by $1 + D + D^2$ and then dividing by $1 + D^2$. One possible implementation is shown in Fig. 4.24. Because the response to a single nonzero input will be of infinite length, the codes produced by feedback encoders are known as *recursive convolutional codes* (RCCs) [24]. If an RCC is systematic it is commonly referred to as *recursive systematic convolutional* (RSC) code [37].

Similar to the LBCs, we can define the parity-check matrix in transform domain $\boldsymbol{H}(D)$ (of size $(n-k) \times k$) that satisfies the following equation:

$$\boldsymbol{G}(D)\boldsymbol{H}^{\mathrm{T}}(D) = \boldsymbol{0}. \tag{4.84}$$

The every codeword $c(D)$ satisfies the following equation:

$$c(D)\boldsymbol{H}^{\mathrm{T}}(D) = \boldsymbol{0}. \tag{4.85}$$

If $\boldsymbol{G}(\boldsymbol{D})$ is the generator matrix of a systematic convolutional code, given by (4.83), the corresponding parity-check matrix can be written in following format:

$$\boldsymbol{H}(D) = \begin{bmatrix}\boldsymbol{P}^{\mathrm{T}}(D) & \boldsymbol{I}\end{bmatrix}, \tag{4.86}$$

where $\boldsymbol{I}$ is $(n-k) \times (n-k)$ identity matrix. For example, the parity-check matrix in transform-domain shown in Fig. 4.24 can be written as

$$\boldsymbol{H}(D) = \begin{bmatrix}\dfrac{1 + D + D^2}{1 + D^2} & 1\end{bmatrix}.$$

For nonsystematic codes, the code sequence on receiver side $c(D)$ should be multiplied by $\boldsymbol{G}^{-1}(D)$ to determine the information sequence:

$$c(D)\boldsymbol{G}^{-1}(D) = m(D)\boldsymbol{G}(D)\boldsymbol{G}^{-1}(D) = m(D)D^l, \tag{4.87}$$

which can be done if the following is valid:

$$\boldsymbol{G}(D)\boldsymbol{G}^{-1}(D) = \boldsymbol{I}D^l, \quad l \geq 0 \tag{4.88}$$

where the multiplication with D^l corresponds to the delay. Therefore, the generator matrix must be invertible. If not, the corresponding convolutional code is *catastrophic*, that is the finite number of channel errors will introduce infinite number of decoding errors. The sufficient condition for the convolutional code to be noncatastrophic is that the greatest common divisor (GCD) of all submatrices of size $k \times k$ be equal to D^l. (The number of these submatrices is $\binom{n}{k}$.) The systematic codes cannot be catastrophic because their generator matrix is invertible. Any noncatastrophic convolutional code can transformed into systematic equivalent. For our example shown in Fig. 4.23, the GCD can be found as

$$\text{GCD}(1 + D^2, 1 + D + D^2) = 1 = D^0.$$

Therefore, for the $(n, 1, M)$ codes it is sufficient not to have any common roots among generator polynomials except probable one of the form D^l.

The convolutional codes can be considered as being a subclass of the so-called trellis codes. The trellis codes can be described as finite state machines, with the output being the function of Mk previous bits and k current bits, which output can be described in the form of trellis. If the trellis encoder is time-invariant and linear, the corresponding code is called the convolutional code.

Similarly as in block codes where the error correction capability is related to the minimum Hamming distance and decoding algorithm, the performance of convolutional codes are determined by decoding algorithm and so called *free distance* of the code, d_{free}. The free distance of convolutional codes is defined as the minimum Hamming distance between any two code words in the code. Because the convolutional codes contain all-zero codeword and belong to the class of linear codes, the free distance can be determined as the minimum weight output of codeword sequence caused by a certain nonzero input sequence. The following three algorithms are commonly used in decoding of convolutional codes: (a) majority logic decoding, (b) sequential decoding, and (c) decoding by Viterbi algorithm. The Viterbi algorithm is the most popular and it is already explained in previous section. The only difference is in trellis, which is in case of convolutional codes time-invariant, so that Viterbi decoder of convolutional codes is simpler to implement than that of block codes. The description of other two decoding algorithms can be found in any textbook on channel coding [24, 26–28]. The calculation of free distance of convolutional codes is described in the next section.

4.7.1 Distance Properties of Convolutional Codes

The calculation of free distance of convolutional codes is closely related to the *generating function* of a convolutional code [22, 24, 26], i.e., the transfer function of the encoder with respect to state transitions. We have shown in Fig. 4.22 that convolutional codes can be represented using the finite state machine (FSM) approach, and its operation can be described using the state diagram. The state of convolutional encoder of rate $1/n$ is defined by the sequence of M most recent bits $(m_{j-M}, m_{j-+1}, \ldots, m_{j-1}, m_j)$ moved into encoder's shift register, and it is denoted by $(m_{j-1}, \ldots, m_{j-M})$, where m_j denotes the current bit. For example, for convolutional encoder shown in Fig. 4.25a with $M = 2$ flip-flops there are $2^M = 4$ possible states denoted by a (00), b (10), c (01) and d (11). The corresponding state diagram of this encoder is shown in Fig. 4.25b, c. The FSM states correspond to the

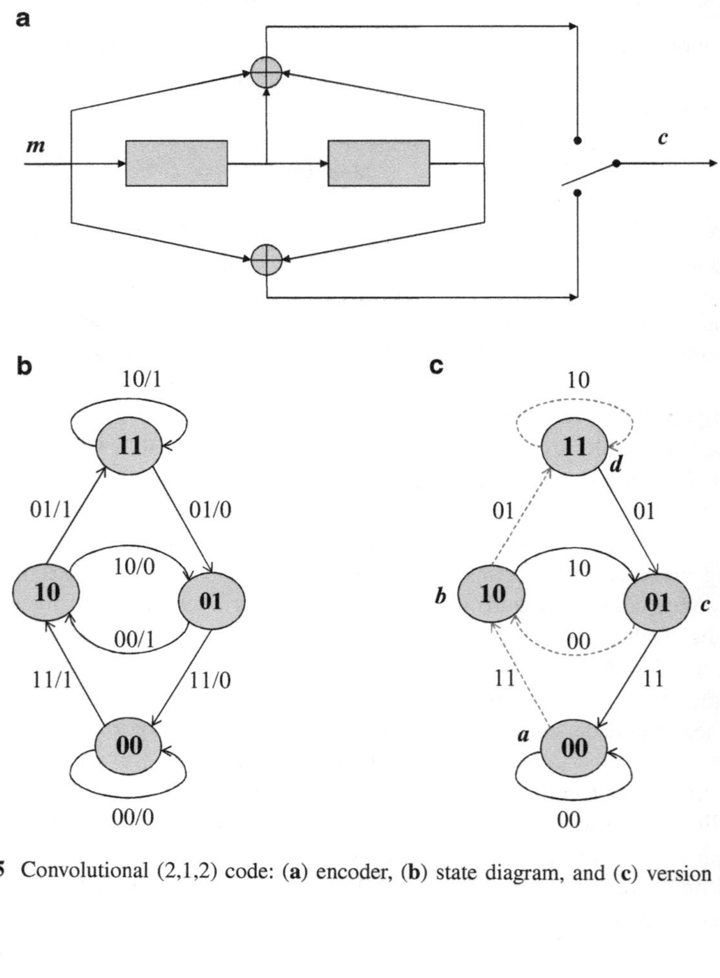

Fig. 4.25 Convolutional (2,1,2) code: (**a**) encoder, (**b**) state diagram, and (**c**) version 2 of state diagram

2-bit states of shift register. The pair of labels above edges, in Fig. 4.25b, represents the outputs of encoder for a given input. For example, if the content of shift register is 10 (we are in state 10) and incoming bit is 1, the generated output sequence is 01, and in the next clock cycle the content of shift register will be 11, meaning that the terminal state is 11. Another equivalent representation is given in Fig. 4.25c, where dashed lines correspond to input bit 1 and the solid lines correspond to input bit 0.

The state diagram can be used as an effective tool to determine the generating function of the code. First of all we have to modify the state diagram as shown in Fig. 4.26. The all-zero state a is split into an initial state a_0 and a final state a_1. The labels of edges are changed as well. The label of each edge is composed of three letters with corresponding exponents. The exponent of I represents the Hamming weight of input (for bit zero we obtain $I^0 = 1$), the exponent of L is the branch length, and the exponent of D represents the Hamming weight of the encoder output corresponding to the input of observed edge. With different paths from initial to final state in modified state diagram we associate the corresponding label by concatenating the edge labels. For example, the path a_0bdca_1 has the label $D^6L^4I^3$. The path starting at a_0 and ending at a_1 is called the fundamental path. If the number of paths starting at a_0 and ending at a_1 with label $D^dL^lI^i$ is denoted by $T_{d,l,i}$ the following generating function will represent the *complete path enumerator* [22, 24, 26]:

$$T(D,L,I) = \sum_{d=1}^{\infty}\sum_{l=1}^{\infty}\sum_{i=1}^{\infty} T_{d,l,i}D^dL^lI^i. \tag{4.89}$$

For example shown in Fig. 4.26, we can determine the $T(D,L,I)$ by using the Mason's gain formula [43] from signal-flaw graph theory to obtain [22]:

$$T(D,L,I) = \frac{D^5L^3I}{1-DLI(1+L)} = D^5L^3I + D^6L^4I^2(1+L) + D^7L^5I^3(1+L)^2 + \cdots \tag{4.90}$$

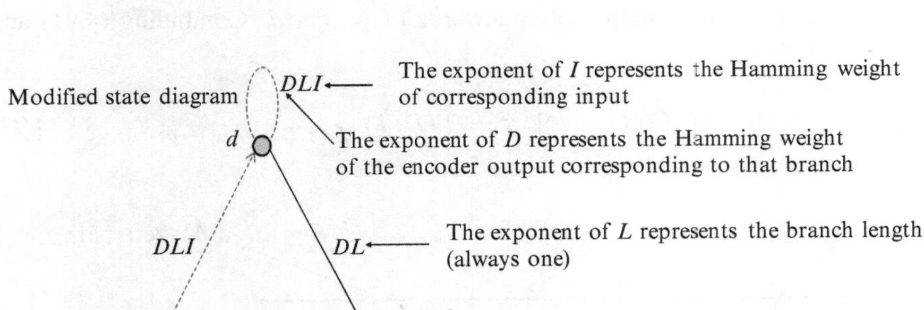

Fig. 4.26 The modified state diagram of convolutional (2,1,2) code from Fig. 4.25a

From (4.90) we conclude that there are no fundamental paths at distance 0, 1, 2, or 3 from the all-zero path; and there is a single fundamental path at distance 5 from all-zero path meaning that the free distance of this convolutional code is $d_{\text{free}} = 5$. This implies that Viterbi decoder can correct up to two errors in decoded sequence.

4.7.2 Bounds on the Bit-Error Ratio of Convolutional Codes

The modified state diagram can also be used to estimate the BERs. Without loss of generality, for an AWGN channel model, we will assume that the all-zero code word was transmitted ($\boldsymbol{c} = \mathbf{0}$). We will say that an *error event* has occur if a nonzero code word survives at state $\mathbf{0}$, therefore eliminating the correct path (all-zero codeword), which is illustrated in Fig. 4.27.

The probability that the first error event E of weight d occurs is given by [26]

$$P(E,d) = \begin{cases} \sum_{i=(d+1)/2}^{d} \binom{d}{i} p^i (1-p)^{d-i}, & d \text{ is odd} \\ \frac{1}{2}\binom{d}{d/2} p^{d/2}(1-p)^{d/2} + \sum_{i=(d/2)+1}^{d} \binom{d}{i} p^i (1-p)^{d-i}, & d \text{ is even} \end{cases} < 2^d p^{d/2}(1-p)^{d/2} \tag{4.91}$$

where p is the crossover probability of BSC. The event error probability upper bound can be determined by [26]

$$P_E < \sum_{d=d_{\text{free}}}^{\infty} T_d P(E,d), \tag{4.92}$$

where T_{d} denotes the number of codewords of weight d. Combining (4.91) and (4.92) we obtain

$$P_E < T_d[4p(1-p)]^{d/2} = T(D,I)\Big|_{I=1,D=\sqrt{4p(1-p)}}. \tag{4.93}$$

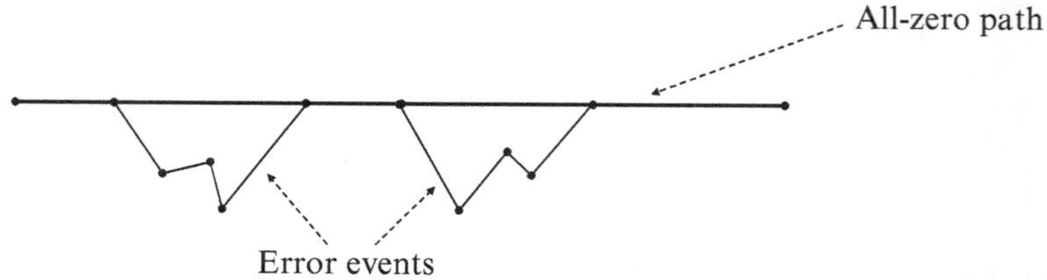

Fig. 4.27 Introducing the concept of error event

The BER of a convolutional code of rate $r = k/n$ is upper bounded by [26]

$$\text{BER} \leq \frac{1}{k} \sum_{d=d_{\text{free}}}^{\infty} \beta_d P(E, d), \tag{4.94}$$

where β_{d} denotes the number of nonzero data bits carried by all T_d paths of weight d. By setting $L = 1$ in (4.89) we obtain

$$T(D, I) = \sum_{d=d_{\text{free}}}^{\infty} \sum_{i=1}^{\infty} T_{d,i} D^d I^i. \tag{4.95}$$

By differentiating $T(D, I)$ with respect to I and by setting $I = 1$ upon differentiation we obtain

$$\left.\frac{\partial T(D, I)}{\partial I}\right|_{I=1} = \sum_{d=d_{\text{free}}}^{\infty} \sum_{i=1}^{\infty} iT_{d,i} D^d = \sum_{d=d_{\text{free}}}^{\infty} \beta_d D^d, \quad \beta_d = \sum_{i=1}^{\infty} iT_{d,i}. \tag{4.96}$$

Finally, by comparing (4.94) and (4.96) we conclude that BER is upper bounded by

$$\text{BER} \leq \frac{1}{k} \left.\frac{\partial T(D, I)}{\partial I}\right|_{I=1} = \frac{1}{k} \sum_{d=d_{\text{free}}}^{\infty} \beta_d D^d. \tag{4.97}$$

For BSC the crossover probability is

$$p = \frac{1}{2} \text{erfc}\left(\sqrt{\frac{E_b}{N_0}}\right) \approx \frac{1}{2\sqrt{\pi}} \exp\left(-\frac{E_b}{N_0}\right), \tag{4.98}$$

where E_b/N_0 is electrical bit energy per power spectral density ratio, and the parameter D from (4.93) can be estimated as

$$D = \sqrt{4p(1-p)} \approx 1.06 \exp\left(-\frac{E_b}{N_0}\right). \tag{4.99}$$

By substituting (4.99) into (4.97) we obtain

$$BER \leq \frac{1}{k} \sum_{d=d_{\text{free}}}^{\infty} \beta_d D^d \approx \frac{1}{k} \sum_{d=d_{\text{free}}}^{\infty} 1.06^d \beta_d \exp\left(-\frac{dRE_b}{2N_0}\right), \tag{4.100}$$

where R is the code rate. For high SNRs only the first term in summation above dominates

$$\mathrm{BER} \approx \frac{1}{k} 1.06^{d_{\mathrm{free}}} \beta_d \exp\left(-\frac{d_{\mathrm{free}} R E_b}{2N_0}\right)\left(\text{for large} \frac{E_b}{N_0}\right). \tag{4.101}$$

The corresponding expression for uncoded BER at high SNR can be estimated by

$$\mathrm{BER}_{\mathrm{unc}} \approx 0.282 \exp\left(-\frac{E_b}{N_0}\right). \tag{4.102}$$

By comparing the exponents in (4.101) and (4.102) we conclude that argument in coded case is $d_{\mathrm{free}} R/2$ times larger, so that corresponding hard decision asymptotic coding gain is

$$G_h = 10 \log_{10}\left(\frac{d_{\mathrm{free}} R}{2}\right) [\mathrm{dB}]. \tag{4.103}$$

For binary input AWGN channel we can repeat the similar procedure and obtain the following asymptotic coding gain for soft decoding

$$G_s = 10 \log_{10}(d_{\mathrm{free}} R)\ [\mathrm{dB}], \tag{4.104}$$

which is 3 dB better than that for hard decision decoding (see (4.103)).

4.8 Summary

Different FEC schemes suitable for use in optical communication systems were classified into two categories, standard block and convolutional codes and codes on graphs. The standard codes are described in this chapter, while the codes on graphs will be described in next chapter. Given the fact that convolutional codes are of low rate, the block codes are described with much more details. Someone may use the puncturing to increase the code of convolutional codes at the expense of BER performance degradation. Nevertheless the convolutional codes are weak, unless they are used in concatenation with RS codes. On the other hand the code rate of this concatenation scheme is too low to be of practical importance for optical communications.

Two classes of standard block codes are described, namely, cyclic codes and BCH codes. The cyclic codes are suitable for use in error detection, while the BCH codes for error correction. The important subclass of BCH codes, RS codes, are standardized for use in optical communications [1]. To deal with simultaneous random errors and burst errors we described the concatenated codes, and interleaved codes. Because the RCCs are used as component codes for turbo codes, and turbo codes are used in deep-space communications, we describe the convolutional codes to a sufficient level so that a potential reader can easily follow the section on turbo codes.

References

1. Forward error correction for submarine systems. Telecommunication Standardization Sector, International Telecommunication Union, Technical Recommendation G.975/G709
2. Forward error correction for high bit rate DWDM submarine systems. Telecommunication Standardization Sector, International Telecommunication Union, G.975.1, Feb. 2004
3. Mizuochi T et al (2004) Forward error correction based on block turbo code with 3-bit soft decision for 10 Gb/s optical communication systems. IEEE/LEOS J Sel Top Quantum Electron 10(2):376–386
4. Mizuochi T et al (2003) Next generation FEC for optical transmission systems. In: Proceedings of the optical fiber communication conference (OFC 2003), vol 2, pp 527–528
5. Sab OA (2001) FEC techniques in submarine transmission systems. In: Proceedings of the optical fiber communication conference, vol 2, pp TuF1-1–TuF1-3
6. Berrou C, Glavieux A, Thitimajshima P (1993) Near Shannon limit error-correcting coding and decoding: Turbo codes. In: Proc 1993 international conference on communication (ICC 1993), vol 2, pp 1064–1070
7. Berrou C, Glavieux A (1996) Near optimum error correcting coding and decoding: turbo codes. IEEE Trans Commun 44(10):1261–1271
8. Pyndiah RM (1998) Near optimum decoding of product codes. IEEE Trans Commun 46: 1003–1010
9. Sab OA, Lemarie V (2001) Block turbo code performances for long-haul DWDM optical transmission systems. In: Proceedings of the optical fiber communication conference, vol 3, pp 280–282
10. Mizuochi T (2006) Recent progress in forward error correction and its interplay with transmission impairments. IEEE/LEOS J Sel Top Quantum Electron 12(4):544–554
11. Gallager RG (1963) Low density parity check codes. MIT Press, Cambridge, MA
12. Djordjevic IB, Sankaranarayanan S, Chilappagari SK, Vasic B (2006) Low-density parity-check codes for 40 Gb/s optical transmission systems. IEEE/LEOS J Sel Top Quantum Electron 12(4):555–562
13. Djordjevic IB, Milenkovic O, Vasic B (2005) Generalized low-density parity-check codes for optical communication systems. IEEE/OSA J Lightwave Technol 23:1939–1946
14. Vasic B, Djordjevic IB, Kostuk R (2003) Low-density parity check codes and iterative decoding for long haul optical communication systems IEEE/OSA J Lightwave Technol 21:438–446
15. Djordjevic IB et al (2004) Projective plane iteratively decodable block codes for WDM high-speed long-haul transmission systems. IEEE/OSA J Lightwave Technol 22:695–702
16. Milenkovic O, Djordjevic IB, Vasic B (2004) Block-circulant low-density parity-check codes for optical communication systems IEEE/LEOS J Sel Top Quantum Electron 10:294–299
17. Vasic B, Djordjevic IB (2002) Low-density parity check codes for long haul optical communications systems. IEEE Photon Technol Lett 14:1208–1210
18. Djordjevic IB, Arabaci M, Minkov L Next generation FEC for high-capacity communication in optical transport networks. IEEE/OSA J Lightwave Technol 27(16):3518–3530 (invited paper)
19. Chung S et al (2001) On the design of low-density parity-check codes within 0.0045 dB of the Shannon Limit. IEEE Commun Lett 5:58–60
20. Djordjevic IB, Minkov LL, Batshon HG (2008) Mitigation of linear and nonlinear impairments in high-speed optical networks by using LDPC-coded turbo equalization. IEEE J Sel Areas Comm, Optical Commun Netw 26(6):73–83
21. Ingels FM (1971) Information and coding theory. Intext Educational, Scranton, PA
22. Haykin S (2004) Communication systems. Wiley, New York
23. Cover TM, Thomas JA (1991) Elements of information theory. Wiley, New York
24. Lin S, Costello DJ (1983) Error control coding: fundamentals and applications. Prentice-Hall, Englewood Cliffs, NJ
25. Elias P (1954) Error-free coding. IRE Trans Inf Theory IT-4:29–37
26. Anderson JB, Mohan S (1991) Source and channel coding: an algorithmic approach. Kluwer Academic, Boston, MA

27. MacWilliams FJ, Sloane NJA (1977) The theory of error-correcting codes. North Holland, Amsterdam, the Netherlands
28. Wicker SB (1995) Error control systems for digital communication and storage. Prentice-Hall, Englewood Cliffs, NJ
29. Forney GD Jr (1966) Concatenated codes. MIT Press, Cambridge, MA
30. Morelos-Zaragoza RH (2002) The art of error correcting coding. Wiley, Boston, MA
31. Drajic DB (2004) (An introduction to information theory and coding, 2nd edn). Akademska Misao, Belgrade, Serbia (in Serbian)
32. Proakis JG (2001) Digital communications. McGaw-Hill, Boston, MA
33. Anderson I (1997) Combinatorial designs and tournaments. Oxford University Press, New York
34. Raghavarao D (1988) Constructions and combinatorial problems in design of experiments. Dover, New York (reprint)
35. Bosco G, Poggiolini P (2006) Long-distance effectiveness of MLSE IMDD receivers IEEE Photon Technol Lett 18(9):1037–1039
36. Bahl LR, Cocke J, Jelinek F, Raviv J (1974) Optimal decoding of linear codes for minimizing symbol error rate IEEE Trans Inf Theory IT-20(2):284–287
37. Ryan WE (2003) Concatenated convolutional codes and iterative decoding. In: Proakis JG (ed) Wiley encyclopedia in telecommunications. Wiley, New York
38. Reed IS, Solomon G (1960) Polynomial codes over certain finite fields. SIAM J Appl Math 8:300–304
39. Wicker SB, Bhargva VK (eds) (1994) Reed-Solomon codes and their applications. IEEE, New York
40. Wolf JK (1978) Efficient maximum likelihood decoding of linear block codes using a trellis. IEEE Trans Inf Theory IT-24(1):76–80
41. Vucetic B, Yuan J (2000) Turbo codes-principles and applications. Kluwer Academic, Boston
42. Ivkovic M, Djordjevic IB, Vasic B (2007) Calculation of achievable information rates of long-haul optical transmission systems using instanton approach. IEEE/OSA J Lightwave Technol 25:1163–1168
43. van Valkenburg ME (1974) Network analysis, 3rd edn. Prentice-Hall, Englewood Cliffs
44. Ivkovic M, Djordjevic I, Rajkovic P, Vasic B (2007) Pulse energy probability density functions for long-haul optical fiber transmission systems by using instantons and edgeworth expansion. IEEE Photon Technol Lett 19(20):1604–1606
45. Divsalar D, Pollara F (1995) Turbo Codes for deep-space communications. TDA progress report 42–120, pp 29–39

Chapter 5
Graph-Based Codes

The 1990s brought about classes of codes that are capable of performance near Shannon's capacity limits for the most common channel models. These code classes include convolutional turbo codes, block turbo codes (BTCs), and low-density parity-check (LDPC) codes [1–6]. Each of these codes can be conveniently modeled by graphs. Such graphical models for codes lead to simplified decoder descriptions which in turn lead to an improved understanding of the performance characteristics of the decoders. A graphical description of a decoder has been well known since around 1970: the Viterbi decoder for decoding convolutional codes is generally described by a graph called a trellis. This chapter gives detailed introductions to parallel and serial convolutional turbo codes, BTCs, and LDPC codes. In discussing the decoding algorithms for each of these code classes, we chose the binary-input additive white Gaussian noise (AWGN) channel model to make our discussions more concrete. This channel model is appropriate for coherent optical communications, but is also often used as an approximation for photodetection receivers. Extensions to other channels require only modified decoder inputs.

5.1 Overview of Graph-Based Codes

A graph is a set of nodes connected by edges. There are several ways to represent a code by a graph, but the one we favor in this chapter is as follows. There are two node types: *bit nodes* (BNs) and *constraint nodes* (CNs); bit nodes are also called *variable nodes* (VNs). Further, bit nodes may only be connected to constraint nodes and vice versa. Such a graph, with two node types and with the restriction that only nodes of different types may be connected, is called a *bipartite graph*. Example bipartite graphs are given in Fig. 5.1, which includes a graph for parallel turbo codes and a graph for LDPC codes.

There is a one-to-one correspondence between the set of all BNs and the set of all bits in a codeword, but we allow selected BNs to be *punctured*, meaning their corresponding code bits are not transmitted. The subset of all BNs connected to a particular CN is permitted to take on only certain bit patterns, as determined by the CN's constraint equations. In this chapter, the CN constraints of interest are those

I. Djordjevic et al., *Coding for Optical Channels*,
DOI 10.1007/978-1-4419-5569-2_5,

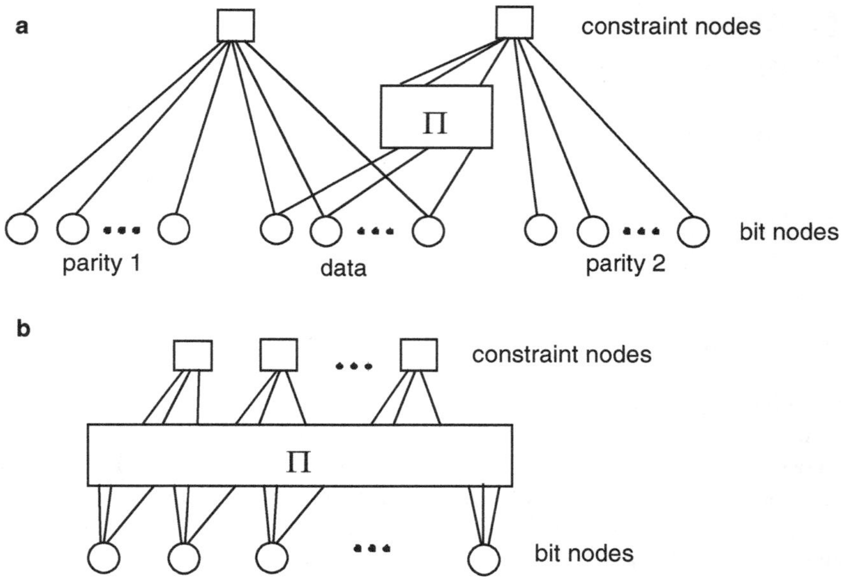

Fig. 5.1 Example bipartite graphs. (**a**) Bipartite graph for a parallel concatenation of codes (constraints are in general any linear code). (**b**) Bipartite graph for an LDPC code (constraints are single parity checks)

of a linear code such as a single-parity-check (SPC) code, a convolutional code, or a general block code. The collection of constraints on the various BN subsets specifies which patterns of bit values constitute codewords.

In the graphical representation of conventional turbo codes, there are two constraint nodes (two constituent codes) each of which are connected to a large number (hundreds or thousands) of BNs. This is demonstrated in Fig. 5.1a for parallel turbo codes whose constituent codes are conventionally convolutional codes. From this, the reader should be able to deduce the graphical representation of serial concatenated code (discussed below). In the graphical representation of an LDPC code, Fig. 5.1b, the CNs represent SPC codes and are typically connected to fewer than about 40 BNs, depending on the code rate and code design. It is occasionally helpful to view the BNs as representations of repetition (REP) codes. From this perspective, an LDPC code graph consists of an interconnection of SPC and REP nodes.

The utility of describing a code by a graph lies in the simplicity with which an iterative decoder may be described and analyzed using the graph. An example of this is the use of the trellis graph in describing the Viterbi algorithm and estimating its performance. The iterative decoders associated with graph-based codes pass messages from node to node in the code's graph and as such are called *message-passing decoders*. The messages being passed within such decoders contain probabilistic information about the values of the code bits associated with the BNs. Details of these decoders are presented in the following sections.

5.2 Convolutional Turbo Codes

Turbo codes, first presented to the coding community in 1993 [1, 2], represent the most important breakthrough in coding since Ungerboeck introduced trellis codes in 1982 [7]. A turbo code encoder comprises a concatenation of two (or more) encoders and its decoder consists of two (or more) "soft" decoders which feed probabilistic information back and forth to each other in a manner that is reminiscent of a turbo engine. This section presents a tutorial exposition of parallel and serial concatenated convolutional codes (PCCCs and SCCCs), which we will also call parallel and serial turbo codes. Also included is a description of their iterative decoders. Later in the chapter we will discuss BTCs.

Figure 5.2 depicts a parallel turbo encoder. As seen in the figure, the encoder consists of two binary rate-1/2 recursive systematic convolutional (RSC) encoders arranged in a so-called *parallel concatenation*, separated by a K-bit pseudorandom interleaver or permuter. Also included is an optional puncturing mechanism to obtain high code rates. Clearly, without the puncturer, the encoder is rate $1/3$, mapping K data bits to $3K$ code bits. With the puncturer, the code rate $R = K/(K + P)$, where P is the number of parity bits remaining after puncturing. As discussed below, the recursive nature of the convolutional encoders is necessary to attain the exceptional performance provided by turbo codes, attributed to so-called interleaver gain. Without any essential loss of generality, we assume that the constituent codes of a PCCC are identical.

Figure 5.3 depicts a serial turbo encoder. As seen in the figure, the serially concatenated convolutional encoders are separated by an interleaver. The inner encoder is required to be an RSC code, whereas the outer encoder need not be recursive. However, RSC inner and outer encoders are occasionally preferred since it is convenient to puncture only parity bits to obtain high code rates. The code rate for the serial turbo encoder is $R = R_i R_o$, where R_i and R_o are the code rates for the inner and outer codes, respectively.

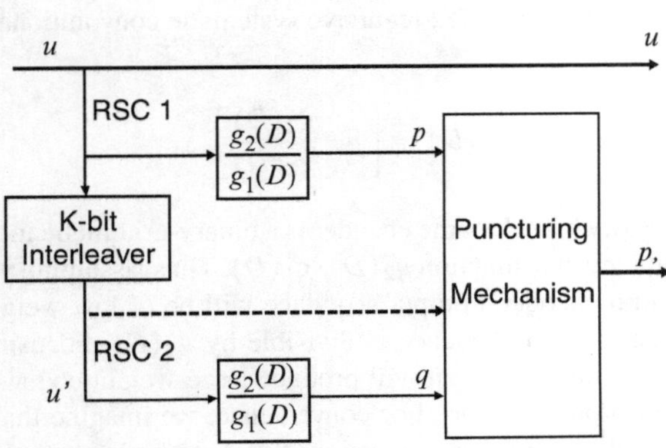

Fig. 5.2 PCCC encoder diagram

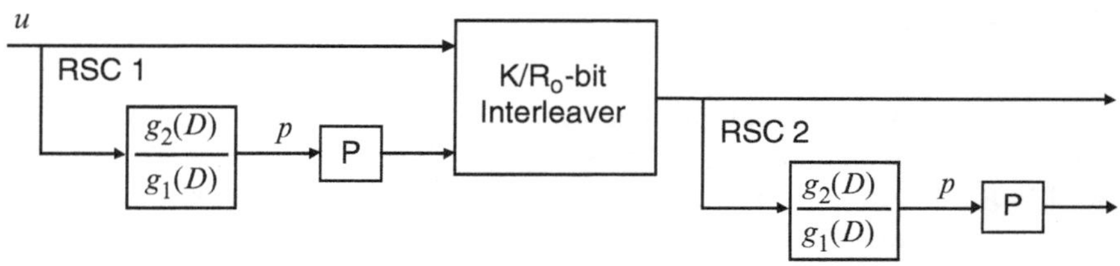

Fig. 5.3 SCCC encoder diagram with RSC component codes. "P" signifies possible puncturing of parity bits

For both parallel and serial turbo codes, the codeword length is $N = K/R$ bits, and we may consider both classes to be (N, K) block codes, particularly since they are generally used as such.

The function of the interleavers in Figs. 5.2 and 5.3 is to take each incoming block of bits and rearrange them in a pseudorandom fashion prior to encoding by the second encoder. For the PCCC, the interleaver permutes K bits and, for the SCCC, the interleaver permutes K/R_o bits. Unlike the classical interleaver (e.g., block or convolutional interleaver), which rearranges the bits in some systematic fashion, it is crucial that this interleaver sort the bits in a manner that lacks any apparent order, although it might be tailored in a certain way for weight-2 and weight-3 inputs as will be made clearer below. The S-random interleaver [8] is quite effective in this regard. This particular interleaver ensures that any two input bits whose positions are within S of each other are separated by an amount greater than S at the interleaver output. S should be selected to be as large as possible for a given value of K. Also, as we shall see, performance increases with K, and so $K \geq 1,000$ is typical.

5.2.1 Performance Characteristics of Parallel and Serial Turbo Codes

The generator matrix for a rate-1/2 recursive systematic convolutional code has the form

$$G(D) = \begin{bmatrix} 1 & \frac{g_2(D)}{g_1(D)} \end{bmatrix}.$$

Observe that the parity branch of the encoder is a binary-arithmetic infinite-impulse-response filter with transfer function $g_2(D)/g_1(D)$. Thus, assuming very long input sequences, the turbo encoder's output sequence will be of low weight if and only if the RSC encoder's input sequence is divisible by $g_1(D)$. Focusing first on the PCCC, a weight-one encoder input will produce large-weight codewords and such codewords are not prone to error. (For convenience, we imagine that the all-zeros codeword is transmitted.) Further, if a weight-two encoder input to the top encoder is divisible by $g_1(D)$, due to the presence of the interleaver, it is unlikely that the input to the bottom encoder will also be divisible. Similar comments hold for inputs of weight three and greater, that is, it is unlikely that the top and bottom encoder will simultaneously produce low-weight codewords. In fact, from a combinatorial perspective, as the encoder input weight increases from two to three, three to four, and so on up to weight-$K/2$, it can be expected that there will be fewer and fewer low-weight codewords at the PCCC encoder output. Because the inner encoder is required to be a RSC code, a similar discussion holds for SCCCs. Note because the input to the inner encoder is an outer code codeword, we can expect the minimum distance of an SCCC to be larger than that of a PCCC (all things being equal).

Although the iterative decoder employed for turbo codes is not maximum-likelihood (ML), it has approximately maximum-likelihood performance. An estimate of the bit-error probability of an ML decoder for turbo codes on the binary-input AWGN channel with power spectral density $N_0/2$ is

$$P_b \simeq \sum_{w=1}^{K} \sum_{v=1}^{\binom{K}{w}} \frac{w}{K} Q\left(\sqrt{\frac{2Rd_{wv}E_b}{N_0}}\right), \tag{5.1}$$

where the first sum is over the weight-w inputs, the second sum is over the $\binom{K}{w}$ different weight-w inputs, and d_{wv} is the weight of the vth codeword produced by a weight-w input. We emphasize that this expression holds for both PCCCs and SCCCs and, in fact, for any linear code. However, what distinguishes PCCCs and SCCCs from other linear codes is the fact that d_{wv} is the smallest for w small (say, $w = 1, 2$, and 3) so that the first few terms in the summation above dominates. In this case, we may write

$$P_b \simeq \sum_{w=1}^{3} \frac{wn_w}{K} Q\left(\sqrt{\frac{2Rd_{w,\min}E_b}{N_0}}\right), \tag{5.2}$$

where $d_{w,\min}$ is the minimum weight of codewords produced by weight-w inputs and n_w is the number of such codewords. Clearly, the $w = 1$ term can be omitted for PCCCs and for SCCCs whose inner code is recursive. Due to the presence of the interleaver, $wn_w \ll K$, and we say that the factor wn_w/K in (5.2) corresponds to *interleaver gain.* If codeword error rate, P_{cw}, is the preferred performance metric, then an estimate of P_{cw} may be obtained from (5.2) by removing the factor w/K. (P_{cw} is also called frame-error rate, FER.)

Example. We consider in this example a PCCC and an SCCC code, both rate 1/2 with parameters $(N, K) = (2048, 1024)$. We use identical four-state RSC encoders in the PCCC encoder whose generators polynomials are, in octal form, $(g_1, g_2) = (7, 5)$, that is, $g_1(D) = 1 + D + D^2$ and $g_2(D) = 1 + D^2$. The outer constituent encoder in the SCCC encoder is this same four-state RSC encoder, and the inner code is a rate-1 differential encoder with transfer function $1/(1 + D)$. The PCCC interleaver is a 1,024-bit S-random interleaver with $S = 16$ and the SCCC interleaver is a 2,048-bit S-random interleaver with $S = 20$. The results are presented

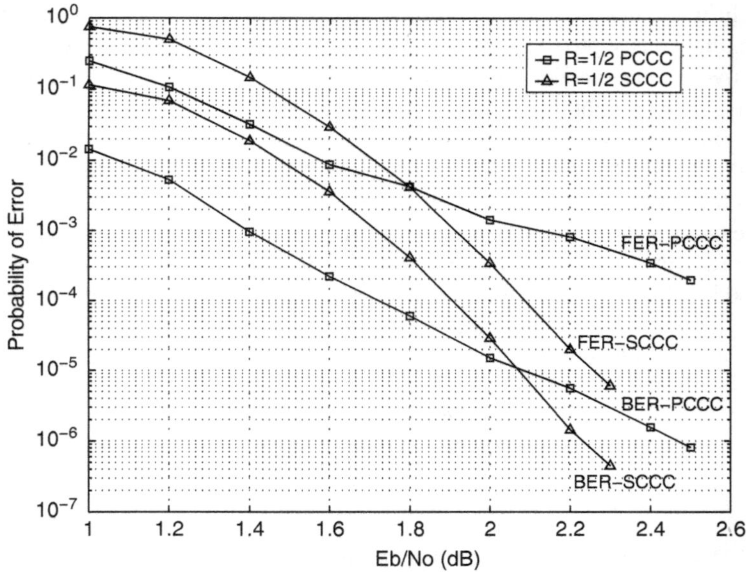

Fig. 5.4 Rate-1/2 PCCC and SCCC bit error rate (BER) and frame error rate (FER) simulation results

in Fig. 5.4 where we observe that the SCCC has a much lower error rate floor, particularly for the FER curves, which we attribute to a larger minimum distance. This floor is at the expense of inferior performance in the waterfall region. This example helps explain the "floor" effect of the error rate curves: it may be interpreted as the usual Q-function shape for a signaling scheme with a modest $d_{\min}$, "pushed down" by the interleaver gain $w^* n_{w^*}/K$, where w^* is the value of w corresponding to the dominant term in (5.2).

5.2.2 The PCCC Iterative Decoder

The key idea behind graph-based codes is that, because a graph-based code consists of an interconnection of two or more *component codes*, it may be decoded (near-optimally) using an interconnection of component decoders matched to the component codes. Thus, multiple low-complexity decoders operating in cooperation replace a single extremely high complexity decoder, with little loss in performance. Further, since a graph-based code comprises multiple component codes, its performance can surpass that of a single code, and can in fact operate near channel capacity. The decoder for a graph-based code is called a turbo decoder or a message-passing decoder, depending on the code, and the component decoders are soft-in/soft-out (SISO) decoders that share probabilistic information cooperatively and iteratively.

The goal of the decoder is to iteratively estimate the a posteriori probabilities (APPs) $\Pr(u_k \mid \mathbf{y})$ where u_k is the kth data bit, $k = 1, 2, \ldots, K$, and $\mathbf{y}$ is the received codeword in noise, $\mathbf{y} = \mathbf{c} + \mathbf{n}$. In this equation, we assume that the components of $\mathbf{c}$ take values in the set $\{\pm 1\}$ (and similarly for $\mathbf{u}$) and $\mathbf{n}$ is a noise word whose components are AWGN samples distributed as $\mathcal{N}(0, \sigma^2)$, with $\sigma^2 = N_0/2$. Knowledge of the APPs allows for optimal decisions on the bits u_k via the maximum a posteriori (MAP) rule[1]

$$\frac{P(u_k = +1 \mid \mathbf{y})}{P(u_k = -1 \mid \mathbf{y})} \underset{-1}{\overset{+1}{\gtrless}} 1$$

or, more conveniently,

$$\hat{u}_k = \text{sign}\left[L(u_k)\right], \tag{5.3}$$

where $L(u_k)$ is the log a posteriori probability (log-APP) ratio defined as

$$L(u_k) \triangleq \log\left(\frac{P(u_k = +1 \mid \mathbf{y})}{P(u_k = -1 \mid \mathbf{y})}\right).$$

We shall use the term log-likelihood ratio (LLR) in place of log-APP ratio for consistency within the chapters. The component SISO decoders compute the LLRs for component code inputs (u_k), component code outputs (c_k), or both.

We introduce the convention that, for PCCCs, the top component encoder is encoder 1 (denoted E1) and the bottom component encoder is encoder 2 (denoted E2). For SCCCs, the outer encoder is encoder 1 (E1) and the inner encoder is encoder 2 (E2). The SISO component decoders matched to E1 and E2 will be denoted by D1 and D2, respectively. Because the SISO decoders D1 and D2 compute $L(u_{ik})$ and/or $L(c_{ik})$, $i = 1, 2$, we will temporarily use the notation $L(b_k)$ where b_k represents either u_{ik} or c_{ik}.

From Bayes' rule, the LLR for an arbitrary SISO decoder can be written as

$$L(b_k) = \log\left(\frac{P(\mathbf{y} \mid b_k = +1)}{P(\mathbf{y} \mid b_k = -1)}\right) + \log\left(\frac{P(b_k = +1)}{P(b_k = -1)}\right) \tag{5.4}$$

with the second term representing a priori information. Since typically $P(b_k = +1) = P(b_k = -1)$, the a priori term is usually zero for conventional decoders. However, for *iterative* decoders (also called *turbo decoders*), each component decoder receives *extrinsic* information for each b_k from its companion decoder which serves as a priori information. The idea behind extrinsic information is that D2 provides soft information to D1 for each b_k using only information not available to D1, and D1 does likewise for D2. For SCCCs, the iterative decoding proceeds as D2→D1→D2→D1→ ⋯, with the previous decoder passing extrinsic information

[1] The MAP rule minimizes the probability of bit error. By contrast, the ML rule, which maximizes the likelihoods $P(\mathbf{y} \mid \mathbf{c})$ over the codewords $\mathbf{c}$, minimizes the probability of codeword error.

along to the next decoder at each half-iteration. For PCCCs, either decoder may initiate the chain of component decodings or, for hardware implementations, D1 and D2 may operate simultaneously. This type of iterative algorithm is known to converge to the true value of the LLR $L(u_k)$ for the concatenated code provided that the graphical representation of this code contains no loops [9–11]. The graph of a turbo code does in fact contain loops [11], but the algorithm nevertheless provides near-optimal performance for virtually all turbo codes.

In the context of PCCCs and SCCCs, each component decoder is a SISO decoder of a single RSC code that uses the BCJR algorithm [12] for computing $L(b_k)$. We now describe the BCJR algorithm after which we describe how it is used in the turbo decoding of PCCCs and SCCCs.

5.2.2.1 The BCJR Algorithm

We consider the BCJR algorithm for decoding a single rate-1/2 RSC code on an AWGN channel. Thus, the transmitted codeword $\mathbf{c}$ will have the form $\mathbf{c} = [c_1, c_2, \ldots, c_K] = [u_1, p_1, u_2, p_2, \ldots, u_K, p_K]$ where $c_k \triangleq [u_k, p_k]$. The received word $\mathbf{y} = \mathbf{c} + \mathbf{n}$ will have the form $\mathbf{y} = [y_1, y_2, \ldots, y_K] = [y_1^u, y_1^p, y_2^u, y_2^p, \ldots, y_K^u, y_K^p]$ where $y_k \triangleq [y_k^u, y_k^p]$ and similarly for $\mathbf{n}$. As above, we assume that our binary variables take values from the set $\{\pm 1\}$.

The BCJR algorithm computes the LLR

$$L(b_k) = \log\left(\frac{P(b_k = +1 \mid \mathbf{y})}{P(b_k = -1 \mid \mathbf{y})}\right)$$

given the received word $\mathbf{y}$, where b_k represents u_k or p_k, depending on the situation. The BCJR algorithm consists primarily of a forward-going and a backward-going Viterbi-like algorithm whose computations are used to arrive at $L(b_k)$. Specifically, we (recursively) define the *forward metric* at state s at time k as

$$\alpha_k(s) \triangleq \max_{s'}{}^* \left[\alpha_{k-1}(s') + \gamma_k(s', s)\right], \tag{5.5}$$

where s' ranges over all possible states at time $(k - 1)$ and where $\max^*$ for two variables is defined as

$$\begin{aligned} \max{}^*(x, y) &\triangleq \log(\mathrm{e}^x + \mathrm{e}^y) \\ &= \max(x, y) + \log\left(1 + \mathrm{e}^{-|x-y|}\right). \end{aligned}$$

This function extends to more than two variables so that, for example, $\max^*(x, y, z) \triangleq \log(\mathrm{e}^x + \mathrm{e}^y + \mathrm{e}^z) = \max^*[\max^*(x, y), z]$. The *branch metric* $\gamma_k(s', s)$ in (5.5) is given by

$$\gamma_k(s', s) = -\frac{\|y_k - c_k\|^2}{2\sigma^2}. \tag{5.6}$$

Note that (5.5) not only defines $\alpha_k(s)$, but also gives its recursion. These forward metrics are initialized as

$$\alpha_0(s) = \begin{cases} 0, & s = 0 \\ -\infty, & s \neq 0 \end{cases}. \tag{5.7}$$

We also define the *backward metric* at state s' at time $(k-1)$ to be

$$\beta_{k-1}(s') \triangleq \max_s{}^* \left[\beta_k(s) + \gamma_k(s', s)\right] \tag{5.8}$$

with initial conditions

$$\beta_K(s) = \begin{cases} 0, & s = 0 \\ -\infty, & s \neq 0 \end{cases} \tag{5.9}$$

under the assumption that the encoder has been terminated to the zero state (by appending appropriate bits to the data word). If the encoder has not been terminated, then $\beta_K(s)$ can be set to the same constant for all s.

Once the forward and backward metrics have been computed, $L(b_k)$ can be computed as

$$\begin{aligned} L(b_k) = & \max_{B+}{}^* \left[\alpha_{k-1}(s') + \gamma_k(s', s) + \beta_k(s)\right] \\ & - \max_{B-}{}^* \left[\alpha_{k-1}(s') + \gamma_k(s', s) + \beta_k(s)\right], \end{aligned} \tag{5.10}$$

where B^+ is the set of state transitions ($s' \to s$) for which $b_k = +1$ and similarly for B^-. Figure 5.5 illustrates pictorially the trellis-based computations that the three equations (5.5), (5.8), and (5.10) represent where we see that analogy with the Viterbi algorithm. We point out that the implementation of the $\max^*(\cdot)$ function involves only a two-input $\max(\cdot)$ function plus a lookup table for the "correction term" $\log\left(1 + e^{-|x-y|}\right)$. Robertson et al. [13] have shown that a table size of eight is usually sufficient. Because the correction term is upper bounded by 0.693, $\max^*(x, y)$ can be approximated by $\max(x, y)$, with some performance degradation, and the recursions (5.5) and (5.8) become forward and reverse Viterbi algorithms, respectively.

5.2.2.2 The Iterative Decoder

We present in this section the iterative decoder for a PCCC consisting of two component rate-1/2 RSC encoders concatenated in parallel. We assume no puncturing so that the overall code rate is 1/3. Block diagrams of the PCCC encoder and its iterative decoder with component SISO decoders are presented in Fig. 5.6. As indicated in Fig. 5.6a, the transmitted codeword $\mathbf{c}$ will have the form $\mathbf{c} = [c_1, c_2, \ldots, c_K] = [u_1, p_1, q_1, \ldots, u_K, p_K, q_K]$ where $c_k \triangleq [u_k, p_k, q_k]$. The received word $\mathbf{y} = \mathbf{c} + \mathbf{n}$ will have the form $\mathbf{y} = [y_1, y_2, \ldots, y_K] = [y_1^u, y_1^p, y_1^q, \ldots, y_K^u, y_K^p, y_K^q]$ where $y_k \triangleq [y_k^u, y_k^p, y_k^q]$ and similarly for $\mathbf{n}$. We denote the codewords produced

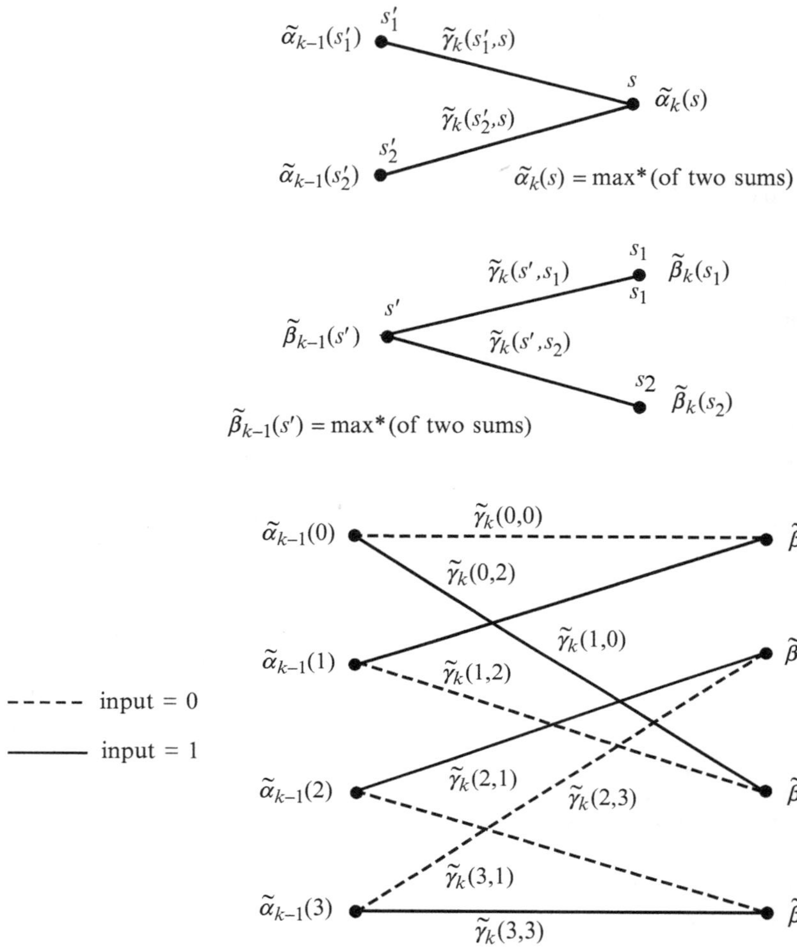

$$L(b_k) = \max{}^*\{\tilde{\alpha}_{k-1}+\tilde{\gamma}_k+\tilde{\beta}_k \text{ for solid lines}\} - \max{}^*\{\tilde{\alpha}_{k-1}+\tilde{\gamma}_k+\tilde{\beta}_k \text{ for dashed lines}\}$$

Fig. 5.5 The *top diagram* depicts the forward recursion in (5.5), the *middle diagram* depicts the backward recursion in (5.8), and the *bottom diagram* depicts the computation of $L(b_k)$ via (5.10)

by E1 and E2 by, respectively, $\mathbf{c}_1 = [c_1^1, c_2^1, \ldots, c_K^1]$ where $c_k^1 \triangleq [u_k, p_k]$ and $\mathbf{c}_2 = [c_1^2, c_2^2, \ldots, c_K^2]$ where $c_k^2 \triangleq [u'_k, q_k]$. Note that $\{u'_k\}$ is a permuted version of $\{u_k\}$ and is not actually transmitted (see Fig. 5.6a). We define the noisy received versions of $\mathbf{c}_1$ and $\mathbf{c}_2$ to be $\mathbf{y}_1$ and $\mathbf{y}_2$, respectively, having components $y_k^1 \triangleq [y_k^u, y_k^p]$ and $y_k^2 \triangleq [y_k^{u'}, y_k^q]$, respectively. Note that $\mathbf{y}_1$ and $\mathbf{y}_2$ can be assembled from $\mathbf{y}$ in an obvious fashion (using an interleaver to obtain $\left\{y_k^{u'}\right\}$ from $\{y_k^u\}$). By doing so, the component decoder inputs are the two vectors $\mathbf{y}_1$ and $\mathbf{y}_2$ as indicated in Fig. 5.6b.

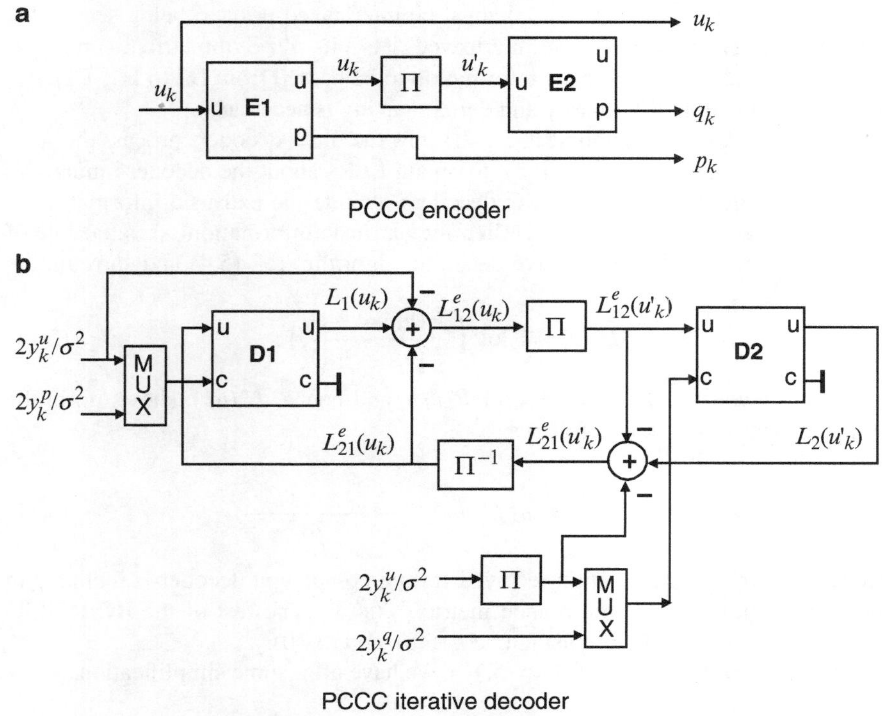

Fig. 5.6 Block diagrams for the PCCC encoder and iterative decoder

In contrast to the BCJR decoder of Sect. 5.2.2.1 whose input was $\mathbf{y} = \mathbf{c} + \mathbf{n}$ and whose output was $\{L(u_k)\}$ (or $\{\hat{u}_k\}$), the SISO decoders in Fig. 5.6b possess two inputs and two outputs. The SISO decoders are essentially the BCJR decoders discussed above, except the SISO decoders have the ability to accept from a companion decoder extrinsic information about its encoder's input (SISO input label "u") and/or about its encoder's output (SISO input label "c"). The SISO decoders also have the ability to produce likelihood information about its encoder's input (SISO output label "u") and/or about its encoder's output (SISO output label "c"). Note that the SISO decoder is to be interpreted as a decoding module not all of whose inputs or outputs need be used [6]. (Note the RSC encoders in Fig. 5.6a have also been treated as modules.) As we will see, the SISO modules are connected in a slightly different fashion for the SCCC case.

Note from Fig. 5.6b that the extrinsic information to be passed from D1 to D2 about bit u_k, denoted $L_{12}^{\mathrm{e}}(u_k)$, is equal to the LLR $L_1(u_k)$ produced by D1 minus the channel likelihood $2y_k^u/\sigma^2$ and the extrinsic information $L_{21}^{\mathrm{e}}(u_k)$ that D1 had just received from D2. The idea is that $L_{12}^{\mathrm{e}}(u_k)$ should indeed be extrinsic (and uncorrelated) with the probabilistic information already possessed by D2. As we will see, $L_{12}^{\mathrm{e}}(u_k)$ is strictly a function of received E1 parity $\{y_k^p\}$ which is not directly sent to D2. Observe that $\{L_{12}^{e}(u_k)\}$ must be interleaved prior to being sent to D2 since E2 and D2 operate on the interleaved data bits u'_k. Symmetrical comments may be made about the extrinsic information to be passed from D2 to D1, $L_{21}^{e}(u_k)$ (e.g., it is a function of E2 parity and deinterleaving is necessary).

We already know from the Sect. 5.2.2.1 how the SISO decoders process the samples from the channel, $\mathbf{y}_i$ $(i = 1, 2)$, to obtain LLRs about the decoder inputs. We now need to discuss how the SISO decoders include the extrinsic information in their computations. As indicated earlier, the extrinsic information takes the role of a priori information in the iterative decoding algorithm [cf. (5.4) and surrounding discussion],

$$L^{e}(u_k) \triangleq \log\left(\frac{P(u_k = +1)}{P(u_k = -1)}\right). \tag{5.11}$$

It can be shown that the a priori term $P(u_k)$, and hence, $L^{e}(u_k)$, shows up in the branch metric expression as follows:

$$\gamma_k(s', s) = u_k L^{e}(u_k)/2 - \frac{\|y_k - c_k\|^2}{2\sigma^2}. \tag{5.12}$$

Thus, the extrinsic information received from a companion decoder is included in the computation through the branch metric $\gamma_k(s', s)$. The rest of the BCJR/SISO algorithm proceeds as before using (5.5), (5.8), and (5.10).

Upon substitution of (5.12) into (5.10), we have after some simplification

$$\begin{aligned} L(u_k) = {} & L^{e}(u_k) + \max_{U+}{}^{*}\left[\alpha_{k-1}(s') + u_k y_k^u/\sigma^2 + p_k y_k^p/\sigma^2 + \beta_k(s)\right] \\ & - \max_{U-}{}^{*}\left[\alpha_{k-1}(s') + u_k y_k^u/\sigma^2 + p_k y_k^p/\sigma^2 + \beta_k(s)\right], \end{aligned} \tag{5.13}$$

where we have used the fact that

$$\begin{aligned} \|y_k - c_k\|^2 &= \left(y_k^u - u_k\right)^2 + \left(y_k^p - p_k\right)^2 \\ &= \left(y_k^u\right)^2 - 2u_k y_k^u + u_k^2 + \left(y_k^p\right)^2 - 2p_k y_k^p + p_k^2 \end{aligned}$$

and only the terms dependent on U^+ or U^-, $u_k y_k^u/\sigma^2$ and $p_k y_k^p/\sigma^2$, survive after the subtraction. Now note that $u_k y_k^u/\sigma^2 = y_k^u/\sigma^2$ under the first $\max^*(\cdot)$ operation in (5.13) (U^+ is the set of state transitions for which $u_k = +1$) and $u_k y_k^u/\sigma^2 = -y_k^u/\sigma^2$ under the second $\max^*(\cdot)$ operation. Using the definition for $\max^*(\cdot)$, it is easy to see that these terms may be isolated out so that

$$\begin{aligned} L(u_k) = {} & 2y_k^u/\sigma^2 + L^{e}(u_k) + \max_{U+}{}^{*}\left[\tilde{\alpha}_{k-1}(s') + p_k y_k^p/\sigma^2 + \tilde{\beta}_k(s)\right] \\ & - \max_{U-}{}^{*}\left[\tilde{\alpha}_{k-1}(s') + p_k y_k^p/\sigma^2 + \tilde{\beta}_k(s)\right]. \end{aligned} \tag{5.14}$$

The interpretation of this new expression for $L(u_k)$ is that the first term is likelihood information received directly from the channel, the second term is extrinsic likelihood information received from a companion decoder, and the third "term" $(\max^*_{U+} - \max^*_{U-})$ is extrinsic likelihood information to be passed to a companion decoder. Note that this third term is likelihood information gleaned from received parity not available to the companion decoder. Thus, specializing to decoder D1, for example, on any given iteration, D1 computes

$$L_1(u_k) = 2y_k^u/\sigma^2 + L_{21}^{\mathrm{e}}(u_k) + L_{12}^{\mathrm{e}}(u_k),$$

where $L_{21}^{\mathrm{e}}(u_k)$ is extrinsic information received from D2, and $L_{12}^{\mathrm{e}}(u_k)$ is the third term $(\max^*_{U+} - \max^*_{U-})$ in (5.14) which is to be used as extrinsic information from D1 to D2.

5.2.3 The SCCC Iterative Decoder

We present in this section the iterative decoder for an SCCC consisting of two component rate-1/2 RSC encoders concatenated in series. We assume no puncturing so that the overall code rate is 1/4. Higher code rates are achievable via puncturing and/or by replacing the inner encoder with a rate 1 differential encoder with transfer function $\frac{1}{1 \oplus D}$. It is straightforward to derive the iterative decoding algorithm for other SCCC codes from the special case that we consider here.

Block diagrams of the SCCC encoder and its iterative decoder with component SISO decoders are presented in Fig. 5.7. We denote by $\mathbf{c}_1 = [c_1^1, c_2^1, \ldots, c_{2K}^1] =$

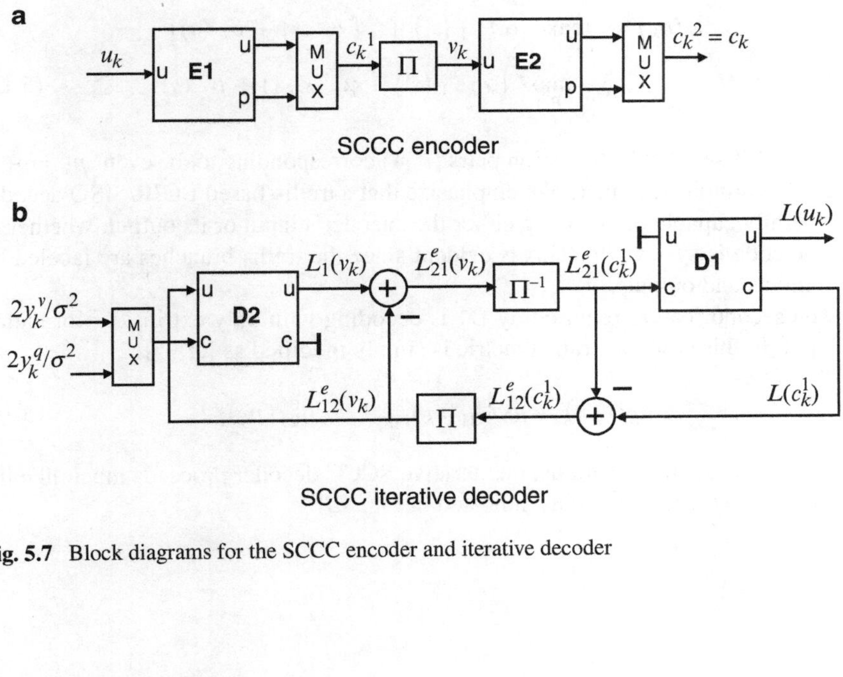

Fig. 5.7 Block diagrams for the SCCC encoder and iterative decoder

$[u_1, p_1, u_2, p_2, \ldots, u_K, p_K]$, where $c_k^1 \triangleq [u_k, p_k]$, the codeword produced by E1 whose input is $\mathbf{u} = [u_1, u_2, \ldots, u_K]$. We denote by $\mathbf{c}_2 = [c_1^2, c_2^2, \ldots, c_{2K}^2] = [v_1, q_1, v_2, q_2, \ldots, v_{2K}, q_{2K}]$, where $c_k^2 \triangleq [v_k, q_k]$, the codeword produced by E2 whose input $\mathbf{v} = [v_1, v_2, \ldots, v_{2K}]$ is the interleaved version of $\mathbf{c}_1$, that is, $\mathbf{v} = \mathbf{c}_1'$. As indicated in Fig. 5.7a, the transmitted codeword $\mathbf{c}$ is the codeword $\mathbf{c}_2$. The received word $\mathbf{y} = \mathbf{c} + \mathbf{n}$ will have the form $\mathbf{y} = [y_1, y_2, \ldots, y_{2K}] = [y_1^v, y_1^q, \ldots, y_{2K}^v, y_{2K}^q]$, where $y_k \triangleq [y_k^v, y_k^q]$ and similarly for $\mathbf{n}$.

The iterative SCCC decoder in Fig. 5.7b employs two SISO decoding modules (described in Sect. 5.2.2.2). Note that these SISO decoders share extrinsic information on the code bits $\{c_k^1\}$ (equivalently, on the E2 input bits $\{v_k\}$) in accordance with the fact that these are the bits known to both encoders. This is in contrast with the PCCC case for which the SISO decoders share extrinsic information only on the input bits of the component encoders. A consequence of this is that D1 must provide likelihood information on E1 *output* bits, whereas D2 produces likelihood information on E2 *input* bits as indicated in Fig. 5.7b. Because LLRs must be obtained on the original data bits u_k so that final decisions may be made, D1 must also compute likelihood information on E1 input bits. Note also that, because E1 sends no bits directly to the channel, D1 receives no samples directly from the channel. Instead, the only input to D1 is the extrinsic information it receives from D2.

Thus, the SISO module D1 requires two features that we have not discussed in any detail to this point. The first is providing likelihood information on the encoder's input *and* output. However, since we assume the component codes are systematic, we need to only compute LLRs on the encoder's output bits $[u_1, p_1, u_2, p_2, \ldots, u_K, p_K]$. Doing this is a simple matter of modifying the max* indices in (5.10) to those relevant to the output bit of interest. For example, the LLR $L(p_k)$ for the E1 parity bit p_k is obtained via

$$\begin{aligned} L(b_k) = & \max_{P+}{}^* \left[\alpha_{k-1}(s') + \gamma_k(s', s) + \beta_k(s)\right] \\ & - \max_{P-}{}^* \left[\alpha_{k-1}(s') + \gamma_k(s', s) + \beta_k(s)\right], \end{aligned} \tag{5.15}$$

where P^+ is set of state transition pairs (s', s) corresponding to the event $p_k = +1$, and P^- is similarly defined. We emphasize that a trellis-based BCJR/SISO decoder is generally capable of decoding either the encoder's input or its output, whether or not the code is systematic. This is evident since the trellis branches are labeled by both inputs and outputs.

The second feature required by D1 is decoding with only extrinsic information as input. In this case the branch metric is simply modified as [cf. (5.12)]

$$\tilde{\gamma}_k(s', s) = u_k L_{21}^{\mathrm{e}}(u_k)/2 + p_k L_{21}^{\mathrm{e}}(p_k)/2. \tag{5.16}$$

Other than these modifications, the iterative SCCC decoder proceeds much like the PCCC iterative decoder and as indicated in Fig. 5.7b.

5.3 Block Turbo Codes

A *block turbo code* (BTC) [14], or *turbo product code* (TPC), is the well-known product code together with soft iterative decoding. Each codeword is an $n_1 \times n_2$ rectangular array in which the columns consists of codewords from the first constituent code, which has parameters (n_1, k_1), and the rows consist of codewords from the second constituent code, which has parameters (n_2, k_2). The product code array is constructed from a $k_1 \times k_2$ subarray that is a k_1k_2-bit data word as follows. The columns of this subarray are encoded by the "column code," after which the rows of the resulting $n_1 \times k_2$ matrix are encoded by the "row code." Alternatively, row encoding may occur first, followed by column encoding. Because the codes are linear, the resulting codeword is independent of the encoding order. In particular, the "checks on checks" submatrix will be unchanged.

The aggregate code rate for this product code is $R = R_1R_2 = (k_1k_2)/(n_1n_2)$, where R_1 and R_2 are the code rates of the individual codes. The minimum distance of the product code is $d_{\min} = d_{\min,1}d_{\min,2}$, where $d_{\min,1}$ and $d_{\min,2}$ are the minimum distances of the individual codes. The constituent codes are typically extended BCH codes (including extended Hamming codes). The extended BCH codes are particularly advantageous because the extension beyond the nominal BCH code increases the (design) minimum distance of each constituent code by one, at the expense of only one extra parity bit, while achieving an increase in aggregate minimum distance by $d_{\min,1} + d_{\min,2} + 1$.

The P_{b} and P_{cw} expressions for ML decoding performance of a BTC on the AWGN channel is no different from any other code:

$$P_{\mathrm{b}} \sim \frac{w_{\min}}{k_1k_2} Q\left(\sqrt{\frac{2Rd_{\min}E_{\mathrm{b}}}{N_0}}\right),$$

$$P_{\mathrm{cw}} \sim A_{\min} Q\left(\sqrt{\frac{2Rd_{\min}E_{\mathrm{b}}}{N_0}}\right),$$

where $w_{\min}$ is the total information weight corresponding to all of the $A_{\min}$ BTC codewords at the minimum distance $d_{\min}$. We note that, unlike PCCCs and SCCCs, $w_{\min}$ and $A_{\min}$ are quite large for BTCs.

5.3.1 Overview of Turbo Decoding of BTCs

Product codes were invented (in 1954 [15]) long before TPCs, but the qualifier "turbo" refers to the iterative decoder which comprises two SISO constituent decoders [14]. Such a turbo decoder is easy to derive if we recast a product code as a serial concatenation of block codes. Under this formulation, the codes in Fig. 5.7a are block codes and the interleaver is a deterministic "column–row" interleaver. A column–row interleaver can be represented as a rectangular array whereby the array is written column-wise and the bits are read out row-wise. The corresponding iterative (turbo) decoder would be that in Fig. 5.7b with SISO convolutional decoders replaced by SISO block decoders. The challenge then would be to design constituent SISO block decoders. One approach is BCJR decoders based on the BCJR trellises of each block code [15, 16]. However, except for very short codes, this approach leads to a high-complexity decoder because the maximum number of states in the time-varying BCJR trellis of an (n, k) block code is 2^{n-k}. The most common approach for its low-complexity is a SISO Chase decoder [14], which is at the expense of some performance loss.

We shall now describe the BTC decoder based on the SISO Chase decoder. We assume knowledge of turbo decoding as described in the previous sections. Each SISO Chase decoder performs the following steps (see [14, 16] for details):

1. Produce a list $\mathcal{L}$ of candidate codewords of size $2^p - 1$. This is done by varying the values of the bits in the p least reliable positions of the received word and algebraically (hard) decoding each time.
2. From the list $\mathcal{L}$, make a decision on which codeword was transmitted. This is done by choosing the codeword in the list that is the closest to the received word in terms of minimum distance.
3. For each code bit in the selected codeword, compute soft-outputs and identify extrinsic information. This is done by starting with the LLR

$$L(c_k) = \ln\left(\frac{\sum_{\mathbf{c}\in\mathcal{L}_k^+} p(\mathbf{y} \mid \mathbf{c})}{\sum_{\mathbf{c}\in\mathcal{L}_k^-} p(\mathbf{y} \mid \mathbf{c})}\right), \tag{5.17}$$

where $\mathcal{L}_k^+$ represents the set of codewords in $\mathcal{L}$ for which $c_k = +1$ and let $\mathcal{L}_k^-$ represent the set of codewords in $\mathcal{L}$ for which $c_k = -1$ (we use here the correspondence $0 \leftrightarrow +1$ and $1 \leftrightarrow -1$). Under the Gaussian assumption, this becomes

$$L(c_k) = \max_{\mathbf{c}\in\mathcal{L}_k^+}{}^* \left(-\|\mathbf{y}-\mathbf{c}\|^2/2\sigma^2\right) - \max_{\mathbf{c}\in\mathcal{L}_k^-}{}^* \left(-\|\mathbf{y}-\mathbf{c}\|^2/2\sigma^2\right). \tag{5.18}$$

The preceding expression may be used to compute the reliabilities $|L(c_k)|$, but an easier-to-implement approximation is often used in practice [14]. For example, with max* replaced by max in (5.18).

Given the SISO constituent Chase algorithm block decoders, the BTC turbo decoder operates as follows. Let $L^{\mathrm{e}}_{rc,k}$ be the extrinsic information computed by the row decoder be passed on to the column decoder, and let $L^{\mathrm{e}}_{cr,k}$ be the extrinsic information received from the column decoder. Similarly, let $L^{\mathrm{e}}_{cr,k}$ be the extrinsic information computed by the column decoder to be passed on to the row decoder, and $L^{\mathrm{e}}_{rc,k}$ be the extrinsic information received from the row decoder. Then the row decoder takes as inputs y_k and $L^{\mathrm{e}}_{cr,k}$ and processes them to produce an output LLR that can be put in the form

$$L_{\mathrm{row}}(c_k) = \frac{2y_k}{\sigma^2} + L^{\mathrm{e}}_{rc,k} + L^{\mathrm{e}}_{cr,k}. \tag{5.19}$$

If an approximation is used, $L_{\mathrm{row}}(c_k)$ has the form

$$L_{\mathrm{row}}(c_k) = y_k + a_l L^{\mathrm{e}}_{rc,k} + a_l L^{\mathrm{e}}_{cr,k}, \tag{5.20}$$

where the scale factors $a_l \in [0, 1]$ are chosen to attenuate the (approximate) extrinsic LLRs which are typically too large. Similarly, the column decoder takes as inputs y_k and $L^{\mathrm{e}}_{rc,k}$ and processes them to produce an output LLR that can be put in the form

$$L_{\mathrm{col}}(c_k) = \frac{2y_k}{\sigma^2} + L^{\mathrm{e}}_{cr,k} + L^{\mathrm{e}}_{rc,k}. \tag{5.21}$$

When an approximation is used, this is

$$L_{\mathrm{col}}(c_k) = y_k + a_l L^{\mathrm{e}}_{cr,k} + a_l L^{\mathrm{e}}_{rc,k}. \tag{5.22}$$

Then, the BTC turbo decoder, depicted in Fig. 5.8, performs the following steps (assuming the row decoder decodes first):

1. **initialize:** $L^{\mathrm{e}}_{cr,k} = L^{\mathrm{e}}_{rc,k} = 0$ for all k.
2. **row decoder:** Run the SISO Chase algorithm with inputs $\{y_k\}$ and $\{L^{\mathrm{e}}_{cr,k}\}$ to obtain $\{L_{\mathrm{row}}(c_k)\}$ and $\{L^{\mathrm{e}}_{rc,k}\}$. Send extrinsic information $\{L^{\mathrm{e}}_{rc,k}\}$ to the column decoder.
3. **column decoder:** Run the SISO Chase algorithm with inputs $\{y_k\}$ and $\{L^{\mathrm{e}}_{rc,k}\}$ to obtain $\{L_{\mathrm{col}}(c_k)\}$ and $\{L^{\mathrm{e}}_{cr,k}\}$. Send extrinsic information $\{L^{\mathrm{e}}_{cr,k}\}$ to the row decoder.

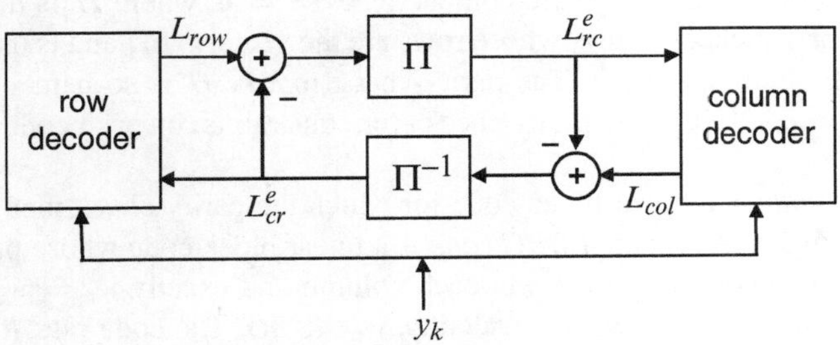

Fig. 5.8 Block turbo code decoder. Π and Π^{-1} represent the implicit row–column interleaving that exists between the row and column encoders (and hence the row and column decoders)

4. **bit decisions:** Repeat steps 2 and 3 until the preset maximum number of iterations is reached or some other stopping criterion is satisfied. Make decisions on bits according to sign$[L_{\text{row}}(c_k)]$ or sign$[L_{\text{col}}(c_k)]$.

5.4 LDPC Codes

LDPC codes are a class of linear block codes which provide near-capacity performance on a large collection of data transmission and storage channels while simultaneously admitting implementable decoders. LDPC codes were first proposed by Gallager in his 1960 doctoral dissertation [17] and was scarcely considered in the 35 years that followed. One notable exception is the important work of Tanner in 1981 [18] in which Tanner generalized LDPC codes and introduced a graphical representation of LDPC codes, now called Tanner graphs. The study of LDPC codes was resurrected in the mid-1990s with the work of MacKay, Luby, and others [19–21], who noticed, apparently independently of the work of Gallager, the advantages of linear block codes which possess sparse (low-density) parity-check matrices. This section provides a brief introduction to LDPC codes.

5.4.1 Matrix Representation

Although LDPC codes can be generalized to nonbinary alphabets, we shall consider only binary LDPC codes for the sake of simplicity. Because LDPC codes form a class of linear block codes, they may be described as a k-dimensional subspace $\mathcal{C}$ of the vector space $\mathbb{F}_2^n$ of binary n-tuples over the binary field $\mathbb{F}_2$. Given this, we may find a basis $B = \{\mathbf{g}_0, \mathbf{g}_1, \ldots, \mathbf{g}_{k-1}\}$ which spans $\mathcal{C}$ so that each $\mathbf{c} \in \mathcal{C}$ may be written as $\mathbf{c} = u_0\mathbf{g}_0 + u_1\mathbf{g}_1 + \cdots + u_{k-1}\mathbf{g}_{k-1}$ for some $\{u_i\}$; more compactly, $\mathbf{c} = \mathbf{u}G$ where $\mathbf{u} = [u_0\ u_1\ \cdots\ u_{k-1}]$ and G is the so-called $k \times n$ generator matrix whose rows are the vectors $\{\mathbf{g}_i\}$ (as is conventional in coding, all vectors are row vectors). The $(n-k)$-dimensional null space $\mathcal{C}^\perp$ of G comprises all vectors $\mathbf{x} \in \mathbb{F}_2^n$ for which $\mathbf{x}G^{\mathrm{T}} = \mathbf{0}$ and is spanned by the basis $B^\perp = \{\mathbf{h}_0, \mathbf{h}_1, \ldots, \mathbf{h}_{n-k-1}\}$. Thus, for each $\mathbf{c} \in \mathcal{C}$, $\mathbf{c}\mathbf{h}_i^{\mathrm{T}} = 0$ for all i or, more compactly, $\mathbf{c}H^{\mathrm{T}} = \mathbf{0}$, where H is the so-called $(n-k) \times n$ parity-check matrix whose rows are the vectors $\{\mathbf{h}_i\}$ and is the generator matrix for the null space $\mathcal{C}^\perp$. The parity-check matrix H is so named because it contains $m = n - k$ separate parity checks (or constraints) on each codeword in the code $\mathcal{C}$.

A *LDPC code* is a linear block code for which the parity-check matrix H has a low density of 1's. A *regular LDPC code* is a linear block code whose parity-check matrix H contains exactly w_{c} 1's in each column and exactly $w_{\mathrm{r}} = w_{\mathrm{c}}(n/m)$ 1's in each row, where $w_{\mathrm{c}} \ll m$ (equivalently, $w_{\mathrm{c}} \ll m$). The code rate $R = k/n$ is related to these parameters via $R = 1 - w_{\mathrm{c}}/w_{\mathrm{r}}$ (this assumes H is full rank). If H is low density, but the number of 1's in each column or row is not constant, then the code is an *irregular LDPC code*. It is easiest to see the sense in which an LDPC code is regular or irregular through its graphical representation.

5.4.2 Graphical Representation

Tanner considered LDPC codes (and a generalization) and showed how they may be represented effectively by a so-called bipartite graph, now called a Tanner graph [18]. The Tanner graph for an LDPC code is shown in Fig. 5.1b. In the context of an LDPC code, the constraint nodes are also called *check nodes* (c-nodes) in accordance with the fact that each constraint is a parity-check constraint. We will also write v-nodes as a shorthand for variable nodes. The Tanner graph of a code is drawn according to the following rule: check node j is connected to variable node i whenever element h_{ji} in H is a 1. One may deduce from this that there are $m = n - k$ check nodes, one for each check equation, and n variable nodes, one for each code bit c_i. Further, the m rows of H specify the m check-node connections, and the n columns of H specify the n variable-node connections.

Example. Consider a (10, 5) linear block code with $w_c = 2$ and $w_r = w_c(n/m) = 4$ with the following H matrix:

$$H = \begin{bmatrix} 1 & 1 & 1 & 1 & 0 & 0 & 0 & 0 & 0 & 0 \\ 1 & 0 & 0 & 0 & 1 & 1 & 1 & 0 & 0 & 0 \\ 0 & 1 & 0 & 0 & 1 & 0 & 0 & 1 & 1 & 0 \\ 0 & 0 & 1 & 0 & 0 & 1 & 0 & 1 & 0 & 1 \\ 0 & 0 & 0 & 1 & 0 & 0 & 1 & 0 & 1 & 1 \end{bmatrix}.$$

The Tanner graph corresponding to H is depicted in Fig. 5.9. Observe that v-nodes c_0, c_1, c_2, and c_3 are connected to c-node f_0 in accordance with the fact that, in the zeroth row of H, $h_{00} = h_{01} = h_{02} = h_{03} = 1$ (all others are zero). Observe that analogous situations holds for c-nodes f_1, f_2, f_3, and f_4 which corresponds to rows 1, 2, 3, and 4 of H, respectively. Note, as follows from the fact that $\mathbf{c}H^{\mathrm{T}} = \mathbf{0}$, the bit values connected to the same check node must sum to zero. We may also proceed along columns to construct the Tanner graph. For example, note that v-node

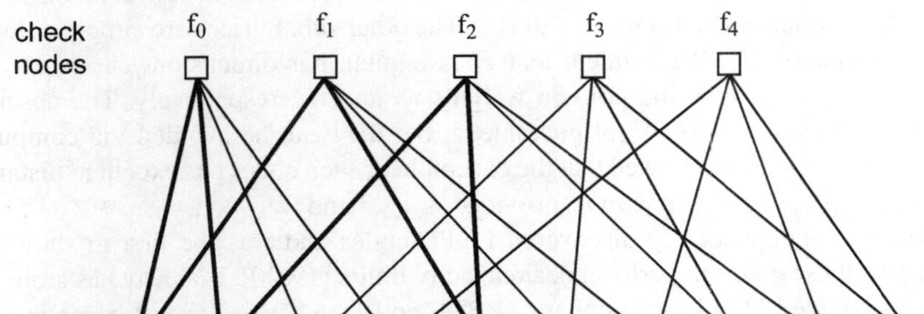

Fig. 5.9 Tanner graph for example code

c_0 is connected to c-nodes f_0 and f_1 in accordance with the fact that, in the zeroth column of H, $h_{00} = h_{10} = 1$.

Note that the Tanner graph in this example is regular: each v-node has two edge connections and each c-node has four edge connections (that is, the *degree* of each v-node is 2 and the degree of each c-node is 4). This is in accordance with the fact that $w_c = 2$ and $w_r = 4$. Is is also clear from this example that $mw_r = nw_c$. For irregular LDPC codes, the parameters w_c and w_r are functions of the column and row numbers.

A *cycle* (or *loop*) of length ν in a Tanner graph is a path comprising ν edges which closes back on itself. The Tanner graph in the above example possesses a length-6 cycle as exemplified by the six bold edges in the figure. The *girth* γ of a Tanner graph is the minimum cycle length of the graph. The shortest possible cycle in a bipartite graph is clearly a length-4 cycle, and such cycles manifest themselves in the H matrix as four 1's that lie on the corners of a submatrix of H. We are interested in cycles, particularly short cycles, because they degrade the performance of the iterative decoding algorithm used for LDPC codes. This fact will be made evident in the discussion of the iterative decoding algorithm.

5.4.3 LDPC Code Design Approaches

Clearly, the most obvious path to the construction of an LDPC code is via the construction of a LDPC matrix with certain desired properties. A large number of design techniques exist in the literature, and we introduce some of the more prominent ones in this section, albeit at a superficial level.

Gallager [17] gave the first code design technique in his seminal work on LDPC codes. In this design approach, the $m \times n$ parity-check matrix H matrix consists of w_c submatrices H_d of size $\mu \times n$ stacked on top of each other. The submatrices H_d have the following structure: For any integers μ and w_r greater than 1, each submatrix H_d is $\mu \times \mu w_r$ with row weight w_r and column weight 1. The submatrix H_1 has the following specific form: for $i = 1, 2, \ldots, \mu$, the ith row contains all of its w_r 1's in columns $(i-1)w_r + 1$ to iw_r. The other submatrices are simply column permutations of H_1. It is evident that H is regular, has dimension $\mu w_c \times \mu w_r = \mu w_c \times n$, and has row and column weights w_r and w_c, respectively. The absence of length-4 cycles in H is not guaranteed, but they can be avoided via computer design of H. Gallager showed that the ensemble of such codes has excellent distance properties (d_{min} increases with n) provided $w_c \geq 3$ and $w_r > w_c$.

MacKay independently discovered LDPC codes and was the first to show the ability of these codes to perform near capacity limits [19,20]. MacKay has archived on a Web page [22] a large number of LPDC codes he has designed for application to data communication and storage, most of which are regular. The algorithms to semirandomly generate sparse H matrices involve simply populating a matrix by placing 1's in it column-by-column to ensure a desired row and column weight distributions and to avoid 4-cycles.

One drawback of Gallager and MacKay codes is that they lack sufficient structure to enable low-complexity encoding. Encoding is performed by putting H in the form $[P^{\mathrm{T}}\ I]$ via Gauss–Jordan elimination, from which the generator matrix can be put in the systematic form $G = [I\ P]$. The problem with encoding via G is that the submatrix P is generally not sparse so that, for codes of length $n = 1,000$ or more, encoding complexity is high. An efficient encoding technique employing only the H matrix was proposed in [23].

Since these early works on LDPC codes, a large number of papers have been published on the design and analysis of LDPC codes. Most of the design technique focus on the encoder complexity issue by incorporating structure in H and on the code performance. The reader is referred to the following books for coverage of the some of the most well-known approaches [16, 24, 25]. These approaches involve accumulators, algebra, combinatorics, and many others.

5.4.4 LDPC Decoding Algorithms

In addition to introducing LDPC codes in his seminal work in 1960 [17], Gallager also provided a decoding algorithm that is generally near optimal. The algorithm has been called the sum–product algorithm (SPA), the belief propagation algorithm (BPA), and the message passing algorithm (MPA). The term "message passing" usually refers to all such iterative algorithms, including the SPA and its approximations.

Much like optimal [maximum a posteriori (MAP)] symbol-by-symbol decoding of trellis-based codes, we are interested in computing the a posteriori probability that a given bit in the transmitted codeword $\mathbf{c} = [c_0\ c_1\ \cdots\ c_{n-1}]$ equals 1, given the received word $\mathbf{y} = [y_0\ y_1\ \cdots\ y_{n-1}]$. Without loss of generality, let us focus on the decoding of bit c_i so that we are interested in computing the LLR

$$L(c_i) \triangleq \log\left(\frac{\Pr(c_i = 0 \mid \mathbf{y})}{\Pr(c_i = 1 \mid \mathbf{y})}\right).$$

The SPA for the computation of $L(c_i)$ is an iterative algorithm which is based on the code's Tanner graph. Specifically, we imagine that the v-nodes represent processors of one type, c-nodes represent processors of another type, and the edges represent message (LLR) paths. In one half-iteration, each v-node processes its input messages and passes its resulting output messages *up* to neighboring c-nodes (two nodes are said to be *neighbors* if they are connected by an edge). This is depicted in Fig. 5.10 for the message $m_{\uparrow 02}$ from v-node c_0 to c-node f_2 (the subscripted arrow indicates the direction of the message, keeping in mind that our Tanner graph convention places c-nodes above v-nodes). Note in the figure that the information passed to c-node f_2 is all the information available to v-node c_0 from the channel and through its neighbors, excluding c-node f_2; that is, only *extrinsic information* is passed. Such extrinsic information $m_{\uparrow ij}$ is computed for each connected

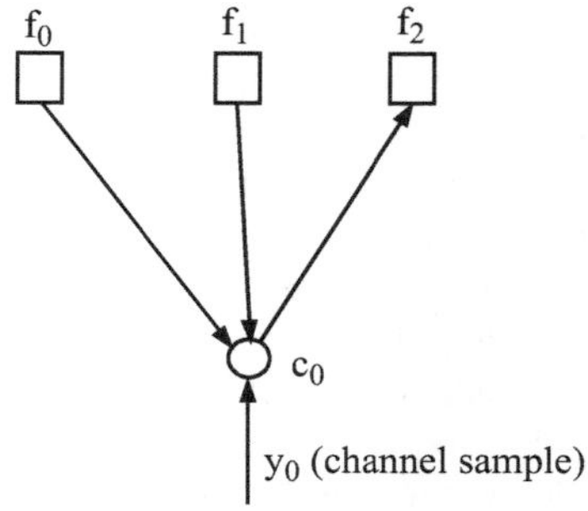

Fig. 5.10 Subgraph of a Tanner graph corresponding to an H matrix whose zeroth column is $[1\ 1\ 1\ 0\ 0\ \cdots\ 0]^{\mathrm{T}}$. The *arrows* indicate message passing from node c_0 to node f_2

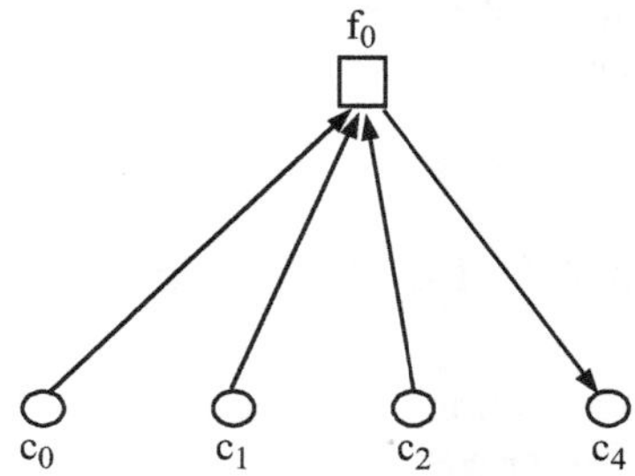

Fig. 5.11 Subgraph of a Tanner graph corresponding to an H matrix whose zeroth row is $[1\ 1\ 1\ 0\ 1\ 0\ 0\ \cdots\ 0]^{\mathrm{T}}$. The *arrows* indicate message passing from node f_0 to node c_4

v-node/c-node pair c_i/f_j at each half-iteration. Specifically, the messages sent from v-node c_i to c-node f_j is given by

$$m_{\uparrow ij} = 2y_i/\sigma^2 + \sum_{N_i\backslash\{j\}} m_{\downarrow ji}, \tag{5.23}$$

where N_i is the set of neighbors of c_i, σ^2 is the variance of each AWGN sample, and $m_{\downarrow ji}$ is the message from f_j to c_i. This expression derives from the bit-wise MAP decoder expression for a repetition code and the fact that each v-node represents a repetition code.

In the other half-iteration, each c-node processes its input messages and passes its resulting output messages *down* to its neighboring v-nodes. This is depicted in Fig. 5.11 for the message $m_{\downarrow 04}$ from c-node f_0 down to v-node c_4. Note that only extrinsic information is passed to v-node c_4. Such extrinsic information $m_{\downarrow ji}$ is computed for each connected c-node/v-node pair f_j/c_i at each half-iteration. Specifically, the message sent from c-node f_j to v-node c_i is given by

$$m_{\downarrow ji} = \underset{N_j\backslash\{i\}}{\boxplus} m_{\uparrow ij}, \tag{5.24}$$

where N_j is the set of neighbors of f_j and the operator $\boxplus$ for two messages (two LLRs) can be defined by any of the three following expressions

$$L_1 \boxplus L_2 = \max{}^*(L_1 + L_2, 0) - \max{}^*(L_1, L_2,) \tag{5.25}$$

$$= \prod_{k=1}^{2} \alpha_k \cdot \phi\left(\sum_{k=1}^{2} \phi(\beta_k)\right) \tag{5.26}$$

$$= \prod_{k=1}^{2} \alpha_k \cdot \min(\beta_1, \beta_2) + s(L_1, L_2,), \tag{5.27}$$

where

$$\alpha_k = \operatorname{sign}(L_k),$$
$$\beta_k = |L_k|,$$
$$\phi(x) = -\log\tanh(x/2),$$
$$s(x,y) = \log\left(1 + \mathrm{e}^{-|x+y|}\right) - \log\left(1 + \mathrm{e}^{-|x-y|}\right).$$

This expression derives from the bit-wise MAP decoder expression for a SPC code and the fact that each c-node represents a repitition code. Any of these pairwise expressions can be used repeatedly when more than two LLRs are involved. For example, $L_1 \boxplus L_2 \boxplus L_3$ can be computed as $(L_1 \boxplus L_2) \boxplus L_3$. Also, for ℓ LLRs, (5.26) becomes

$$L_1 \boxplus L_2 \boxplus \cdots \boxplus L_\ell = \prod_{k=1}^{\ell} \alpha_k \cdot \phi\left(\sum_{k=1}^{\ell} \phi(\beta_k)\right).$$

After a preset maximum number of iterations or after some stopping criterion has been met, the decoder computes the LLR from which decisions on the bits c_i are made. One example of stopping criterion is to stop iterating when $\hat{\mathbf{c}}H^T = \mathbf{0}$, where $\hat{\mathbf{c}}$ is a tentatively decoded codeword. The bit decisions $\hat{c}_i$ in $\hat{\mathbf{c}}$ are made by adding up all of the incoming message in v-node c_i and comparing that sum to zero as detailed below.

The MPA assumes that the messages passed are statistically independent throughout the decoding process. When the y_i are independent, this independence assumption would hold true if the Tanner graph possessed no cycles. Further, the MPA would yield exact LLRs in this case [9]. However, for a graph of girth γ, the independence assumption is only true up to the $\gamma/2$th iteration, after which messages start to loop back on themselves in the graph's various cycles. Still, simulations have shown that the MPA is generally very effective provided length-4 cycles are avoided.

Given (5.23) and (5.24) and the decision step described above, what remains in the description of the SPA is the initialization step. The messages $m_{\downarrow ji}$ from the c-nodes are all initialized to zero and the messages $m_{\uparrow ij}$ from the v-nodes are set to $2y_i/\sigma^2$, for all j. We may now present the SPA decoding algorithm for LDPC codes.

SUM–PRODUCT DECODING ALGORITHM

1. **initialization:** For $i = 0, 1, \ldots, n-1$, set $m_{\uparrow ij} = 2y_i/\sigma^2$ for all j for which $h_{ij} = 1$.
2. **c-node update:** For $j = 0, 1, \ldots, m-1$, compute $m_{\downarrow ji} = \underset{N_j \backslash \{i\}}{\boxplus} m_{\uparrow ij}$ for all i for which $h_{ij} = 1$.
3. **v-node update:** For $i = 0, 1, \ldots, n-1$, set $m_{\uparrow ij} = 2y_i/\sigma^2 + \sum_{N_i \backslash \{j\}} m_{\downarrow ji}$ for all j for which $h_{ij} = 1$
4. **bit decisions:** Update $\{L(c_i)\}$ using $L(c_i) = 2y_i/\sigma^2 + \sum_{N_i} m_{\downarrow ji}$. For $i = 0, 1, \ldots, n-1$, set

$$\hat{c}_i = \begin{cases} 1 \text{ if } L(c_i) < 0 \\ 0 \text{ else} \end{cases} .$$

If $\hat{\mathbf{c}}\mathbf{H}^{\mathrm{T}} = \mathbf{0}$ or the number of iterations equals the maximum limit, stop; else, go to Step 2.

Remark. This algorithm can be applied to other channels simply by changing the initialization step. Also, modifications of steps 2–4 are possible. For example, it is possible to combine steps 3 and 4 since very similar calculations are made.

5.4.5 Reduced Complexity Decoders

Implementation of the $\boxplus$ operator is quite complex as it involves log and tanh functions. A look-up table is one possible route, but sometimes even a very large table can produce an error-rate floor in a code's performance curve. Other techniques such as a piecewise linear approximation are possible. Two common approximations are presented below, each of which can lower the error-rate floor, but at the expense of some performance loss in the performance curve's waterfall region. The extent of the performance loss depends on the code.

5.4.5.1 The Min-Sum Decoder [11]

Consider the c-node update equations (5.24) and (5.26), in particular, the expression $\phi\left(\sum_k \phi(\beta_k)\right)$. Note from the shape of $\phi(x)$, shown in Fig. 5.12, that the term corresponding to the smallest β_k in the summation dominates, so that

$$\begin{aligned} \phi\left(\sum_k \phi(\beta_k)\right) &\simeq \phi\left(\phi\left(\min_k \beta_k\right)\right) \\ &= \min_k \beta_k \, . \end{aligned}$$

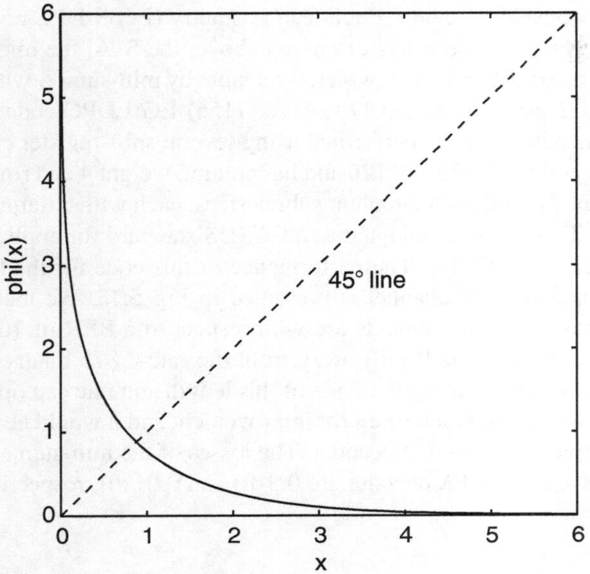

Fig. 5.12 Plot of the $\phi(x)$ function

Thus, the min-sum algorithm is simply the log-domain SPA with c-node update equation replaced by

$$m_{\downarrow ji} = \prod_{N_j \backslash \{i\}} \text{sign}(m_{\uparrow ij}) \cdot \min_{N_j \backslash \{i\}} \left| m_{\uparrow ij} \right|.$$

It can also be shown that the initialization $m_{\uparrow ij} = 2y_i/\sigma^2$ may be replaced by $m_{\uparrow ij} = y_i$ when the min-sum algorithm is employed. The advantage, of course, is that knowledge of the noise power σ^2 is unnecessary in this case.

5.4.5.2 The Min-Sum-Plus-Correction-Factor Decoder [26]

Consider the c-node update equations (5.24) and (5.27). Because $|s(x, y)| \leq 0.693$, the term $s(x, y)$ in (5.27) can (sometimes) be ignored so that $L_1 \boxplus L_2$ is approximated as

$$L_1 \boxplus L_2 \simeq \text{sign}(L_1)\, \text{sign}(L_2) \min\left(|L_1|, |L_2|\right). \tag{5.28}$$

Of course, this is the min-sum algorithm. Alternatively, the complex function $s(x, y)$ can be replaced by the approximation $\tilde{s}(x, y)$ given by

$$\tilde{s}(x, y) = \begin{cases} c & \text{if } |x+y| < 2 \text{ and } |x-y| > 2|x+y| \\ -c & \text{if } |x-y| < 2 \text{ and } |x+y| > 2|x-y| \\ 0 & \text{otherwise} \end{cases}$$

and where c on the order of 0.5 is typical.

Example. We consider a regular Euclidean geometry (EG) LDPC code and its performance with the three decoders discussed above: the SPA, the min-sum, and the min-sum with a correction factor (which we denote by min-sum-c, with c set to 0.5). This code is a quasicyclic rate-0.875 (8176, 7156) EG LDPC code. Because it is quasicyclic, encoding may be performed using several shift-register circuits. The H matrix for this code is $1,022 \times 8,176$ and has column weight 4 and row weight 32. It comprises eight $511 \times 2,044$ circulant submatrices, each with column weight 2 and row weight 8. This has been adopted as a CCSDS standard for application to satellite communications [27, 28]. The performance of this code for the three decoders on a binary-input AWGN channel is presented in Fig. 5.13. We make the following observations (all measurements are with respect to a BER of 10^{-5}). The SPA performance of this code is 0.9 dB away from the rate-0.875 binary-input AWGN capacity limit. An irregular regular code of this length and rate can operate closer to the limit, but there is not much room for improvement and it would be at the expense of a more complex encoder and decoder. The losses of the min-sum and min-sum-c decoders relative to the SPA decoder are 0.3 dB and 0.01 dB, respectively.

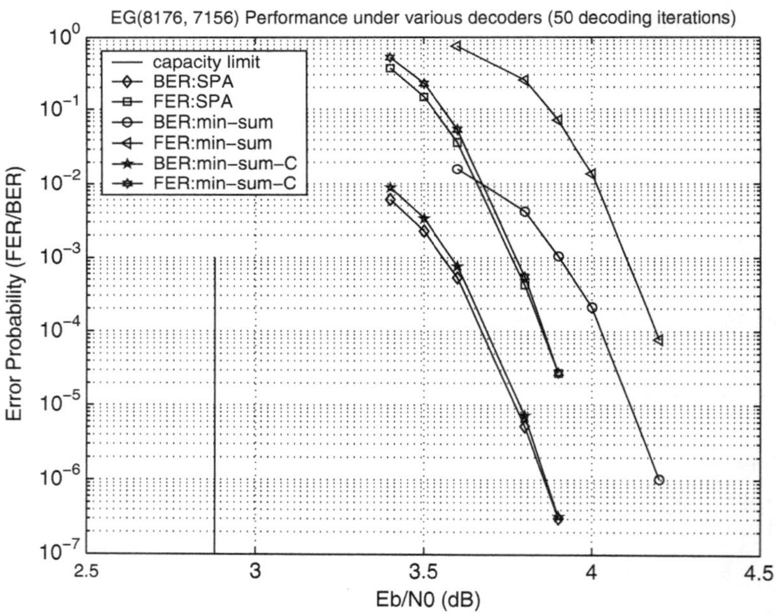

Fig. 5.13 Performance of a quasicyclic EG(8176, 7156) LDPC code on a binary-input AWGN channel and three decoding algorithms

5.5 Concluding Remarks

This chapter presented an introduction to parallel and serial turbo codes, block turbo codes, and LDPC codes. Each of these has their advantages and disadvantages, but one notable disadvantage for parallel and serial turbo codes is that they are intrinsically low-rate codes, making them an imperfect match for the optical communications application. Higher rates are possible by puncturing, but it would be at the expense of decreasing minimum distance, raising the error-rate floor. By contrast, block turbo codes are intrinsically high rate codes because decoding-complexity issues dictate that their component codes be high rate. Further, these codes admit relatively low complexity (soft) list decoding, which explains why an experimental demonstration has already been given for a BTC on the optical communication channel [29]. LDPC codes have the potential for the best performance because their design space has the most degrees of freedom. Further, they are capable of high-speed operation because of the highly parallel nature of the encoder and decoder. However, LDPC encoder and decoder complexities are greater than that of BTCs.

References

1. Berrou C, Glavieux A, Thitimajshima P (1993) Near Shannon limit error-correcting coding and decoding: turbo codes. In: Proceedings of 1993 International conference on communication, pp 1064–1070
2. Berrou C, Glavieux A (1996) Near optimum error correcting coding and decoding: turbo-codes. IEEE Trans Commun 44(10):1261–1271
3. Benedetto S, Montorsi G (1996) Unveiling turbo codes: some results on parallel concatenated coding schemes. IEEE Trans Inform Theory 40(2):409–428
4. Benedetto S, Montorsi G (1996) Design of parallel concatenated codes. IEEE Trans Commun 44(5):591–600
5. Benedetto S, Divsalar D, Montorsi G, Pollara F (1998) Serial concatenation of interleaved codes: performance analysis, design, and iterative decoding. IEEE Trans Inform Theory 44(3):909–926
6. Benedetto S, Divsalar D, Montorsi G, Pollara F (1996) A soft-input soft-output maximum a posteriori (MAP) module to decode parallel and serial concatenated codes. TDA Progress Report 42–127, Nov. 15, 1996
7. Ungerboeck G (1982) Channel coding with multilevel/phase signals. IEEE Trans Inform Theory 28(1):55–67
8. Divsalar D, Pollara F (1995) Multiple turbo codes for deep-space communications. JPL TDA Progress Report 42–121, May 15, 1995
9. Pearl J (1988) Probabilistic reasoning in intelligent systems. Morgan Kaufmann, San Mateo, CA
10. Frey B (1998) Graphical models for machine learning and digital communication. MIT, Cambridge, MA
11. Wiberg N (1996) Codes and decoding on general graphs. PhD dissertation, U. Linköping, Sweden
12. Bahl L, Cocke J, Jelinek F, Raviv J (1974) Optimal decoding of linear codes for minimizing symbol error rate. IEEE Trans Inform Theory 20(2):284–287

13. Robertson P, Villebrun E, Hoeher P (1995) A comparison of optimal and suboptimal MAP decoding algorithms operating in the log domain. In: Proceedings of 1995 international conference on communication, pp 1009–1013
14. Pyndiah R (1998) Near-optimum decoding of product codes: block turbo codes. IEEE Trans Commun 46(8):1003–1010
15. Elias P (1954) Error-free coding. IEEE Trans Inform Theory 4(4):29–37
16. Ryan WE, Lin S (2009) Channel codes: classical and modern. Cambridge University Press, Cambridge
17. Gallager R (1962) Low-density parity-check codes. IRE Trans Inform Theory 8(1):21–28
18. Tanner RM (1981) A recursive approach to low complexity codes. IEEE Trans Inform Theory 27(5):533–547
19. MacKay D, Neal R (1995) Good codes based on very sparse matrices. In Boyd C (ed) Cryptography and coding, 5th IMA conference, Lecture notes in computer science. Springer, Berlin, pp 100–111
20. MacKay D (1999) Good error correcting codes based on very sparse matrices. IEEE Trans Inform Theory 45(2):399–431
21. Alon N, Luby M (1996) A linear time erasure-resilient code with nearly optimal recovery. IEEE Trans Inform Theory 42(6):1732–1736
22. Mackay DJC, http://wol.ra.phy. cam.ac.uk/mackay
23. Richardson TJ, Urbanke R (2001) Efficient encoding of low-density parity-check codes. IEEE Trans Inform Theory 47:638–656
24. Lin S, Costello D (2004) Error-control coding, 2nd edn. Prentice-Hall, Upper Saddle River, NJ
25. Richardson T, Urbanke R (2008) Modern coding theory. Cambridge University Press, Cambridge
26. Hu X-Y, Eleftherious E, Arnold DM, Dholakia A (2001) Efficient implementation of the sum–product algorithm for decoding LDPC codes. In: Proceedings of 2001 IEEE GlobeCom conference, pp 1036–1036E
27. Fong W (2002) White paper for LDPC codes for CCSDS channel Blue Book. NASA GSFC White paper submitted to the Panel 1B CCSDS standards committee, Oct. 2002
28. CCSDS draft recommendation for space data systems standards: TM synchronization and channel coding, Pink Book, August 2009
29. Mizuochi T et al (2004) Forward error correction based on block turbo code with 3-bit soft decision for 10 Gb/s optical communication systems. IEEE J Sel Top Quantum Electron 10(2):376–386

Chapter 6
Coded Modulation

In this chapter, we describe how to optimally combine modulation with channel coding, and describe several coded-modulation schemes: (a) multilevel coding (MLC) [1–4], (b) multidimensional coded modulation [5, 6], and (c) coded orthogonal frequency division multiplexing (OFDM) [7–12]. Using this approach, modulation, coding, and multiplexing are performed in a unified fashion so that, effectively, the transmission, signal processing, detection, and decoding are done at much lower symbol rates. At these lower rates, dealing with the nonlinear effects and polarization mode dispersion (PMD) is more manageable, while the aggregate data rate per wavelength is maintained above 100 Gb/s.

The chapter is organized as follows. In Sect. 6.1 we describe M-ary quadrature amplitude multiplexing (QAM) [13, 14], M-ary phase shift keying (PSK), and an optimum signal constellation in minimum mean-square error (MMSE) sense [15]. In Sect. 6.2 we describe MLC, in which the component block codes have different code rates, and bit-interleaved coded modulation (BICM), in which all component codes have the same code rate. In Sect. 6.3 we describe multidimensional coded modulation as an efficient way to increase the spectral efficiency. In Sect. 6.4 we describe coded OFDM as an efficient way to compensate simultaneously for chromatic dispersion (CD) and PMD, and to increase the spectral efficiency.

6.1 Multilevel Modulation Schemes

M-ary QAM is a two-dimensional (2D) generalization of M-ary pulse-amplitude modulation (PAM), and its formulation involves two orthogonal passband basis functions:

$$\phi_I(t) = \sqrt{\frac{2}{T}}\cos(2\pi f_c t), \quad 0 \le t \le T,$$
$$\phi_Q(t) = \sqrt{\frac{2}{T}}\sin(2\pi f_c t), \quad 0 \le t \le T, \tag{6.1}$$

I. Djordjevic et al., *Coding for Optical Channels*,
DOI 10.1007/978-1-4419-5569-2_6,

where f_c denotes the carrier frequency and T denotes the symbol duration. The message point $\boldsymbol{s}_i$ can be described in terms of basis functions by $(I_k d_{\min}, Q_k d_{\min}/2)$, where $d_{\min}$ is related to the energy of the signal with smallest amplitude E_0 by $d_{\min}/2 = \sqrt{E_0}$. The coordinates of kth signal constellation point for square M-ary QAM signal constellation are given by

$$\{I_k, Q_k\} = \begin{bmatrix} (-\sqrt{M}+1, \sqrt{M}-1) & (-\sqrt{M}+3, \sqrt{M}-1) & \ldots & (\sqrt{M}-1, \sqrt{M}-1) \\ (-\sqrt{M}+1, \sqrt{M}-3) & (-\sqrt{M}+3, \sqrt{M}-3) & \ldots & (\sqrt{M}-1, \sqrt{M}-3) \\ \ldots & \ldots & \ldots & \ldots \\ (-\sqrt{M}+1, -\sqrt{M}+1) & (-\sqrt{M}+3, -L+1) & \ldots & (\sqrt{M}-1, -\sqrt{M}+1) \end{bmatrix}. \tag{6.2}$$

For example, for $M = 16$ signal constellation points are given by

$$\{I_k, Q_k\} = \begin{bmatrix} (-3,3) & (-1,3) & (1,3) & (3,3) \\ (-3,1) & (-1,1) & (1,1) & (3,1) \\ (-3,-1) & (-1,-1) & (1,-1) & (3,-1) \\ (-3,-3) & (-1,-3) & (1,-3) & (3,-3) \end{bmatrix}.$$

The transmitted M-ary QAM signal for kth symbol can, therefore, be expressed as

$$s_k(t) = \sqrt{\frac{2E_0}{T}} I_k \cos(2\pi f_c t) - \sqrt{\frac{2E_0}{T}} Q_k \sin(2\pi f_c t), \quad 0 < t \le T; \quad k = 1, \ldots, M \tag{6.3}$$

with constellation points coordinates given by (6.2). On the other hand, using the same basis functions, the transmitted M-ary PSK signal for kth symbol can be expressed as

$$s_k(t) = \begin{cases} \sqrt{\frac{2E}{T}} \cos\left(2\pi f_c t + (k-1)\frac{2\pi}{M}\right), & 0 \le t \le T, \\ 0, & \text{otherwise}, \end{cases} \quad k = 1, 2, \ldots, M \tag{6.4}$$

The star-QAM [14] can be considered as generalization of M-ary PSK given by (6.4), in which the points are placed at circles of different radius.

As an illustration, the 8-PSK, 8-QAM, and 64-QAM constellation diagrams are given in Fig. 6.1. The corresponding Gray-mapping rule, for 8-PSK and 8-QAM, is given in Table 6.1.

In [15] we designed an optimum signal constellation based on iterative polar quantization (IPQ) procedure. The IPQ constellation was determined based on the minimum mean-square quantization error, and it is optimum for the ASE noise-dominated scenario. This scheme can be considered as a generalization of star-QAM

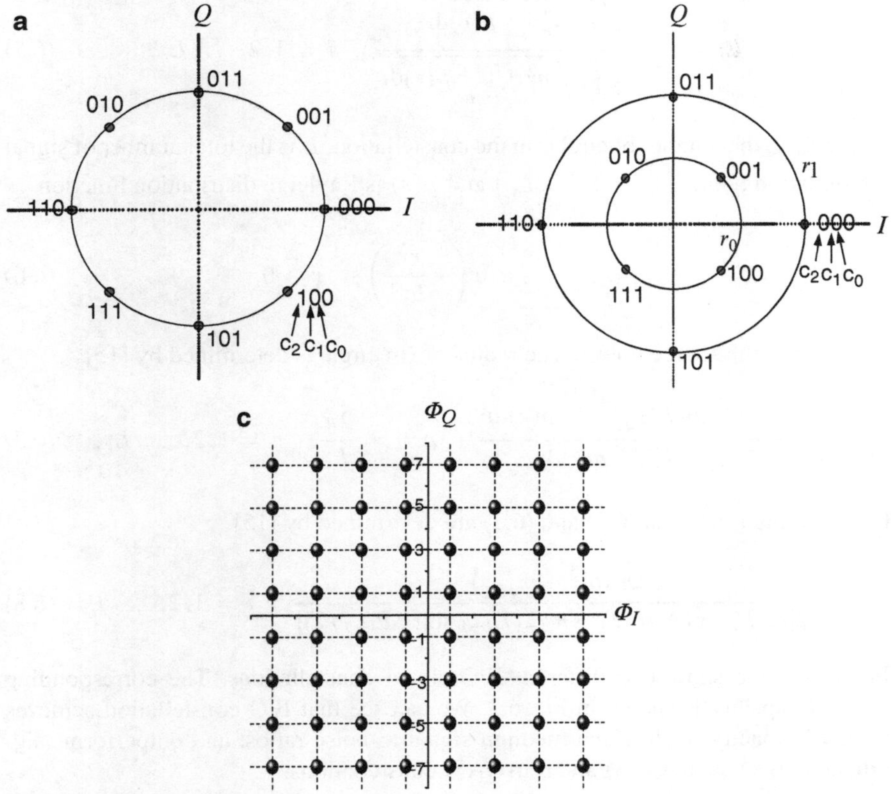

Fig. 6.1 The constellation diagrams for: (**a**) 8-PSK, (**b**) 8-QAM, and (**c**) 64-point square QAM

Table 6.1 8-PSK/8-QAM gray-mapping rule

Input bits ($c_2c_1c_0$)	ϕ_k	8-PSK		8-QAM	
		I_k	Q_k	I_k	Q_k
000	0	1	0	$1+\sqrt{3}$	0
001	$\pi/4$	$\sqrt{2}/2$	$\sqrt{2}/2$	1	1
011	$\pi/2$	0	1	0	$1+\sqrt{3}$
010	$3\pi/4$	$-\sqrt{2}/2$	$\sqrt{2}/2$	-1	1
110	π	-1	0	$-(1+\sqrt{3})$	0
111	$5\pi/4$	$-\sqrt{2}/2$	$-\sqrt{2}/2$	-1	-1
101	$3\pi/2$	0	-1	0	$-(1+\sqrt{3})$
100	$7\pi/4$	$\sqrt{2}/2$	$-\sqrt{2}/2$	1	-1

(SQAM), in which the constellation points are distributed over the circles of radius determined by Rayleigh distribution. Let L_i denote the number of constellation points per circle of radius m_i. The optimum number of constellation points is determined by [15]

$$L_i = \frac{\sqrt[3]{m_i^2 \int_{r_i}^{r_i+1} p(r)\mathrm{d}r}}{\sum_{i=2}^{L_r} \frac{1}{L} \sqrt[3]{m_i^2 \int_{r_i}^{r_i+1} p(r)\mathrm{d}r}}, \quad i = 1, 2, \ldots, L_r, \tag{6.5}$$

where L_r is the number of circles in the constellation, L is the total number of signal constellation points $\left(L = \sum_{i=2}^{L_r} L_i\right)$ and $p(r)$ is Rayleigh distribution function

$$p(r) = \frac{r}{\sigma^2} \exp\left(-\frac{r^2}{2\sigma^2}\right), \quad r \geq 0, \tag{6.6}$$

where σ^2 is the source power. The radius of ith circle is determined by [15]

$$m_i = \frac{2\sin(\Delta\theta_i/2) \int_{r_i}^{r_i+1} rp(r)\mathrm{d}r}{\Delta\theta_i \int_{r_i}^{r_i+1} p(r)\mathrm{d}r}, \quad \Delta\theta_i = \frac{2\pi}{L_i}, \quad i = 1, 2, \ldots, L_r. \tag{6.7}$$

Limits of integration in (6.6) and (6.7) are determined by [15]

$$r_i = \frac{\pi\left(m_i^2 - m_{i-1}^2\right)}{2[m_i L_i \sin(\Delta\theta_i/2) - m_{i-1} L_{i-1} \sin(\Delta\theta_{i-1}/2)]}, \quad i = 1, 2, \ldots, L_r \tag{6.8}$$

In Fig. 6.2 we show two different IPQ signal constellations. The corresponding channel capacity is shown in Fig. 6.3. We can see that IPQ constellation achieves channel capacity for low and medium signal-to-noise ratios, and outperforms significantly the square-QAM and star-QAM constellations.

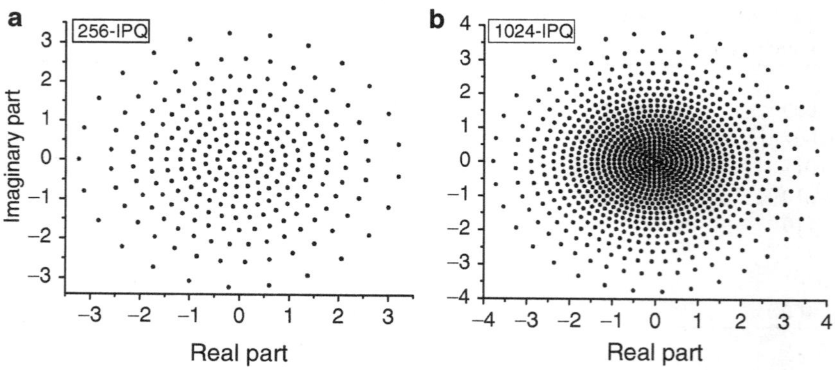

Fig. 6.2 IPQ signal constellations: (**a**) $M = 256$, and (**b**) $M = 1024$

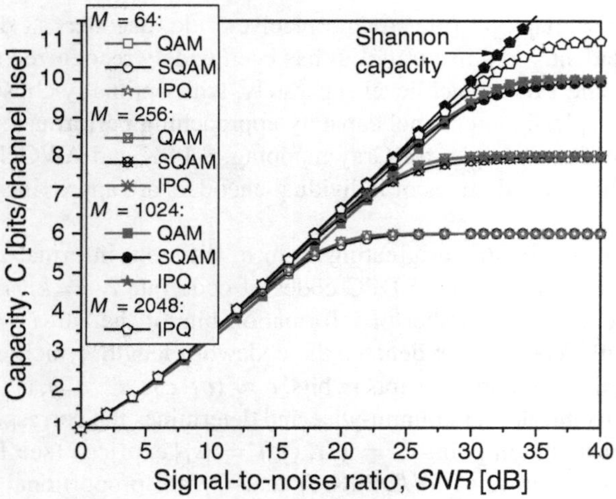

Fig. 6.3 Channel capacity of IPQ

6.2 Single-Carrier-Coded Modulation Schemes

M-ary PSK, M-ary QAM, and M-ary DPSK achieve the transmission of $\log_2 M(= m)$ bits per symbol, providing bandwidth-efficient communication. In coherent detection, for M-ary PSK, the data phasor $\phi_l \in \{0, 2\pi/M, \ldots, 2\pi(M-1)/M\}$ is sent at each lth transmission interval. In direct detection for M-ary DPSK, the modulation is differential, the data phasor $\phi_l = \phi_{l-1} + \Delta\phi_l$ is sent instead, where $\Delta\phi_l \in \{0, 2\pi/M, \ldots, 2\pi(M-1)/M\}$ is determined by the sequence of $\log_2 M$ input bits using an appropriate mapping rule. Let us now introduce the transmitter architecture employing LDPC codes as channel codes. We have two options: (a) to use the same LDPC code for different data streams, this scheme is known as bit-interleaved coded modulation (BICM) [2]; and (b) to use different code rate LDPC codes (of the same length), this scheme is known as multilevel coding (MLC) [1].

The MLC, proposed by Imai and Hirakawa in 1977 [1], is a powerful coded modulation scheme capable of achieving the bandwidth-efficient communication [16,17]. The key idea behind the MLC is to protect the individual bits using different binary codes and use M-ary signal constellation. The decoding is based on the so-called multistage decoding (MSD) algorithm, in which decisions from prior (lower) decoding stages are passed to next (higher) stages (see [17]). Despite its attractiveness with respect to large coding gain, the MLC with the MSD algorithm has serious limitation for use in high-speed applications, such as optical communications, that is due to inherently large delay of the MSD algorithm. One possible solution is to use the *parallel independent decoding* (PID) [16]. In MLC/PID scheme the information bit stream is split into L different levels, and the corresponding bits at different levels are encoded using different encoders, and then combined into a signal point using appropriate mapping rule. At the receiver side, decoders at different levels operate independently and in parallel. It has been widely recognized that the use of Gray mapping and PID at each level separately, with optimally chosen component codes [16, 18, 19] leads to channel capacity approaching performance, and as such is adopted here. For example, for Gray mapping, 8-PSK and AWGN, it was found in [15] that optimum code rates of individual encoders are approximately 0.75, 0.5, and 0.75.

In MLC, the bit streams originating from m different information sources are encoded using different (n, k_i) LDPC codes of code rate $r_i = k_i/n$, as shown in Fig. 6.4a. k_i denotes the number of information bits of the ith ($i = 1, 2, \ldots, m$) component LDPC code, and n denotes the codeword length, which is the same for all LDPC codes. The mapper accepts m bits, $c = (c_1, c_2, \ldots, c_m)$, at time instance i from the $(m \times n)$ interleaver column-wise and determines the corresponding M-ary ($M = 2^m$) constellation point $\boldsymbol{s}_i = (I_i, Q_i) = |\boldsymbol{s}_i| \exp(\mathrm{j}\phi_i)$ (see Fig. 6.4a). The mapper outputs for M-ary PSK/DPSK I_i and Q_i are proportional to $\cos\phi_i$ and $\sin\phi_i$, respectively. The use of MLC allows optimally to allocate the code rates. It is straightforward to show [16] that the spectral efficiency R_s, expressed in number of bits/symbol, is equal to the sum of the individual code rates $R_i = k_i/n$

$$R_s = \sum_{i=0}^{L-1} R_i = \frac{1}{n}\sum_{i=0}^{L-1} k_i = \frac{k}{n}, \quad k = \sum_{i=0}^{L-1} k_i. \tag{6.9}$$

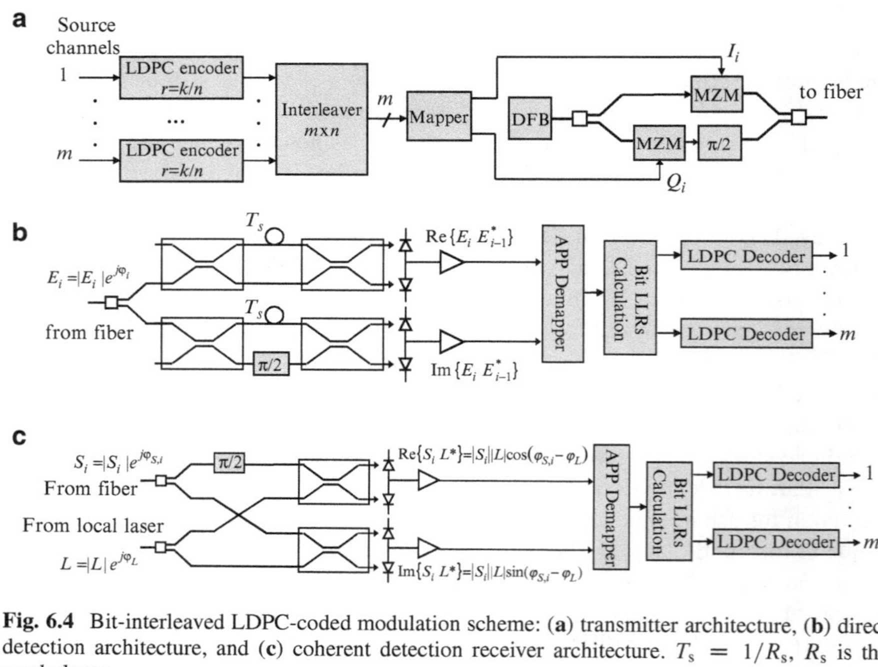

Fig. 6.4 Bit-interleaved LDPC-coded modulation scheme: (**a**) transmitter architecture, (**b**) direct detection architecture, and (**c**) coherent detection receiver architecture. $T_s = 1/R_s$, R_s is the symbol rate

The receiver input electrical field at time instance i for an optical M-ary DPSK receiver configuration from Fig. 6.4b is denoted by $E_i = |E_i| \exp(j\varphi_i)$. The outputs of I- and Q-branches (upper and lower-branches in Fig. 6.4b) are proportional to $\mathrm{Re}\{E_i E_{i-1}^*\}$ and $\mathrm{Im}\{E_i E_{i-1}^*\}$, respectively. The phase difference $\Delta\varphi_k = \varphi_k - \varphi_{k-1}$ can be determined from the ratio of the lower- and upper-branch outputs by

$$\frac{\mathrm{Im}\{E_k E_{k-1}^*\}}{\mathrm{Re}\{E_k E_{k-1}^*\}} = \frac{|E_k|\,|E_{k-1}| \sin \Delta\varphi_k}{|E_k|\,|E_{k-1}| \cos \Delta\varphi_k} = \tan \Delta\varphi_k. \tag{6.10}$$

Notice that the phase difference can be determined from the lower- and upper-branch signals even when the amplitude of pulses into two consecutive intervals $|E_k|$ and $|E_{k-1}|$ are different. Therefore, the arbitrary QAM constellation, including square constellation can be applied.

As an illustration, the eye diagrams of a transmitted and a received RZ-8-DPSK signal at 40 Giga symbols/s (40 GS/s) are shown in Fig. 6.5. The average power spectral densities of RZ-DPSK and RZ-8-DPSK are shown in Fig. 6.6. The eye diagrams of a transmitted and a received 8-DQAM signals are shown in Fig. 6.7a, b, while the power spectral density is given in Fig. 6.7c. Notice that the main lobe widths of 8-DPSK and 8-DQAM spectra are about three times lower than that of DPSK.

The outputs at I- and Q-branches (in either coherent or direct detection case), are sampled at the symbol rate, while the symbol log-likelihood ratios (LLRs) are calculated in a posteriori probability (APP) demapper block as follows:

$$\lambda(s) = \log \frac{P(s = s_0|\boldsymbol{r})}{P(s \neq s_0|\boldsymbol{r})}, \tag{6.11}$$

where $P(s|\boldsymbol{r})$ is determined by using Bayes' rule

$$P(s|\boldsymbol{r}) = \frac{P(\boldsymbol{r}|s)P(s)}{\sum_{s'} P(\boldsymbol{r}|s')P(s')}. \tag{6.12}$$

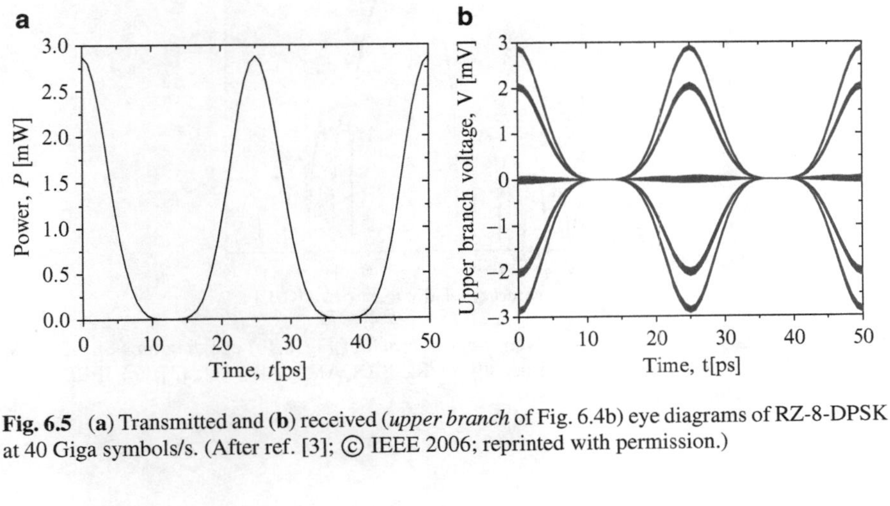

Fig. 6.5 (**a**) Transmitted and (**b**) received (*upper branch* of Fig. 6.4b) eye diagrams of RZ-8-DPSK at 40 Giga symbols/s. (After ref. [3]; © IEEE 2006; reprinted with permission.)

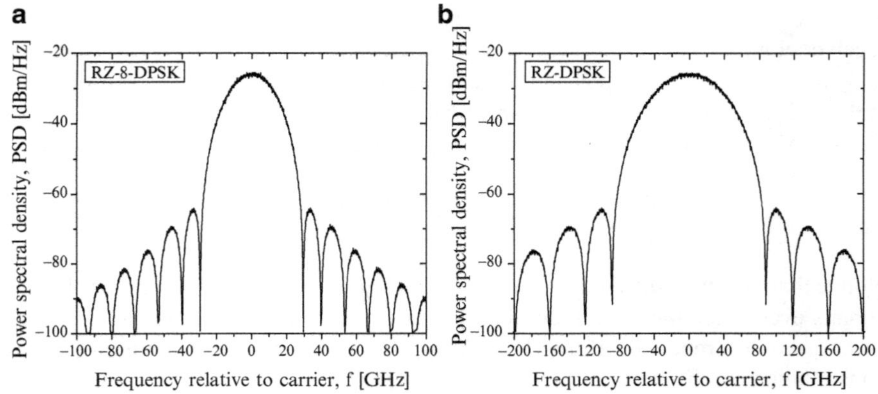

Fig. 6.6 Power spectral densities of (**a**) RZ-8-DPSK, and (**b**) RZ-DPSK at 40 Giga symbols/s (duty cycle: 0.33). (After ref. [3]; © IEEE 2006; reprinted with permission.)

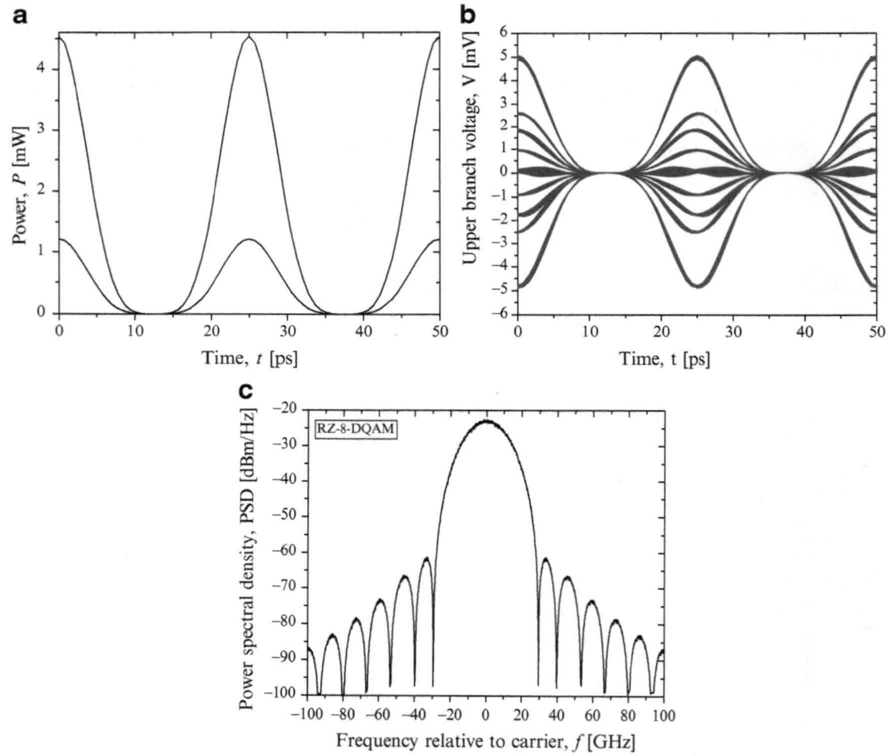

Fig. 6.7 (**a**) Transmitted and (**b**) received (*upper branch* of Fig. 6.4b) eye diagrams of 8-DQAM at 40 Gsymbols/s. (**c**) Power spectral density of RZ-8-DQAM. (After ref. [3]; © IEEE 2006; reprinted with permission.)

$\boldsymbol{s} = (I_i, Q_i)$ denotes the transmitted signal constellation point at time instance i, while $\boldsymbol{r} = (r_{\mathrm{I}}, r_{\mathrm{Q}})$ denotes the received point. ($r_{\mathrm{I}} = v_{\mathrm{I}}(t = iT_{\mathrm{s}})$, and $r_{\mathrm{Q}} = v_Q(t = iT_{\mathrm{s}})$ are the samples of I- and Q-detection branches from Fig. 6.4b, c.) $P(\boldsymbol{r}|\boldsymbol{s})$ from (6.12) is estimated by evaluation of histograms, employing sufficiently long training sequence. With $P(\boldsymbol{s})$ we denoted the a priori probability of symbol $\boldsymbol{s}$, while $\boldsymbol{s}_0$ is a referent symbol. The bit LLRs $c_j (j = 1, 2, \ldots, m)$ are determined from symbol LLRs of (6.11) as

$$L(\hat{c}_j) = \log \frac{\sum_{\boldsymbol{s}:c_j=0} \exp[\lambda(\boldsymbol{s})]}{\sum_{\boldsymbol{s}:c_j=1} \exp[\lambda(\boldsymbol{s})]}. \tag{6.13}$$

The APP demapper extrinsic LLRs (the difference of demapper bit LLRs and LDPC decoder LLRs from previous step) for LDPC decoders become

$$L_{M,e}(\hat{c}_j) = L(\hat{c}_j) - L_{D,e}(c_j). \tag{6.14}$$

With $L_{D,e}(c)$ we denoted LDPC decoder extrinsic LLRs, which is initially set to zero value. The LDPC decoder is implemented by employing the sum–product algorithm. The LDPC decoders extrinsic LLRs (the difference between LDPC decoder output and the input LLRs), $L_{D,e}$, are forwarded to the APP demapper as a priori bit LLRs ($L_{M,a}$), so that the symbol a priori LLRs are calculated as

$$\lambda_a(\boldsymbol{s}) = \log P(\boldsymbol{s}) = \sum_{j=0}^{m-1} (1 - c_j) L_{D,e}(c_j). \tag{6.15}$$

By substituting (6.15) into (6.12), and then (6.11), we are able to calculate the symbol LLRs for the subsequent iteration. The iteration between the APP demapper and LDPC decoder is performed until the maximum number of iterations is reached, or the valid codewords are obtained.

For convergence behavior analysis, the EXIT chart analysis should be performed [20]. In order to determine the mutual information (MI) transfer characteristics of the demapper, we model the a priori input LLR, $L_{\mathrm{M,a}}$, as a conditional Gaussian random variable [20]. The MI between c and $L_{\mathrm{M,a}}$ is determined numerically as explained in [20–23]. Similarly, the MI $I_{LM,e}$ between c and $L_{\mathrm{M,e}}$ is calculated numerically, but with the p.d.f. of c and $L_{\mathrm{M,e}}$ determined form histogram obtained by Monte Carlo simulation, as explained in [20]. By observing the $I_{LM,e}$ as a function of the MI of $I_{LM,a}$ and optical signal-to-noise ratio, *OSNR*, in dB, the demapper EXIT characteristic (denoted as T_{M}) is given by

$$I_{LM,e} = T_M(I_{LM,a}, \mathrm{OSNR}). \tag{6.16}$$

The EXIT characteristic of LDPC decoder (denoted as T_{D}) is defined in a similar fashion as

$$I_{LD,e} = T_D(I_{LD,a}). \tag{6.17}$$

The "turbo" demapping-based receiver operates by passing extrinsic LLRs between demapper and LDPC decoder. The iterative process starts with an initial demapping in which $L_{M,a}$ is set to zero, and as a consequence, $I_{LM,a}$ becomes zero as well. The demapper output LLRs, described by

$$I_{LM,e} = I_{LD,a} \tag{6.18}$$

are fed to LDPC decoder. The LDPC decoder output LLRs, described by

$$I_{LD,e} = I_{LM,a} \tag{6.19}$$

are fed to the APP demaper. The iterative procedure is repeated until the convergence or the maximum number of iterations has been reached. This procedure is illustrated in Fig. 6.8, where the APP demapper and LDPC decoder EXIT charts are shown together on the same graph. Three modulation formats (8-PSK, 16-PSK, 16-QAM) are observed, as well as the following mappings: natural, Gray, and anti-Gray. The EXIT curves have different slopes for different mappings. The existence of "tunnel" between corresponding demapping and decoder curves indicates that iteration between demapper and decoder will be successful. The smallest OSNR, at which iterative scheme start to converge, is known as threshold (pinch-off) limit [20]. The threshold limit in the case of 16-PSK (Fig. 6.8b) is about 3 dB worse as compared to 8-PSK (Fig. 6.8a). The 16-QAM mapping curve is well above the 16-PSK curve (see Fig. 6.8b), indicating that 16-QAM scheme is going to significantly outperform the 16-PSK one.

The results of simulations for 30 iterations in the sum–product algorithm and 10 APP demapper-LDPC decoder iterations for an AWGN channel model are shown in Fig. 6.9a. The information symbol rate is set to 40 GS/s, while 8-PSK is employed, so that the aggregate bit rate becomes 120 Gb/s. Two different mappers are considered: Gray and natural mapping.

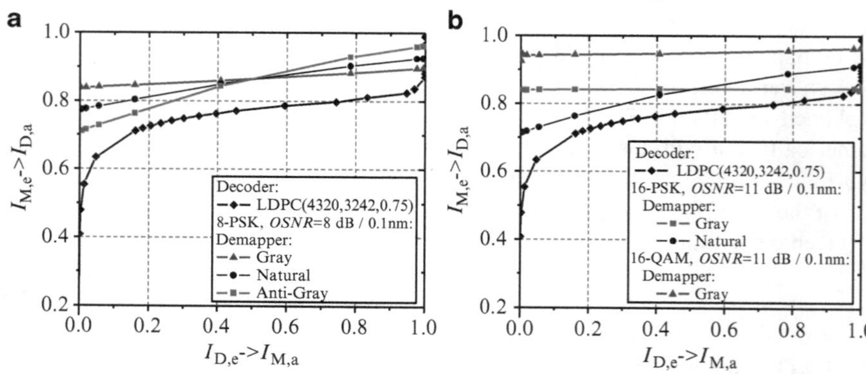

Fig. 6.8 The EXIT chart for different mappings and modulations. (After ref. [4]; © IEEE 2007; reprinted with permission.)

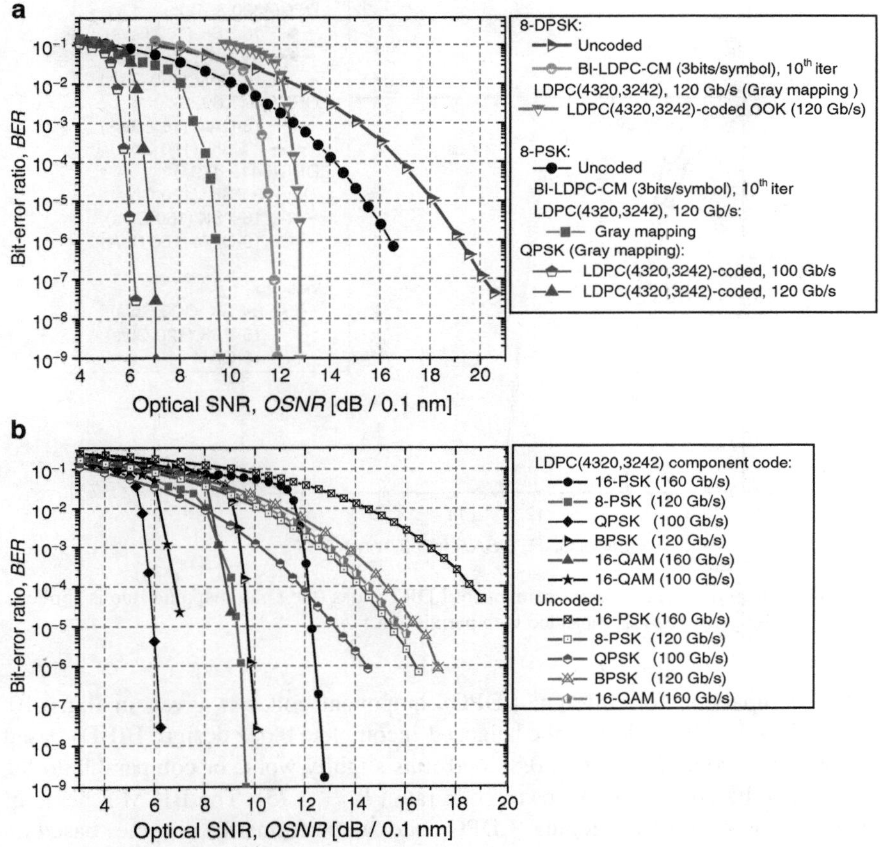

Fig. 6.9 (**a**) BER performance of coherent detection bit interleaved LDPC-coded modulation scheme versus direct detection one on an AWGN channel model, and (**b**) performance comparison for different modulation schemes (the Gray mapping rule is applied). (After ref. [4]; © IEEE 2007; reprinted with permission.)

The coding gain for 8-PSK at bir-error ratio (BER) of 10^{-9} is about 9.5 dB, and much larger coding gain is expected at BER below 10^{-12}. The coherent detection scheme offers an improvement of at least 2.3 dB as compared to the corresponding direct detection scheme. The BER performance of coherent BICM with LDPC(4320,3242) code, employed as component code for different modulations, is shown in Fig. 6.9b. We can see that 16-QAM (with an aggregate rate of 160 Gb/s) outperforms 16-PSK by more than 3 dB. It is also interesting that 16-QAM slightly outperforms 8-PSK scheme of lower aggregate data rate (120 Gb/s). The 8-PSK scheme of aggregate rate of 120 Gb/s outperforms BPSK scheme of data rate 120 Gb/s. Moreover, since the transmission symbol rate for 8-PSK is 53.4 GS/s, the impact of PMD and intrachannel nonlinearities is much less important than that at 120 G/s. Consequently, for 100 Gb/s Ethernet transmission, it is better to multiplex two 50 Gb/s channels than four 25 Gb/s channels.

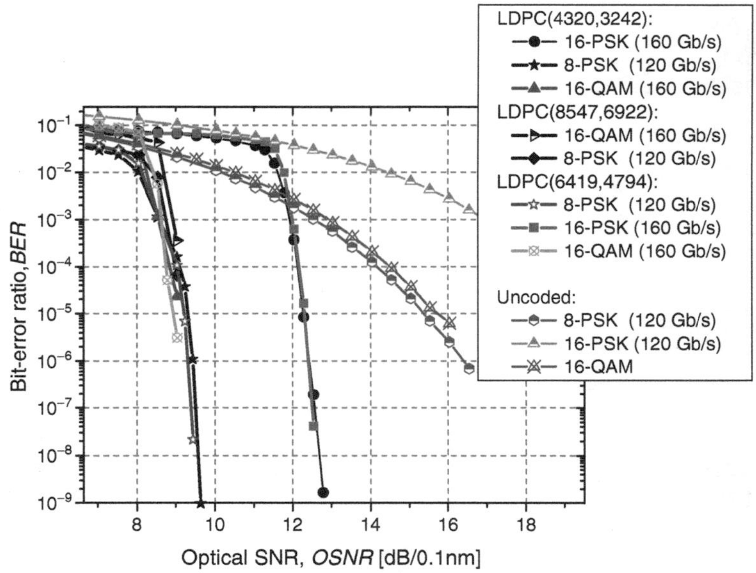

Fig. 6.10 Comparison for different component LDPC codes (the Gray mapping rule is applied). (After ref. [4]; © IEEE 2007; reprinted with permission.)

The comparison for different LDPC component codes is given in Fig. 6.10. The BICM scheme employing the balanced-incomplete block design (BIBD)-based girth-8 LDPC code [21] of rate 0.81 performs slightly worse or comparable to the quasi-cyclic-based scheme of lower code rate ($R = 0.75$). The BICM scheme of rate 0.75, based on PBD irregular LDPC code, outperforms the schemes based on regular LDPC codes.

6.3 Multidimensional Coded Modulation Schemes

In this section, we describe another approach to increase the spectral efficiency based on multilevel multidimensional (N-dimensional) coded modulation [5, 6]. By increasing the number of dimensions (i.e., the number of orthonormal basis functions) we can increase the aggregate rate of the system, while enabling reliable transmission at these higher speeds using LDPC codes at each level. The transmitter and receiver block diagrams are shown in Fig. 6.11. As shown in the setup, N different bit streams coming from different information sources are encoded using identical LDPC codes. The outputs of the encoders are interleaved by the ($N \times n$) block interleaver. The block interleaver accepts data from the encoders row-wise, and outputs the data column-wise to the mapper that accepts N bits at time instance i. The mapper determines the corresponding M-ary ($M = 2^N$) signal constellation point

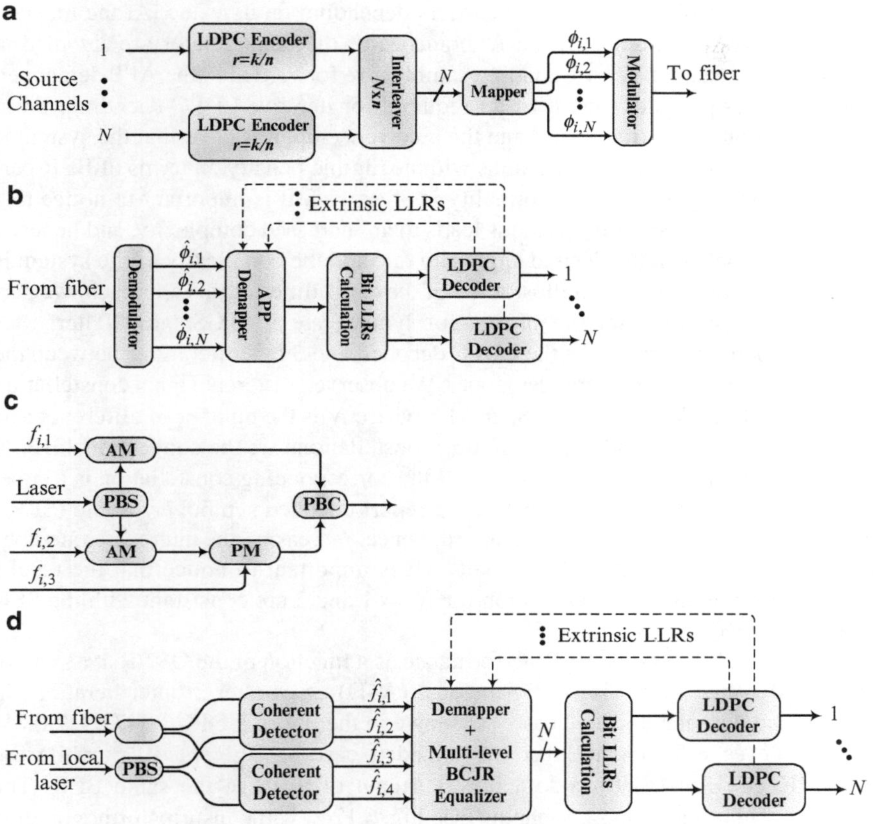

Fig. 6.11 Multidimensional LDPC-coded modulation scheme: (**a**) transmitter and (**b**) receiver configurations. (**c**) HAPP modulator, (**d**) HAPP receiver configuration

$$s_i = N^{-1/2} \sum_{j=1}^{N} \varphi_{i,j} \Phi_j, \tag{6.20}$$

as shown in Fig. 6.11a. After that the signals are modulated and sent over the fiber. In (6.20), which represents the general formula applicable to any N-dimensional constellation, the set $\{\Phi_1, \Phi_2, \ldots, \Phi_N\}$ represents a set of N orthonormal basis functions. The number N is determined by the desired final rate and the availability of orthogonal functions. If we take, for example, the desired aggregate rate to be 200 Gb/s with symbol rate of 50 GS/s, then we have to find four orthogonal functions to achieve 4 bits/symbol. For instance, phase provides the in-phase and the quadrature components; hence, the first two bits define the mapping of the signal onto a conventional QPSK, while the third bit defines the polarization. The mapper will map a 1 in the third bit to the x-polarization and 0 to the y-polarization. The fourth component can be the frequency. In this example, the fourth bit is mapped to one of the two available orthogonal subcarriers depending on its value. At the receiver side, Fig. 6.8b, the outputs of the N branches of the demodulator are sampled at the symbol rate and the corresponding samples are forwarded to the APP demapper. The demapper provides the bit LLRs required for iterative LDPC decoding. From the description of the transmitter and the receiver setup, it is clear that the system is scalable to any number of dimensions with negligible penalty, in terms of BER performance, as long as the orthonormality is preserved. It is important to notice that increasing the number of dimensions leads to an increased complexity, and hence, a compromise between the desired aggregate rate and the complexity of the system is to be made in practice. As an illustration of the capabilities of this approach, we perform simulations assuming the information symbol rate of 40 GS/s, for 20 iterations of sum–product algorithm for the LDPC decoder, and 3 outer iterations between the LDPC decoder and the natural demapper. We observed different signal constellation formats, such as: $N = 1, 2, 3, 4, 8$, and 10, where N is the number of different basis functions. For $N = 1$ and 2, the resulting constellations are the conventional BPSK and QPSK, respectively, while for $N = 3$ the corresponding constellation is a cube. As noticed from the Fig. 6.12a, where we report uncoded symbol error rate (SER), increasing the number of dimensions and, hence, increasing the aggregate rate does not introduce significant OSNR penalty. (It is important to notice that the results achieved for the uncoded modulation for $N = 1$ and 2 are consistent with the SER reported in [13].)

Figure 6.12b shows the BER performance as a function of the OSNR at a symbol rate of 40 GS/s for the six N-dimensional (N-D) cases after 3 outer iterations in comparison with the uncoded case. As shown in the figure, LDPC(8547,6922,0.81) code achieves 8.5 dB gain over the uncoded case at BER of 10^{-8} while the LDPC(16935,13550,0.80) code achieves a gain of 9 dB at the same BER. The improvement of the proposed scheme (see Fig. 6.12c), while insuring orthogonality, is computed at BER of 10^{-9} to be: 3 dB over 8-QAM, 9.75 dB over 3D-constellation with 256 constellation points we introduced in [5], and 14 dB over 256-QAM.

Another example of multidimensional coded modulation to be described here is a particular instance of hybrid amplitude/phase/polarization (HAPP) coded modulation. Using transmission equipment operating at 40 GS/s, the HAPP scheme achieves $N \times 40\,\text{Gb/s}$ aggregate rates where $N = 3, 4, \ldots$ represents the total number of source channels connected to the transmitter. The transmitter of the HAPP coded-modulation scheme accepts input from N different bit streams coming from different information sources and feeds them into N identical encoders. The number N is determined based on the desired final rate. The encoding process is similar to that explained above. The mapper accepts N bits at a time and maps them into a 2^N-ary signal constellation point on a vertex of a regular polyhedron inscribed in a Poincaré sphere (see Fig. 6.13) based on a lookup table. The signal is then modulated by the HAPP modulator and sent to the fiber, as shown in Fig. 6.11a. The lookup table maps each N bits into a triple $(f_{1,i}, f_{2,i}, f_{3,i})$ where i is the time index and f_j $(j = 1, 2, 3)$ are the voltages needed to control a set of modulators. As, the polyhedrons used are inscribed in a Poincaré sphere, Stokes parameters [24]

$$s_1 = a_x^2 - a_y^2, \quad s_2 = 2a_x a_y \cos(\delta), \quad s_3 = 2a_x a_y \sin(\delta), \quad \delta = \varphi_x - \varphi_y \tag{6.21}$$

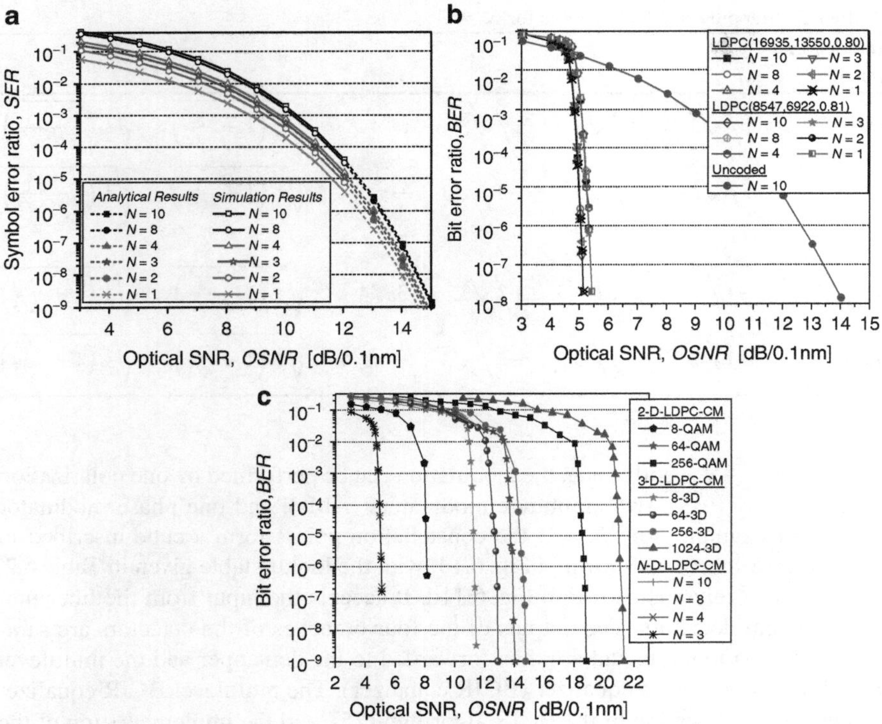

Fig. 6.12 (**a**) SER performance of uncoded modulation for different values of N (*solid lines* simulation results, *dashed lines* analytical results); (**b**) BER performance of N-D-LDPC-CM schemes for different LDPC codes, in comparison with the uncoded case; (**c**) BER performance of the 2D, 3D, and N-D-LDPC-CM schemes for different constellations

Fig. 6.13 Signal constellation in Stokes coordinates

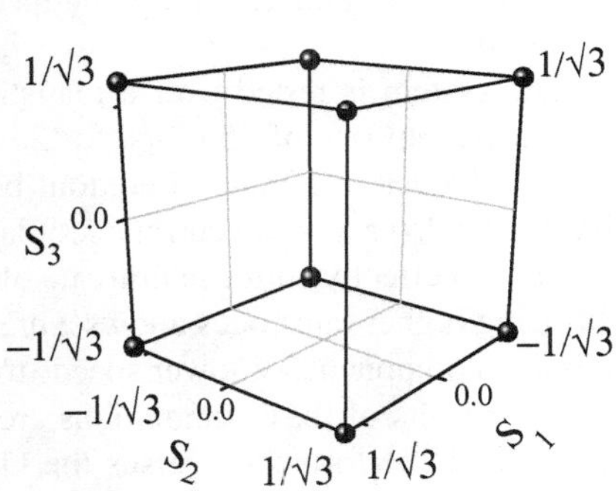

are used for the design of the polyhedron. In (6.21), $a_x(a_y)$ corresponds to the amplitude of electrical field in x-polarization (y-polarization) and $\phi_x(\phi_y)$ is the corresponding phase. The HAPP modulator is shown in Fig. 6.11c. By setting

Table 6.2 Mapping rule lookup table for $N = 3$

Interleaver output	s_1	s_2	s_3	δ	a_x	a_y
000	$1/\sqrt{3}$	$1/\sqrt{3}$	$1/\sqrt{3}$	$\pi/4$	$\sqrt{\frac{1}{2}\left(1+\frac{1}{\sqrt{3}}\right)}$	$\sqrt{\frac{1}{2}\left(1-\frac{1}{\sqrt{3}}\right)}$
001	$1/\sqrt{3}$	$1/\sqrt{3}$	$-1/\sqrt{3}$	$-\pi/4$	$\sqrt{\frac{1}{2}\left(1+\frac{1}{\sqrt{3}}\right)}$	$\sqrt{\frac{1}{2}\left(1-\frac{1}{\sqrt{3}}\right)}$
				$\vdots$		
110	$-1/\sqrt{3}$	$-1/\sqrt{3}$	$1/\sqrt{3}$	$3\pi/4$	$\sqrt{\frac{1}{2}\left(1-\frac{1}{\sqrt{3}}\right)}$	$\sqrt{\frac{1}{2}\left(1+\frac{1}{\sqrt{3}}\right)}$
111	$-1/\sqrt{3}$	$-1/\sqrt{3}$	$-1/\sqrt{3}$	$-3\pi/4$	$\sqrt{\frac{1}{2}\left(1-\frac{1}{\sqrt{3}}\right)}$	$\sqrt{\frac{1}{2}\left(1+\frac{1}{\sqrt{3}}\right)}$

$\phi_x = 0$, $\delta = -\phi_y$, and hence, the modulation can be performed by one polarization beam splitter (PBS), two amplitude modulators (AMs), and one phase modulator (PM). For example, for $N = 3$ the constellation points form a cube inscribed in the Poincaré sphere, as shown in Fig. 6.13, with the lookup table given in Table 6.2. The HAPP receiver is shown in Fig. 6.11d. It accepts the input from the fiber into two coherent detectors. The outputs of the four branches of the detectors are sampled at the symbol rate and are then forwarded to the demapper and the multilevel BCJR algorithm-based equalizer (BCJR equalizer). The multilevel BCJR equalizer is a generalized version of the BCJR algorithm [25], and the implementation of the equalizer will be described later in Chap. 7. The output of the equalizer is then forwarded to the bit LLRs calculator which provides the LLRs required for the LDPC decoding process. The LDPC decoder forwards the extrinsic LLRs to the BCJR equalizer, and the extrinsic information is iterated back and forth between the decoder and the equalizer for a predefined number of iterations unless convergence is achieved.

The system is tested over a realistic model using VPITransmisionMaker [26], with a symbol rate of 45 GS/s, for 25 iterations of sum–product algorithm for the LDPC decoder, and 3 outer iterations between the LDPC decoder and the multilevel BCJR equalizer. For the current simulation, LDPC(4320,3242) of rate 0.75 is used. The actual effective information rate of the system is $3 \times 45 \times 0.75 = 101.25\,\text{Gb/s}$. Utilizing higher rate codes allows a higher actual transmission rate, or allows transmission components of lower speeds to achieve the current transmission rate.

The results of these simulations are summarized in Fig. 6.14. We show the uncoded BER performance versus the OSNR per bit for different differential group delays (DGDs) and for different memories m, where m represents the number of symbols preceding and succeeding the symbol of concern. As noticed from the figure, the back-to-back configuration reaches the BER of 10^{-6} at OSNR of 12 dB, and the configuration with DGD of $0.6T$ (where T is the symbol-time) reaches the BER of 10^{-6} beyond OSNR of 19 dB. As the DGD value increases above $1T$, the equalizer memory needs to be extended in order to avoid error floor. This is obvious from the figure, where we show that for DGDs $1.2T$, and $2T$, the symbol-by-symbol

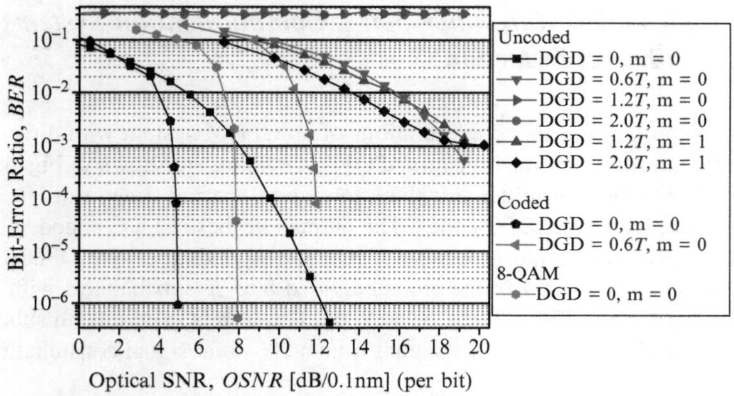

Fig. 6.14 BER performance versus the OSNR per bit for both uncoded and LDPC-coded data

equalizer enters the error floor at BER of 0.33. On the other hand, for DGD of $1.2T$, the BCJR equalizer does not enter the error floor for $m = 1$. For DGD of $2.0T$ BCJR equalizer enters the error floor around BER of 1×10^{-3} for $m = 1$, which can be avoided by increasing the equalizer memory. Since the BCJR equalizer error floor is below the BER threshold of the LDPC code (above 10^{-2}), then $m = 1$ is sufficient for the coded case. Notice that configuration for DGD of $2T$ outperforms that of DGD $1.2T$ which is in agreement with the results published in [27] for binary direct detection systems. For the coded case, we show in Fig. 6.14 the BER performance for the proposed scheme in the back-to-back configuration and for DGD of $0.6T$, in addition to 8-QAM for comparison purposes. As noticed, the proposed scheme outperforms 8-QAM by 3 dB at BER 10^{-6}.

6.4 Coded OFDM in Fiber-Optics Communication Systems

OFDM [28, 29] is an efficient approach to deal with intersymbol interference (ISI) due to CD and PMD [29–42]. By providing that the guard interval is larger than the combined delay spread due to CD and maximum DGD, the ISI can be eliminated successfully. However, the four-wave mixing (FWM) between different subcarriers and its interplay with CD and PMD will result in different subcarriers being affected differently. Even though that most of the subcarriers are without errors, the overall BER will be dominated by BER of the worst subcarriers. In order to avoid this problem, the use of forward error correction (FEC) is essential. The use of advanced FEC schemes is needed to provide that BER performance of an OFDM system that is determined by the average received power rather than by the power of the weakest subcarrier. In this section, we describe coded OFDM with both direct detection (Sect. 6.4.1) and coherent detection (Sect. 6.4.2).

6.4.1 Coded OFDM in Fiber-optics Communication Systems with Direct Detection

The transmitter and receiver configurations of an OFDM system with direct detection [7, 29] and format of the transmitted OFDM symbol are shown in Fig. 6.15a–c, respectively. On the transmitter side the information-bearing streams at 10 Gb/s are encoded using identical LDPC codes. The outputs of these LDPC encoders are demultiplexed and parsed into groups of B bits corresponding to one OFDM frame. The B bits in each OFDM frame are subdivided into K subchannels with the ith subcarrier carrying b_i bits, so that $B = \sum b_i$. The b_i bits from the ith subchannel are mapped into a complex-valued signal from a 2^{b_i}-point signal constellation such

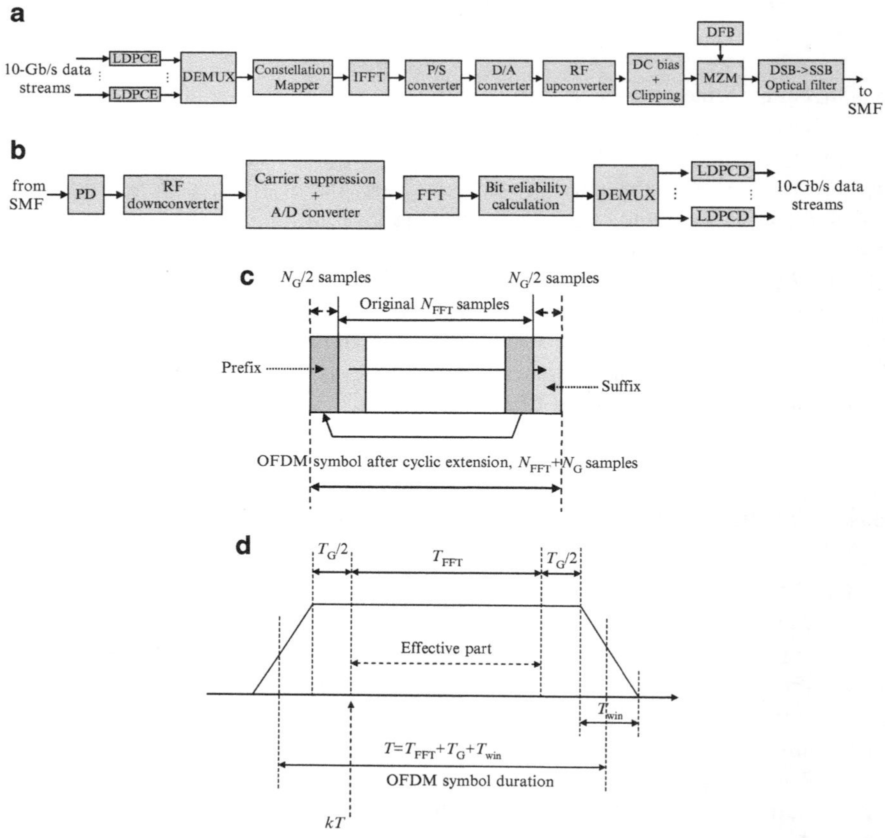

Fig. 6.15 (**a**) Transmitter configuration, (**b**) receiver configuration, (**c**) OFDM symbol after cyclic extension, and (**d**) OFDM symbol after windowing. *LDPCE* LDPC encoder, *LDPCD* LDPC decoder, *S/P* serial-to-parallel converter, *MZM* Mach–Zehnder modulator, *SMF* single-mode optical fiber, *PD* photodetector, *DSB* double-sideband, *SSB* single-sideband

as M-ary QAM and M-ary PSK. For example, $b_i = 2$ for QPSK and $b_i = 4$ for 16-QAM. The complex-valued signal points from subchannels are considered to be the values of the fast Fourier transform (FFT) of a multicarrier OFDM signal. By selecting the number of subchannels K, sufficiently large, OFDM symbol interval can be made significantly larger than the dispersed pulse-width of an equivalent single-carrier system, resulting in ISI reduction. The OFDM symbol, shown in Fig. 6.15c, is generated as follows: $N_{\text{QAM}}(= K)$ input QAM symbols are zero-padded to obtain N_{FFT} input samples for IFFT, N_{G} nonzero samples are inserted (as explained below) to create the guard interval, and the OFDM symbol is multiplied by the window function. The OFDM symbol windowing is illustrated in Fig. 6.15d. The purpose of cyclic extension is to preserve the orthogonality among subcarriers even when the neighboring OFDM symbols partially overlap due to dispersion, and the role of windowing is to reduce the out-of-band spectrum. For efficient chromatic dispersion and PMD compensation, the length of cyclically extended guard interval should be larger than the spread due to chromatic dispersion and PMD.

The cyclic extension, illustrated in Fig. 6.15c, is accomplished by repeating the last $N_{\text{G}}/2$ samples of the effective OFDM symbol part (N_{FFT} samples) as a prefix, and repeating the first $N_{\text{G}}/2$ samples as a suffix. After D/A conversion and RF up-conversion, the RF signal can be mapped to the optical domain using one of two possible options: (1) the OFDM signal can directly modulate a DFB laser, or (2) the OFDM signal can be used as the RF input of a Mach–Zehnder modulator (MZM). A *DC bias component* is added to the OFDM signal in order to enable recovery of the QAM symbols incoherently. In what follows, three different OFDM schemes are described. The first scheme is based on direct modulation, and shall be referred to as the "biased-OFDM" (B-OFDM) scheme. Because bipolar signals cannot be transmitted over an IM/DD link, it is assumed that the bias component is sufficiently large so that when added to the OFDM signal the resulting sum is nonnegative. The main disadvantage of the B-OFDM scheme is its poor power efficiency. To improve the OFDM power efficiency, two alternative schemes can be used. The first scheme, which we shall refer to as the "clipped-OFDM" (C-OFDM) scheme, is based on single-sideband (SSB) transmission, and clipping of the OFDM signal after bias addition. The bias is varied in order to find the optimum one for fixed optical launched power. It was found that the optimum bias is one in which $\sim$50% of the total electrical signal energy before clipping is allocated for transmission of a carrier. The second power-efficient scheme, which we shall refer to as the "unclipped-OFDM" (U-OFDM) scheme, is based on SSB transmission using a $LiNbO_3$ MZM. To avoid distortion due to clipping at the transmitter, the information is mapped into the optical domain by modulating the electrical field of the optical carrier (instead of intensity modulation employed in the B-OFDM and C-OFDM schemes). In this way, both positive and negative portions of the electrical OFDM signal are transmitted to the photodetector. Distortion introduced by the photodetector, caused by squaring, is successfully eliminated by proper filtering, and recovered signal does not exhibit significant distortion. It is important to note, however, that the U-OFDM scheme is slightly less power efficient than the C-OFDM scheme. The SSB modulation can be achieved either by appropriate optical filtering the double-side band signal at MZM output (see Fig. 6.15a) or by using the Hilbert transformation of in-phase component of OFDM RF signal. The first version requires the use of only in-phase component of RF OFDM signal, providing that zero-padding is done in the middle of OFDM symbol rather than at the edges.

The transmitted OFDM signal can be written as

$$s(t) = s_{\mathrm{OFDM}}(t) + D, \tag{6.22}$$

where

$$s_{\mathrm{OFDM}}(t) = \mathrm{Re}\left\{ \sum_{k=-\infty}^{\infty} w(t-kT) \sum_{i=-N_{\mathrm{FFT}}/2}^{N_{\mathrm{FFT}}/2-1} X_{i,k} \cdot \exp\left(\mathrm{j}2\pi \frac{i}{T_{\mathrm{FFT}}}(t-kT) \right) \mathrm{e}^{\mathrm{j}2\pi f_{\mathrm{RF}} t} \right\} \tag{6.23}$$

is defined for $t \in [kT - T_{\mathrm{G}}/2 - T_{\mathrm{win}}, kT + T_{\mathrm{FFT}} + T_{\mathrm{G}}/2 + T_{\mathrm{win}}]$. In the above expression, $X_{i,k}$ denotes the ith subcarrier of the kth OFDM symbol, $w(t)$ is the window function, and f_{RF} is the RF carrier frequency. T denotes the duration of the OFDM symbol, T_{FFT} denotes the FFT sequence duration, T_{G} is the guard interval duration (the duration of cyclic extension), and T_{win} denotes the windowing interval duration. D denotes the DC bias component, which is introduced to enable the OFDM demodulation using the direct detection.

The PIN photodiode output current can be written as

$$i(t) = R_{\mathrm{PIN}} \left\{ [s_{\mathrm{OFDM}}(t) + D] * h(t) + N(t) \right\}^2, \tag{6.24}$$

where $s_{\mathrm{OFDM}}(t)$ denotes the transmitted OFDM signal in RF domain given by (6.23). D is introduced above, while R_{PIN} denotes the photodiode responsivity. The impulse response of the optical channel is represented by $h(t)$. The signal after RF down-conversion and appropriate filtering, can be written as

$$r(t) = [i(t) k_{\mathrm{RF}} \cos(\omega_{\mathrm{RF}} t)] * h_e(t) + n(t), \tag{6.25}$$

where $h_{\mathrm{e}}(t)$ is the impulse response of the low-pass filter, $n(t)$ is electronic noise in the receiver, and k_{RF} denotes the RF down-conversion coefficient. Finally, after the A/D conversion and cyclic extension removal, the signal is demodulated by using the FFT algorithm. The soft outputs of the FFT demodulator are used to estimate the bit reliabilities that are fed to identical LDPC iterative decoders implemented based on the sum–product with correction term algorithm as we explained above.

For the sake of illustration, let us consider the signal waveforms and power-spectral densities (PSDs) at various points in the OFDM system given in Fig. 6.15. These examples were generated using SSB transmission in a back-to-back configuration. The bandwidth of the OFDM signal is set to B GHz, and the RF carrier to $0.75B$; where B denotes the aggregate data rate. The number of OFDM subchannels is set to 64, the OFDM sequence is zero-padded, and the FFT is calculated using 128 points. The guard interval is obtained by a cyclic extension of 2×16 samples. The

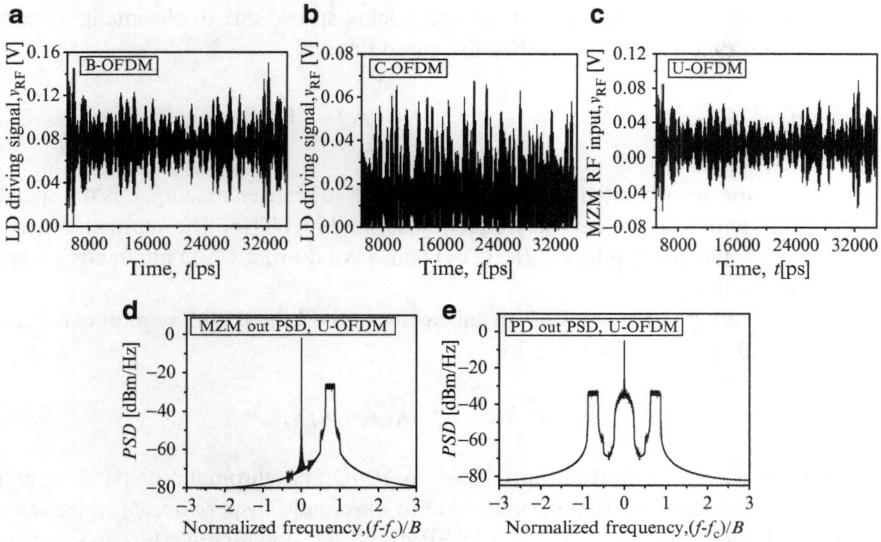

Fig. 6.16 Waveforms and PSDs of SSB QPSK-OFDM signal at different points during transmission for electrical SNR (per bit) of 6 dB. (f_c the optical carrier frequency, LD the laser diode.)

average transmitted launch power is set to 0 dBm. The OFDM transmitter parameters are carefully chosen such that RF driver amplifier and MZM operate in linear regime (see Fig. 6.16a–c). The PSDs of MZM output signal, and the photodetector output signal are shown in Fig. 6.16d, e, respectively. The OFDM term after beating in the PD, the low-pass term, and the squared OFDM term can easily be identified.

The received electrical field, at the input of the transimpedance amplifier (TA), in the presence of chromatic dispersion and first-order PMD, can be represented by

$$E(t) = \mathrm{FT}^{-1}\left\{\mathrm{FT}\left\{E_0\begin{bmatrix}\sqrt{1-\kappa}\\ \sqrt{k}\mathrm{e}^{\mathrm{j}\delta}\end{bmatrix}[s_{\mathrm{OFDM}}(t)+D]\right\}\right.$$
$$\left.\times\exp\left[\mathrm{j}\left(\frac{\beta_2\omega^2}{2}-\frac{\beta_3\omega^3}{6}\right)L_{\mathrm{tot}}\right]\right\}+\begin{bmatrix}N_x(t)\\ N_y(t)\end{bmatrix}, \quad (6.26)$$

where β_2 and β_3 represent the group-velocity dispersion (GVD) and second-order GVD parameters, L_{tot} is the total SMF length, k is the splitting ratio between two principal states of polarization (PSPs), δ is the phase difference between PSPs, E_0 is transmitted laser electrical field amplitude, and N_x and N_y represent x- and y-polarization ASE noise components. With FT and FT^{-1} we denoted the Fourier transform and inverse Fourier transform, respectively. The TA output signal can be represented by $v(t) = R_{\mathrm{F}}R_{\mathrm{PIN}}|E(t)|^2 + n(t)$, where R_{PIN} is the photodiode responsivity, R_{F} is the TA feedback resistor, and $n(t)$ is TA thermal noise.

For complete elimination of ISI, the total delay spread due to chromatic dispersion and DGD should be shorter than the guard interval:

$$|\beta_2|\, L_{\text{tot}}\Delta\omega + \text{DGD}_{\max} = \frac{c}{f^2}\,|D_t|\, N_{\text{FFT}}\Delta f + \text{DGD}_{\max} \le T_G, \qquad (6.27)$$

where D_t is the accumulated dispersion, Δf is the subcarrier spacing, c is the speed of the light, and f is the central frequency set to 193.1 THz. The number of subcarriers N_{FFT}, the guard interval T_{G}, GVD and second-order GVD parameters were introduced earlier.

The received QAM symbol of ith subcarrier in the kth OFDM symbol is related to transmitted QAM symbol $X_{i,k}$ by

$$Y_{i,k} = h_i e^{j\theta_i} e^{j\phi_k} X_{i,k} + n_{i,k}, \qquad (6.28)$$

where h_i is channel distortion introduced by PMD and chromatic dispersion, and θ_i is the phase shift of ith subcarrier due to chromatic dispersion. ϕ_k represents the OFDM symbol phase noise due to SPM and RF down-converter, and can be eliminated by pilot-aided channel estimation. Notice that in direct detection case, the laser phase noise is completely cancelled by photodetection. To estimate the channel distortion due to PMD, h_i and phase shift due to chromatic dispersion θ_i, we need to pretransmit the training sequence. Because in ASE noise dominated scenario (considered here) the channel estimates are sensitive to ASE noise, the training sequence should be sufficiently long to average the noise. For DGDs up to 100 ps, the training sequence composed of several OFDM symbols is sufficient. For larger DGDs longer OFDM training sequence is required; alternatively, the channel coefficients can be chosen to maximize the LLRs or someone can use the PBS to separate the x- and y-polarization components, and consequently process them. The phase shift of ith subcarrier due to chromatic dispersion can be determined from training sequence as difference of transmitted and received phase averaged over different OFDM symbols. Once the channel coefficients and phase shifts due to PMD and chromatic dispersion are determined, in a decision-directed mode, the transmitted QAM symbols are estimated by

$$\hat{X}_{i,k} = \left(h_i^* / |h_i|^2\right) e^{-j\theta_i} e^{-j\phi_k} Y_{i,k}. \qquad (6.29)$$

The symbol LLRs $\lambda(q)(q = 0, 1, \ldots, 2^b - 1)$ can be determined by

$$\lambda(q) = -\frac{\left(\text{Re}\left[\hat{X}_{i,k}\right] - \text{Re}\left[\text{QAM}(\text{map}(q))\right]\right)^2}{N_0} - \frac{\left(\text{Im}[\hat{X}_{i,k}] - \text{Im}\left[\text{QAM}(\text{map}(q))\right]\right)^2}{N_0}, \quad q = 0, 1, \ldots, 2^b - 1 \qquad (6.30)$$

where Re[] and Im[] denote the real and imaginary part of a complex number, QAM denotes the QAM-constellation diagram, N_0 denotes the PSD of an equivalent Gaussian noise process, and map(q) denotes a corresponding mapping rule (Gray mapping is applied here). (b denotes the number of bits per constellation point.) Let us denote by v_j the jth bit in an observed symbol q binary representation $\boldsymbol{v} = (v_1, v_2, \ldots, v_b)$. The bit LLRs needed for LDPC decoding are calculated from symbol LLRs by

$$L(\hat{v}_j) = \log \frac{\sum_{q:v_j=0} \exp[\lambda(q)]}{\sum_{q:v_j=1} \exp[\lambda(q)]}. \tag{6.31}$$

Therefore, the jth bit reliability is calculated as the logarithm of the ratio of a probability that $v_j = 0$ and probability that $v_j = 1$. In the nominator, the summation is done over all symbols q having 0 at the position j, while in the denominator over all symbols q having 1 at the position j.

The results of simulation, for ASE noise-dominated scenario and single-wavelength channel transmission, are shown in Figs. 6.17–6.19, for the LDPC-coded SSB OFDM system with aggregate rate of 10 Gb/s, 512 subcarriers, RF carrier frequency of 10 GHz, oversampling factor of 2, and cyclic extension with 512 samples. The modulation format being applied is QPSK. The LDPC(16935,13550) code of girt-10, code rate 0.8, and column-weight 3 is used. In Fig. 6.17, we show the BER performance for DGD of 100 ps, without residual chromatic dispersion. We see that uncoded case faces significant performance degradation at low BERs. On the other hand, the LDPC-coded case has degradation of 1.1 dB at BER of 10^{-9} (when compared to the back-to-back configuration). In Fig. 6.18, we show the BER performance after 6,500 km of SMF (without optical dispersion compensation),

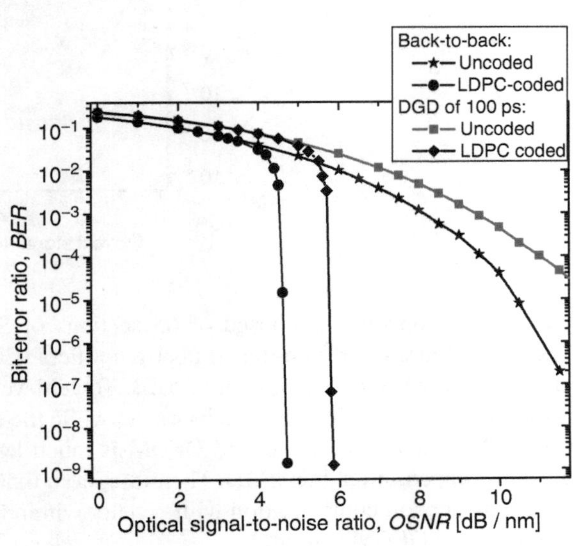

Fig. 6.17 BER performance of LDPC-coded OFDM system with aggregate rate of 10 Gb/s, for DGD of 100 ps

Fig. 6.18 BER performance of LDPC-coded OFDM system with aggregate rate of 10 Gb/s, after 6,500 km of SMF

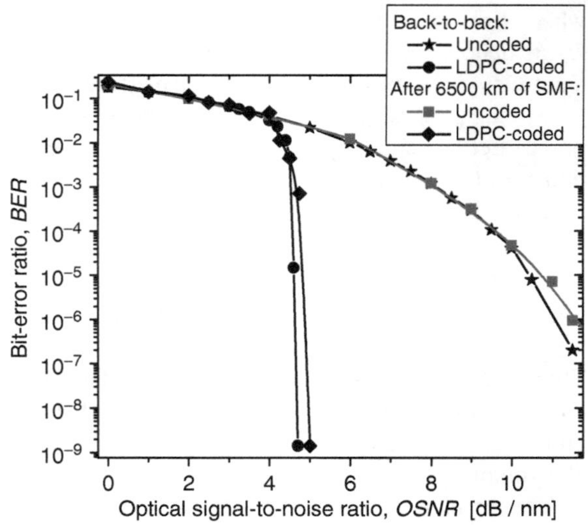

Fig. 6.19 BER performance of LDPC-coded OFDM system with aggregate rate of 10 Gb/s, after 6,500 km of SMF and for DGD of 100 ps

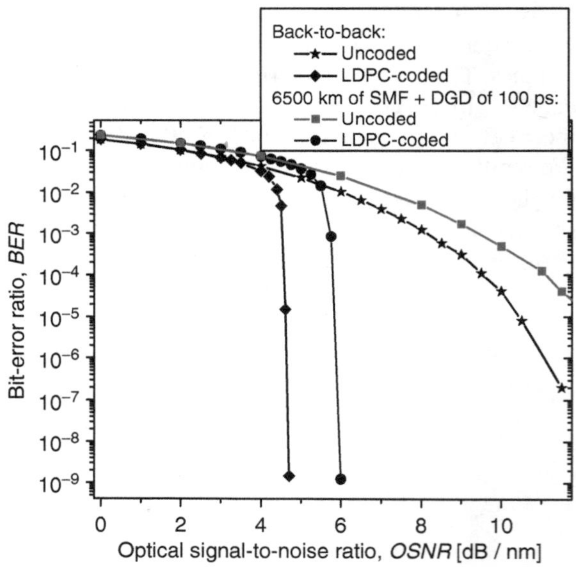

for a dispersion map composed of 65 sections of SMF with 100 km in length. The noise figure of erbium-doped fiber amplifiers (EDFAs), deployed periodically after every SMF section, was set to 5 dB. To achieve the desired OSNR, the ASE noise loading was applied on receiver side, while the launch power was kept below 0 dBm. We see that LDPC-coded OFDM is much less sensitive to chromatic dispersion compensation than PMD. Therefore, even 6,500 km can be reached without optical dispersion compensation with penalty within 0.4 dB at BER of 10^{-9}, when LDPC-coded OFDM is used.

In Fig. 6.19, the efficiency of LDPC-coded OFDM in simultaneous chromatic dispersion and PMD compensation is studied. After 6,500 km of SMF (without optical dispersion compensation) and for DGD of 100 ps, the LDPC-coded OFDM has the penalty within 1.5 dB. Notice that coded turbo equalization cannot be used at all for this level of residual chromatic dispersion and DGD. It can also be noticed that, from numerical results presented here, that the major factor of performance degradation in LDPC-coded OFDM with direct detection, in ASE noise-dominated scenario, is PMD. To improve the tolerance to PMD someone may use longer training sequences and redistribute the transmitted information among the subcarriers less affected by DGD, or to use the PBS and separately process x- and y-PSPs, in a fashion similar to that proposed for OFDM with coherent detection as described in next section; however, the complexity of such a scheme would be at least two times higher. Notice that for this level of DGD, the redistribution of power among subcarriers not being faded away is not needed. For larger values of DGDs, the penalty due to DGD grows as DGD increases, if the redistribution of subcarriers is not performed.

6.4.2 Coded OFDM in Fiber-Optics Communication Systems with Coherent Detection

The transmitter and receiver configurations for coherent detection and format of the transmitted OFDM symbol are shown in Fig. 6.20a–c, respectively. The bit streams originating from m different information sources are encoded using different (n, k_l) LDPC codes of code rate $r_l = k_l/n$. k_l denotes the number of information bits of lth ($l = 1, 2, \ldots, m$) component LDPC code, and n denotes the codeword length, which is the same for all LDPC codes. The outputs of m LDPC encoders are written row-wise into a block–interleaver block. The mapper accepts m bits at time instance i from the $(m \times n)$ interleaver column-wise and determines the corresponding M-ary ($M = 2^m$) signal constellation point $(\phi_{\mathrm{I},i}, \phi_{\mathrm{Q},i})$ in 2D constellation diagram such as M-ary PSK or M-ary QAM. The coordinates correspond to in-phase (I) and quadrature (Q) components of M-ary 2D constellation. The 2D constellation points, after serial-to-parallel (S/P) conversion, are used as the data-values for the FFT of a multicarrier OFDM signal. By selecting the number of subcarriers K, sufficiently large, OFDM symbol interval can be made significantly larger than the dispersed pulse-width of an equivalent single-carrier system, resulting in ISI reduction. The OFDM symbol, shown in Fig. 6.20c, is generated as follows: $N_{\mathrm{QAM}}(\leq K)$ input 2D symbols are zero-padded to obtain N_{FFT} input samples for IFFT, N_{G} nonzero samples are inserted (as explained below) to create the guard interval, and the OFDM symbol is multiplied by the window function. The purpose of cyclic extension is to preserve the orthogonality among subcarriers even when the neighboring OFDM symbols partially overlap due to dispersion, and the role of windowing is to reduce the out-of-band spectrum. For efficient chromatic dispersion and PMD compensation, the length of cyclically extended guard interval should be larger than the spread due to chromatic dispersion and PMD. The cyclic extension, illustrated in Fig. 6.20c, is

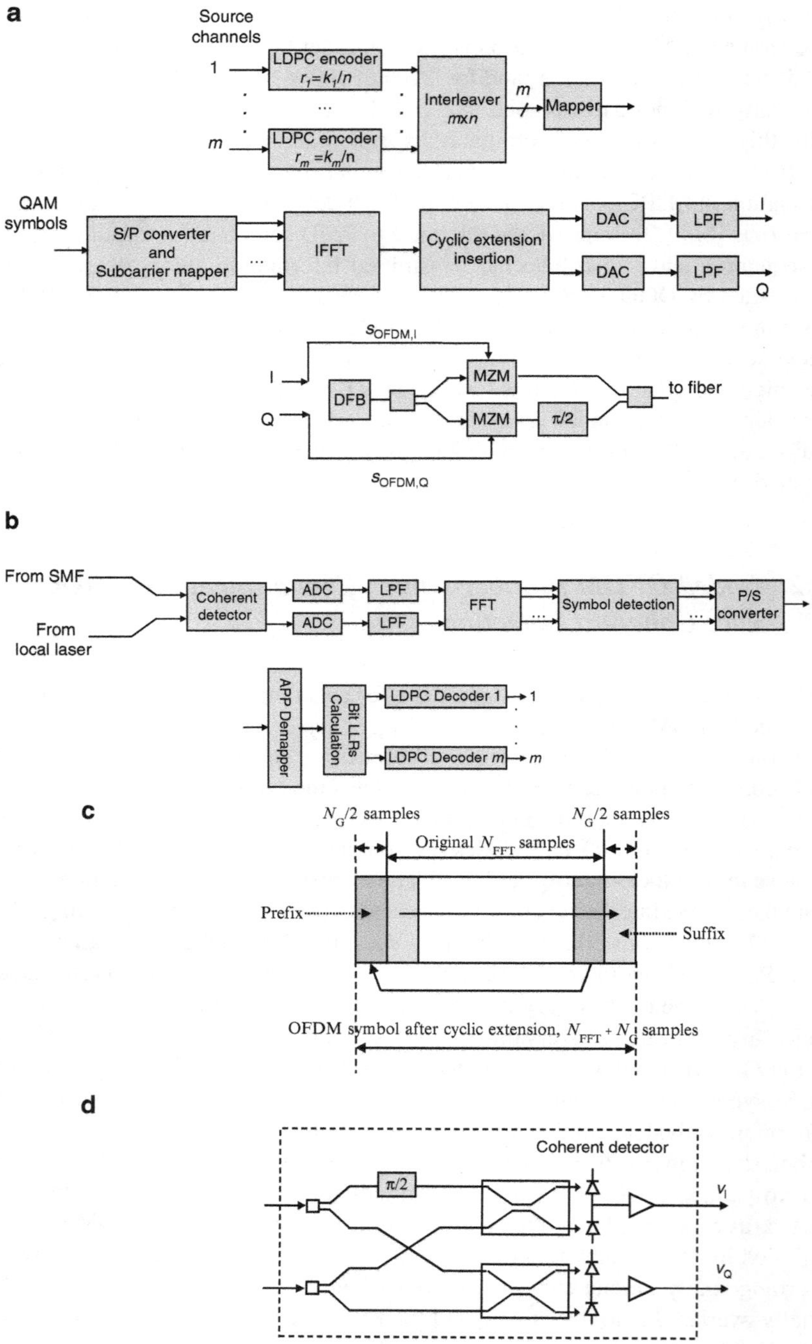

Fig. 6.20 (**a**) Transmitter configuration, (**b**) receiver configuration, (**c**) OFDM symbol after cyclic extension, and (**d**) coherent detector architecture. *S/P* serial-to-parallel converter, *MZM* Mach–Zehnder modulator, *SMF* single-mode optical fiber, *DAC* D/A converter

accomplished by repeating the last $N_G/2$ samples of the effective OFDM symbol part (N_{FFT} samples) as a prefix, and repeating the first $N_G/2$ samples as a suffix. After D/A conversion and low-pass filtering, the real-part of OFDM signal (I) and the imaginary-part of OFDM signal (Q) are mapped to the optical domain using two single-drive MZMs.

The complex envelope of a transmitted OFDM signal can be written as

$$s_{OFDM}(t) = \sum_{k=-\infty}^{\infty} w(t-kT) \sum_{i=-N_{FFT}/2}^{N_{FFT}/2-1} X_{i,k} \cdot e^{j2\pi \frac{i}{T_{FFT}} \cdot (t-kT)}, \tag{6.32}$$

and it is defined for $t \in [kT - T_G/2 - T_{win}, kT + T_{FFT} + T_G/2 + T_{win}]$. In the above expression, $X_{i,k}$ denotes the ith subcarrier of the kth OFDM symbol, $w(t)$ is the window function, T denotes the duration of the OFDM symbol, T_{FFT} denotes the FFT sequence duration, T_G is the guard interval duration (the duration of cyclic extension), and T_{win} denotes the windowing interval duration. One DFB laser is used as CW source, and 3-dB coupler is used to split the CW signal for both MZMs. After the appropriate sampling, the sum in (6.32) (obtained by ignoring the windowing function) can be written as follows:

$$x_{m,k} = \sum_{i=-N_{FFT}/2}^{N_{FFT}/2-1} X_{i,k} \exp\left(j2\pi \frac{im}{N_{FFT}}\right), \quad m = 0, 1, \ldots, N_{FFT}-1, \tag{6.33}$$

and corresponds to the discrete Fourier transform (DFT) (except for the normalization factor $1/N$).

The received electrical field, in the presence of chromatic dispersion, can be written as

$$E_s = \exp\left[j(2\pi f_{LD}{}^t + \phi_{PN,S})\right] \sum_{k=1}^{N_{FFT}} X_k e^{j2\pi f_k{}^t} e^{j\phi_{CD}(k)}, \tag{6.34}$$

where $\phi_{CD}(k)$ denotes the phase factor of kth subcarrier

$$\phi_{CD}(k) = \frac{\omega_k^2 |\beta_2| L}{2} = \frac{4\pi^2 f_k^2}{2} \frac{\lambda_{LD}^2}{2\pi c} DL = \frac{\pi c}{f_{LD}^2} D_t f_k^2, \tag{6.35}$$

with f_k being the kth subcarrier frequency (corresponding angular frequency is $\omega_k = 2\pi f_k$), f_{LD} being the transmitting laser-emitting frequency (the corresponding wavelength is λ_k), D_t being the total dispersion coefficients ($D_t = DL$, D – dispersion coefficient, L– the fiber length), and c being the speed of light. (β_2 denotes the GVD parameter.) The coherent detector output (see Fig. 6.20d), in complex notation, can be written as

$$v(t) \simeq R_{PIN} r_F \exp\left\{j\left[2\pi\left(f_{LD} - f_{LO}\right)t + \phi_{PN,S} - \phi_{PN,LO}\right]\right\} \sum_{k=1}^{N_{FFT}} X_k e^{j2\pi f_k{}^t} e^{j\phi_{CD}(k)} + N, \tag{6.36}$$

where f_{LO} denotes the local laser-emitting frequency, R_{PIN} is the photodetector responsivity, r_{F} is feedback resistor from trans-impedance amplifier configuration, while $\varphi_{\mathrm{PN,S}}$ and $\varphi_{\mathrm{PN,LO}}$ represent the laser phase noise of transmitting and receiving (local) laser, respectively. These two noise sources are commonly modeled as Wiener–Lévy process [43], which is a zero-mean Gaussian process with variance $2\pi(\Delta\nu_{\mathrm{S}} + \Delta\nu_{\mathrm{L}})|t|$, where $\Delta\nu_{\mathrm{S}}$ and $\Delta\nu_{\mathrm{L}}$ are the laser linewidths of transmitting and receiving laser, respectively. $N = N_{\mathrm{I}} - \mathrm{j}N_{\mathrm{Q}}$ represent the noise process, mostly dominated by ASE noise.

The kth subcarrier received symbol Y_k can, therefore, be represented by

$$Y_k = X_k \exp\left[\mathrm{j}\left(\phi_{\mathrm{PN,S}} - \phi_{\mathrm{PN,LO}}\right)\right] \exp\left[\mathrm{j}\phi_{\mathrm{CD}}(k)\right] + N_k, \quad N_k = N_{k,I} + N_{k,Q}, \tag{6.37}$$

where N_k represents the circular complex Gaussian process due to ASE noise. The transmitted symbol on kth subcarrier can be estimated by

$$\tilde{X}_k = Y_k \exp\left[-\mathrm{j}\left(\phi_{\mathrm{PN,S}} - \phi_{\mathrm{PN,LO}}\right)\right] \exp\left[-\mathrm{j}\phi_{\mathrm{CD}}(k)\right], \tag{6.38}$$

where the chromatic dispersion phase factor of kth subcarrier $\phi_{\mathrm{CD}}(k)$ is estimated by training based channel estimation, while the phase factor $\varphi_{\mathrm{PN,S}} - \varphi_{\mathrm{PN,LO}}$ is estimated by pilot-aided channel estimation. The soft outputs of the FFT demodulator are used to estimate the bit reliabilities that are fed to identical LDPC iterative decoders implemented based on the sum-product with correction term algorithm as we explained earlier. The chromatic dispersion is, therefore, quite straightforward to compensate for, while for PMD compensation we have different options described in incoming subsections. Before we turn our attention to different PMD compensation schemes of high spectral efficiency, in next section we briefly describe the PMD channel model.

6.4.2.1 Description of PMD Channel Model

For the first-order PMD study, the Jones matrix, neglecting the polarization dependent loss and depolarization effects, can be represented by [44]

$$\boldsymbol{H} = \begin{bmatrix} h_{xx}(\omega) & h_{xy}(\omega) \\ h_{yx}(\omega) & h_{yy}(\omega) \end{bmatrix} = \boldsymbol{R}\boldsymbol{P}(\omega)\boldsymbol{R}^{-1}, \quad \boldsymbol{P}(\omega) = \begin{bmatrix} e^{-\mathrm{j}\omega\tau/2} & 0 \\ 0 & e^{\mathrm{j}\omega\tau/2} \end{bmatrix}, \tag{6.39}$$

where τ denotes DGD, ω is the angular frequency, and $R = R(\theta, \varepsilon)$ is the rotational matrix [44]

$$\boldsymbol{R} = \begin{bmatrix} \cos\left(\frac{\theta}{2}\right) \mathrm{e}^{\mathrm{j}\varepsilon/2} & \sin\left(\frac{\theta}{2}\right) \mathrm{e}^{-\mathrm{j}\varepsilon/2} \\ -\sin\left(\frac{\theta}{2}\right) \mathrm{e}^{\mathrm{j}\varepsilon/2} & \cos\left(\frac{\theta}{2}\right) \mathrm{e}^{-\mathrm{j}\varepsilon/2} \end{bmatrix} \tag{6.40}$$

with θ being the polar angle, and ε being the azimuth angle. For OFDM with coherent detection, the received symbol vector of kth subcarrier in ith OFDM symbol $\boldsymbol{r}_{i,k} = [r_{x,i,k} r_{y,i,k}]^{\mathrm{T}}$ can be represented by

$$\boldsymbol{r}_{i,k} = \boldsymbol{H}_k \boldsymbol{s}_{i,k} e^{\mathrm{j}[\phi_{\mathrm{CD}}(k)+\phi_{\mathrm{T}}-\phi_{\mathrm{LO}}]} + \boldsymbol{n}_{i,k}, \tag{6.41}$$

where $\boldsymbol{s}_{i,k} = [s_{x,i,k} s_{y,i,k}]^{\mathrm{T}}$ denotes the transmitted symbol vector of kth subcarrier in ith OFDM symbol, for both polarizations, $\boldsymbol{n}_{i,k} = [n_{x,i,k} n_{y,i,k}]^{\mathrm{T}}$ denotes the noise vector dominantly determined by the amplified spontaneous emission (ASE) noise; ϕ_{T} and ϕ_{LO} denote the laser phase noise processes of transmitting and local lasers, $\phi_{\mathrm{CD}}(k)$ denotes the phase distortion of kth subcarrier due to chromatic dispersion, and the Jones matrix of kth subcarrier $\boldsymbol{H}_k$ is already introduced in (6.39). The transmitted/received symbols are complex-valued, with real part corresponding to the in-phase coordinate and imaginary part corresponding to the quadrature coordinate. The equivalent mathematical model is similar to that for wireless communications, and it is shown in Fig. 6.21.

Figure 6.22 shows the magnitude responses of h_{xx} and h_{xy} coefficients of Jones channel matrix against normalized frequency $f\tau$ (the frequency is normalized with DGD τ so that the conclusions are independent on the data rate) for two different cases: (a) $\theta = \pi/2$ and $\varepsilon = 0$, and (b) $\theta = \pi/3$ and $\varepsilon = 0$. In the first case channel coefficient h_{xx} completely fades away for certain frequencies, while in the second case it never completely fades away; suggesting that the first case represents the worst case scenario. To avoid this problem, in OFDM systems someone can redistribute the transmitted power among subcarriers not being under fading, or use the polarization diversity. We describe several alternative approaches instead in the following that can be used for a number of modulation formats including M-ary PSK, M-ary QAM, and IPQ.

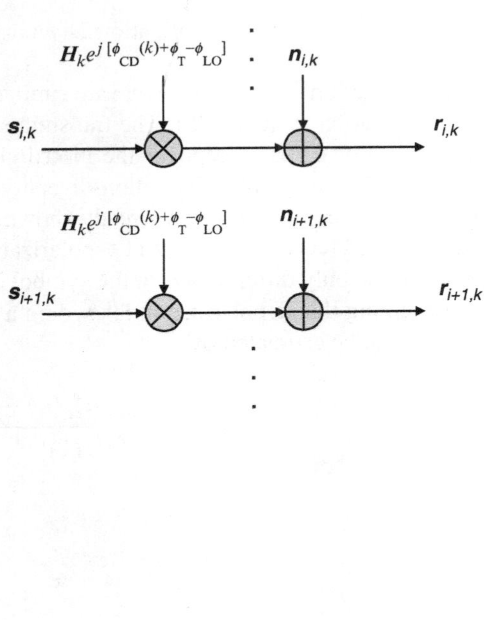

Fig. 6.21 Equivalent OFDM channel model

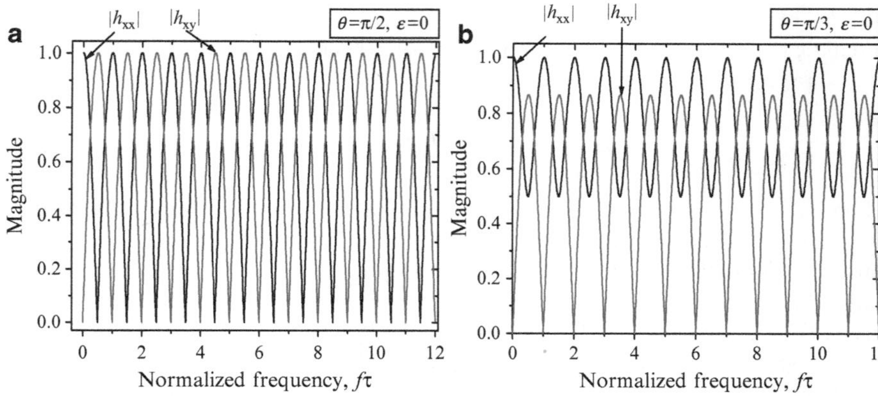

Fig. 6.22 Magnitude response of h_{xx} and h_{xy} Jones matrix coefficients against the normalized frequency for: (**a**) $\theta = \pi/2$ and $\varepsilon = 0$, and (**b**) $\theta = \pi/3$ and $\varepsilon = 0$

6.4.2.2 PMD Compensation by Coded OFDM in Fiber-Optics Communication Systems with Coherent Detection

In this section, we describe several PMD compensation schemes based on coded OFDM: (a) polarization diversity coded OFDM (described in the first subsection below, (b) BLAST-type polarization interference cancellation (described in the second subsection below), (c) iterative polarization cancellation (described in the second subsection below), and (d) Alamouti-type polarization-time (PT) coding (described in the third subsection below). The performance assessment of different PMD compensation schemes is given in the fourth subsection below.

PMD Compensation by Polarization-Diversity-Coded OFDM

The polarization diversity receivers are similar to diversity receivers used in wireless communication systems [45]. The transmitter configuration is essentially the same as that in Fig. 6.20a, except for the insertion of PBS after the DFB laser. The receiver configuration of polarization-diversity-coded OFDM is shown in Fig. 6.23, while the coherent detector is already shown in Fig. 6.20d. The operational principle of OFDM receivers, for x- and y-polarizations, is similar to that of single OFDM receiver, the only difference is in the symbol detector.

Assuming that x-polarization is used on a transmitter side, the transmitted symbol $s_{k,i}$ can be estimated by

$$\tilde{s}_{k,i} = \frac{r_{i,k,x} h_{xx}^{*}(k) + r_{i,k,y} h_{xy}^{*}(k)}{|h_{xx}(k)|^2 + |h_{xy}(k)|^2}, \tag{6.42}$$

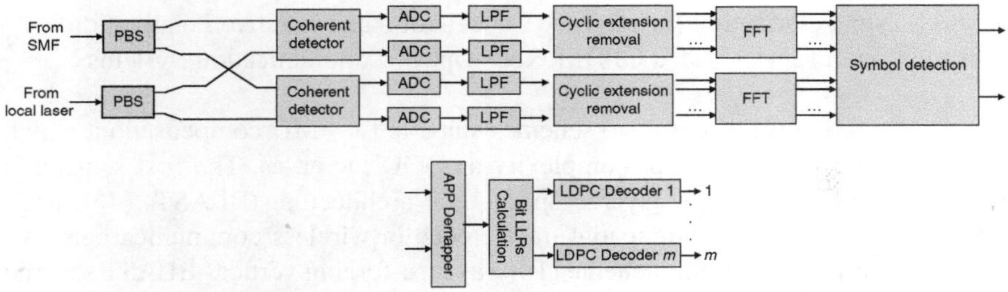

Fig. 6.23 The receiver configuration for polarization-diversity-coded OFDM with coherent detection. *PBS* polarization beam splitter

where h_{xx} and h_{xy} are channel coefficients introduced already in (6.39); while $r_{i,k,x}$ and $r_{i,k,y}$ represent the received QAM symbol in kth subcarrier of ith OFDM symbol corresponding to x- and y-polarizations, respectively.

The symbol-detector soft estimates of symbols carried by kth subcarrier in ith OFDM symbol are forwarded to the APP demapper, which determines the symbol LLRs $\lambda(q)(q = 0, 1, \ldots, 2^b - 1)$ by

$$\lambda(q) = -\frac{\left(\mathrm{Re}[\tilde{s}_{i,k}] - \mathrm{Re}\,[\mathrm{QAM}(\mathrm{map}(q))]\right)^2}{2\sigma^2} - \frac{\left(\mathrm{Im}[\tilde{s}_{i,k}] - \mathrm{Im}[\mathrm{QAM}(\mathrm{map}(q))]\right)^2}{2\sigma^2}, \quad q = 0, 1, \ldots, 2^b - 1 \qquad (6.43)$$

where Re[] and Im[] denote the real and imaginary part of a complex number, QAM denotes the QAM-constellation diagram, σ^2 denotes the variance of an equivalent Gaussian noise process originating from ASE noise, and map(q) denotes a corresponding mapping rule (Gray mapping is applied here). (b denotes the number of bits per constellation point.) Let us denote by v_j the jth bit in an observed symbol q binary representation $\boldsymbol{v} = (v_1, v_2, \ldots, v_b)$. The bit LLRs needed for LDPC decoding are calculated from symbol LLRs by

$$L(\hat{v}_j) = \log \frac{\sum_{q:v_j=0} \exp[\lambda(q)]}{\sum_{q:v_j=1} \exp[\lambda(q)]}. \qquad (6.44)$$

Therefore, the jth bit reliability is calculated as the logarithm of the ratio of a probability that $v_j = 0$ and probability that $v_j = 1$. In the nominator, the summation is done over all symbols q having 0 at the position j, while in the denominator over all symbols q having 1 at the position j. The extrinsic LLRs are iterated backward and forward until convergence or pre-determined number of iterations is reached. The LDPC code used here belongs to the class of quasi-cyclic codes of large girth ($g \geq 10$) described earlier, so that the corresponding decoder complexity is low compared to random LDPC codes, and does not exhibit the error floor phenomena in the region of interest in fiber-optics communications ($\leq 10^{-15}$) [51].

BLAST-Type Polarization Interference Cancellation and Iterative Polarization Cancellation in LDPC-Coded OFDM Fiber-Optics Communication Systems

In this subsection we discuss two schemes suitable for PMD compensation, which do not require the increase of complexity as DGD increases. The first scheme is based on Bell Laboratories layered space–time architecture (BLAST) [46], originally proposed to deal with spatial interference in wireless communications. We consider two versions of this scheme [10]: (a) zero-forcing vertical-BLAST scheme (ZF V-BLAST), and (b) minimum-mean-square-error vertical-BLAST (MMSE V-BLAST) scheme. Because the ZF V-BLAST scheme is derived by ignoring the influence of ASE noise, we proposed the second scheme that uses the output of ZF V-BLAST scheme as starting point and removes the remaining polarization interference in an iterative fashion. This approach also leads to reducing the influence of ASE noise. We evaluate the performance of those schemes when used in combination with coherent detection based OFDM. We describe how to use those schemes together with multilevel modulation and FEC. The arbitrary FEC scheme can be used with proposed PMD compensation schemes; however, the use of LDPC codes leads to channel capacity achieving performance.

The polarization interference cancellation scheme based on V-BLAST algorithm, which uses an LDPC code as channel code, is shown in Fig. 6.24. The 2D signal constellation points, generated in fashion similar to that reported in Sect. 6.1, are split into two streams for OFDM transmitters (see Fig. 6.24a) corresponding to the x- and y-polarizations. The QAM constellation points are considered to be the values of the FFT of a multicarrier OFDM signal. The OFDM symbol is generated similarly as reported above: N_{QAM} input QAM symbols are zero-padded to obtain N_{FFT} input samples for inverse FFT (IFFT), N_{G} nonzero samples are inserted to create the guard interval, and the OFDM symbol is multiplied by the window function. For efficient chromatic dispersion and PMD compensation, the length of cyclically extended guard interval should be smaller than the total spread due to chromatic dispersion and DGD. The cyclic extension is accomplished by repeating the last $N_{\mathrm{G}}/2$ samples of the effective OFDM symbol part (N_{FFT} samples) as a prefix, and repeating the first $N_{\mathrm{G}}/2$ samples as a suffix. After DAC, the RF OFDM signal is converted into the optical domain using the dual-drive MZM. Two MZMs are needed, one for each polarization. The outputs of MZMs are combined using the polarization beam combiner (PBC). One DFB laser is used as CW source, with x- and y-polarization separated by a PBS.

The receiver architecture is shown in Fig. 6.24b. The configuration of polarization interference cancellation scheme by BLAST-algorithm is shown in Fig. 6.25a. The received symbol vector in kth subcarrier of ith OFDM symbol in both polarization is linearly processed, the processing is described by matrix $\boldsymbol{C}_k$ related to channel matrix $\boldsymbol{H}_k$ as shown below, and the estimate of polarization interference obtained from preliminary decisions $\tilde{\boldsymbol{s}}_{i,k}$, denoted as $\boldsymbol{D}_k\tilde{\boldsymbol{s}}_{i,k}$, is removed from received symbol $\boldsymbol{r}_{i,k}$. The Euclidean detector can be used to create the preliminary decisions. When the presence of ASE noise is ignored, the zero-forcing V-BLAST

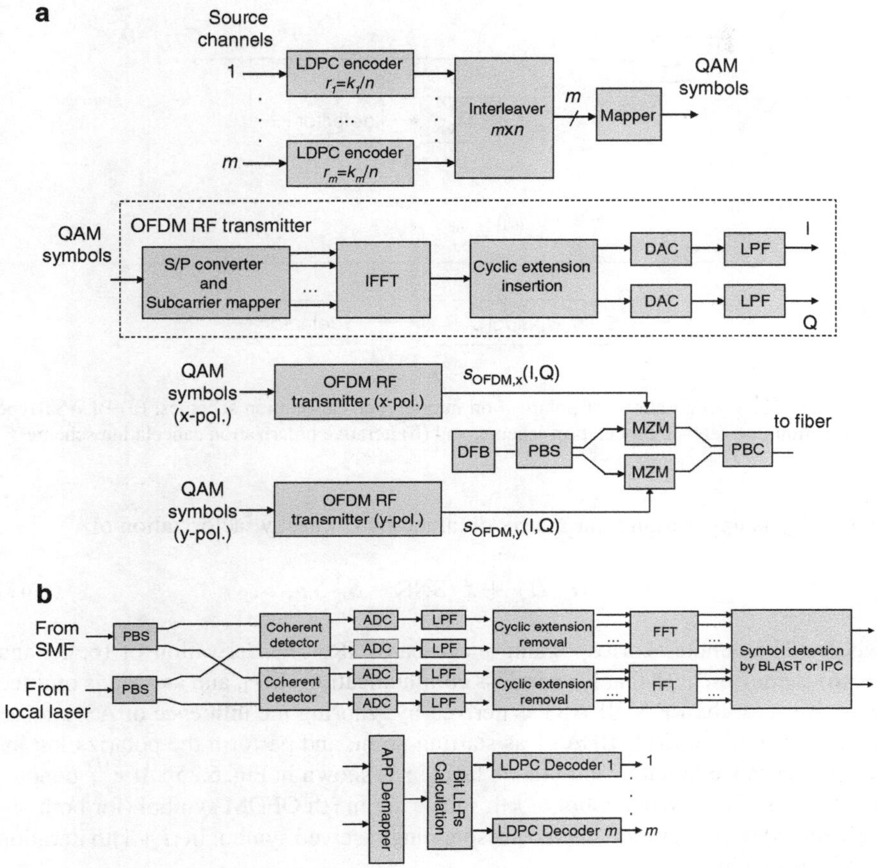

Fig. 6.24 The architecture of polarization interference cancelation scheme in combination with LDPC-coded OFDM: (**a**) transmitter architecture, and (**b**) receiver architecture. *DFB* distributed feedback laser, *PBS(C)* polarization beam splitter (combiner), *MZM* dual-drive Mach–Zehnder modulator, *APP* a posteriori probability, *LLRs* log-likelihood ratios

polarization interference cancellation scheme results. The matrices $\boldsymbol{C}_k$ and $\boldsymbol{D}_k$ can be determined from QR-factorization of channel matrix $\mathbf{H}_k = \boldsymbol{Q}_k \boldsymbol{R}_k$, as follows [10, 47]:

$$\boldsymbol{C}_k = \text{diag}^{-1}(\boldsymbol{R}_k)\boldsymbol{Q}_k^{\dagger}, \quad \boldsymbol{D}_k = \text{diag}^{-1}(\boldsymbol{R}_k)\boldsymbol{R}_k - \boldsymbol{I}, \tag{6.45}$$

where $\boldsymbol{I}$ is identity matrix, and with diag() we denoted the diagonal elements of $\boldsymbol{R}_k$. Notice that elements at main diagonal in $\boldsymbol{D}_k$ are set to zero in order to have only polarization interference be removed. (We use † to denote the simultaneous transposition and complex-conjugation.) In the presence of ASE noise, the matrices $\boldsymbol{C}_k$ and $\boldsymbol{D}_k$ can be determined by minimizing the MSE, which leads to

$$\boldsymbol{C}_k = \text{diag}^{-1}(\boldsymbol{S}_k)(\boldsymbol{S}_k^{\dagger})^{-1}\boldsymbol{H}_k^{\dagger}, \quad \boldsymbol{D}_k = \text{diag}^{-1}(\boldsymbol{S}_k)\boldsymbol{S}_k - \boldsymbol{I}, \tag{6.46}$$

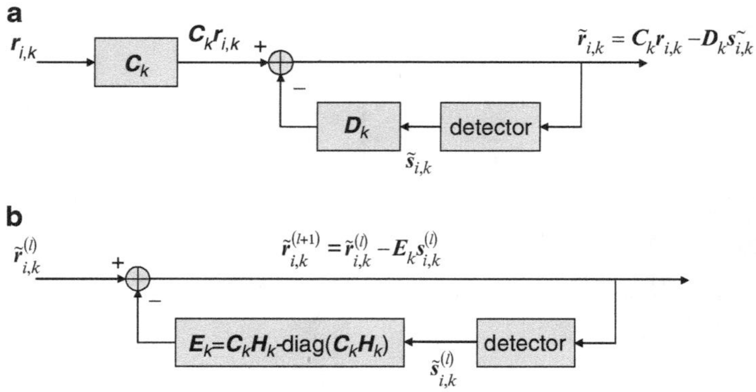

Fig. 6.25 The configurations of polarization interference cancelation schemes: (**a**) BLAST-type polarization interference cancelation scheme, and (**b**) iterative polarization cancelation scheme

where S_k is upper triangular matrix obtained by Cholesky factorization of

$$H_k^{\dagger} H_k + I/\mathrm{SNR} - S_k^{\dagger} S_k, \tag{6.47}$$

where SNR denotes corresponding electrical SNR. The derivation of (6.45) and (6.46) is equivalent to that for wireless communications [47], and as such is omitted here. Because the ZF V-BLAST is derived by ignoring the influence of ASE noise, we propose to use ZF V-BLAST as starting point, and perform the polarization interference cancellation in an iterative fashion as shown in Fig. 6.25b. If $\tilde{r}^{(l)}_{i,k}$ denotes the processed received symbol of kth subcarrier in ith OFDM symbol (for both polarizations) in lth iteration, then corresponding received symbol in $(l + 1)$th iteration can be found by

$$\tilde{r}^{(l+1)}_{i,k} = \tilde{r}^{(l)}_{i,k} - [C_k H_k - \mathrm{diag}(C_k H_k)]\, \tilde{s}_{i,k}{}^{(l)}, \tag{6.48}$$

where $\tilde{s}^{(l)}_{i,k}$ denotes the transmitted symbol (of kth subcarrier in ith OFDM symbol (for both polarizations)) estimate in lth iteration. The matrices C_k and D_k are already introduced above. Notice that different matrix operations applied above are trivial because the dimensionality of matrices is small, 2×2.

The BLAST-detector soft estimates of symbols carried by kth subcarrier in ith OFDM symbol, $s_{i,k,x(y)}$, are forwarded to the APP demapper, which determines the symbol LLRs $\lambda_{x(y)}(q)(q = 0, 1, \ldots, 2^b - 1)$ of x- (y-) polarization by

$$\lambda_{x(y)}(q) = -\frac{\left(\mathrm{Re}[\tilde{s}_{i,k,x(y)}] - \mathrm{Re}[\mathrm{QAM}(\mathrm{map}(q))]\right)^2}{2\sigma^2} - \frac{(\mathrm{Im}[\tilde{s}_{i,k,x(y)}] - \mathrm{Im}[\mathrm{QAM}(\mathrm{map}(q))])^2}{2\sigma^2}, \quad q = 0, 1, \ldots, 2^b - 1 \tag{6.49}$$

where Re[] and Im[] denote the real and imaginary part of a complex number, QAM denotes the QAM-constellation diagram, σ^2 denotes the variance of an equivalent Gaussian noise process originating from ASE noise, and map(q) denotes a corresponding mapping rule (Gray mapping is applied here). (b denotes the number of bits per constellation point.) Let us denote by $v_{j,x(y)}$ the jth bit in an observed symbol q binary representation $v = (v_1, v_2, \ldots, v_b)$ for x- (y-) polarization. The bit LLRs needed for LDPC decoding are calculated from symbol LLRs by (6.44).

PMD Compensation by Polarization-Time Coding

In this section, we describe the PT-encoder and PT-decoder based on Alamouti-type scheme, and describe how to interplay them with coded OFDM. The PT-encoder operates as follows [11]. In the first half of ith time instance ("the first channel use") it sends symbol s_x to be transmitted using x-polarization channel and symbol s_y to be transmitted using y-polarization channel. In the second half of ith time instance ("the second channel use") it sends symbol $-s_y^*$ to be transmitted using x-polarization channel, and symbol $-s_x^*$ to be transmitted using y-polarization. Therefore, the PT-coding procedure is similar to the Alamouti-scheme [48]. Notice that Alamouti-type PT-coding scheme has the spectral efficiency comparable to coherent OFDM with polarization diversity scheme. When the channel is used twice during the same symbol period, the spectral efficiency of this scheme is twice higher than that of polarization diversity OFDM. Notice that the hardware complexity of PT-encoder/decoder is trivial compared to that of OFDM. The transmitter complexity is slightly higher than that that reported in Sect. 6.4.2.2; it requires additional PT-encoder, a PBS and a PBC. On the receiver side, we have the option to use only one polarization or to use both polarizations. The receiver architecture employing both polarizations is shown in Fig. 6.26b.

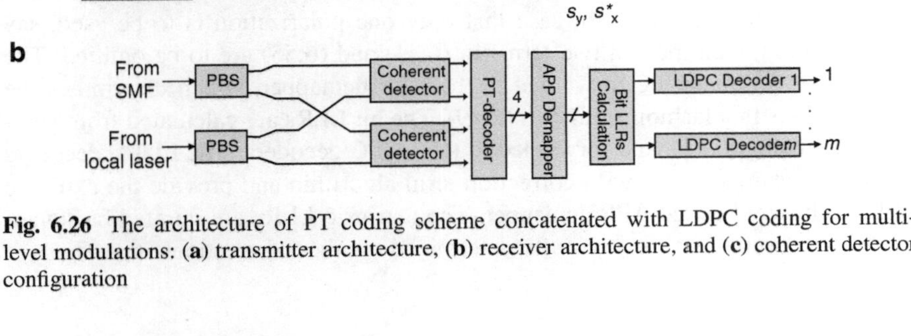

Fig. 6.26 The architecture of PT coding scheme concatenated with LDPC coding for multilevel modulations: (**a**) transmitter architecture, (**b**) receiver architecture, and (**c**) coherent detector configuration

The operations of all blocks, except the PT-decoder, are similar to those we reported above. Hence, we concentrate only on description of PT-decoder. The received symbol vectors in ith time instance for the first ($\boldsymbol{r}_{i,1}$) and second channel use ($\boldsymbol{r}_{i,2}$) can be written, respectively, as follows:

$$\boldsymbol{r}_{i,1} = \boldsymbol{H}\boldsymbol{s}_{i,1}\mathrm{e}^{\mathrm{j}(\phi_{\mathrm{T}}-\phi_{\mathrm{LO}})+\mathrm{j}\phi_{\mathrm{CD}}(k)} + \boldsymbol{n}_{i,1}, \tag{6.50}$$

$$\boldsymbol{r}_{i,2} = \boldsymbol{H}\boldsymbol{s}_{i,2}\mathrm{e}^{\mathrm{j}(\phi_{\mathrm{T}}-\phi_{\mathrm{LO}})+\mathrm{j}\phi_{\mathrm{CD}}(k)} + n_{i,2}, \tag{6.51}$$

where the Jones (channel) matrix $\boldsymbol{H}$, $\boldsymbol{r}_{i,1(2)} = [r_{x,i,1(2)} r_{y,i,1(2)}]^{\mathrm{T}}$ denotes the received symbol vector in the first (second) channel use of ith time instance, while $\boldsymbol{n}_{i,1(2)} = [n_{x,i,1(2)} n_{y,i,1(2)}]^{\mathrm{T}}$ is the ASE noise vector corresponding the first (second) channel use in ith time instance. We use $\boldsymbol{s}_{i,1} = [s_{x,i} s_{y,i}]^{\mathrm{T}}$ to denote the symbol transmitted in the first channel use of ith time instance, and $\boldsymbol{s}_{i,2} = [-s_{y,i}^* \; s_{x,i}^*]^{\mathrm{T}}$ to denote the symbol transmitted in the second channel use (of the same time instance). Because the symbol vectors transmitted in the first and the second channel use of ith time instance are orthogonal, $\boldsymbol{s}_{i,1}^{+}\boldsymbol{s}_{i,2} = \boldsymbol{0}$, (6.50) and (6.51) can be rewritten by separately grouping x- and y-polarizations as follows:

$$\begin{bmatrix} r_{x,i,1} \\ r_{x,i,2}^* \end{bmatrix} = \begin{bmatrix} h_{xx} & h_{xy} \\ h_{xy}^* & -h_{xx}^* \end{bmatrix} \begin{bmatrix} s_{x,i} \\ s_{y,i} \end{bmatrix} e^{\mathrm{j}(\phi_T-\phi_{\mathrm{LO}})+j\phi_{\mathrm{CD}}(k)} + \begin{bmatrix} n_{x,i,1} \\ n_{x,i,2}^* \end{bmatrix}, \tag{6.52}$$

$$\begin{bmatrix} r_{y,i,1} \\ r_{y,i,2}^* \end{bmatrix} = \begin{bmatrix} h_{yx} & h_{yy} \\ h_{yy}^* & -h_{yx}^* \end{bmatrix} \begin{bmatrix} s_{x,i} \\ s_{y,i} \end{bmatrix} e^{\mathrm{j}(\phi_T-\phi_{\mathrm{LO}})+\mathrm{j}\phi_{\mathrm{CD}}(k)} + \begin{bmatrix} n_{y,i,1} \\ n_{y,i,2}^* \end{bmatrix}. \tag{6.53}$$

If only one polarization is to be used, we can solve either (6.52) or (6.53). However, the use of only one polarization results in 3 dB penalty with respect to the case when both polarizations are used. Following the derivation similar to that performed by Alamouti [48], it can be shown that the optimum estimates of transmitted symbols at the output of PT-decoder (for ASE noise dominated scenario) can be obtained as follows:

$$\tilde{s}_{x,i} = h_{xx}^* r_{x,1} + h_{xy} r_{x,2}^* + h_{yx}^* r_{y,1} + h_{yy} r_{y,2}^*, \tag{6.54}$$

$$\tilde{s}_{y,i} = h_{xy}^* r_{x,1} - h_{xx} r_{x,2}^* + h_{yy}^* r_{y,1} - h_{yx} r_{y,2}^*, \tag{6.55}$$

where $\tilde{s}_{x,i}$ and $\tilde{s}_{y,i}$ denote the PT-decoder estimates of symbols $s_{x,i}$ and $s_{y,i}$ transmitted in ith time instance. In case that only one polarization is to be used, say x-polarization, then the last two terms in (6.54) and (6.55) are to be omitted. The PT-decoder estimates are forwarded to the APP demapper, which determines the symbol LLRs in a fashion similar to (6.49). The bit LLRs are calculated from symbol LLRs using (6.44), and forwarded to the LDPC decoders. The LDPC decoders employ the sum–product-with-correction term algorithm and provide the extrinsic LLRs to be used in the APP demapper. The extrinsic LLRs are iterated backward and forward until convergence or predetermined number of iterations is reached.

Evaluation of Different PMD Compensation Schemes

We are turning our attention to the BER performance evaluation of the schemes described in previous sections. The results of simulation for uncoded OFDM for different PMD compensation schemes are shown in Fig. 6.27. The OFDM system parameters are chosen as follows: $N_{\mathrm{QAM}} = 512$, oversampling is two times, OFDM signal bandwidth is set to 10 GHz, and $N_{\mathrm{G}} = 256$ samples. The MMSE V-BLAST and iterative polarization cancellation schemes (with ZF V-BLAST as starting point) perform identically (only MMSE curve is shown because the curves overlap each other), while ZF V-BLAST is slightly worse. Polarization diversity OFDM outperforms MMSE V-BLAST at low BERs, but performs comparable at BERs above 10^{-2}, which is the threshold region of girth-10 LDPC codes employed here. Moreover, the spectral efficiency of MMSE V-BLAST is twice higher because in polarization diversity OFDM only one polarization is used for transmission. The MMSE V-BLAST OFDM outperforms the PT-coding-based OFDM at both low and high BERs, and has two times higher spectral efficiency. The MMSE V-BLAST OFDM scheme is able to compensate for even 1,200 ps of DGD with negligible penalty. Notice that for corresponding turbo equalization [49] or maximum-likelihood sequence estimation schemes, the detector complexity grows

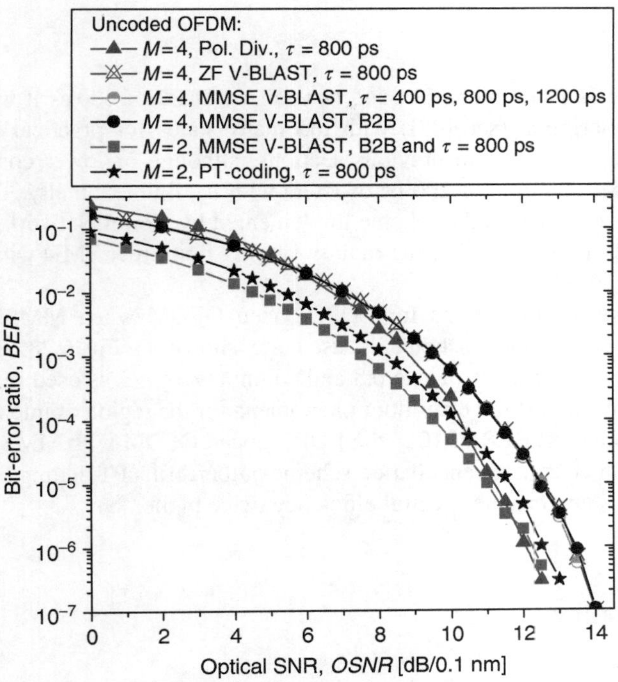

Fig. 6.27 BER performance of BLAST algorithm-based PMD compensation schemes against polarization diversity OFDM, and PT-coding-based OFDM. *B2B* back-to-backs

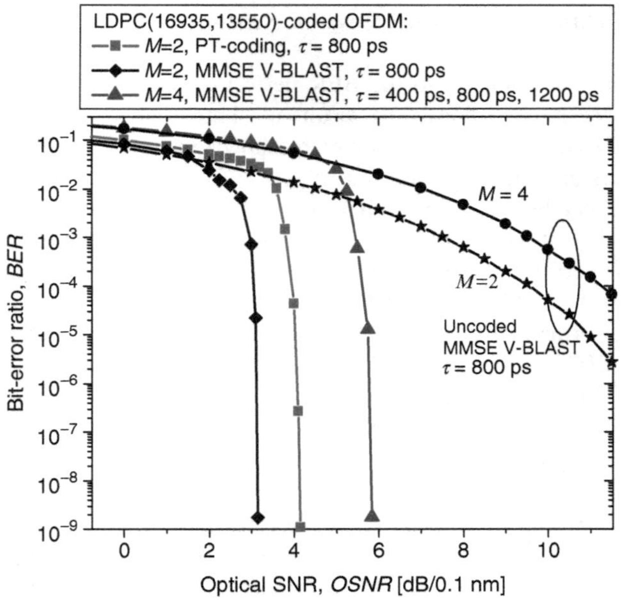

Fig. 6.28 BER performance of the polarization interference cancelation schemes for LDPC-coded OFDM

exponentially as DGD increases, and normalized DGD of 800 ps it would require the trellis description (see [49]) with too many states for practical implementation. Schemes described in previous sections, although of lower complexity, are able to compensate up to 1,200 ps of DGD with negligible penalty. The proposed schemes also outperform the scheme implemented by Nortel Networks researchers [50], capable of compensating the rapidly varying first-order PMD with peak DGD of 150 ps.

The results of simulations for LDPC-coded OFDM when MMSE V-BLAST polarization cancellation scheme is used are shown in Fig. 6.28. The girth-10 LDPC(16935,13550) code of rate 0.8 and column weight 3, is used in simulations. This code does not exhibit error floor phenomena for the region of interest in optical communications. At BER of 10^{-9} the LDPC-coded OFDM with MMSE V-BLAST polarization interference cancellation scheme outperforms PT-coding based OFDM by about 1 dB, and has the spectral efficiency twice higher.

6.5 Summary

In this chapter, we described several coded-modulation schemes: (1) MLC (Sect. 6.2), (2) multidimensional coded modulation (Sect. 6.3), and (3) coded OFDM (Sect. 6.4). In Sect. 6.1 we described different M-ary modulation formats, suitable for use in optical coded modulation schemes to achieve high spectral-efficiency. We described several coded OFDM schemes suitable for use in polarization multiplexed OFDM with coherent detection: polarization-time coding, BLAST algorithm based OFDM and polarization interference cancellation scheme.

References

1. Imai H, Hirakawa S (1977) A new multilevel coding method using error correcting codes. IEEE Trans Inform Theory IT-23(3):371–377
2. Caire G, Taricco G, Biglieri E (1998) Bit-interleaved coded modulation. IEEE Trans Inform Theory 44:927–946
3. Djordjevic IB, Vasic B (2006) Multilevel coding in *M-ary* DPSK/differential QAM high-speed optical transmission with direct detection IEEE/OSA J Lightwave Technol 24:420–428
4. Djordjevic IB, Cvijetic M, Xu L, Wang T (2007) Using LDPC-coded modulation and coherent detection for ultra high-speed optical transmission. IEEE/OSA J Lightwave Technol 25: 3619–3625
5. Batshon HG, Djordjevic IB, Xu L, Wang T, Cvijetic M (2008) Proposal to achieve 1 Tb/s per wavelength transmission using 3-dimensional LDPC-coded modulation. IEEE Photon Technol Lett 20(9):721–723
6. Batshon HG, Djordjevic IB, Minkov LL, Xu L, Wang T (2009) Multidimensional LDPC-coded modulation for beyond 400 Gb/s per wavelength optical transmission. IEEE Photon Technol Lett 21:1139–1141
7. Djordjevic IB, Vasic B (2006) Orthogonal frequency-division multiplexing for high-speed optical transmission. Opt Express 14:3767–3775
8. Djordjevic IB, Vasic B (2006) 100 Gb/s transmission using orthogonal frequency-division multiplexing. IEEE Photon Technol Lett 18(15):1576–1578
9. Djordjevic IB (2007) PMD compensation in fiber-optic communication systems with direct detection using LDPC-coded OFDM. Opt Express 15:3692–3701
10. Djordjevic IB, Xu L, Wang T (2008) PMD compensation in multilevel coded-modulation schemes with coherent detection using BLAST algorithm and iterative polarization cancellation. Opt Express 16(19):14845–14852
11. Djordjevic IB, Xu L, Wang T (2008) PMD compensation in coded-modulation schemes with coherent detection using Alamouti-type polarization-time coding. Opt Express 16(18): 14163–14172
12. Djordjevic IB, Xu L, Wang T (2009) Beyond 100 Gb/s optical transmission based on polarization multiplexed coded-OFDM with coherent detection. IEEE/OSA J Opt Commun Netw 1(1):50–56
13. Proakis JG (2001) Digital communications. McGaw-Hill, Boston, MA
14. Webb WT, Steele R (1995) Variable rate QAM for mobile radio. IEEE Trans Comm 43: 2223–2230
15. Peric ZH, Djordjevic IB, Bogosavljevic SM, Stefanovic MC (1998) Design of signal constellations for Gaussian channel by iterative polar quantization. In: Proceedings of the 9th mediterranean electrotechnical conference, vol 2, pp 866–869, Tel-Aviv, Israel, 18–20 May 1998
16. Hou J, Siegel PH, Milstein LB, Pfitser HD (2003) Capacity-approaching bandwidth-efficient coded modulation schemes based on low-density parity-check codes. IEEE Trans Inform Theory 49(9):2141–2155
17. Wachsmann U, Fischer RFH, Huber JB (1999) Multilevel codes: theoretical concepts and practical design rules. IEEE Trans Inform Theory 45:1361–1391

18. Hou J, Lee MH (2004) Multilevel LDPC codes design for semi-BICM. IEEE Comm Lett 8:674–676
19. Limpaphayom P, Winick KA (2004) Power- and bandwidth-efficient communications using LDPC codes. IEEE Trans Comm 52:350–354
20. ten Brink S (2000) Designing iterative decoding schemes with the extrinsic information transfer chart. AEÜ Int J Electron Comm 54:389–398
21. Vasic B, Djordjevic IB, Kostuk R (2003) Low-density parity check codes and iterative decoding for long haul optical communication systems IEEE/OSA J Lightwave Technol 21:438–446
22. Hou J, Siegel PH, Milstein LB (2005) Design of multi-input multi-output systems based on low-density parity-check codes IEEE Trans Comm 53:601–611
23. Tan J, Stüber GL (2005) Analysis of symbol mappers for iteratively decoded BICM. IEEE Trans Wireless Comm 4:662–672
24. Benedetto S, Poggiolini P (1992) Theory of polarization shift keying modulation. IEEE Trans Comm 4:708–721
25. Bahl L, Cocke J, Jelinek F, Raviv J (1974) Optimal decoding of linear codes for minimizing symbol error rate. IEEE Trans Inform Theory IT-20(2):284–287
26. http://www.vpiphotonics.com
27. Jäger M, Rankl T, Speidel J, Bülow H, Buchali F (2006) Performance of turbo equalizers for optical PMD channels. IEEE/OSA J Lightwave Technol 24(3):1226–1236
28. Prasad R (2004) OFDM for wireless communications systems. Artech House, Boston
29. Shieh W, Djordjevic IB (2009) OFDM for optical communications. Academic, Burlington, MA
30. Djordjevic IB, Vasic B, Neifeld MA (2006) LDPC coded orthogonal frequency division multiplexing over the atmospheric turbulence channel. In: Proceedings of the CLEO/QELS 2006, paper CMDD5
31. Djordjevic IB, Vasic B, Neifeld MA (2007) LDPC coded OFDM over the atmospheric turbulence channel. Opt Express 15(10):6332–6346
32. Shieh W, Athaudage C (2006) Coherent optical frequency division multiplexing. Electron Lett 42:587–589
33. Lowery AJ, Du L, Armstrong J (2006) Orthogonal frequency division multiplexing for adaptive dispersion compensation in long haul WDM systems. In: Proceedings of the OFC postdeadline papers, March 2006, paper no. PDP39
34. Lowery AJ, Armstrong J (2005) 10 Gb/s multimode fiber link using power-efficient orthogonal-frequency-division multiplexing. Opt Express 13(25):10 003–10 009
35. Djordjevic IB (2007) LDPC-coded OFDM transmission over graded-index plastic optical fiber links. IEEE Photon Technol Lett 19(12):871–873
36. Shieh W (2006) PMD-supported coherent optical OFDM systems. IEEE Photon Technol Lett 19:134–136
37. Shieh W, Yi X, Ma Y, Tang Y (2007) Theoretical and experimental study on PMD-supported transmission using polarization diversity in coherent optical OFDM systems. Opt Express 15:9936–9947
38. Jansen SL, Morita I, Takeda N, Tanaka H (2007) 20-Gb/s OFDM transmission over 4,160-km SSMF enabled by RF-pilot tone phase compensation. In: Proceedings of the OFC/ NFOEC 2007 postdeadline papers, Anaheim, CA, March 25–29, 2007, paper no. PDP15
39. Schmidt BJ, Lawery AJ, Amstrong J (2007) Experimental demonstration of 20 Gbit/s direct-detection optical OFDM and 12 Gbit/s with a colorless transmitter. In: Proceedings of the OFC/ NFOEC 2007 postdeadline papers, Anaheim, CA, March 25–29, 2007, paper no. PDP18
40. Du LB, Lowery AJ (2009) Fiber nonlinearity compensation for CO-OFDM systems with periodic dispersion maps. In: Proceedings of the optical fiber telecommunications (OFC 2009), San Diego, CA, paper no. OTuO1
41. Du LB, Lowery AJ (2008) Improved nonlinearity precompensation for long-haul high-data-rate transmission using coherent optical OFDM. Opt Express 16(24):19920–19925
42. Du LB, Lowery AJ (2008) Improving nonlinear precompensation in direct-detection optical OFDM communications systems. In: Proceedings of the european conference on optical communications (ECOC), Brussels, September 2008, paper no. P. 4.08

43. Cvijetic M (1996) Coherent and nonlinear lightwave communications. Artech House, Boston
44. Penninckx D, Morenás V (1999) Jones matrix of polarization mode dispersion. Opt Lett 24:875–877
45. Goldsmith A (2005) Wireless communications. Cambridge University Press, Cambridge
46. Foschini GJ (1996) Layered space-time architecture for wireless communication in a fading environment when using multi-element antennas. Bell Labs Tech J 1:41–59
47. Biglieri E, Calderbank R, Constantinides A, Goldsmith A, Paulraj A, Poor HV (2007) MIMO wireless communications. Cambridge University Press, Cambridge
48. Alamouti S (1998) A simple transmit diversity technique for wireless communications. IEEE J Sel Areas Commun 16:1451–1458
49. Djordjevic IB, Batshon HG, Cvijetic M, Xu L, Wang T (2007) PMD compensation by LDPC-coded turbo equalization. IEEE Photon Technol Lett 19(15):1163–1165
50. Sun H, Wu K-T, Roberts K (2008) Real-time measurements of a 40 Gb/s coherent system. Opt Express 16:873–879
51. Djordjevic IB, Arabaci M, Minkov L (2009) Next generation FEC for high-capacity communication in optical transport networks. IEEE/OSA J Lightwave Technol 27(16):3518–3530 (invited paper)

Chapter 7
Turbo Equalization in Fiber-Optics Communication Systems

In order to adapt to the ever-increasing demands of telecommunication needs, the network operators already consider 100 Gb/s per dense wavelength division multiplexing (DWDM) channel transmission. At those data rates, the performance of fiber-optic communication systems is degraded significantly due to intra- and interchannel fiber nonlinearities, polarization mode dispersion (PMD), and chromatic dispersion [1–11]. In order to mitigate the signal distortions at ultra-high bit rates, some new technologies have been proposed and deployed in optical systems, and they represent a distinctive new trend in optical fiber communications.

These new technologies include digital signal processing (DSP)-aided optical channel equalization, digital coherent receiving, multilevel modulations and optical polarization multiplexing (or optical multiple-input multiple-output technologies) [1–6]. To deal with chromatic dispersion and PMD a number of channel equalization techniques have been proposed recently including digital-filtering approach [1] and maximum likelihood sequence detection/estimation (MLSD/E) [2]. To simultaneously suppress chromatic dispersion and PMD, orthogonal frequency division multiplexing (OFDM) has been proposed [4, 5]. On the other hand, to deal with intrachannel nonlinearities someone may use either constrained coding [6].

In several recent publications [3, 6–11], we have shown that fiber nonlinearities, chromatic dispersion, and PMD can simultaneously be compensated for by using low-density parity-check (LDPC)-coded turbo equalization. This chapter is therefore devoted to the compensation of different linear and nonlinear channel impairments by turbo equalization. The chapter is organized as follows. In Sect. 7.1, we provide some basic concepts of both linear and nonlinear equalization. In Sect. 7.2, we describe the use of turbo equalization for compensation of different linear and nonlinear channel impairments in fiber-optics communication systems with direct detection. In Sect. 7.3, we describe the *multilevel* maximum a posteriori probability (MAP) turbo equalization scheme based on multilevel Bahl–Cocke–Jelinek–Raviv (BCJR) algorithm (called here multilevel BCJR equalizer) and a soft-iterative decoder. This scheme is suitable for compensation of fiber nonlinearities, chromatic dispersion, and PMD compensation in polarization-multiplexed optical communication systems using multilevel modulations and coherent detection.

I. Djordjevic et al., *Coding for Optical Channels*,
DOI 10.1007/978-1-4419-5569-2_7,

7.1 Channel Equalization Preliminaries

To compensate for linear distortion effects, such as chromatic dispersion, we may use the circuit known as equalizer, connected in cascade with the system under study as shown in Fig. 7.1a. The overall frequency response is equal to $H_c(\omega)H_{eq}(\omega)$, where $H_c(\omega)$ is the system (transfer) function and $H_{eq}(\omega)$ is the equalizer transfer function. In this section, we assume that balanced coherent detection is used. For distortionless transmission we require:

$$H_c(\omega)H_{eq}(\omega) = e^{-j\omega t_0}, \tag{7.1}$$

so that the frequency response of the equalizer is reversely proportional to that of the system as follows:

$$H_{eq}(\omega) = \frac{e^{-j\omega t_0}}{H_c(\omega)}. \tag{7.2}$$

A system that is well suited for equalization is *tapped-delay-line (transversal) equalizer* or the finite impulse response (FIR) filter, which is shown in Fig. 7.1b. The equalizer design can be done by using the window method, which can be described as follows. Start with a specified order M (M is an even integer). For a given sampling interval T_s, proceed as follows: (1) set the constant time delay $t_0 = (M/2)/T_s$; (2) take the inverse Fourier transform of $H_{eq}(\omega)$ to obtain a desired impulse response $h_{eq}(t)$; (3) set $h(n) = w(n)h_{eq}(nT_s)$, where $w(n)$ is a window of length $(M+1)$. The simplest windowing function is the "rectangular window":

$$w(n) = \begin{cases} 1, & n = 0, 1, \ldots, M-1, \\ 0, & \text{otherwise}. \end{cases} \tag{7.3}$$

For example, the influence of chromatic dispersion can be described by the following transfer function

$$H_c(\omega) = \exp\left[j\left(\frac{\beta_2}{2}\omega^2 + \frac{\beta_3}{6}\omega^3\right)L\right], \tag{7.4}$$

where β_2 and and β_3 are group-velocity dispersion (GVD) and second-order GVD parameters, introduced in Chap. 3; and L is the fiber length. From (7.3) it is straightforward to design an FIR equalizer to compensate for chromatic dispersion as shown

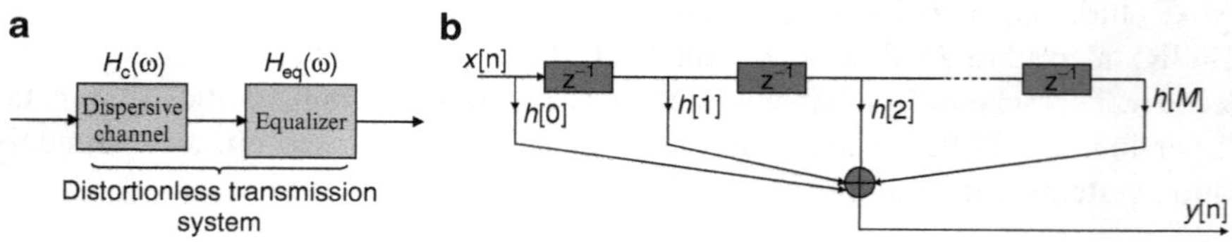

Fig. 7.1 (**a**) Equalization principle, (**b**) tapped-delay-line equalizer (FIR filter)

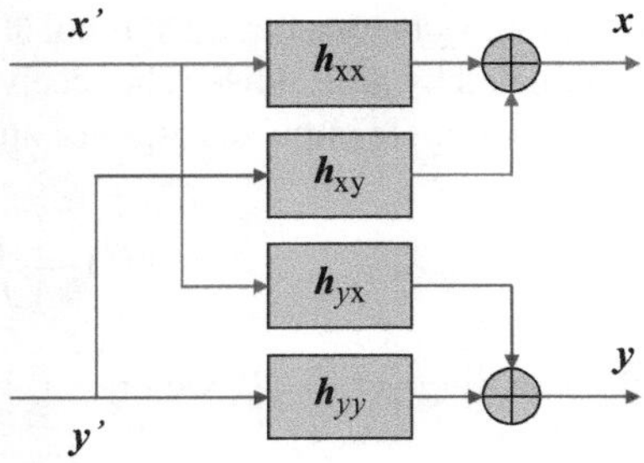

Fig. 7.2 The compensation of polarization-dependent impairments by FIR filters

in [1]. If the second-order GVD parameter can be neglected the FIR equalizer coefficients can even be determined analytically [1]:

$$h[k] = \sqrt{\frac{jcT^2}{D\lambda^2 L}} \exp\left(-\mathrm{j}\frac{\pi c T^2}{D\lambda^2 L}k^2\right), \quad -\left\lfloor \frac{N}{2} \right\rfloor \le k \le \left\lfloor \frac{N}{2} \right\rfloor,$$
$$N = 2\left\lfloor \frac{|D|\,\lambda^2 L}{2cT^2} \right\rfloor + 1, \tag{7.5}$$

where D is dispersion parameter related to GVD by $D = -(2\pi c/\lambda^2)\,\beta_2$. The first-order PMD can also be compensated for by using a series of FIR filters as shown in Fig. 7.2, adopted from [1]. The output symbols in x-polarization can be determined by [1]

$$x[k] = \boldsymbol{h}_{xx}^T \boldsymbol{x}' + \boldsymbol{h}_{xy}^T \boldsymbol{y}' = \sum_{m=0}^{M-1} \{h_{xx}[m]x'[k-m] + h_{xy}[m]y'[k-m]\}, \tag{7.6}$$

where $\boldsymbol{h}_{ij}(i, j \in \{x, y\})$ are FIR filters with M taps. The corresponding equation to determine the symbols in y-polarization can be written in a similar fashion.

The objective of FIR equalizers is to make the intersymbol interference (ISI) to be zero at all instances $t = nT_s$, except for $n = 0$, and are known as the *zero-forcing equalizers*. However, the amplified-spontaneous emission (ASE) noise, chromatic dispersion, and PMD act together, affecting the behavior of a transmission system in a combined manner. Transversal equalizer ignores the effect of channel noise, which leads to the noise enhancement phenomenon. The better approach for receiver design would be to use the *minimum-mean square error* (MMSE) criterion to determine the equalizer coefficients, which provides a balanced solution to the problem of reducing the effects of both channel noise and ISI [12–15]. Let $c(t)$ be the receiver filter impulse response and $x(t)$ be the channel output determined by

$$x(t) = \sum_k s_k q(t - kT_s) + w(t), \quad q(t) = g(t)*h_c(t) \tag{7.7}$$

where $g(t)$ is the transmit filter impulse response, $h_c(t)$ is the channel impulse response (due to chromatic dispersion and optical/electrical filters), T_s is the symbol duration, s_k is transmitted symbol at kth time instance, and $w(t)$ is the channel noise dominated by ASE noise. The receive filter output can be determined by convolution of receive filter impulse response and corresponding input by

$$y(t) = \int_{-\infty}^{\infty} c(\tau)x(t-\tau)\mathrm{d}\tau. \tag{7.8}$$

By sampling at $t = iT_s$ we obtain

$$\begin{aligned} y(iT_s) &= \xi_i + n_i, \quad \xi_i = \sum_k s_k \int_{-\infty}^{\infty} c(\tau)q(iT_s - kT_s - \tau)\mathrm{d}\tau, \\ w_i &= \int_{-\infty}^{\infty} c(\tau)w(iT_s - \tau)\mathrm{d}\tau. \end{aligned} \tag{7.9}$$

The error signal can be defined as the difference of receive sample and transmitted symbol as follows

$$e_i = y(iT_s) - s_i = \xi_i + w_i - s_i. \tag{7.10}$$

The corresponding mean-square error is

$$\mathrm{MSE} = \frac{1}{2}E\left[e_i^2\right] = \frac{1}{2}E\left[\xi_i^2\right] + \frac{1}{2}E\left[w_i^2\right] + \frac{1}{2}E[s_i^2] + E[\xi_i w_i] - E[w_i s_i] - E[\xi_i s_i]. \tag{7.11}$$

Assuming the stationary environment (7.11) can be rewritten as

$$\begin{aligned} \mathrm{MSE} = \frac{1}{2} + \frac{1}{2}\int_{-\infty}^{\infty}\int_{-\infty}^{\infty}\left[R_q(t-\tau) + \frac{N_0}{2}\delta(t-\tau)\right]c(t)c(\tau)\mathrm{d}t\,\mathrm{d}\tau \\ - \int_{-\infty}^{\infty} c(t)q(-t)\mathrm{d}t, \end{aligned} \tag{7.12}$$

where $R_q(\tau)$ is the autocorrelation function of $q(t)$ and N_0 is the power spectral density of ASE noise. To determine the optimum filter in MMSE sense, we perform the derivation of MSE with respect to c(t) and set the corresponding derivative to zero to obtain:

$$\int_{-\infty}^{\infty}\left[R_q(t-\tau) + \frac{N_0}{2}\delta(t-\tau)\right]c(\tau)\mathrm{d}\tau = q(-t). \tag{7.13}$$

By applying the Fourier transform (FT) of (7.13) and solving for receive filter transfer function $C(f) = \mathrm{FT}[c(t)]$ we obtain

$$C(f) = \frac{Q^*(f)}{S_q(f) + \frac{N_0}{2}}. \tag{7.14}$$

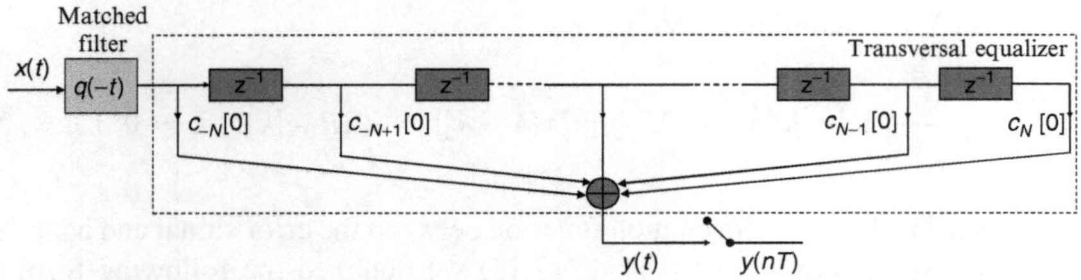

Fig. 7.3 Optimum linear receiver in MMSE sense

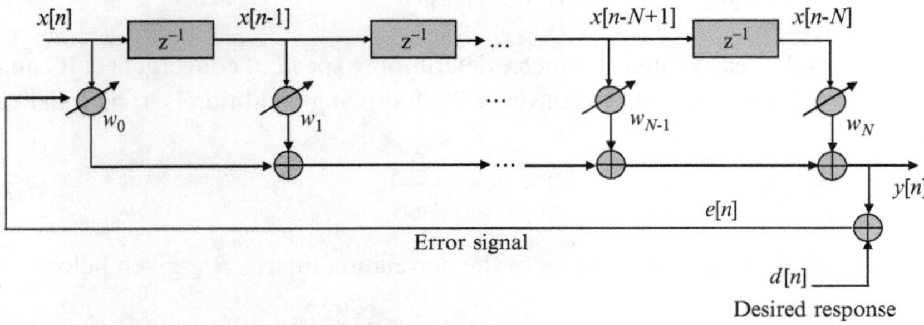

Fig. 7.4 Adaptive equalizer

Therefore, the optimum linear receiver, in MMSE sense, consists of the cascade connection of matched filter and transversal equalizer, as shown in Fig. 7.3.

So far we have assumed that different channel impairments in fiber-optics communication systems are time-invariant, which is not quite true, especially for PMD. To solve for this problem the *adaptive filtering* offers an attractive solution [12], [13]. An adaptive filter has a set of adjustable filter coefficients, as shown in Fig. 7.4, which are adjusted in accordance with an algorithm. Very popular adaptive filter algorithms are steepest descent algorithm and least-mean-square (LMS) algorithm [12, 13].

According to the *steepest descent algorithm*, we update the kth filter coefficient w_k, shown in Fig. 7.4, by making correction of the present value in direction opposite to the gradient ∇_k (in the direction of steepest descent of the error-performance surface) [14]:

$$w_k[n+1] = w_k[n] + \frac{1}{2}\mu\left\{-\nabla_k[n]\right\}, \quad k = 0, 1, \ldots, N. \tag{7.15}$$

If the error signal is defined as difference of desired signal (commonly a training sequence) and corresponding FIR filter output $y[n]$ as follows:

$$e[n] = \mathrm{d}[n] - \sum_{k=0}^{N} w_k[n]x[n-k], \tag{7.16}$$

the gradient ∇_k is determined by

$$\nabla_k[n] = \frac{\partial}{\partial w_k[n]} E\left[e^2[n]\right] = -2E\left[e[n]x[n-k]\right] = -2R_{\text{EX}}[k], \quad k = 0, 1, \ldots, N, \tag{7.17}$$

where $R_{\text{EX}}[k]$ is the cross-correlation function between the error signal and adaptive filter input. By substituting (7.17) into (7.15) we obtained the following form of steepest descent algorithm:

$$w_k[n+1] = w_k[n] + \mu R_{\text{EX}}[k], \quad k = 0, 1, \ldots, N. \tag{7.18}$$

In (7.18) μ is the real-valued parameter determining speed of convergence. It can be shown that for this algorithm to converge the following condition is to be satisfied:

$$0 < \mu < \frac{2}{\lambda_{\max}}, \tag{7.19}$$

where $\lambda_{\max}$ is the largest eigenvalue of the correlation matrix $\boldsymbol{R}_X$ given below

$$\boldsymbol{R}_X = \begin{bmatrix} R_X(0) & R_X(1) & \cdots & R_X(N) \\ R_X(1) & R_X(0) & \cdots & R_X(N-1) \\ \cdots & \cdots & \cdots & \cdots \\ R_X(N) & R_X(N-1) & \cdots & R_X(0) \end{bmatrix}, \quad R_X(l) = E[x[n]x[n-l]] \tag{7.20}$$

A drawback of the steepest descent algorithm is that it requires the knowledge of gradient ∇_k in each iteration. The key idea in LMS algorithm is to approximate the operator of averaging E[.] in (7.17) by its instantaneous value $e[n]x[n-k]$ so that the update r is simply:

$$\hat{w}_k[n+1] = \hat{w}_k[n] + \mu e[n]x[n-k], \quad k = 0, 1, \ldots, N \tag{7.21}$$

Another interesting equalizer is the decision-feedback equalizer (DFE), shown in Fig. 7.5. The key idea of decision-feedback equalization is to use decision made on the basis of precursors of the channel impulse response to take care of postcurosrs,

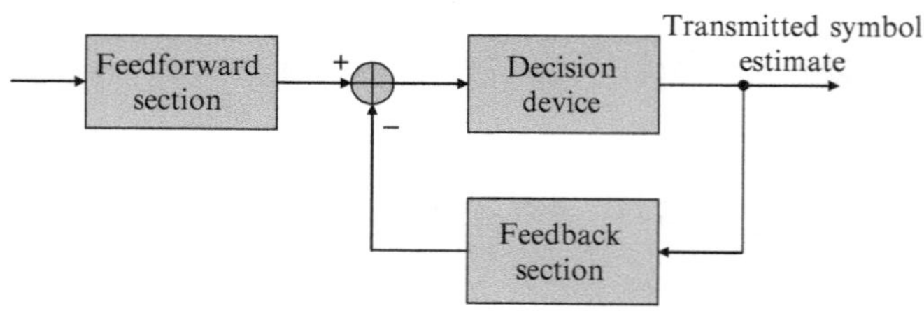

Fig. 7.5 Decision-feedback equalizer (DFE)

assuming that decisions were correct. Let the channel impulse response in discrete form be denoted by $h[n]$. The response of the channel, in absence of noise, to an input sequence $x[n]$ is as follows:

$$y[n] = \sum_k h[k]x[n-k] = h[0]x[n] + \underbrace{\sum_{k<0} h[k]x[n-k]}_{\text{precursors}} + \underbrace{\sum_{k>0} h[k]x[n-k]}_{\text{postcursors}} \tag{7.22}$$

The first term $h[0]x[n]$ represents the desired data symbol; the second term is a function of previous samples only, with the channel coefficients being known as precursors; and the third term is a function of incoming samples, with channel coefficients being known as postcursors. The DFE is composed of feed-forward section, a feedback section, and a decision device, as shown in Fig. 7.5. The feedforward and feedback sections can be implemented as FIR filters, and can be adaptive. Because the transmitted symbol estimate is dependent on previous decisions, this equalizer is nonlinear. When previous decisions are in error, the *error propagation* effect occurs. The errors, however, do not persist indefinitely, rather they occur in bursts.

Another very important equalizer is based on *maximum likelihood sequence detection (estimation)* (MLSD or MLSE) [2, 15–22]. Because this method estimates the sequence of transmitted symbols, it avoids the problem of noise enhancement. The MLSE chooses an input sequence of transmitted symbols $\{s_k\}$ that maximizes the likelihood of received signal $y(t)$, see (7.8). By using Gram–Schmidt orthonormalization procedure we can express the received signal $y(t)$ by using a complete set of orthonormal basis functions $\{\phi_n(t)\}$ as follows [12–15]:

$$y(t) = \sum_{n=1}^{N} y_n \phi_n(t), \tag{7.23a}$$

where

$$y_n = \sum_{k=-\infty}^{\infty} s_k h_{nk} + v_n = \sum_{k=0}^{L} s_k h_{nk} + v_n, \quad h_{nk} = \int_0^{LT_s} h(t-kT_s)\phi_n^*(t)\mathrm{d}t,$$
$$v_n = \int_0^{LT_s} w(t_s)\phi_n^*(t)\mathrm{d}t, \quad h(t) = g(t)^* h_c(t)^* c(t). \tag{7.23b}$$

N in (7.23) denotes the cardinality of basis set, L is the channel memory and $h(t)$ is the overall system impulse response (the convolution of transmit, channel, and receive filters impulse responses). Because v_n are Gaussian random variables, the distribution of $\mathbf{y}^N = [y_1\ y_2\ \ldots\ y_N]^{\mathrm{T}}$ is a multivariate Gaussian:

$$p(y^N|d^L, h(t)) = \prod_{n=1}^{N} \left\{ \frac{1}{\pi N_0} \exp\left[-\frac{1}{N_0} \left| y_n - \sum_{k=0}^{L} d_k h_{nk} \right|^2 \right] \right\}, \tag{7.24}$$

where N_0 is the PSD of ASE noise (see Chap. 3). The MLSE decides in favor of the symbol sequence $\boldsymbol{s}^L$ that maximizes the likelihood function above ((7.24)):

$$\hat{\boldsymbol{s}}^L = \arg\max_{s^L} p\left(y^N|s^L, h(t)\right) = \arg\max_{s^L}\left[-\left|y_n - \sum_{k=0}^{L} s_k h_{nk}\right|^2\right]$$
$$= \arg\max_{s^L}\left\{2\,\mathrm{Re}\left[\sum_k s_k^* \sum_{n=1}^{N} y_n h_{nk}^*\right] - \sum_k\sum_m s_k s_m^* \sum_{n=1}^{N} h_{nk}h_{nm}^*\right\}. \quad (7.25)$$

By noticing that the following is valid:

$$\sum_{n=1}^{N} y_n h_{nk}^* = \int_{-\infty}^{\infty} y(\tau)h^*(\tau - kT_s)\mathrm{d}\tau = s[k], \quad \sum_{n=1}^{N} h_{nk}h_{nm}^*$$
$$= \int_{-\infty}^{\infty} h(\tau - kT_s)\,h^*(\tau - mT_s)\mathrm{d}\tau = f[k-m], \quad (7.26)$$

(7.25) can be simplified as follows:

$$\hat{\boldsymbol{s}}^L = \arg\max_{s^L}\left\{2\,\mathrm{Re}\left[\sum_k s_k^* y[k]\right] - \sum_k\sum_m s_k s_m^* u[k-m]\right\},$$
$$u[k-m] = u(nT_s - kT_s),\; u(t) = h(t) * h^*(-t). \quad (7.27)$$

Equation (7.27) can efficiently calculated by Viterbi algorithm [13]. The Viterbi algorithm provides the hard decisions and as such is not suitable for use with soft decision decoding schemes. To fully exploit the advantages of iterative decoding, the soft reliabilities are needed that can be obtained by soft-output Viterbi algorithm (SOVA) [23], BCJR algorithm [24], or by Monte Carlo equalization [25]. To further improve the BER performance, perform the iteration of extrinsic information between soft equalizer and soft decoder, the procedure known as *turbo equalization* [3, 7–11, 26–28], which is described in the next two sections. The turbo equalization can be used to simultaneously compensate for chromatic dispersion, PMD and fiber nonlinearities, as we have shown in [3]. To compensate for nonlinear impairments we can also use the nonlinear filtering methods [29].

In this section, we described different equalization methods assuming that coherent balanced detection is used. For direct detection, however, the Gaussian assumption in (7.27) is not more valid. We have shown in [30] that for direct detection systems the distribution is rather exponential than Gaussian. The description of equalization of different channel impairments for direct detection systems can be found in [16–22]. In the next section, we describe how LDPC-coded turbo equalization can be used to simultaneously compensate for both linear and nonlinear channel impairments in fiber-optic communication systems with direct detection.

7.2 Turbo Equalization in Fiber-Optics Communication Systems with Direct Detection

As mentioned in Introduction, we will study LDPC-coded turbo equalization scheme, as a universal detection-decoding scheme that can be used simultaneously for: (1) suppression of fiber nonlinearities [7], (2) PMD compensation [8–10], and (3) chromatic dispersion compensation [3]. Before we describe a dynamical nonlinear optical channel model, we provide the basic concepts of optimum detection in minimum probability of error sense. Let $\boldsymbol{x}$ denote the transmitted sequence and $\boldsymbol{y}$ the received sequence. The optimum receiver assigns $\hat{x}_k$ to the value $x \in \{0,1\}$ that maximizes the a posteriori probability (APP) $P(x_k = x|\boldsymbol{y})$ given the received sequence $\boldsymbol{y}$

$$\hat{x}_k = \arg\max_{x\in\{0,1\}} P(x_k = x|\boldsymbol{y}). \tag{7.28}$$

The corresponding algorithm is commonly referred to as a maximum a posteriori probability (MAP) algorithm. In practice, it is common to use the logarithmic version of equation (7.28) as follows:

$$\hat{x}_k = \begin{cases} 0, & L(x_k|\boldsymbol{y}) \geq 0, \\ 1, & \text{otherwise}, \end{cases} \qquad L(x_k|\boldsymbol{y}) = \log\left[\frac{P(x_k = 0|\boldsymbol{y})}{P(x_k = 1|\boldsymbol{y})}\right], \tag{7.29}$$

where $L(x_k|\boldsymbol{y})$ is the conditional log-likelihood ratio (LLR). To calculate the $P(x_k = x|\boldsymbol{y})$ needed in either equation above we invoke the Bayes' rule:

$$P(x_k = x|\boldsymbol{y}) = \sum_{\forall \boldsymbol{x}:\, x_k = x} P(\boldsymbol{x}|\boldsymbol{y}) = \sum_{\forall \boldsymbol{x}:\, x_k = x} \frac{P(\boldsymbol{y}|\boldsymbol{x})P(\boldsymbol{x})}{P(\boldsymbol{y})}, \tag{7.30}$$

where $P(\boldsymbol{y}|\boldsymbol{x})$ is conditional probability density function (PDF), and $P(\boldsymbol{x})$ is the a priori probability of input sequence $\boldsymbol{x}$, which when the symbols are independent factors as $P(\boldsymbol{x}) = \prod_{n=1}^{N} P(x_n)$. By substituting (7.30) into (7.29), the conditional LLR can be written as:

$$L(x_k|\boldsymbol{y}) = \log\left[\frac{\sum\limits_{\forall\, \boldsymbol{x}: x_k=0} p(\boldsymbol{y}|\boldsymbol{x}) \prod\limits_{n=1}^{N} P(x_n)}{\sum\limits_{\forall\, \boldsymbol{x}: x_k=1} p(\boldsymbol{y}|\boldsymbol{x}) \prod\limits_{n=1}^{N} P(x_n)}\right] = L_{\text{ext}}(x_k|\boldsymbol{y}) + L(x_k), \tag{7.31}$$

where the extrinsic information about x_k contained in y $L_{\text{ext}}(x_k|\boldsymbol{y})$ and the a priori LLR $L(x_k)$ are defined, respectively, as

$$L_{\text{ext}}(x_k|\boldsymbol{y}) = \log\left[\frac{\sum\limits_{\forall\, x:x_k=0} p(\boldsymbol{y}|\boldsymbol{x}) \prod\limits_{n=1,n\neq k}^{N} P(x_n)}{\sum\limits_{\forall\, x:x_k=1} p(\boldsymbol{y}|\boldsymbol{x}) \prod\limits_{n=1,n\neq k}^{N} P(x_n)}\right],\ L(x_k) = \log\left[\frac{P(x_k=0)}{P(x_k=1)}\right]. \tag{7.32}$$

From (7.31) and (7.32), it is clear that computation of conditional LLRs can be computationally extensive. One possible computation is based on BCJR algorithm [24], with log-domain version described below.

7.2.1 Description of LDPC-Coded Turbo Equalizer

The LDPC-coded turbo equalizer (TE), shown in Fig. 7.6a, is composed of two ingredients: (1) the BCJR algorithm based equalizer, and (2) the LDPC decoder. Transmitter (shown in Fig. 7.6b) is based on standard RZ transmitter and an LDPC encoder.

BCJR equalizer serves as nonlinear ISI canceller, reduces the BER down to the forward error correction (FEC) threshold, and provides accurate estimates of LLRs for LDPC decoder. To further improve BER performance we allow for iteration of extrinsic information between BCJR equalizer and LDPC decoder. The BCJR equalizer operates on a discrete dynamical trellis description of the

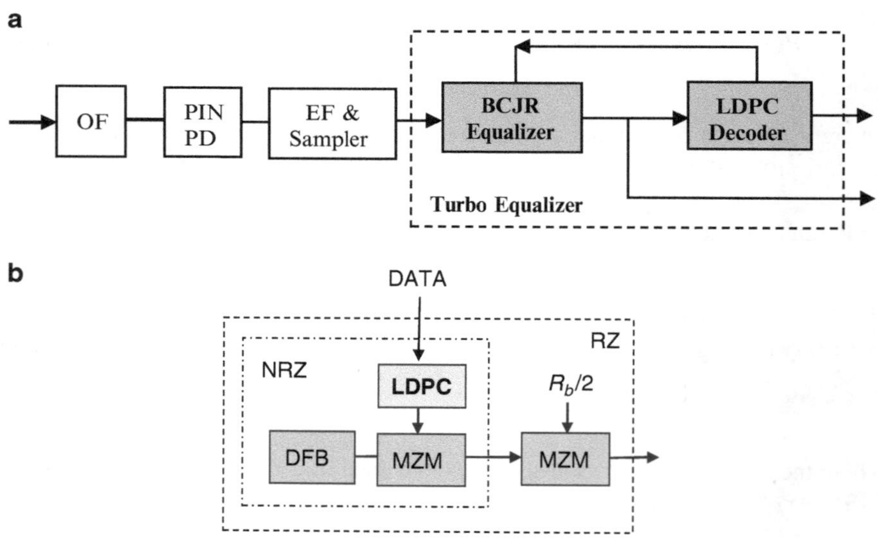

Fig. 7.6 (**a**) Receiver and LDPC-coded turbo equalizer architectures; and (**b**) transmitter configuration. *MZM* Mach–Zhender modulator, *DFB* distributed feedback laser, *LDPC* LDPC encoder, *OF* optical filter, *PD* photodetector, *EF* electrical filter

Fig. 7.7 Trellis description for $2m + 1 = 5$

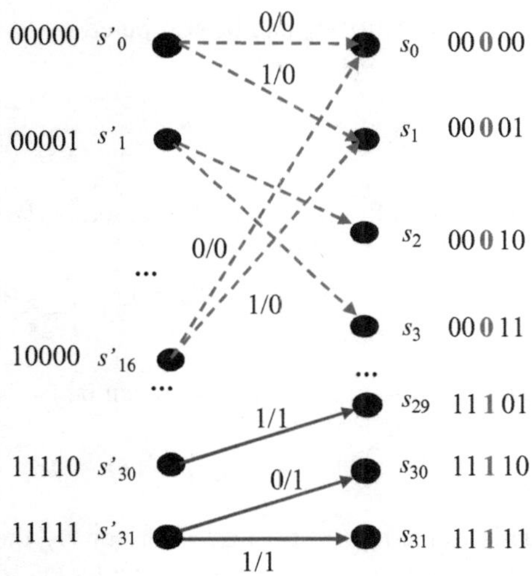

optical channel, with memory equal to $2m + 1$, with $2m$ being the number of bits that influence the observed bit from both sides. This dynamical trellis is uniquely defined by the set of previous state, the next state, in addition to the channel output. The state (the bit-pattern configuration) in the trellis is defined as $\boldsymbol{s}_j = (x_{j-m}, x_{j-m+1}, \ldots, x_j, x_{j+1}, \ldots, x_{j+m}) = \boldsymbol{x}[j-m, j+m]$, where $x_k \in \boldsymbol{X} = \{0, 1\}$. An example trellis of memory $2m + 1 = 5$ is shown in Fig. 7.7. The trellis has $2^5 = 32$ states $(\boldsymbol{s}_0, \boldsymbol{s}_1, \ldots, \boldsymbol{s}_{31})$, each of which corresponds to a different 5-bit pattern. For the complete description of the trellis, the transition probability density functions (PDFs) $p(y_j|x_j) = p(y_j|\boldsymbol{s}), \boldsymbol{s} \in \boldsymbol{S}$ can be determined from *collected histograms* (y_j represents the sample at the input of the BCJR equalizer that corresponds to the transmitted bit u_j, and $\boldsymbol{S}$ is the set of states in the trellis). Because the collection of histograms is time consuming and sensitive to the precision, a better option would be to estimate conditional PDFs using *instantons* (or method of optimal fluctuations) to estimate far tails, and *Edgeworth expansion* to refine the middle part, as we proposed in [30].

Let $\mathbf{s}' = \boldsymbol{x}[k-m-1, k+m-1]$ represent the previous state, $\boldsymbol{s} = \boldsymbol{x}[k-m, k+m]$ the present state, $\boldsymbol{x} = (x_1, x_2, \ldots, x_N)$ the transmitted codeword (or sequence when FEC is not used), and $\boldsymbol{y} = (y_1, y_2, \ldots, y_N)$ the received sequence of samples. To efficiently compute $P(x_k|\boldsymbol{y})$, our starting point is to calculate $P(\boldsymbol{s}', \boldsymbol{s}, \boldsymbol{y})$ as follows

$$p(\boldsymbol{s}', \boldsymbol{s}, \boldsymbol{y}) = P(\boldsymbol{s}', \boldsymbol{s}|\boldsymbol{y})p(\boldsymbol{y}). \tag{7.33}$$

The sequence $\boldsymbol{y}$ can be interpreted as $\boldsymbol{y} = (y_1 \ldots y_{k-1})y_k(y_{k+1} \ldots y_N) = \boldsymbol{y}[1, k-1]y_k\boldsymbol{y}[k+1, N]$. By using the joint probability in terms of conditional probability rule $P(a, b) = P(b|a)P(a)$, the previous equation can be rewritten as

$$\begin{aligned} p(\boldsymbol{s}', \boldsymbol{s}, \boldsymbol{y}) &= \alpha'_k(\boldsymbol{s})\gamma'_k(\boldsymbol{s}', \boldsymbol{s})\beta'_k(\boldsymbol{s}); \quad \alpha'_k(\boldsymbol{s}) = p(\boldsymbol{s}, \boldsymbol{y}[1, k-1]), \\ \gamma'_k(\boldsymbol{s}', \boldsymbol{s}) &= p(\boldsymbol{s}, y_k|\boldsymbol{s}']), \quad \beta'_k(\boldsymbol{s}) = p(\boldsymbol{y}[k+1, N]|\boldsymbol{s}). \end{aligned} \tag{7.34}$$

The probability α'_k can be computed by *forward recursion*

$$\alpha'_k(s) = \sum_{\forall s' \in S} \gamma'_k(s', s)\beta'_k(s). \tag{7.35}$$

The probability β'_k can be computed by *backward recursion*

$$\beta'_k(s') = \sum_{\forall s \in S} \alpha'_k(s)\gamma'_k(s', s). \tag{7.36}$$

The transition probability term can be further decomposed as

$$\gamma'_k(s', s) = p(s|s')p(y_k|s', s). \tag{7.37}$$

This transition probability is driven by the input symbol $x_{(s',s)}$ (that causes s' to s transition, denoted as (s', s)) and $p(y_k|s', s)$ is dependent on the corresponding output symbol $y_{(s',s)}$ so that it can be rewritten as follows:

$$\gamma'_k(s', s) = \begin{cases} p(x_k = x_{(s',s)})p(y_k|y_{(s',s)}), & (s', s) \in \boldsymbol{T}, \\ 0, & \text{otherwise}, \end{cases} \tag{7.38}$$

where $\boldsymbol{T}$ is the set of all possible transitions in trellis channel description. For example, for ASE noise-dominated scenario and balanced coherent detection the $p(y_k|y_{(s',s)})$ is given by

$$p(y_k|y_{(s',s)}) = \frac{1}{\sigma\sqrt{2\pi}} \exp\left[-\frac{\left(y_k - y_{(s',s)}\right)^2}{2\sigma^2}\right], \tag{7.39}$$

where σ^2 is the electrical domain variance due to ASE noise.

What remains now is to calculate the APP $P(x_k|\boldsymbol{y})$, which can be obtained by summing (7.34) over all branches corresponding to input symbol x_k:

$$P(x_k = x|\boldsymbol{y}) = \sum_{\forall (s',s) \in T: x_k = x} p(s', s, \boldsymbol{y}) = \sum_{\forall (s',s) \in T: x_k = x} \alpha'_k(s)\gamma'_k(s', s)\beta'_k(s). \tag{7.40}$$

The corresponding conditional LLR $L(x_k|\boldsymbol{y})$ can be obtained from (7.29) by using (7.40) as follows:

$$L(x_k|\boldsymbol{y}) = \log\left[\frac{P(x_k = 0|\boldsymbol{y})}{P(x_k = 1|\boldsymbol{y})}\right] = \log\left[\frac{\sum_{\forall (s',s) \in T: x_k = 0} \alpha'_k(s)\gamma'_k(s', s)\beta'_k(s)}{\sum_{\forall (s',s) \in T: x_k = 1} \alpha'_k(s)\gamma'_k(s', s)\beta'_k(s)}\right]. \tag{7.41}$$

The conditional LLR in (7.41) can be simplified by defining the log-domain *forward metric* as $\alpha_k(s) = \log \alpha'_k(s)$ and the *backward metric* as $\beta_k(s) = \log \beta'_k(s)$, and it is given by

$$L(x_k|\boldsymbol{y}) = \max_{\forall (s',s)\in \boldsymbol{T}: x_k=0}{}^{*} \left[\alpha_{k-1}(s') + \gamma_k(s',s) + \beta_k(s)\right] - \max_{\forall (s',s)\in \boldsymbol{T}: x_k=1}{}^{*} \left[\alpha_{k-1}(s') + \gamma_k(s',s) + \beta_k(s)\right], \quad (7.42)$$

where max*-operator is defined by $\max^*(x,y) = \log(\mathrm{e}^x + \mathrm{e}^y)$, and it can be efficiently calculated by $\max^*(x,y) = \max(x,y) + c(x,y)$, where $c(x,y)$ is the correction factor, $c(x,y) = \log[1 + \exp(-|x-y|)]$, commonly approximated or implemented using a look-up table. The forward and backward recursion steps of BCJR equalizer are illustrated in Figs. 7.8a, b, respectively. For log-domain the branch metric can be defined as

$$\gamma_k(s',s) = \log \gamma'_k(s',s) = \begin{cases} \log\left[p(x_k = x_{(s',s)})p(y_k|y_{(s',s)})\right], & (s',s) \in \boldsymbol{T}, \\ -\infty, & \text{otherwise}, \end{cases} \quad (7.43)$$

$p(y_k|y_{(s',s)}) = p(y_k|\boldsymbol{x}[k-m,k+m])$ is obtained, as already explained above, by either collecting the histograms or by instanton–Edgeworth expansion method, and $P(x_k)$ represents a priori probability of transmitted bit x_j, which is 1/2 when equally probable transmission is observed.

The dashed lines in Fig. 7.8 corresponds to transitions $(s',s) : x_k = 0$, and the solid lines to transitions $(s',s) : x_k = 1$. The initial forward and backward metrics values are set to

$$\alpha_0(s) = \begin{cases} 0, & s = s_0 \\ -\infty, & s \neq s_0 \end{cases} \quad \text{and} \quad \beta_n(s) = \begin{cases} 0, & s = s_0, \\ -\infty, & s \neq s_0, \end{cases} \quad (7.44)$$

where s_0 is an initial state. A key difference with respect to turbo equalization employed in magnetic recording systems [20] is that the BCJR equalizer operates

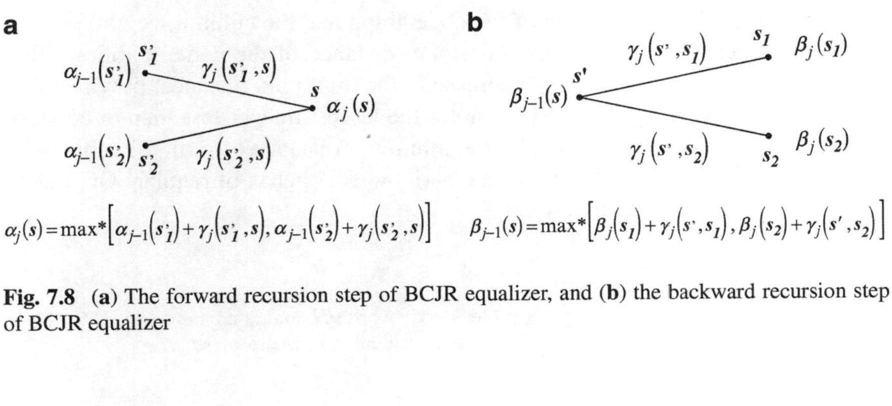

Fig. 7.8 (**a**) The forward recursion step of BCJR equalizer, and (**b**) the backward recursion step of BCJR equalizer

on a trellis that includes both pre- and postcursor ISI. The forward recursion step (Fig. 7.8a), and backward recursion step (Fig. 7.8b) are similar to that in the original log-domain BCJR decoding algorithm described in Chap. 5.

The BCJR equalizer LLRs are forwarded to the soft-decoding LDPC decoder implemented based on sum-product-with-correction-term algorithm [15]. To improve the overall performance of LDPC-coded turbo equalizer we allow the iteration of *extrinsic* LLRs between LDPC decoder and BCJR equalizer. For example, extrinsic LLRs for BCJR equalizer in kth iteration are determined by

$$L_{\mathrm{LDPC},e}\left(x_j^{(k)}\right) = L_{\mathrm{LDPC}}\left(x_j^{(k)}\right) - L_{\mathrm{LDPC}}\left(x_j^{(k-1)}\right), \tag{7.45}$$

where $L_{\mathrm{LDPC}}(x_j)$ is corresponding LLR of bit x_j, while indices k and $k-1$ are used to denote the current and previous iterations. The extrinsic LLRs from (7.45) are used as a priori LLRs in (7.43), which is calculated as follows:

$$L_{\mathrm{BCJR},a}(x_j) = \log[P(x_j)] = L_{\mathrm{LDPC},e}(x_j). \tag{7.46}$$

The LDPC codes suitable for use in turbo equalization can be selected using the concept of EXIT charts [32], in a fashion similar to that we reported in Chap. 6. To facilitate the implementation at high speed we prefer the use of structured quasi-cyclic LDPC codes of large girth. Such codes lead to encoders that can be implemented based on shift-register and modulo-2 adders, and the complexity of decoder is low. One such class of LDPC codes is proposed in the next section.

7.2.2 Large-Girth LDPC Codes

Based on Tanner's bound for the minimum distance of an LDPC code [39]

$$d \geq \begin{cases} 1 + \frac{w_c}{w_c-2}\left((w_c-1)^{\lfloor (g-2)/4 \rfloor} - 1\right), & g/2 = 2m+1, \\ 1 + \frac{w_c}{w_c-2}\left((w_c-1)^{\lfloor (g-2)/4 \rfloor} - 1\right) + (w_c-1)^{\lfloor (g-2)/4 \rfloor}, & g/2 = 2m, \end{cases} \tag{7.47}$$

where g and w_c denote the girth of the code graph and the column weight, respectively, and where d stands for the minimum distance of the code; it follows that large girth leads to an exponential increase in the minimum distance, provided that the column weight is at least 3. ($\lfloor\rfloor$ denotes the largest integer less than or equal to the enclosed quantity.) For example, the minimum distance of girth-10 codes with column weight $r = 3$ is at least 10. The parity-check matrix of regular[1] QC LDPC codes [31] can be represented by

[1] A (w_c, w_r)-*regular LDPC code* is a linear block code whose H-matrix contains exactly w_c 1's in each column and exactly $w_r = w_c n/(n-k)$ 1's in each column, where $w_c \ll n-k$.

$$H = \begin{bmatrix} I & I & I & \dots & I \\ I & P^{S[1]} & P^{S[2]} & \dots & P^{S[c-1]} \\ I & P^{2S[1]} & P^{2S[2]} & \dots & P^{2S[c-1]} \\ \dots & \dots & \dots & \dots & \dots \\ I & P^{(r-1)S[1]} & P^{(r-1)S[2]} & \dots & P^{(r-1)S[c-1} \end{bmatrix}, \tag{7.48}$$

where I is $p \times p$ (p is a prime number) identity matrix, P is $p \times p$ permutation matrix given by $P = (p_{ij})_{p\times p}$, $p_{i,i+1} = p_{p,1} = 1$ (zero otherwise), and where r and c represent the number of block-rows and block-columns in (7.48), respectively. The set of integers S are to be carefully chosen from the set $\{0, 1, \dots, p-1\}$ so that the cycles of short length, in the corresponding Tanner (bipartite) graph representation of (7.48), are avoided. According to Theorem 2.1 in [31], we have to avoid the cycles of length $2k$ ($k = 3$ or 4) defined by the following equation

$$S[i_1]j_1 + S[i_2]j_2 + \dots + S[i_k]j_k = S[i_1]j_2 + S[i_2]j_3 + \dots + S[i_k]j_1 \bmod p, \tag{7.49}$$

where the closed path is defined by (i_1, j_1), (i_1, j_2), (i_2, j_2), (i_2, j_3), ..., (i_k, j_k), (i_k, j_1) with the pair of indices denoting row–column indices of permutation-blocks in (7.48) such that $l_m \neq l_{m+1}$, $l_k \neq l_1$ ($m = 1, 2, \dots, k; l \in \{i, j\}$). Therefore, we have to identify the sequence of integers $S[i] \in \{0, 1, \dots, p-1\}(i = 0, 1, \dots, r-1; r < p)$ not satisfying (7.49), which can be done either by computer search or in a combinatorial fashion. For example, to design the QC LDPC codes in [40], we introduced the concept of the cyclic-invariant difference set (CIDS). The CIDS-based codes come naturally as girth-6 codes, and to increase the girth we had to selectively remove certain elements from a CIDS. The design of LDPC codes of rate above 0.8, column weight 3 and girth 10 using the CIDS approach is a very challenging and is still an open problem. Instead, in our recent paper [41], we solved this problem by developing an efficient computer search algorithm. We add an integer at a time from the set $\{0, 1, \dots, p-1\}$ (not used before) to the initial set S and check if (7.49) is satisfied. If (7.49) is satisfied, we remove that integer from the set S and continue our search with another integer from set $\{0, 1, \dots, p-1\}$ until we exploit all the elements from $\{0, 1, \dots, p-1\}$. The code rate of these QC codes, R, is lower-bounded by

$$R \geq \frac{|S|p - rp}{|S|p} = \frac{1-r}{|S|}, \tag{7.50}$$

and the codeword length is $|S|p$, where $|S|$ denotes the cardinality of set S. For a given code rate R_0, the number of elements from S to be used is $\lfloor r/(1-R_0) \rfloor$. With this algorithm, LDPC codes of arbitrary rate can be designed.

Example 7.1. By setting $p = 2,311$, the set of integers to be used in (7.48) is obtained as $S = \{1, 2, 7, 14, 30, 51, 78, 104, 129, 212, 223, 318, 427, 600, 808\}$. The corresponding LDPC code has rate $R_0 = 1 - 3/15 = 0.8$, column weight 3, girth 10, and length $|S|p = 15 \cdot 2,311 = 34,665$. In the above example, the initial set of integers was $S = \{1, 2, 7\}$, and the set of row to be used in (7.48) is $\{1, 3, 6\}$. The use of a different initial set will result in a different set from that obtained above.

Example 7.2. By setting $p = 269$, the set S is obtained as $S = \{0, 2, 3, 5, 9, 11, 12, 14, 27, 29, 30, 32, 36, 38, 39, 41, 81, 83, 84, 86, 90, 92, 93, 95, 108, 110, 111, 113, 117, 119, 120, 122\}$. If 30 integers are used, the corresponding LDPC code has rate $R_0 = 1 - 3/30 = 0.9$, column weight 3, girth 8, and length $30 \cdot 269 = 8,070$.

The results of simulations for an additive white Gaussian noise (AWGN) channel model are given in Fig. 7.9, where we compare the large-girth LDPC codes (Fig. 7.9a) against RS codes, concatenated RS codes, TPCs, and other classes of LDPC codes. In optical communications, it is a common practice to use the Q-factor as a figure of merit of binary modulation schemes instead of signal-to-noise ratio.[2] In all simulation results in this paper, we maintained the double precision. For the LDPC(16935,13550) code, we also provided 3- and 4-bit fixed-point simulation results (see Fig. 7.9a). Our results indicate that the 4-bit representation performs comparable to the double-precision representation whereas the 3-bit representation performs 0.27 dB worse than the double-precision representation at the BER of $2 \cdot 10^{-8}$. The girth-10 LDPC(24015,19212) code of rate 0.8 outperforms the concatenation $RS(255,239) + RS(255,223)$ (of rate 0.82) by 3.35 dB and RS(255,239) by 4.75 dB both at BER of 10^{-7}. The same LDPC code outperforms projective

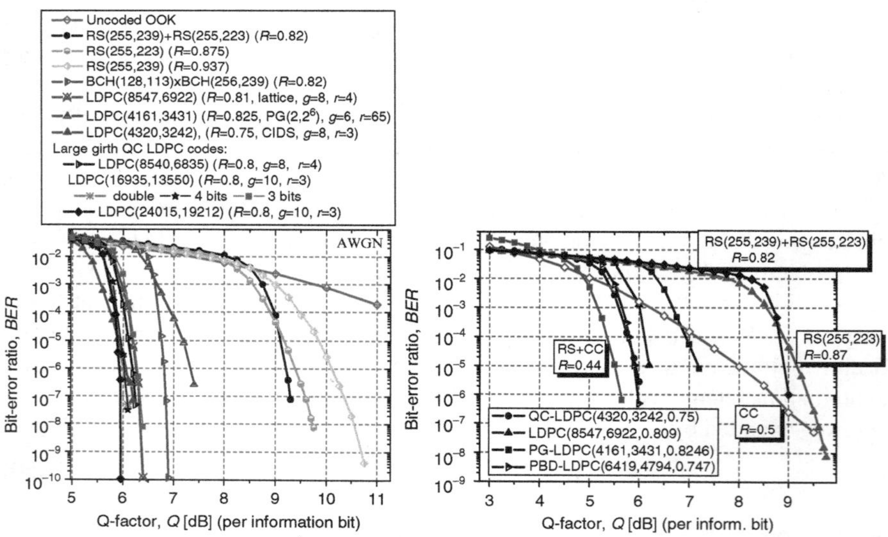

Fig. 7.9 (**a**) Large-girth QC LDPC codes against RS codes, concatenated RS codes, TPCs, and previously proposed LDPC codes on an AWGN channel model, and (**b**) LDPC codes versus convolutional, concatenated RS, and concatenation of convolutional and RS codes on an AWGN channel. Number of iterations in sum–product-with-correction-term algorithm was set to 25 (After ref. [36]; © IEEE 2009; reprinted with permission.)

[2] The Q-factor is defined as $Q = (\mu_1 - \mu_0)/(\sigma_1 + \sigma_0)$, where μ_j and $\sigma_j\,(j = 0, 1)$ represent the mean and the standard deviation corresponding to the bits $j = 0, 1$.

geometry (PG) $(2, 2^6)$-based LDPC(4161,3431) (of rate 0.825) of girth 6 by 1.49 dB at BER of 10^{-7}, and outperforms CIDS-based LDPC(4320,3242) of rate 0.75 and girth-8 LDPC codes by 0.25 dB. At BER of 10^{-10}, it outperforms lattice-based LDPC(8547,6922) of rate 0.81 and girth-8 LDPC code by 0.44 dB, and $\text{BCH}(128, 113) \times \text{BCH}(256, 239)$ TPC of rate 0.82 by 0.95 dB. The net effective coding gain (NECG) at BER of 10^{-12} is 10.95 dB.

In Fig. 7.9b, different LDPC codes are compared against RS(255,223) code, concatenated RS code of rate 0.82 and convolutional code (CC) (of constraint length 5). It can be seen that LDPC codes, both regular and irregular, offer much better performance than hard-decision codes. It should be noticed that pairwised balanced design (PBD)-based irregular LDPC code of rate 0.75 is only 0.4-dB away from the concatenation of convolutional-RS codes (denoted in Fig. 7.9b as RS + CC) with significantly lower code rate $R = 0.44$ at BER of 10^{-6}. As expected, irregular LDPC codes outperform regular LDPC codes.

7.2.3 Suppression of Intrachannel Nonlinearities by LDPC-Coded Turbo Equalization

We are turning our attention to the description how to deal with intra-channel nonlinearities via LDPC-coded turbo equalization. For this purpose, we developed a realistic fiber-optic communication system model based on the nonlinear Schrödinger equation that was solved using the split-step Fourier method. This model takes into account Kerr nonlinearities, stimulated Raman scattering, dispersion effects, ASE noise, filtering effects, ISI, and linear crosstalk effects (for more details an interested reader is referred to Chap. 2). It is well known that at 40-Gb/s and above the intrachannel nonlinearities, such as intrachannel four-wave mixing (IFWM) and intrachannel cross-phase modulation (IXPM), dominate over inter-channel nonlinearities. IXPM effects can be controlled by a proper dispersion map design, while IFWM cannot be eliminated using this approach.

The dispersion map shown in Fig. 7.10 is selected in such a way that IFWM is the predominant intrachannel nonlinear effect. The span length is set to $L = 120$ km, and each span consists of $2L/3$ km of D_+ fiber followed by $L/3$ km of D_- fiber. Precompensation of $-1{,}600$ ps/nm and corresponding postcompensation are also applied. The parameters of D_+ and D_- fibers, used in simulations, are given in

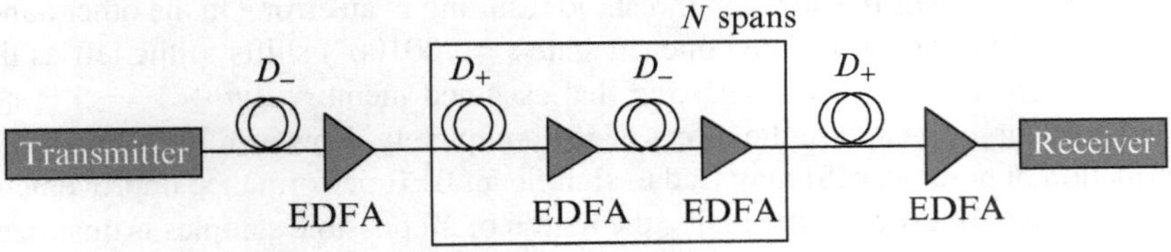

Fig. 7.10 Dispersion map under study

Table 7.1 Fiber parameters

Parameters	D_+ Fiber	D_- Fiber
Dispersion [ps/(nm km)]	20	−40
Dispersion slope [ps/(nm^2 km)]	0.06	−0.12
Effective cross-sectional area (μm^2)	110	50
Nonlinear refractive index (m^2/W)	2.6×10^{-20}	2.6×10^{-20}
Attenuation coefficient (dB/km)	0.19	0.25

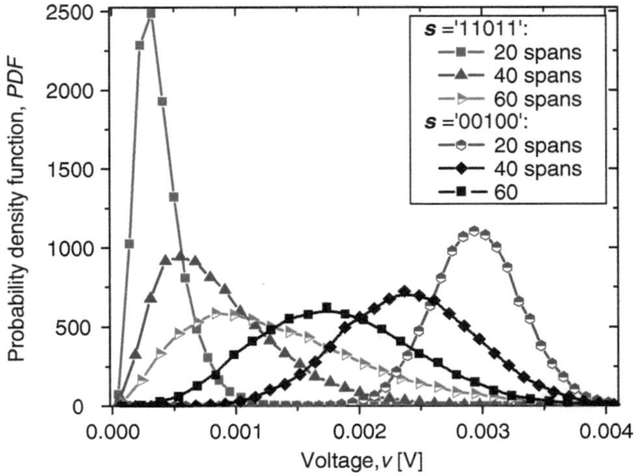

Fig. 7.11 Conditional PDFs $p(y_j|\boldsymbol{x}[j-m, j+m])$ against different number of spans estimated from simulator for RZ OOK

Table 7.1. RZ modulation format of a duty cycle of 33% is observed, the extinction ratio is 14 dB, and the launched power is set to 0 dBm. EDFAs with noise figure of 6 dB are deployed after every fiber section, the bandwidth of optical filter (modeled as super-Gaussian filter of eight order) is set to $3R_l$ and the bandwidth of electrical filter (modeled as Gaussian filter) to $0.7R_l$, with R_l being the line rate (defined as the bit rate (40 Gb/s) divided by a code rate). The line rate (defined above) is appropriately chosen so that the effective information rate is 40 Gb/s.

Figure 7.11 shows the conditional PDFs $p(y_j|\boldsymbol{x}[j-m, j+m])$ obtained for dispersion map described above for $m = 2$. As expected, by increasing the number of spans, the ghost pulse at the central bit position for the state $\boldsymbol{s}$ = "11,011" grows, hence shifting the mean of the PDF to the right. After certain number of spans the mean of PDF exceeds the decision threshold resulting in an error. On the other hand, the mean of PDF for an isolated one (in state $\boldsymbol{s}$ = "00100") shifts to the left as the number of spans increases, suggesting that assumed memory $2m + 1 = 5$ is not sufficiently large to capture the effect of ISI completely. However, even the partial elimination of nonlinear ISI may lead to significant BER performance improvement, as shown later. To estimate the PDFs, the region of all possible samples is quantized in 64 bins, and the number of occurrences of samples in a given bin is counted and normalized with total number of samples in all bins.

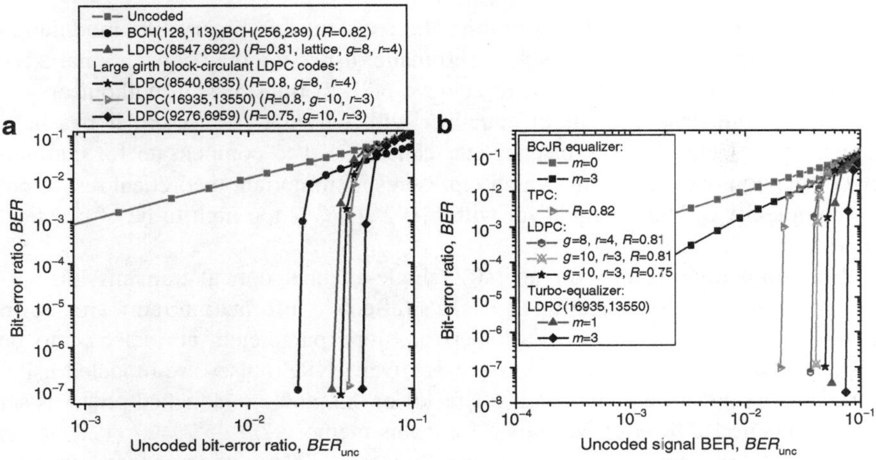

Fig. 7.12 BER performance of different coding schemes at 40 Gb/s for dispersion map from Fig. 7.10: (**a**) BER performance of large-girth block-circulant LDPC codes, (**b**) LDPC-coded TE against LDPC codes and TPC (After ref. [36]; © IEEE 2009; reprinted with permission.)

The results of simulations for a single-channel optical transmission system operating at 40 Gb/s, with dispersion map described above, are shown in Fig. 7.12. The number of spans was changed from 20 to 60 in step of 2, the uncoded BER at 40 Gb/s (denoted as BER_{unc}) and BER after iterative decoding at line rate $R_l = R_b/R(R_b = 40\,\text{Gb/s}$, R is the code rate) were calculated and given in Fig. 7.12 as x- and y-axis, respectively. In Fig. 7.12a we omitted the BCJR equalizer in order to be able to evaluate different classes of LDPC codes introduced in the previous section in the presence of bursts of errors due to intrachannel nonlinearities. The large girth LDPC codes outperform the TPC even in the presence of simultaneous random and bursts errors. The BER performance comparison of LDPC-coded TE against large-girth LDPC codes and turbo-product codes is given in Fig. 7.12b, for different trellis memories. LDPC-coded TE with memory $2m + 1 = 7$ provides almost 12 dB improvement over memoryless BCJR equalizer (assumed memory is $m = 0$) at BER of 10^{-8}.

7.2.4 Chromatic Dispersion Compensation

The chromatic dispersion is time invariant so that it is quite straightforward to compensate for by using the advanced detection and coding techniques. Several of such techniques based on Viterbi equalizer have been recently proposed [2, 16].

The Viterbi equalizer is a maximum-likelihood sequence detection technique whose complexity grows exponentially with residual group-velocity dispersion. In order to reduce the complexity of Viterbi equalizer a narrowband electrical filtering has been proposed in [2], by exploiting the fact that duo-binary like modulation formats are more immune to residual chromatic dispersion than conventional NRZ modulation format. Nevertheless, to achieve 600 km transmission without optical dispersion compensation Viterbi equalizer with 64 states was used. It has been shown in [16] that the Viterbi equalizer can be used to compensate for cumulative dispersion over 1,000 km, which represents an important theoretical result, but the complexity of Viterbi equalizer with 8,192 states is too high to be of practical importance.

The simulations results for an NRZ single-channel optical transmission system given in Fig. 7.13, operating at 10 Gb/s (effective information rate), are shown in Fig. 7.14. The dispersion and dispersion slope parameters are selected to be 16 ps/nm km, and 0.08 ps/(nm^2 km), respectively. NRZ pulses are modeled using raised-cosine pulse shape with roll-off factor of 0.5, and the launched power is set to 0 dBm. The BCJR equalizer alone for trellis memory $2m + 1 = 7$ (128 states) enters the error floor, while TE provides more than 700 km of error-free transmission. Moreover, here we employed conventional NRZ instead of narrowband filtered NRZ used in [2].

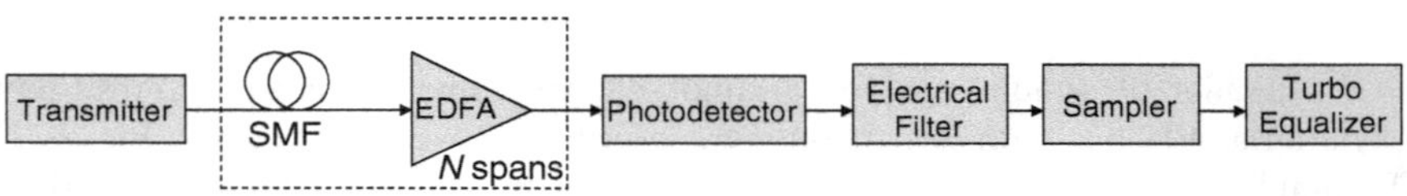

Fig. 7.13 System description for study of chromatic dispersion compensation based on LDPC-coded turbo equalization

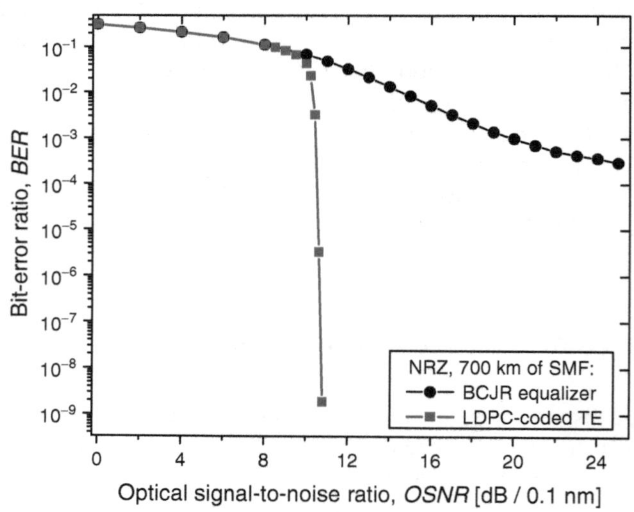

Fig. 7.14 BER performance of LDPC(16935,13550)-coded chromatic dispersion turbo equalizer of trellis memory $2m + 1 = 7$ for SMF length of 700 km

7.2.5 PMD Compensation

The nonlinear ISI TE introduced in Sect. 7.2 can also be used as a PMD compensator. As explained earlier, the TE (see Fig. 7.6a) is composed of two components: the BCJR equalizer, and LDPC decoder. The purpose of BCJR equalizer is to deal with ISI due to PMD, and to provide soft bit reliabilities (LLRs) for LDPC decoder.

The results of simulations, for 10 Gb/s (effective information rate) transmission and ASE noise-dominated scenario, are shown in Fig. 7.15 for DGD $\Delta\tau = 100$ ps and girth-10 LDPC code of rate 0.81. RZ-OOK of a duty cycle of 33% is observed. The bandwidth of optical filter (modeled as super-Gaussian of order 8) is set to $3R_l$, and the bandwidth of electrical filter (modeled as Gaussian) to $0.7R_l$, with R_l being the line rate. For DGD of 100 ps, the $R = 0.81$ LDPC-coded turbo equalizer (for trellis memory $2m + 1 = 7$) has penalty of only 2 dB with respect to the back-to-back configuration. The turbo equalizer employing the code of girth 10 provides more than 1 dB improvement with respect to girth-8 LDPC code.

We further demonstrate the capability of LDPC-coded turbo equalizer in simultaneous chromatic dispersion and PMD compensation. The results of simulation after 700 km of SMF and for DGD of 50 ps, assuming NRZ transmission at 10 Gb/s, are given in Fig. 7.16. The BCJR equalizer enters BER floor that is so high that even concatenated RS code is not able to handle, while LDPC-coded turbo equalizer is able to operate properly.

In the rest of this section we turn our attention to the experimental verification. The experimental setup for PMD compensation study by LDPC-coded turbo equalization is shown in Fig. 7.17. The LDPC-encoded sequence is uploaded into Anritsu pattern generator via GPIB card controlled by a PC. A zero-chirp Mach–Zehnder

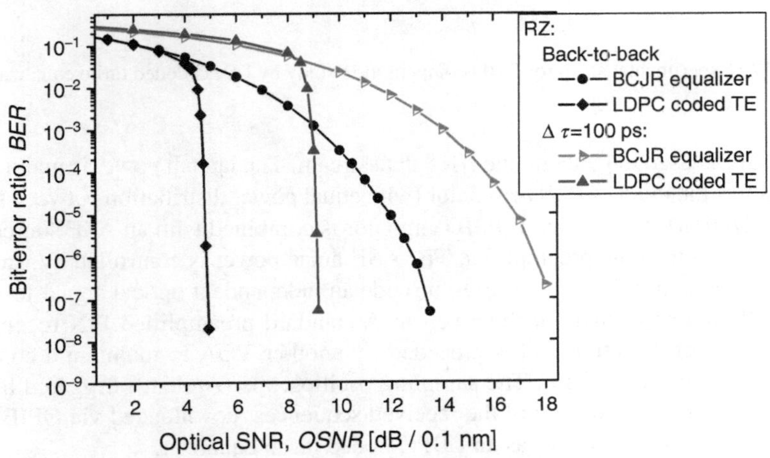

Fig. 7.15 BER performance of LDPC(16935,13550)-coded PMD TE with trellis memory $2m + 1 = 7$ (After ref. [36]; © IEEE 2009; reprinted with permission.)

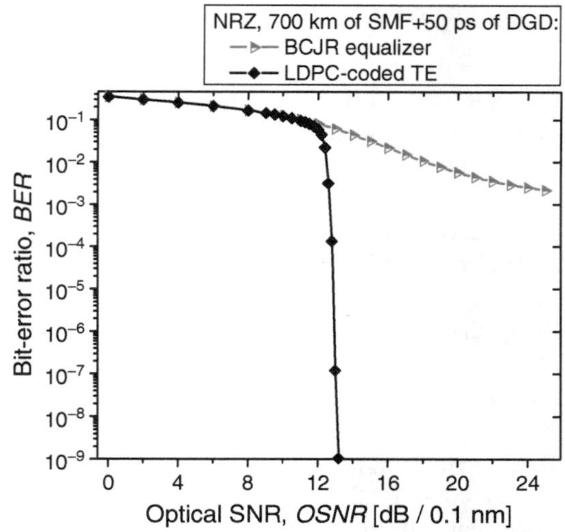

Fig. 7.16 BER performance of LDPC(16935,13550)-coded turbo equalizer of trellis memory $2m + 1 = 7$ for simultaneous chromatic dispersion and PMD compensation

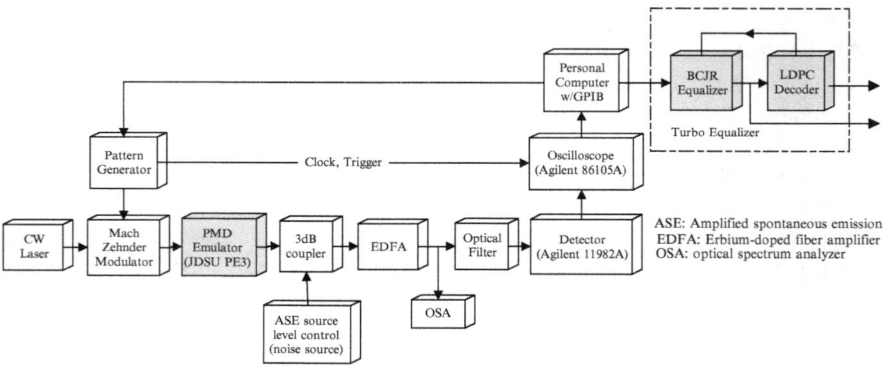

Fig. 7.17 Experimental setup for PMD compensation study by LDPC-coded turbo equalization

modulator is used to generate the NRZ data stream. The launch power is maintained at 0 dBm at the input of PMD emulator (with equal power distribution between states of polarization). The output of PMD emulator is combined with an ASE source immediately prior to the preamplifier. The ASE noise power is controlled by variable optical attenuator (VOA) in order to provide an independent optical signal-to-noise ratio (OSNR) adjustment at the receiver. A standard preamplified PIN receiver is used for direct detection and is preceded by another VOA to maintain a constant received power of −6 dBm. The sampling oscilloscope (Agilent), triggered by the data pattern, is used to acquire the received sequences, downloaded via GPIB card back to the PC which serves as an LDPC-coded turbo equalizer.

The experimental results for 10 Giga symbols/s (GS/s) NRZ transmission are shown in Fig. 7.18, for different DGD values. The TE is based on quasi-cyclic

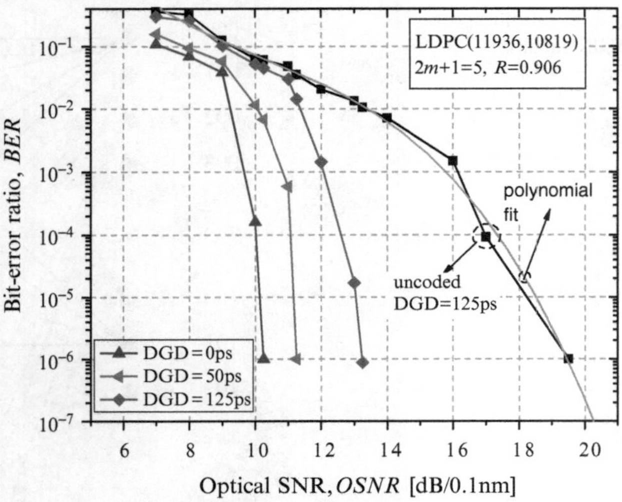

Fig. 7.18 Experimentally obtained BERs of LDPC(11936,10819)-coded PMD compensator (After ref. [36]; © IEEE 2009; reprinted with permission.)

LDPC(11936,10819) code of code rate 0.906 and girth 10, with 5 outer and 25 sum–product algorithm iterations. The OSNR penalty for DGD of 125 ps is about 3 dB at BER $= 10^{-6}$, while the coding gain improvement over BCJR equalizer (with memory $2m + 1 = 5$) for DGD $=$ 125 ps is 6.25 dB at BER $= 10^{-6}$. Larger coding gains are expected at lower BERs.

7.3 Multilevel Turbo Equalization in Fiber-Optics Communication Systems with Coherent Detection

7.3.1 Description of Multilevel Turbo Equalizer

The multilevel BCJR equalizer operates on a discrete dynamical trellis description of the optical channel. This dynamical trellis is uniquely defined by the following triplet: the previous state, the next state, and the channel output. The state in the trellis is defined as $\boldsymbol{s}_j = (x_{j-m}, x_{j-m+1}, \ldots, x_j, x_{j+1}, \ldots, x_{j+m}) = \boldsymbol{x}[j-m, j+m]$, where x_k denotes the index of the symbol from the following set of possible indices $\boldsymbol{X} = \{0, 1, \ldots, M-1\}$, with M being the number of points in corresponding M-ary signal constellation such as M-ary phase-shift keying (PSK), M-ary quadrature-amplitude modulation (QAM) or M-ary polarization-shift keying (PolSK). Every symbol carries $l = \log_2 M$ bits, using the appropriate mapping rule (natural, Gray, anti-Gray, etc.) For example, for QPSK and Gray mapping the following sequences of bits 00, 01, 11, and 10 are mapped to the following signal

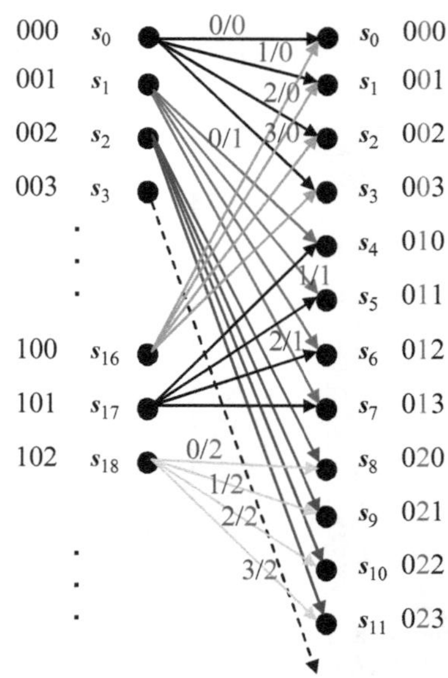

Fig. 7.19 A portion of trellis for four-level BCJR equalizer with memory $2m + 1 = 3$

constellation points QPSK(0) = (1, 0), QPSK(1) = (0, 1), QPSK(2) = (−1, 0) and QPSK(3) = (0, −1), respectively. The memory of the state is equal to $2m + 1$, with $2m$ being the number of *symbols* that influence the observed symbol from both sides. An example trellis of memory $2m+1 = 3$ for 4-ary modulation formats (such as QPSK) is shown in Fig. 7.19. The trellis has $M^{2m+1} = 64$ states $(s_0, s_1, \ldots, s_{63})$, each of which corresponds to different three-symbol patterns (configurations). The state index is determined by considering $(2m + 1)$ symbols as digits in numerical system with the base M. For example, in Fig. 7.19, the quaternary numerical system (with the base 4) is used. (In this system, 18 is represented by $(102)_4$). The left column in the dynamic trellis represents the current states and the right column denotes the terminal states. The branches are labeled by two symbols, the input symbol and the output symbol is the central symbol of terminal state. Therefore, the main difference with respect to other channels (such as magnetic recording channel), where the current symbol is affected only by the previous symbols, is that in optical channel current symbol is affected by both previous and incoming symbols. For the complete description of the dynamical trellis, the transition probability density functions (PDFs) $p(y_j|x_j) = p(y_j|s), s \in \boldsymbol{S}$ are needed; where $\boldsymbol{S}$ is the set of states in the trellis, and y_j is the complex number (corresponding to the transmitted symbol index x_j) with real part being the in-phase channel sample and imaginary part being the quadrature channel sample. The conditional PDFs can be determined from *collected histograms* or by using *instanton-Edgeworth expansion* method [30]. The number of edges originating in any of the left-column states is M, and the number of merging edges in arbitrary terminal state is also M. Notice that samples $y_j (j = 1, \ldots, n)$ are correlated, and to decorrelate them we perform symbol-level interleaving as shown in Fig. 7.20.

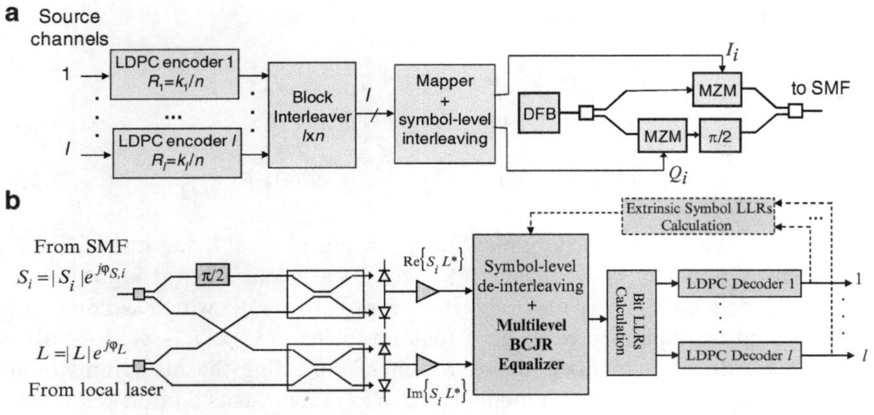

Fig. 7.20 Proposed coded-modulation and turbo equalization schemes: (**a**) transmitter configuration; and (**b**) receiver configuration. *MZM* Mach-Zehnder modulator, *DFB* distributed feedback laser

The transmitter configuration, for multilevel coding (MLC), is shown in Fig. 7.20a. The bit streams originating from l different information sources are encoded using different (n, k_i) LDPC codes of code rate $R_i = k_i/n.k_i$ denotes the number of information bits of ith $(i = 1, 2, \ldots, l)$ component LDPC code, and n denotes the codeword length, which is the same for all LDPC codes. The use of different LDPC codes allows us optimally to allocate the code rates. The bit-interleaved coded modulation (BICM) scheme can be considered as a special MLC scheme in which all of the component codes are of the same rate. The outputs of l LDPC encoders are written row-wise into a block-interleaver block. The mapper accepts l bits at time instance j from the $(l \times n)$ interleaver column-wise and determines the corresponding M-ary $(M = 2^l)$ signal constellation point $s_j = (\phi_{1,j}, \phi_{2,j}) = |s_j| \exp(j\varphi_j)$. The coordinates correspond to in-phase and quadrature components of M-ary QAM constellation, after appropriate mapping. The receiver configuration is shown in Fig. 7.20b. The received electrical field at the ith transmission interval is denoted by $S_i = |S_i| \exp(\mathrm{j}\varphi_{\mathrm{S}i})$, $\varphi_{\mathrm{S}i} = \varphi_i + \varphi_{S,PN}$, where φ_i denotes the data phasor and $\varphi_{S,PN}$ denotes the laser phase noise process of transmitting laser. The local laser electrical field is denoted by $L = |L| \exp(\mathrm{j}\varphi_L)$, where φ_L denotes the laser phase noise process of the local laser. The outputs of upper- and lower-balanced branches, proportional to $\mathrm{Re}\{S_i L^*\}$ and $\mathrm{Im}\ \{S_i L^*\}$, respectively, are used as inputs of multilevel BCJR equalizer.

Before we explain how the symbol LLRs are calculated in multilevel BCJR equalizer block, let us introduce the following notation, which has already been used in Section 7.2.1. The *forward metric* is defined as $\alpha_j(s) = \log\{p(s_j = s, y[1, j])\}(j = 1, 2, \ldots, n)$; the *backward metric* is defined as $\beta_j(s) = \log\{p(y[j+1, n]|s_j = s)\}$; and the *branch metric* is defined as $\gamma_j(s', s) = \log[p(s_j = s, y_j, s_{j-1} = s')]$. The corresponding metrics can be calculated iteratively as follows:

$$\alpha_j(s) = \max_{s'}{}^{*}\left[\alpha_{j-1}(s') + \gamma_j(s', s)\right], \tag{7.51a}$$

$$\beta_{j-1}(s') = \max_{s}{}^{*}\left[\beta_j(s) + \gamma_j(s', s)\right], \tag{7.51b}$$

$$\gamma_j(s', s) = \log\left[p\left(y_j|\boldsymbol{x}[j-m, j+m]\right)P(x_j)\right]. \tag{7.51c}$$

The max*-operator is defined by $\max^*(x, y) = \log(e^x + e^y)$, and it is efficiently calculated by $\max^*(x, y) = \max(x, y) + c_f(x, y)$, where $c_f(x, y)$ is the correction factor, defined as $c_f(x, y) = \log[1 + \exp(-|x - y|)]$, which is commonly approximated or implemented using a look-up table. $p(y_j|\boldsymbol{x}[j-m, j+m])$ is obtained, as already explained above, by either collecting the histograms or by instanton–Edgeworth expansion method, and $P(x_j)$ represents a priori probability of transmitted symbol x_j. In the first outer iteration $P(x_j)$ is set to either $1/M$ (because equally probable transmission is observed) for an existing transition from trellis given in Fig. 7.19, or to zero for a nonexisting transition.

The outer iteration is defined as the calculation of symbol LLRs in multilevel BCJR equalizer block, the calculation of corresponding bit LLRs needed for LDPC decoding, the LDPC decoding, and the calculation of extrinsic symbol LLRs needed for the next iteration. The iterations within LDPC decoder, based on min-sum-with-correction-term algorithm [35], are called here inner iterations. The initial forward and backward metrics values are set to

$$\alpha_0(s) = \begin{cases} 0, s = s_0 \\ -\infty, s \neq s_0 \end{cases} \quad \text{and } \beta_n(s) = \begin{cases} 0, s = s_0 \\ -\infty, s \neq s_0, \end{cases} \tag{7.52}$$

where s_0 is an initial state.

Let $\mathbf{s}' = \boldsymbol{x}\ [j-m-1, j+m-1]$ represent the previous state, $s = \boldsymbol{x}[j-m, j+m]$ the present state, $\boldsymbol{x} = (x_1, x_2, \ldots, x_n)$ the transmitted word of symbols, and $\boldsymbol{y} = (y_1, y_2, \ldots, y_n)$ the received sequence of samples. The LLR, denoting the reliability, of symbol $x_j = \delta (j = 1, 2, \ldots, n)$ can be calculated by

$$\begin{aligned} \Lambda\left(x_j = \delta\right) = &\max_{(s',s):x_j=\delta}{}^{*}\left[\alpha_{j-1}(s') + \gamma_j(s', s) + \beta_j(s)\right] \\ &- \max_{(s',s):x_j=\delta_0}{}^{*}\left[\alpha_{j-1}(s') + \gamma_j(s', s) + \beta_j(s)\right], \end{aligned} \tag{7.53}$$

where δ represents the observed symbol ($\delta \in \{0, 1, \ldots, M-1\}\backslash\{\delta_0\}$) and δ_0 is the referent symbol. The forward and backward metric is calculated using the (7.51). The forward and backward recursion steps of four-level BCJR MAP detector are illustrated in Figs.7.21a and 7.21b, respectively. In Fig. 7.21a s denotes an arbitrary terminal state, which has $M = 4$ edges originating from corresponding initial states, denoted as s'_1, s'_2, s'_3, and s'_4. Notice that the first term in branch metric (see (7.51)) is calculated only once, before the detection/decoding takes place, and stored. The

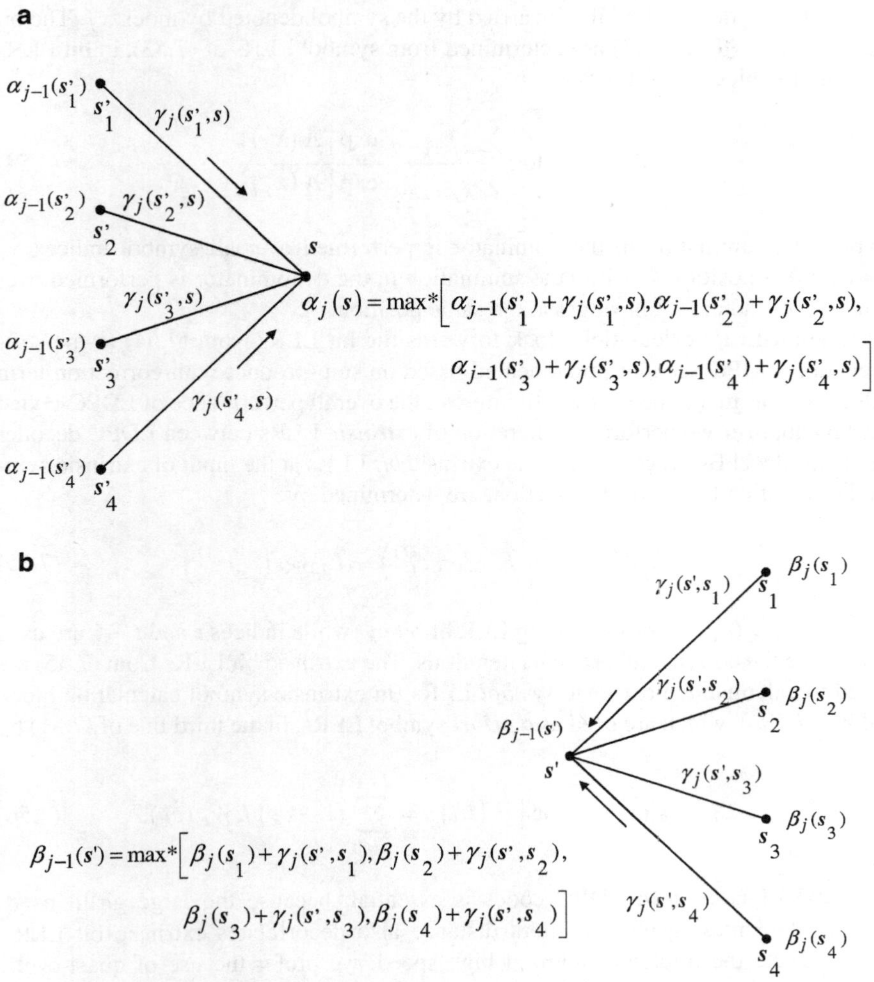

Fig. 7.21 Forward/backward recursion steps for M = four-level BCJR equalizer: (**a**) the forward recursion step, and (**b**) the backward recursion step

second term, $\log(P(x_j))$, is recalculated in every outer iteration. The forward metric of state s in jth step ($j = 1, 2, \ldots, n$) is updated by preserving the maximum term (in max*-sense) $\alpha_{j-1}(s'_k) + \gamma_j(s, s'_k)(k = 1, 2, 3, 4)$. The procedure is repeated for every state in column of terminal states of jth step. The similar procedure is used to calculate the backward metric of state s', $\beta_{j-1}(s')$, (in $(j-1)$th step), as shown in Fig. 7.21b, but now proceeding in backward direction ($j = n, n-1, \ldots, 1$).

Let the c_k denote the kth bit carried by the symbol denoted by index x_j. The bit LLRs $c_k (k = 1, 2, \ldots, l)$ are determined from symbol LLRs of (7.53), in bit LLRs calculation block, as follows:

$$L(\hat{c}_k) = \log \frac{\sum_{x_j : c_k = 0} \exp\left[\Lambda\left(x_j\right)\right]}{\sum_{x_j : c_k = 1} \exp\left[\Lambda\left(x_j\right)\right]}, \tag{7.54}$$

where the summation in the nominator is performed over all symbol indices x_j having 0 at position k, while the summation in the denominator is performed over all symbol indices x_j having 1 at the same position.

The bit LLRs calculation block forwards the bit LLRs from (7.54) to the soft-decoding LDPC decoder implemented based on sum-product-with-correction-term algorithm, as mentioned above. To improve the overall performance of LDPC-coded turbo equalizer we perform the iteration of *extrinsic* LLRs between LDPC decoder and multilevel BCJR equalizer. The extrinsic *bit* LLRs at the input of extrinsic symbol calculation block, in tth iteration, are determined by

$$L_{\text{LDPC},e}\left(c_k^{(t)}\right) = L_{\text{LDPC}}\left(c_k^{(t)}\right) - L_{\text{LDPC}}\left(c_k^{(t-1)}\right), \tag{7.55}$$

where $L_{\text{LDPC}}(c_k)$ is corresponding LLR of bit c_k, while indices t and $t-1$ are used to denote the current and previous iterations. The extrinsic bit LLRs from (7.55) are used to calculate the extrinsic *symbol* LLRs (in extrinsic symbol calculation block of Fig. 7.20b), which are used as *a priori* symbol LLRs, in the third line of (7.51) by

$$L_{\text{BCJR},a}\left(x_j\right) = \log\left[P\left(x_j\right)\right] = \sum_{k=0}^{l-1} (1 - c_k) L_{D,e}(c_k). \tag{7.56}$$

The use of large-girth LDPC codes is essential, because the large girth in addition of increasing the minimum distance also decorrelates extrinsic bit LLRs. To facilitate the implementation at high speed, we prefer the use of quasi-cyclic codes rather than random LDPC codes; and corresponding design is briefly given in Sect. 7.2.2. To optimally match the multilevel BCJR equalizer and LDPC decoder, quasi-cyclic LDPC codes are selected using the concept of EXIT charts as explained in [32].

Notice that complexity of dynamic trellis grows exponentially, because the number of states is determined by M^{2m+1}, so that the increase in signal constellation leads to increase of the base, while the increase in state memory assumption $(2m+1)$ leads to the increase of exponent. We will show in case of QPSK transmission, that even small state memory assumption $(2m + 1 = 3)$ leads to significant performance improvement with respect to the case $m = 0$. For larger constellations and/or larger memories the reduced-state BCJR algorithm [42] is to be used instead. Moreover, we have shown in [7] that approximating $\max^*(x, y)$ by $\max(x, y)$ in the regime when intrachannel nonlinearities dominate leads to insignificant performance degradation, while the forward and the backward steps in BCJR equalizer become the forward and the backward Viterbi equalizers, respectively; which are already implemented at 10 Gb/s. Another interesting approach to reduce the channel memory, introduced recently [33, 34], is the backpropagation method. Namely, in point-to-point links, the receiver knows the dispersion map configuration and can propagate the received signal through the dispersion map with fiber parameters (group velocity dispersion (GVD), second-order GVD, and nonlinearity coefficient) of opposite signs to that used in original map. However, the nonlinear interaction of ASE noise and Kerr nonlinearities cannot be compensated for. Moreover, the complexity of this approach is about two orders of magnitude higher that that of linear equalizer as shown in [34]. The best strategy would be to use the coarse backpropagation (with reasonable small number of coefficients) to reduce the channel memory, and compensate for remained channel distortions by turbo equalization scheme introduced in this section. Another advantage of turbo equalization is that it can operate even when the conditional estimates obtained from instanton–Edgeworth expansion method are not perfect, because we iterate extrinsic information between soft-decoders and BCJR equalizer until the valid codewords are obtained (or predetermined number of iteration has been reached), which improves the system performance.

7.3.2 *Mitigation of Intrachannel Nonlinearities by LDPC-Coded Turbo Equalization Based on Multilevel BCJR Algorithm*

We are turning our attention to the evaluation of proposed scheme in suppression of intrachannel nonlinearities, which are the most challenging impairments to mitigate. For this purpose, we developed a realistic fiber-optic communication system model based on the nonlinear Schrödinger equation that was solved using the split-step Fourier method. This model takes into account Kerr nonlinearities, nonlinear phase noise, stimulated Raman scattering, dispersion effects, amplified-spontaneous emission (ASE) noise, linear filtering effects, ISI, and linear crosstalk effects.

The submarine-like dispersion map under study is shown in Fig. 7.10, which is chosen on such a way that the presence of intrachannel nonlinearities is predominant. The span length is set to $L = 120$ km, and each span consists of $2L/3$ km of D_+ fiber followed by $L/3$ km of D_- fiber. Precompensation of $-1{,}600$ ps/nm and corresponding postcompensation are also applied. The parameters of D_+ and D_- fibers, used in simulations, are given in Table 7.1. The QPSK modulation format is observed with RZ pulses having a duty cycle of 33% ($\sqrt{(P_p)}\cos\{(\pi/2)\sin[\pi R_s(t - 0.5/R_s)]\}$, R_s the symbol rate, P_p the peak power), and the averaged launched power being set to 0 dBm. EDFAs with noise figure of 5 dB are deployed after every fiber section, the bandwidth of optical filter is set to $3R_l$ and the bandwidth of electrical filter to $0.7R_l$, with R_l being the line rate that is defined as the symbol rate of 50 GS/s divided by a code rate. The line rate is appropriately chosen so that the effective aggregate information rate is 100 Gb/s. The scheme is therefore evaluated for use in 100 Gb/s transmission and 100 G Ethernet.

The results of Monte Carlo simulations (performed by parallel computing on several Dual Intel Quad-Core Xeon CPUs) for a single-channel single-polarization optical QPSK transmission system with Gray mapping operating at 50 GS/s, with dispersion map described above, are shown in Fig. 7.22. The number of spans was changed from 4 to 84, the uncoded BER at 50 GS/s and BER after iterative decoding at line rate $R_l = R_s/R(R_s = 50\,\mathrm{GS/s}$, R is the code rate) were calculated and given in Fig. 7.22 against the number of spans. The number of inner (LDPC decoder) iterations was set to 25, and the number of outer (multilevel BCJR–LDPC decoder) iterations to 3. We can see that for 22 spans, four-level BCJR equalizer with memory $2m + 1 = 3$ provides more than one order in magnitude improvement in BER over memoryless case ($2m + 1 = 1$). For the turbo equalization scheme based on four-level BCJR equalizer of memory $2m + 1 = 1$ and the LDPC(16935, 13550) code of girth 10 and column weight 3 (described in Sect. 7.2.2), we achieve transmission over 55 spans (6,600 km) without any error. On the other hand, for the turbo equalization scheme based on four-level BCJR equalizer of memory $2m + 1 = 3$ and the same LDPC code, we are able to achieve even 8,160 km of error-free transmission at an aggregate rate of 100 Gb/s. Because we are concerned with 100 Gb/s per wavelength transmission only QPSK was studied (Fig. 7.22).

In Fig. 7.22 we also report the results obtained by turbo equalization with backpropagation. The dispersion map was composed of standard SMF only with EDFAs of noise figure 6 dB being deployed every 100 km. We can see that backpropagation can indeed reduce the channel memory because the improvement for $m = 1$ is small compared to case $m = 0$.

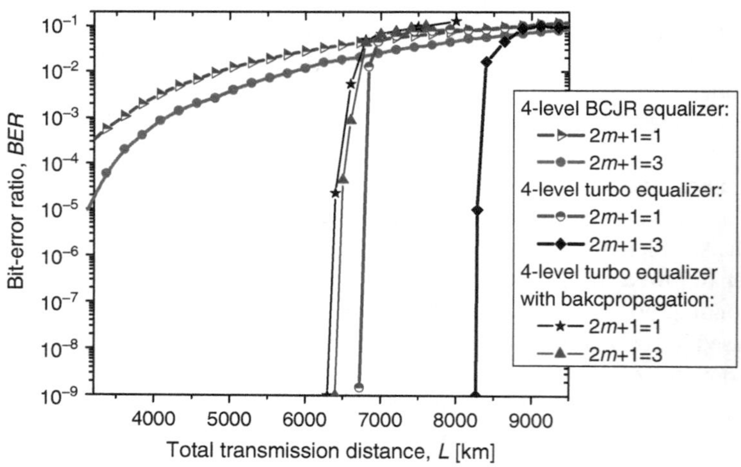

Fig. 7.22 BER performance of turbo equalizer based on four-level BCJR equalizer for QPSK with Gray mapping (After ref. [38]; © IEEE 2009; reprinted with permission.)

7.3.3 *PMD Compensation in Polarization Multiplexed Multilevel Modulations by Turbo Equalization*

In this section, we demonstrate the efficiency of multilevel TE in PMD compensation. The multilevel turbo equalizer described in previous section requires that PMD is compensated for by digital filtering, before decoding takes place. In this section we describe an approach to turbo equalization which does not require previous PMD compensation [37]. This scheme considers the independent symbols transmitted in two orthogonal polarization states as a *super-symbol*, which is a symbol vector composed of two components, corresponding to two orthogonal polarization states. The multilevel turbo equalizer is universal and applicable to any two-dimensional signal constellation such as M-ary phase-shift keying (PSK), M-ary quadrature amplitude modulation (QAM) or M-ary pulse-amplitude modulation (PAM), and both coherent and direct detections. The scheme can be used in combination with an arbitrary soft-decoding process. The transmitter and receiver configurations are shown in Figs. 7.23 (a,b), respectively. The transmitted super-symbol sequence and the received sequences are denoted by $\boldsymbol{X}$ and $\boldsymbol{Y}$, respectively. The super-symbol $x \in \boldsymbol{X}$ has two components $x = (\boldsymbol{s}_{\mathrm{H}}, \boldsymbol{s}_{\mathrm{V}})$, where $\boldsymbol{s}_{\mathrm{H}}(\boldsymbol{s}_{\mathrm{V}})$ denotes the symbol transmitted over horizontal (vertical) polarization. The symbols transmitted over either polarization channel originate from M-ary QAM, M-ary PSK, or M-ary PAM; and every symbol carries $\log_2 M = b$ bits. The received symbol $y \in \boldsymbol{Y}$ has four components $\boldsymbol{y} = (\mathrm{Re}(y_{\mathrm{H}}), \mathrm{Im}(y_{\mathrm{H}}), \mathrm{Re}(y_{\mathrm{V}}), \mathrm{Im}(y_{\mathrm{V}}))$, where $\mathrm{Re}(y_{\mathrm{H}})$ and $\mathrm{Im}(y_{\mathrm{H}})$ denote the samples corresponding to I- and Q-channels in horizontal polarization, while $\mathrm{Re}(y_{\mathrm{V}})$ and $\mathrm{Im}(y_{\mathrm{V}})$ denote the samples corresponding to I- and Q-channels in vertical polarization. y_j is the vector of samples that correspond to the transmitted symbol x_j for the jth discrete time period. The memory assumption indicates that a

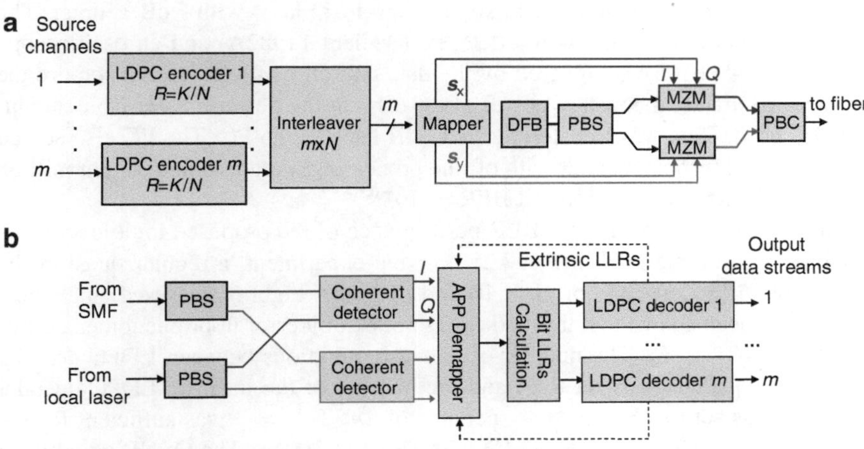

Fig. 7.23 The super-symbol based architectures of: (**a**) transmitter and (**b**) turbo equalizer

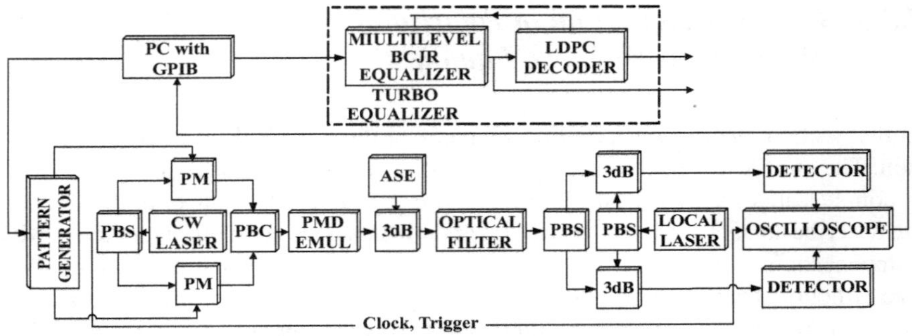

Fig. 7.24 Experimental setup for polarization multiplexed BPSK. CW Laser: continuous wave laser, PM: phase modulator, ASE: amplified spontaneous emission noise source, 3dB: 3dB coupler

super-symbol x_i is influenced by the preceding $m(x_{i-m}, x_{i-m+1}, \ldots, x_{i-1})$ super-symbols and the next $m(x_{i+1}, \ldots, x_{i+m})$ super-symbols in the sequence and the memory is denoted by $2m + 1$.

Figure 7.24 shows the experimental setup. The two orthogonal polarizations of a continuous wave laser source are separated by a polarization beam splitter and are modulated by two phase modulators (Covega) driven at 10 Gb/s (Anritsu MP1763C). A precoded test pattern was loaded into the pattern generator via personal computer with GPIB interface. A polarization beam combiner was used to combine the two modulated signals, followed by a PMD emulator (JDSU PE3), which introduced controlled amount of DGD to the signal. Then the signal distorted by PMD was mixed with controlled amount of ASE noise with 3 dB coupler. Modulated signal level was maintained at 0 dB while the ASE power level was changed to obtain different OSNRs. Next, the optical signal was preamplified, filtered (JSDU 2 nm band-pass filter) and coherently detected. The coherent detection is performed by mixing the received signal with signal from local laser with 3 dB coupler. The resulting signal is detected with a detector (Agilent 11982A) and an oscilloscope (Agilent DCA 86105A), triggered by the data pattern that was used to acquire the samples. To maintain constant power of −6 dBm at the detector, a variable attenuator was used. Data was transferred via GPIB back to the PC. The PC also served as a multilevel turbo equalizer with offline processing. Pattern length of ten million bits was used for measurements at BER $= 10^{-6}$.

The experimental results for BER performance of the proposed multilevel turbo equalizer are summarized in Fig. 7.25. For the experiment, a regular quasi-cyclic LDPC(16935,13550) code of girth 10 and column weight 3 was used as channel code. It has been shown that this code does not exhibit error floor phenomena down to BER $= 10^{-9}$ [36]. The number of extrinsic iterations between LDPC decoder and BCJR equalizer was set to 3, and the number of the intrinsic LDPC decoder iterations was set to 25. The state memory of $2m + 1 = 3$ was sufficient for the compensation of the first-order PMD with DGD of 100 ps. The OSNR penalty for 100 ps of DGD is 1.5 dB at BER of 10^{-6}. Coding gain for DGD of 0 ps is 7.5 dB

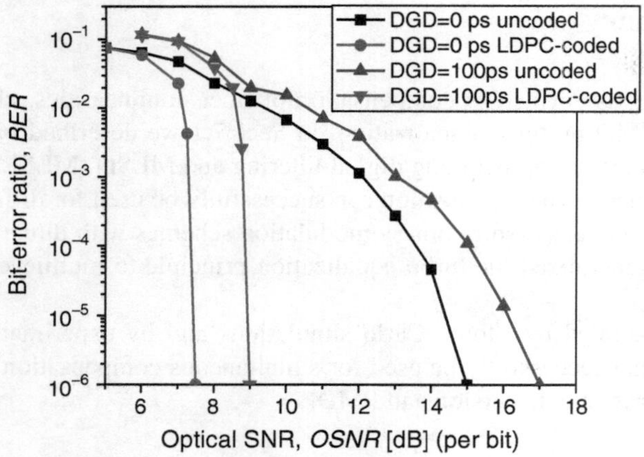

Fig. 7.25 BER performance of super-symbol based multilevel turbo equalization scheme. (After ref. [49]; © IEEE 2009; reprinted with permission.)

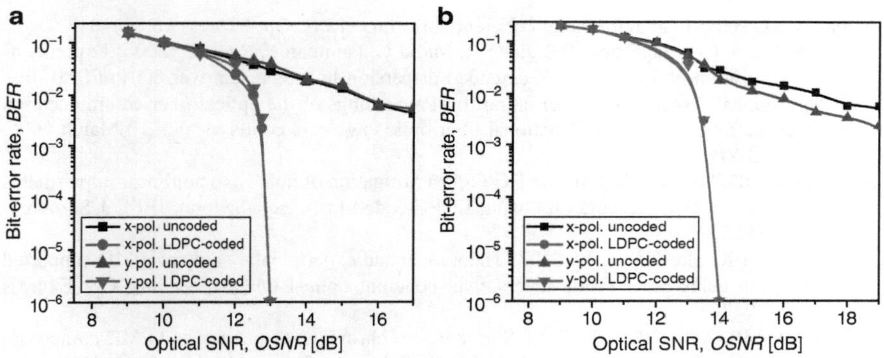

Fig. 7.26 BER performance of conventional turbo equalization scheme: (**a**) DGD = 0 ps, and (**b**) DGD = 100 ps

at BER of 10^{-6}, and the coding gain for DGD of 100 ps is 8 dB. The BER performance, when polarization channels were considered independently, for DGD of 0 ps, is shown in Fig. 7.26a, while that for DGD of 100 ps is shown in Fig. 7.26b. The uncoded curves for DGD of 100 ps exhibit a BER floor, which is so high that the use of advanced FEC is necessary. Using a hard decision equalizer would result in an error floor while the proposed equalization scheme eliminates the error floor and allows for implementation of FEC codes other than LDPC codes. The proposed multilevel turbo equalization scheme outperforms the corresponding scheme that considers the polarization channels independently by even 5 dB at BER of 10^{-6} for DGD of 100 ps and by 5.5 dB at BER of 10^{-6} for DGD of 0 ps.

7.4 Summary

This Chapter was devoted to compensation of fiber nonlinearities, chromatic dispersion and PMD by turbo equalization. In Sect. 7.1 we described basic concepts of channel equalization, including digital filtering and MLSD (MLSE). In Sect. 7.2 we described how turbo equalization can successfully be used for different channel impairments compensation in binary modulation schemes with direct detection. In Sect. 7.3 we generalized the turbo equalization principle to multilevel modulation schemes.

We demonstrated by Monte Carlo simulations and by experiments that turbo equalization can successfully be used for simultaneous compensation of fiber nonlinearities, chromatic dispersion and PMD.

References

1. Savory SJ (2008) Digital filters for coherent optical receivers. Opt Express 16:804–817
2. Alic N, Papen GC, Saperstein RE, Jiang R, Marki C, Fainman Y, Radic S (2006) Experimental demonstration of 10 Gb/s NRZ extended dispersion-limited reach over 600 km-SMF link without optical dispersion compensation. In: Proceedings of the optical fiber communication conference, 2006 and the 2006 national fiber optic engineers conference, 5–10 March 2006, paper no. OWB7
3. Djordjevic IB, Minkov LL, Batshon HG (2008) Mitigation of linear and nonlinear impairments in high-speed optical networks by using LDPC-coded turbo equalization. IEEE J Sel Areas Comm 26(6):73–83
4. Shieh W, Yi X, Ma Y, Tang Y (2007) Theoretical and experimental study on PMD-supported transmission using polarization diversity in coherent optical OFDM systems. Opt Express 15:9936–9947
5. Djordjevic IB, Xu L, Wang T (2008) Simultaneous chromatic dispersion and PMD compensation by using coded-OFDM and girth-10 LDPC codes. Opt Express 16(14):10269–10278
6. Djordjevic IB (2007) Suppression of intrachannel nonlinearities in high-speed WDM systems. In: Xu L (ed) Advanced technologies for high-speed optical communications. Research Signpost, Trivandrum–Kerala, India, pp 247–277
7. Djordjevic IB, Vasic B (2006) Nonlinear BCJR equalizer for suppression of intrachannel nonlinearities in 40 Gb/s optical communication systems. Opt Express 14:4625–4635
8. Djordjevic IB, Batshon HG, Cvijetic M, Xu L, Wang T (2007) PMD compensation by LDPC-coded turbo equalization. IEEE Photon Technol Lett 19(15):1163–1165
9. Minkov LL, Djordjevic IB, Batshon HG, Xu L, Wang T, Cvijetic M, Kueppers F (2007) Demonstration of PMD compensation by LDPC-coded turbo equalization and channel capacity loss characterization due to PMD and quantization. IEEE Photon Technol Lett 19(22): 1852–1854
10. Minkov LL, Djordjevic IB, Xu L, Wang T, Kueppers F (2008) Evaluation of large girth LDPC codes for PMD compensation by turbo equalization. Opt Express 16(17):13450–13455
11. Djordjevic IB, Vasic B (2006) Noise-predictive BCJR equalization for suppression of intrachannel nonlinearities. IEEE Photon Technol Lett 18(12):1317–1319
12. Haykin S (2001) Communication systems. Wiley, Hoboken, NJ
13. Proakis JG (2001) Digital communications. McGrawHill, Boston
14. Haykin S (1988) Digital communications. Wiley, Hoboken, NJ
15. Goldsmith A (2005) Wireless communications. Cambridge University Press, New York

16. Poggiolini P, Bosco G, Savory S, Benlachtar Y, Killey RI, Prat J (2006) 1,040 km uncompensated IMDD transmission over G.652 fiber at 10 Gbit/s using a reduced-state SQRT-metric MLSE receiver. In: Proceedings of the ECOC 2006, post-deadline paper Th4.4.6, Cannes, France, September 2006
17. Bülow H, Buchali F, Franz B (2005) Enhancement of tolerances in transmission systems by analog and digital signal processing. Asia-Pacific optical communications conference (APOC 2005) Shanghai, China, November 2005. In: Proceedings of the SPIE, vol 6021, paper 6021–6163
18. Bülow H, Lanne S (2005) PMD compensation techniques. In: Galtarossa A Menyuk CR (eds) Polarization mode dispersion. Springer, New York
19. Agazzi OE, Hueda MR, Carrer HS, Crivelli DE (2005) Maximum-likelihood sequence estimation in dispersive optical channels. IEEE/OSA J Lightwave Technol 23:749–763
20. Alic N, Papen GC, Saperstein RE, Milstein LB, Fainman Y (2005) Signal statistics and maximum likelihood sequence estimation in intensity modulated fiber optic links containing a single optical preamplifier. Opt Express 13:4568–4579
21. Colavolpe G, Foggi T, Forestieri E, Prati G (2008) Multilevel optical systems with MLSD receivers insensitive to GVD and PMD. IEEE/OSA J Lightwave Technol 26:1263–1273
22. Ivkovic M, Djordjevic IB, Vasic B (2006) Hard decision error correcting scheme based on LDPC codes for long-haul optical transmission. In: Proceedings of the optical transmission systems and equipment for networking V-SPIE optics east conference, Boston, Massachusetts, 1–4 October 2006 vol 6388, pp 63880F.1–63880F.7
23. Hagenauer J, Hoeher P (1989) A Viterbi algorithm with soft-decision outputs and its applications. In: Proceedings of the IEEE Globecom conference, Dallas, TX, November 1989, pp 1680–1686
24. Bahl LR, Cocke J, Jelinek F, Raviv J (1974) Optimal decoding of linear codes for minimizing symbol error rate. IEEE Trans Inf Theory IT-20:284–287
25. Wymeersch H Win MZ (2008) Soft electrical equalization for optical channels. In: Proceedings of the ICC'08, 19–23 May 2008, pp 548–552
26. Douillard C, Jézéquel M, Berrou C, Picart A, Didier P, Glavieux A (1995) Iterative correction of intersymbol interference: turbo equalization. Eur Trans Telecommun 6:507–511
27. Tüchler M, Koetter R, Singer AC (2002) Turbo equalization: principles and new results. IEEE Trans Commun 50(5):754–767
28. Jäger M, Rankl T, Spiedel J, Bulöw H, Buchali F (2006) Performance of turbo equalizers for optical PMD channels. IEEE/OSA J Lightwave Technol 24(3):1226–1236
29. Xia C, Rosenkranz W (2005) Performance enhancement for duobinary modulation through nonlinear electrical equalization. In: Proceedings of the 31st european conference on optical communications (ECOC 2005) vol. 2, Kiel University, Germany, 25–29 September 2005, pp 255–256
30. Ivkovic M, Djordjevic I, Rajkovic P, Vasic B (2007) Pulse energy probability density functions for long-haul optical fiber transmission systems by using instantons and Edgeworth expansion. IEEE Photon Technol Lett 19(20):1604–1606
31. Fossorier MPC (2004) Quasi-cyclic low-density parity-check codes from circulant permutation matricies. IEEE Trans Inf Theory 50:1788–1794
32. ten Brink S (2001) Convergence behavior of iteratively decoded parallel concatenated codes. IEEE Trans Commun 40:1727–1737
33. Essiambre R-J, Foschini GJ, Kramer G, Winzer PJ (2008) Capacity limits of information transport in fiber-optic networks. Phys Rev Lett 101:163901-1–163901-4
34. Ip E, Kahn JM (2008) Compensation of dispersion and nonlinear effects using digital backpropagation. J Lightwave Technol 26(20):3416–3425, October 15, 2008
35. Xiao-Yu H, Eleftheriou E, Arnold D-M, Dholakia A (2001) Efficient implementations of the sum-product algorithm for decoding of LDPC codes. In: Proceedings of the IEEE Globecom, vol 2, November 2001, pp 1036–1036E
36. Djordjevic IB, Arabaci M, Minkov L (2009) Next generation FEC for high-capacity communication in optical transport networks. IEEE/OSA J Lightwave Technol 27(16):3518–3530

37. Minkov LL, Djordjevic IB, Xu L, Wang T (2009) PMD compensation in polarization multiplexed multilevel modulations by turbo equalization. IEEE Photon Technol Lett 21(23): 1773–1775
38. Djordjevic IB, Minkov LL, Xu L, Wang T (2009) Suppression of fiber nonlinearities and PMD in coded-modulation schemes with coherent detection by using turbo equalization IEEE/OSA J Opt Commun Netw 1(6):555–564
39. Tanner RM (1981) A recursive approach to low complexity codes. IEEE Trans Inf Theory IT-27(5):533–547, September 1981
40. Milenkovic O, Djordjevic IB, Vasic B (2004) Block-circulant low-density parity-check codes for optical communication systems. IEEE J Sel Top Quantum Electron 10(2):294–299, March/April 2004
41. Djordjevic IB, Xu L, Wang T, Cvijetic M (2008) Large girth low-density parity-check codes for long-haul high-speed optical communications, in Proc OFC/NFOEC, (IEEE/OSA, San Diego, CA, 2008), Paper no. JWA53
42. Colavolpe G, Ferrari G, Raheli R (2001) Reduced-sate BCJR-type algorithms. IEEE Sel Areas Comm 19:848–858, May 2001

Chapter 8
Constrained Coding for Optical Communication

Anantha Raman Krishnan and Shiva K. Planjery
Department of Electrical & Computer Engineering, University of Arizona,
Tucson, AZ 85721, USA

8.1 Introduction

The increase in speed of optical communication systems has brought new technical challenges to the fore. For example, combatting *intrachannel cross-phase modulation* (IXPM) and *intrachannel four-wave mixing* (IFWM) in high-speed time division multiplexing (TDM) systems has been a focus of considerable research [1–3]. IXPM and IFWM are interactions among neighboring bits (pulses) that are a result of fiber nonlinearities. They can limit system performance by causing energy transfer between interacting pulses, thereby resulting in the formation of *ghost pulses* or shadow pulses. Moreover, these interactions can also lead to timing and amplitude jitters in the system. Though ghost pulses occur due to interactions between pulses in varied positions, it has been observed that certain "resonance" positions lead to more energy transfers than others [2]. In particular, pulses at positions k, l, and m lead to ghost pulse at position $(k + l - m)$. Also, it has been observed that this interaction is the most for pulses that are close to each other. Figure 8.1 illustrates this phenomenon. The figure depicts the creation of a ghost pulse at position 0 as a result of pulses at positions -1, 2, and 3.

In the past, novel modulation formats have been relied upon to suppress ghost pulses [2,3]. An alternative scheme for reduction of ghost pulses relies on the use of constrained coding [4,5]. Constrained codes can be used to map user bit-streams to channel bit-stream with reduced IFWM, hence reducing the extent of ghost pulses. To illustrate this, consider the channel bit-stream obtained by randomly traversing the states of the finite state transition diagram (FSTD) shown in Fig. 8.2 and reading the bits off the edges traversed. It is easily seen that the resultant bit-stream contains no more than two consecutive 1s. This bit-stream avoids the most harmful patterns, namely, a sequence of three or more 1s. Suppose we could design a code based on the FSTD shown here, then the resulting channel bits would also have no more than two consecutive 1s (note that this is merely a toy example and that with more ingenuity, FSTDs that better suppress ghost pulses can be devised. This will be discussed in Sect. 8.5). A typical constrained coding problem can be formulated as follows: Given a FSTD, can a code be constructed that maps blocks of user-bits to blocks of channel-bits? If yes, then what is the maximum achievable rate of such

I. Djordjevic et al., *Coding for Optical Channels*,
DOI 10.1007/978-1-4419-5569-2_8,

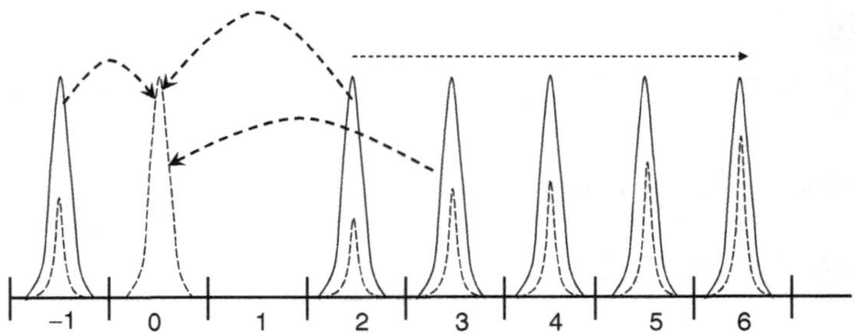

Fig. 8.1 Formation of ghost pulses

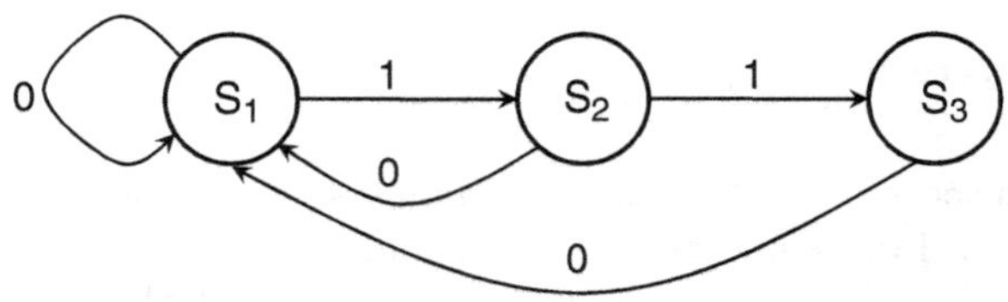

Fig. 8.2 FSTD that avoids three consecutive 1s

a code? Given the code, can a "good" decoder be designed for this code? Some excellent references to learn more about constrained coding are [6] and [7].

This chapter is devoted to explaining the fundamentals of constrained coding and describing how it can be applied to optical communications. We shall begin by providing a general exposition on the concepts of constrained coding. Then, we will focus on constrained coding from the perspective of optical communication. The rest of the chapter is organized as follows. Section 8.2 provides a brief introduction to the fundamentals of constrained systems and its related definitions and terminologies. In Sect. 8.3, we discuss design of encoders from a given FSTD, and in Sect. 8.4, we discuss the decoder design. We shall then describe in detail how constrained coding can be used to suppress the nonlinear intrachannel effects present in high-speed optical transmission in Sect. 8.5. Finally, we provide some concluding remarks in Sect. 8.6.

8.2 Fundamentals of Constrained Systems

Consider a communication system with input alphabet $\mathcal{X}$ (throughout this chapter, we assume $\mathcal{X}$ to be $\{0, 1\}$ or $\{-1, +1\}$, depending on the context. However, the definitions that follow are easily extended for any input alphabet). The system is said to be unconstrained if there are no restrictions on what input sequences the channel allows. Next, suppose that there exists a set of constraints that sequences must satisfy in order to be valid. For example, consider the well-known family of constraints known as (d,k) or runlength-limited (RLL) constraints. The (d,k) constraint allows only sequences with runs of 0 having length at least d and no more than k (a run of 0 with length d has d consecutive 0s between two 1s). For instance, the following sequence satisfies the (2, 4) constraint.

$$\ldots, 1, 0, 0, 0, 1, 0, 0, 1, 0, 0, 0, 1, 0, 0, 0, 0, 1, \ldots$$

Another well-known constraint is the DC-free constraint which suppresses the spectral content at the zero frequency. The DC-free constraint allows all sequences $\mathbf{s} = \ldots, s_{-1}, s_0, s_1, \ldots, s_i \in \{-1, +1\}$ for all i such that for all k

$$\left| \sum_{-\infty}^{k} s_i \right| < D,$$

where D is appropriately chosen positive integer.

It is desirable to represent diagrammatically the process of generating all valid sequences. For example, Fig. 8.3a and b show the diagrammatic representation of processes that generate sequences satisfying the $(1, \infty)$ RLL constraint and the DC-free constraint, respectively. The figure represents an FSTD with directed edges. All the edges in Fig. 8.3a and b are labeled from $\{0, 1\}$ and $\{-1, +1\}$, respectively. Sequences are drawn from such an FSTD by traversing the states along the edges, reading the labels off the edges traversed. It can be easily seen that all valid sequences of length, say n (an n-tuple), can be represented as a path of length n in the FSTD. The set of all finite-length strings generated by an FSTD constitute the constrained system.

Such a diagrammatic representation paves way for a thorough theoretical analysis of constrained systems. We can now draw from varied fields of research like information theory, symbolic dynamics, and theory of nonnegative matrices to yield a near-complete characterization of these systems. But before we do this, we provide some definitions that we shall use throughout this chapter.

An FSTD is said to have a *local anticipation* of a if a is the smallest nonnegative integer such that for any state i, all paths of length $(a + 1)$ that generate identical sequences begin with the same edge. That is, given $(a + 1)$-tuple, the initial state is sufficient to determine the initial edge. Similarly, an FSTD is said to have a *local memory* of m, if m is the smallest nonnegative integer such that for any state i, all paths of length $(m + 1)$ that generate identical sequences end with the same edge. An FSTD is said to be *deterministic* if its local anticipation is 0. Equivalently, an

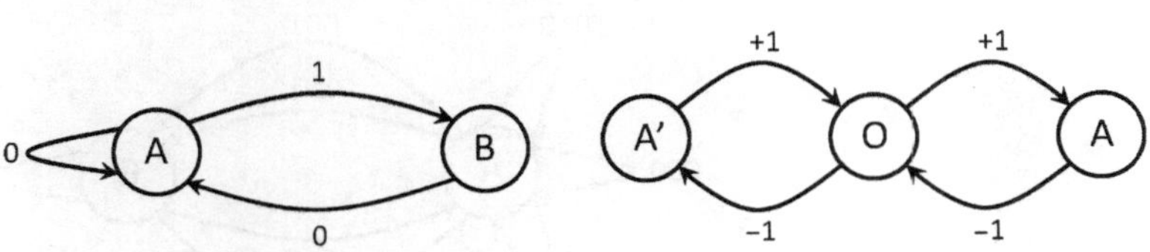

Fig. 8.3 FSTDs for various constraints: (**a**) $(1, \infty)$ RLL constraint and (**b**) DC-free constraint

FSTD is deterministic if for any state i, all its outgoing edges are labeled distinctly. For example, the FSTDs shown in Fig. 8.3 are deterministic. Figure 8.4 shows an FSTD with anticipation of 1.

An FSTD is said to be *irreducible* if for all states i and j, there exists a path from i to j. An FSTD is said to be reducible if it is not irreducible. For example, the FSTDs shown in Fig. 8.3 are irreducible. Figure 8.5 shows an example of a reducible FSTD. An FSTD is said to be *disconnected* if the set of states of the FSTD S can be divided into two subsets S_1 and S_2 such that there is no edge that begins in a state in S_1 and ends in a state in S_2.

We are interested in designing block codes that map blocks of user-symbols to channel-symbols. Hence, it is worthwhile to define the nth power of an FSTD. Let S be an FSTD. Then, the nth power of S (denoted as S^n) is constructed as follows: The states of S^n are the same as those of S. However, each directed edge from state i, to another, say j, represents a path of length n from i to j. The edge-label is the n-tuple obtained by reading off the edges traversed. The third power of the FSTD representing the $(1, \infty)$ constraint is shown in Fig. 8.6. It is interesting to note that if S is an irreducible FSTD, then for all n, S^n is either irreducible or consists of disjoint, irreducible components.

At this point, we observe that a simple block encoder can be readily designed using such an FSTD. As an example, consider the FSTD shown in Fig. 8.6. Suppose that this FSTD is used to design a rate $1/3$ code. This can be done by picking two edges for each state and pruning away the remaining edges. For instance, we choose

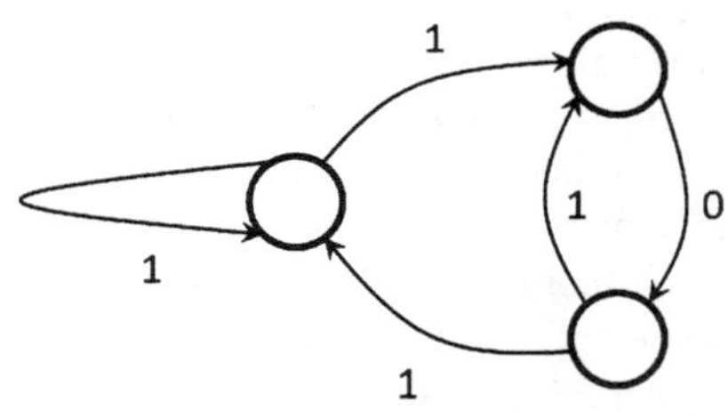

Fig. 8.4 FSTD with anticipation 1

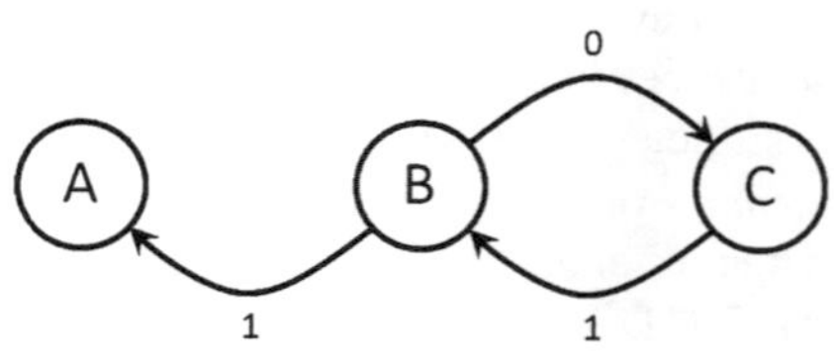

Fig. 8.5 A reducible FSTD

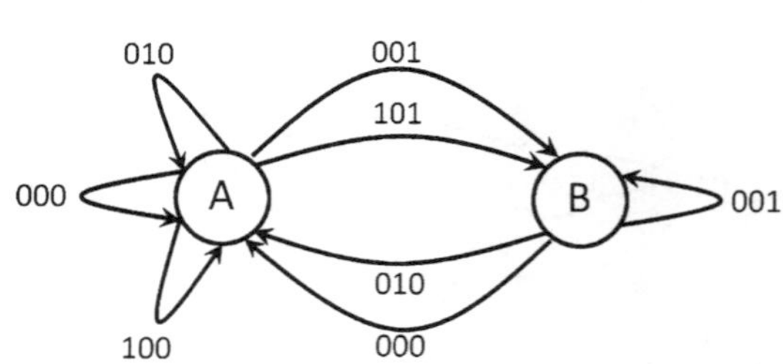

Fig. 8.6 Third power of FSTD of $(1, \infty)$ RLL constraint

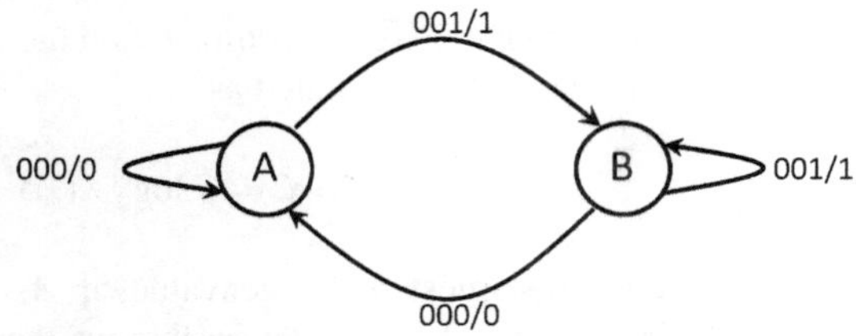

Fig. 8.7 FSTD of a rate 1/3 code corresponding to the $(1, \infty)$ RLL constraint

the edges 100 and 101 for state A and edges 001 and 000 for state B. Next, pick two outgoing edges and assign input-bits to these outgoing edges at each state. For example, we assign 1 to edge 001 and 0 to 000, and prune away the remaining edges. The resulting FSTD is a block encoder shown in Fig. 8.7 that uses the input-bits and the current state of the FSTD to generate channel bits. For example, the input $1, 0, 1, 1, \ldots$ is encoded as $101, 000, 101, 001, \ldots$ (assuming the starting state to be A).

The construction of the encoder raises the following questions. What is the highest achievable rate of a code designed thus? Can such an encoder be realized? In order to answer these questions, we need to form a mathematical framework for the analysis of constrained systems. The beginning of this framework is the definition of the *adjacency matrix* of an FSTD. Let S be an FSTD with s states (ordered from 1 to s). The adjacency matrix A of S is a $s \times s$ matrix. The (i, j)th element of A is the number of edges from i to j. For example, the FSTD shown in Fig. 8.3a has the adjacency matrix

$$A_{(1,\infty)} = \begin{bmatrix} 1 & 1 \\ 1 & 0 \end{bmatrix}.$$

The adjacency matrix of the nth power of an FSTD S with adjacency matrix A can be calculated by taking the nth power of A. It is easy to verify that $A^3_{(1,\infty)}$ is indeed the adjacency matrix of the FSTD given in Fig. 8.6.

We now proceed to find the upper bound on the highest achievable rate of codes based on constrained systems. This is obtained by calculating the *Shannon capacity* or *entropy rate*. The Shannon capacity (or simply "capacity") measures the growth rate of the number of valid n-tuples of a constrained system. The capacity of a constrained system T, C_T is defined as:

$$C_T = \lim_{n \to \infty} \frac{\log_2 N(n, T)}{n}, \tag{8.1}$$

where $N(n, T)$ is the number of valid n-tuples in T. In other words, the number of valid n-tuples of T is well approximated by 2^{nC_T}, for large n. Shannon [8] proved that capacity is the upper bound on the rate achievable by any code based on constrained systems. Moreover, given any integers k, n such that

$$\frac{k}{n} < C_T,$$

Shannon proved that there is an integer l such that there are 2^{kl} valid ln-tuples of T.

Suppose S is the FSTD representing T and has an adjacency matrix A. Then, the capacity can be easily be calculated as:

$$C_S = \log_2 \lambda(A),$$

where $\lambda(A)$ is the largest real eigenvalue of A. The existence of a positive real eigenvalue is guaranteed by Perron–Frobenius theory of nonnegative real matrices. We note that for any FSTD S with capacity C, the capacity of S^n is nC. Finally, it can be shown that if S is a reducible FSTD with capacity C, then S has an irreducible component whose capacity is equal to C. Hence, it is sufficient to consider only the irreducible components of S during code construction.

With the upper bounds on achievable rates established, we now proceed to design block codes with appropriately chosen rates. The methodology of code design is explained in Sect. 8.3.

8.3 Construction of Finite-State Encoders

In order to construct finite-state encoders based on FSTDs, we rely on the following result from [7]:

Theorem 8.1. *Let T be a constrained system with capacity C_T. Let k, n be positive integers satisfying the inequality*

$$k/n \leq C_T.$$

Then, there exists a finite-state encoder with a state-dependent decoder with rate k/n.

The proof of this theorem also provides a systematic way to design the encoder. We provide here a brief description of the proof, which will also serve as a description of the design methodology. For this purpose, we use the example from [7].

Let S be the FSTD representing the constrained system T. In order to design the encoder, we will use the S^n as the starting point. Suppose each of the states in S^n has at least 2^k outgoing edges, we choose exactly 2^k edges in each state and prune away the remaining edges. The resulting FSTD is the required finite-state encoder. In other words, it is equivalent to having the sum of each row of the adjacency matrix being greater than 2^k.

Indeed, it is sufficient to find a sub-FSTD of S^n, S', such that all its states have at least 2^k outgoing edges terminating in S'. In terms of adjacency matrix, it can be expressed as follows: Let A be the adjacency matrix of S^n. Suppose a 0–1 vector $\mathbf{v}$ (not all 0s) can be found such that

$$A^n \mathbf{v} \geq 2^k \mathbf{v}, \tag{8.2}$$

then the FSTD S' constituting the states indexed by $\mathbf{v}$ can be used as an finite-state encoder.

However, finding such a sub-FSTD is not always possible. For instance, consider the problem of designing a finite-state encoder for the $(1, \infty)$ constraint. The capacity of this constraint is 0.69. According to Theorem 8.1, design of a rate $2/3$ code should be possible. However, a cursory analysis of Fig. 8.6 is enough to see that this is not possible. In the FSTD, state B has only three outgoing edges. If state A is chosen as the sub-FSTD (that is, state B is removed), then state A has only three outgoing edges. Similarly, if state A is removed, then B has only one outgoing edge. Equivalently, none of the choices for $\mathbf{v}$, namely, $[1, 1]'$, $[1, 0]'$, and $[0, 1]'$ satisfy (8.2).

When we are unable to find a code by this fashion, one alternative is to look at higher powers of S. However, we can use the technique of "state splitting" [7] to generate a code *without* increasing the block length.

We shall now describe this technique. Let S be an FSTD. Let V be the set of all states in S. Let i belong to V. Let $E = E_1 \bigcup E_2$, E_1 and E_2 disjoint, be the set of outgoing edges from i. Then, splitting of state i results in a new FSTD S', with the set of states $V \backslash \{i\} \bigcup \{i_1, i_2\}$, where i_1, i_2 are referred to as the descendant states of i. The edges of S' are obtained from S as follows:

Case 1. If the edge in S neither begins or ends in i, then replicate it in S'.

Case 2. Suppose the edge begins in state $j \neq i$ and terminates in i. If the edge is in E_1 or E_2, then replicate the edge so that there is an edge from j to i_1, and from j to i_2.

Case 3. Suppose the edge begin in i and ends in $j \neq i$. If it belongs to E_l, then there is an edge in S' from i_l to j.

Case 4. Suppose the edge is a self-loop that begins and ends in i and belongs to E_1. Then there are two edges, beginning in E_1 and ending in i_1 and i_2, respectively. This rule is congruently applied to self-loops belonging to E_2.

Figure 8.8 shows the FSTD resulting from the splitting of state A in Fig. 8.6. In the figure, the outgoing edges are split as $E_1 = \{000, 100\}$ and

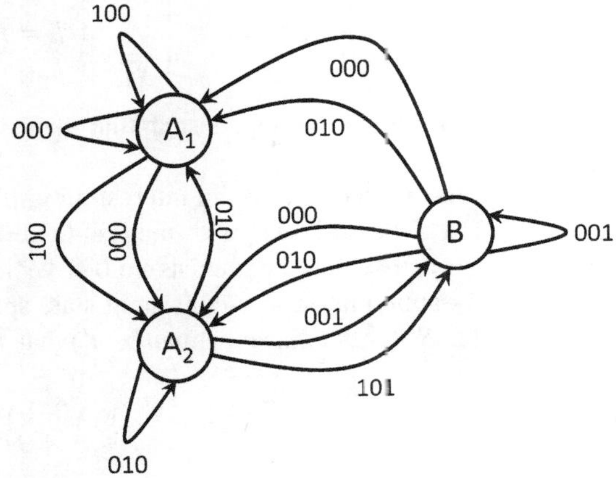

Fig. 8.8 State splitting of FSTD shown in Fig. 8.6

$E_2 = \{010, 001, 101\}$. Note that a generalized notion of state splitting that is the splitting of a state into multiple descendants can be defined as well. However, this is equivalent to a series of splittings into two descendant.

We observe that this transformation retains the properties of the original FSTD in that all the sequences generated by S are generated by S' as well. Also, it can be proved that if S has an anticipation of a, S' has an anticipation of at most $(a + 1)$.

For the purpose of code construction, a specialized state splitting called **v**-consistent state splitting is used. However, before this is described, we introduce *approximate eigenvectors*. An (A, l) approximate eigenvector of a matrix A is defined as a nonzero vector **v** with nonnegative components such that

$$A\mathbf{v} \geq l\mathbf{v}.$$

If $\lambda(A) \geq l$, Perron–Frobenius theory for nonnegative matrices guarantees the existence of an (A, l)-approximate eigenvector.

Recall that for the purpose of code construction, we have chosen the rate $k/n \leq C_T$. Then, the aforementioned result guarantees the existence of a vector **v** such that

$$A^n\mathbf{v} \geq 2^k\mathbf{v}. \tag{8.3}$$

For example, for the $(1, \infty)$ constraint, $v = [2, 1]'$ is an $(A^3, 4)$ approximate eigenvector.

An approximate eigenvector of matrix can be calculated by using Franaszek's algorithm (detailed in [7]). Franaszek's algorithm takes as its input a nonnegative integer matrix A, an integer l, and an integer L. If there is an (A, l)-approximate eigenvector with no element greater than L, it will find one.

A basic **v**-consistent state splitting is explained as follows: Let S be an FSTD with adjacency matrix A. Let V be the set of states of S and let **v** be an (A, l)-approximate eigenvector. Further, let the state $i \in V$ be split into i_1 and i_2, yielding a new FSTD S' with adjacency matrix A'. This split is said to be **v**-consistent if the vector v'

$$v'_j = \begin{cases} v_j & \text{if } j \neq i \\ y_1 & \text{if } j = i_1 \\ y_2 & \text{if } j = i_2 \end{cases},$$

where y_1, y_2 are positive integers such that $y_1 + y_2 = v_i$ is an (A', l)-approximate eigenvector.

The fact that $y_1, y_2 < n$ has an interesting implication. To see this, consider the following: Let S be an FSTD with matrix A. Let k/n be an appropriately chosen rate. Further suppose that there exists no 0–1 vector **v** such that $A^n\mathbf{v} \geq 2^k\mathbf{v}$. Then, by repeatedly applying basic **v**-consistent state splitting, it is possible to construct another FSTD S' (with adjacency matrix A') that satisfies the constraint

$$A'\mathbf{v}' \geq 2^k\mathbf{v}', \ \mathbf{v}' \text{ a 0–1 vector}.$$

In other words, S' (or one of its sub-FSTDs) has at least 2^k outgoing edges from each of its states (sufficient for an encoder). Indeed, it is trivial to show that the number of state-splitting iterations required is at most $\sum v_i - 1$, the sum of all the elements of $\mathbf{v}$.

Finally, to complete the proof of Theorem 8.1, we have the following proposition by [7] (stated without proof).

Proposition 8.2. *Let A be the adjacency matrix of an irreducible FSTD S, and assume that the all-1's vector is not an (A, n)-approximate eigenvector. Let $\mathbf{v}$ be a positive (A, n)-approximate vector. Then, there is a basic $\mathbf{v}$-consistent splitting of H.*

Hence, by using the state-splitting algorithm, a finite-state encoder is constructed for any rates $k/n \leq C_T$. For clarity, we summarize the encoder design procedure as follows:

1. Given the constraint T, generate an FSTD, say S, representing the constrained.
2. From the adjacency matrix of S, say A, calculate the capacity C_S.
3. Select rational rate k/n such that $k/n \leq C_S$.
4. Construct S^n.
5. Use Franaszek's algorithm there exists a $(A^n, 2^k)$-approximate eigenvector whose entries are 0 or 1. If it exists, use the FSTD (or the appropriate sub-FSTD) as the encoder.
6. If no such vectors exist, use Franaszek's algorithm to calculate an $(A^n, 2^k)$-approximate eigenvector. Eliminate all states for which $v_i = 0$.
7. Find a $\mathbf{v}$-consistent partition for some state in S. Find a basic $\mathbf{v}$-consistent state splitting for this state to create a new FSTD S'. The existence of such a split is guaranteed.
8. Repeat step (7) till all states have at least 2^k outgoing edges.

For practical implementations, it is desirable to construct an FSTD of the encoder that has the smallest possible number of states. The number of states can be significantly reduced by using the technique of *state merging*. The technique can be described as follows. Let i and j be states in an encoder FSTD with outer degree n (any extra edges have been deleted from the FSTD). Let $E_i = \{e_1, \ldots, e_n\}$ and $E_j = \{f_1, \ldots, f_n\}$ denote the edge sets corresponding to states i and j, respectively. If the edge sets can be arranged such that for each $k = 1, 2, \ldots, n$, the edges e_k and f_k share the same label and terminating state, then states i and j can be merged to one state, thus reducing the overall number of states by one. Figure 8.9 shows an example of an RLL (0, 1) code, and Fig. 8.10 shows the resulting FSTD of the encoder after merging the states S_1 and S_2.

Since the sequence of state splittings affects the possible state mergings, it would be more efficient if the construction algorithm could identify state mergings on the fly during the state-splitting process rather than performing state merging at the end of the process, in order to construct the FSTD of the encoders with minimal number of states required for representation. Although there is no absolute solution to this problem, certain techniques and heuristics have been proposed in [9] and [10]

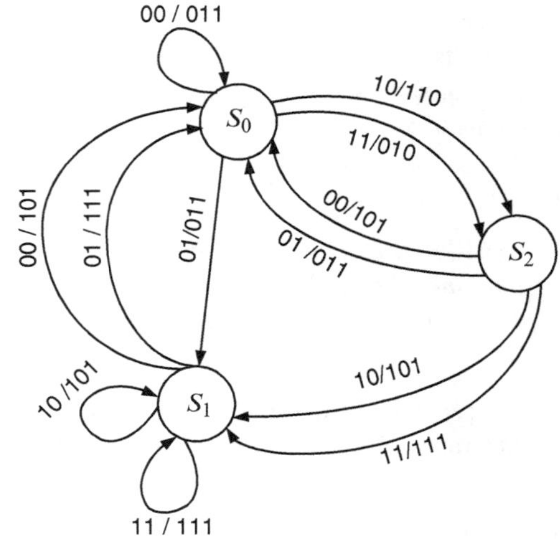

Fig. 8.9 FSTD of the RLL (0, 1) encoder with input tags

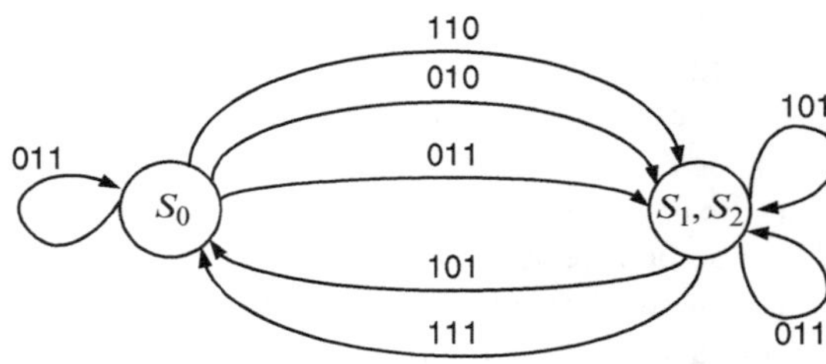

Fig. 8.10 Encoder obtained after stage merging

and have proven to be effective in the construction of codes for recording channels that use peak detection as well for channels using partial response signaling with maximum likelihood detection.

8.4 Decoders for Constrained Codes

The decoders for constrained codes are designed based on the constructed rate k/n finite-state encoders described in Sect. 8.3. We remind that the FSTDs of the finite-state encoders previously described have a finite local anticipation a which is an important property to facilitate decoding. There are mainly two types of decoders: state-dependent decoders and sliding-block decoders. State-dependent decoders are naturally represented by the FSTD of the encoder itself. For these decoders, we can think of the decoding process as exactly the reverse of the encoding process, i.e., the decoder utilizes the current decoder state and received codeword in order to determine the original k-block message and the next decoder state. The initial state of the decoder is matched with the initial state of the encoder at the beginning of the decoding process. The decoding algorithm for state-dependent decoders can be summarized as follows:

1. Set the initial state of the decoder to be the initial state of the encoder.
2. For the current state i, concatenate the current n-block codeword with the next a upcoming codewords to obtain a sequence l of length $a + 1$ (measured in n-blocks). Observe the path corresponding to the sequence l on the FSTD of the encoder starting from the current decoder state i. By definition of the local anticipation, the initial edge e of the path can uniquely be determined. The decoded output will be the input tag assigned to edge e on the FSTD. The next decoder state is the terminal state of edge e.
3. Repeat step 2 until all codewords have been decoded.

Hence as stated in Theorem 8.1, a state-dependent decoder exists for every finite-state k/n encoder, and these decoders are relatively simple and straightforward to implement.

However, state-dependent decoders can lead to the possibility of catastrophic error propagation when the decoders are used in a noisy environment. Catastrophic error propagation means obtaining an unbounded number of errors at the output of the decoder for a finite number of errors at the input of the decoder. Due to the presence of noise, the decoder may have to deal with erroneously detected sequences that may not even be valid sequences. Figure 8.11 shows an example where a single error in the decoder input can cause an unbounded number of errors at the decoder output. If the initial state is S_0 and we encode the input sequence 0000... using the above FSTD, we will obtain the constrained sequence *aaaaa* .... Now assuming there is noise present, if the noise corrupts the first symbol to b, the decoder will receive *baaaa* ..., which is then decoded by the state-dependent decoder as 111111... leading to an unbounded number of errors. Hence, it would be desirable to have a decoder that can bound the number of errors occurring at the output for any given decoder input, and this is possible only for a decoder that is state independent. The sliding-block decoder, which we shall now discuss in detail, is a state-independent decoder than can bound the number of errors occurring at the output.

A sliding-block decoder uses a finite memory ("look-back") m and finite anticipation ("look-ahead") a in order to decode the current codeword. In other words, the decoder uses m previous codewords and a upcoming codewords to decode the

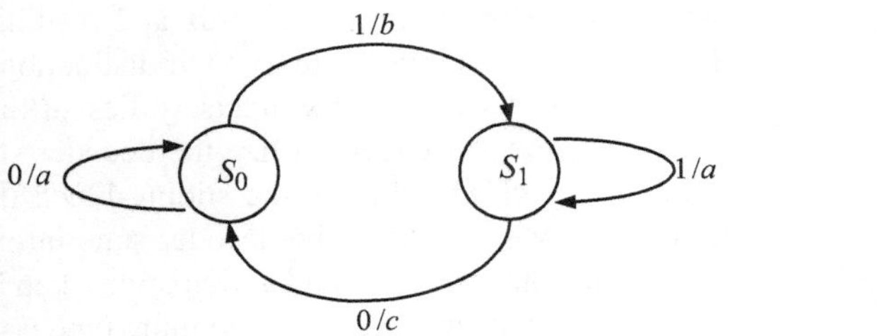

Fig. 8.11 Encoder that can cause catastrophic error propagation

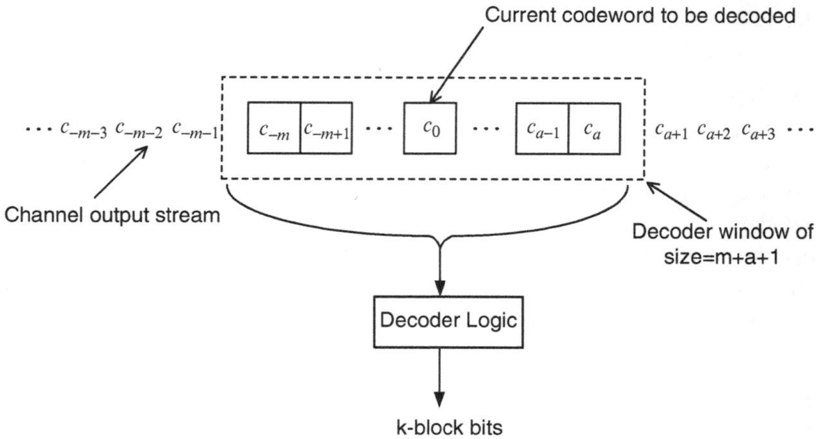

Fig. 8.12 Illustration of sliding-block decoder

current codeword. Figure 8.12 provides a schematic of the sliding-block decoder. Since the decoder window size is $(m + a + 1)$, an error in the detection of the current codeword will only affect at most $(m + a + 1)$ codewords, thus bounding the number of errors occurring at the output of the decoder. In precise terms, a sliding-block decoder for a finite-state rate k/n encoder is a mapping f from $(m + a + 1)$ codewords of length n (n-tuples) to a message block of length k, for some nonnegative integers m and a. Before we delve into the details of these decoders, we must define a new class of constrained systems for which there exist finite-state encoders with sliding-block decoders that achieve rates upto capacity.

An FSTD representing a constrained system T is said to have finite memory m and anticipation a, if for any given sequence of codewords $\mathbf{c} = c_{-m}, \ldots, c_0, \ldots, c_a$ in T, the set of all possible paths $\mathbf{e} = e_{-m}, \ldots, e_0, \ldots, e_a$ that generate $\mathbf{c}$ all contain the same edge e_0. This means that given a codeword c_0, the decoder can uniquely determine the edge e_0 corresponding to the codeword c_0 in the FSTD of the encoder by observing the m previous codewords and the a upcoming codewords. Such an FSTD is said to have finite memory and finite anticipation. Based on the definition of the FSTD, we can see that the sliding-block decoder for an encoder with such an FSTD can simply be considered as a look-up table that determines the edge that generated the current codeword, by looking at the m previous codewords and a upcoming codewords and reading off the input tag on this edge to be the decoder output. Note that the distinction between state-dependent decoders and sliding-block decoders precisely lies in the concept of using finite memory. State-dependent decoders utilize the decoder state information and local anticipation of the FSTD, whereas the sliding-block decoders use finite anticipation and finite memory instead of the decoder state information. It can be proven that if an FSTD has finite memory and anticipation, then it has finite local anticipation [7]. A constrained system T is said to be finite type if there exists an FSTD representation that has finite memory and anticipation. For example, the RLL (d, k) constraint is a finite-type constrained system. It must be noted here that there do exist bad representations of a finite-type system that do not have finite memory and anticipation and we would like to determine a good FSTD representation for a given finite-type constrained system T.

Before one attempts to search for an FSTD representation with finite memory and anticipation for a given constrained system T, one would like to know whether the constrained system T is finite type or not. There are several intrinsic conditions of finite-type constraints that can be useful in making this determination [7, 11]. They are the following:

1. There exists an integer L, and a list $\mathcal{L}$ that contains strings of length L, such that a sequence $\mathbf{c}$ satisfies the specified constraint if and only if each substring of length L in the sequence $\mathbf{c}$ belongs to the list $\mathcal{L}$. We may consider $\mathcal{L}$ to be a valid list. As an example, for the RLL (0, 1) constraint, we can choose $L = 2$ and $\mathcal{L} = \{11, 01, 10\}$.
2. There exists a finite list $\mathcal{F}$ that contains forbidden strings of length L. In other words, a sequence $\mathbf{c}$ satisfies the specified constraint if and only if it contains no substrings of length L that belong to the forbidden list $\mathcal{F}$. Again for the RLl (0, 1) constraint, we may consider $\mathcal{F} = \{00\}$.
3. Let $\mathcal{A}$ be the symbol alphabet. There exists an integer N such that, for a given symbol $a \in \mathcal{A}$, there is a list $T(a)$ that contains strings of length no more than N, with the property that if $\mathbf{w}$ is any constrained sequence, then the concatenation $\mathbf{wa}$ is a sequence that satisfies the constraint if and only if $\mathbf{w}$ ends in a string $\mathbf{u} \in T(a)$. For the RLL (0, 1) constraint, we may choose $N = 1$, $T(0) = \{1\}$, and $T(1) = \{0, 1\}$.

For a given FSTD S that represents a finite-type constrained system T with capacity C_S, it has been proven that there always exists a rate k/n finite-state encoder with a corresponding sliding-block decoder that can achieve rates upto the capacity C_S (see Theorem 2 in [7]). The FSTD of the encoder can be constructed in a manner described in Sect. 8.3, by considering higher powers of any FSTD with finite memory and anticipation and applying the state-splitting algorithm on the resulting FSTD. Just as the state-splitting algorithm preserves local anticipation for the FSTDs considered in Sect. 8.3, the state-splitting algorithm also preserves the finite memory and anticipation when applied to an FSTD that represents the finite-type constrained system T, and only the anticipation may increase by at most one for each state splitting.

From the discussion, it is evident that for the sliding-block decoder of a constraint code under a noisy channel, the propagation of errors in the decoder is controlled by the window size $(m + a + 1)$ of the sliding-block decoder. Therefore, it would be desirable to come up with sliding-block decoders that require the smallest possible window size in order to keep the error propagation to a minimum. Consider the construction of a finite-state encoder represented by a given FSTD S and its nth power S^n, where S^n has memory m and anticipation a. If p is the number of splittings used, then the FSTD of the encoder will have the same memory m but

Table 8.1 Decoding process for RLL (0, 1) encoder

Current codeword	Upcoming codeword	Decoded data
011	110	00
110	111	10
111	101	11
101	011	00
011	010	11
010	–	11

the anticipation cannot be more than $(a + p)$. It follows that an upper bound on the window size can be determined as

$$W \leq m + a + p + 1.$$

In practice, the window size that is realizable can be significantly smaller than the given upper bound. Figure 8.9 shows the example of an FSTD of the RLL (0,1) encoder that has a sliding-block decoder with window size $W = 2$. Table 8.1 illustrates the decoding process for the sequence $\mathbf{c}$ = 011 110 111 101 011 010. From the table, the decoded sequence is 00 10 11 00 11 11. Tighter bounds for the window size of a sliding-block decoder can be obtained by considering the use of a generalized **v**-consistent state-splitting algorithm during construction of the finite-state encoders, which is a slightly modified version of the basic **v**-consistent state-splitting algorithm described in Sect. 8.3. Interested readers can refer to [7] for further details on this algorithm.

In general, the decoder mapping for a given finite-state encoder is fixed by the message(k-block)-to-codeword(n-block) assignment on the FSTD of the encoder and one can find the minimum window size required to correctly decode the valid codeword sequence. The size actually depends on the message-to-codeword assignment on the FSTD and the size can actually increase or decrease for the same FSTD based on this assignment. For example, Fig. 8.13 illustrates two different assignments of the input tags for the same FSTD of a rate 1/2 RLL $(1, 3)$ code. Figure 8.13a has decoder window size of 1, whereas Fig. 8.13b has a window size of 2. Hence, clearly the choice of the assignment plays a role in the required window size of the decoder.

Given a finite-state encoder that has a sliding-block decoder, we would like to address the following question: what choice(s) of the message-to-codeword assignment will minimize the decoder window size W? The window size can be reduced by making consistent assignments of the input tags to edges that have same labels but that are not completely distinguished by the possible contents of the reduced decoder window. Based on the idea of consistent assignment, a general iterative method can be used to determine the assignment that allows the minimum possible decoder window size to be achieved. The method is described as follows. Select targets for the look-back m' and look-ahead a' for the decoder. For each edge e in the FSTD of the encoder, make a list of all the codewords generated by paths $e_{-m'}, \ldots, e_{-1}, e, e_1, \ldots, e_{a'}$ and denote this list as $L(e; m', a')$. In order for

Fig. 8.13 Two different assignments of the input tags

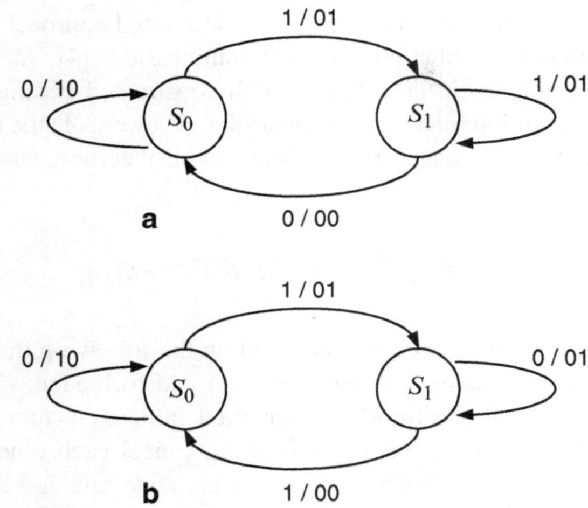

the sliding-block decoder with targets m' and a' to exist, any two edges e and f must be assigned the same input tag whenever the corresponding lists $L(e; m', a')$ and $L(f; m', a')$ are not disjoint. In addition, the assignment must satisfy the criterion that the set of message words assigned to the 2^k edges at each state must encompass the entire list of binary k-tuples. If any condition is not satisfied, then a sliding-block decoder with parameters m' and a' is not realizable and the whole process is repeated for another set of values for m' and a'. Although the approach is straightforward, the problem of obtaining a realizable decoder using this algorithm has been shown to be an NP-hard problem.

So far, we have discussed constrained coding techniques from a general setting where an efficient constrained code is required such that the output sequences that are being transmitted on a channel satisfy a given constrained system which is typically defined based on the application. We shall now discuss constrained coding from the context of optical communications in much greater detail and explain how it can be used to combat the intrachannel nonlinear effects of IXPM and IFWM in high-speed optical transmission.

8.5 Applications to Optical Communications

It was briefly mentioned in Sect. 8.1 as to how high-speed optical transmission can be affected by intrachannel interactions, such as IFWM and IXPM [1]. Approaches for dealing with intrachannel nonlinearities may be classified into two broad categories (1) modulation formats [2, 3] and (2) constrained codes [5, 12]. The first approach deals with removing the short-term phase coherence of the pulses emitted by the optical transmitter since the IFWM interaction is a phase sensitive effect. The second approach involves the use of a constrained code to avoid waveforms in the transmitted signal that will most likely be detected incorrectly. In this section, we shall look more closely at the second approach of using constrained coding in order to combat intrachannel nonlinearities [4]. We shall specifically describe three techniques in conjunction with constrained coding that can be used for suppression of intrachannel nonlinearities. They are (1) the use of constrained encoding, (2) combined constrained and error control coding, and (3) deliberate error insertion.

8.5.1 Use of Constrained Encoding

Constrained encoding is based on the following three principles (1) the most troublesome sequences are identified and forbidden, (2) the zero-symbol in so-called "resonant positions" is converted into one-symbol, and (3) the different contributions to a ghost-pulse creation cancel each other in resonant positions. It also provides flexibility in selecting the code rate and the nonlinear effect suppression factor. We illustrate using three different constraints how constrained encoding can be successful in suppressing the intrachannel nonlinearities.

Constraint A

In this constraint, the resonant sequences such as "1101," "1011," and "11011" are considered as the most troublesome ones that lead to an error-prone channel [5]. The FSTD of the constrained system avoiding those sequences is shown in Fig. 8.14. Valid sequences can be obtained by reading off the edge labels while making transitions from one state to another according to orientation of the edges. The largest code rate permitted by this constraint (which is the capacity) is 0.6942. Since the most troublesome sequences are identified and forbidden, this constraint will result in excellent Q-factor improvement; however, the code rate is unacceptably low for high-speed transmission.

Constraint B

Consider a sequence of length L, $c_i c_{i+1} \cdots c_{i+L-1}$, satisfying the following constraint: if for $k, l, m \in [i, i+1)$ and $k+l-m \in [i, i+L)$ (k and l not necessarily

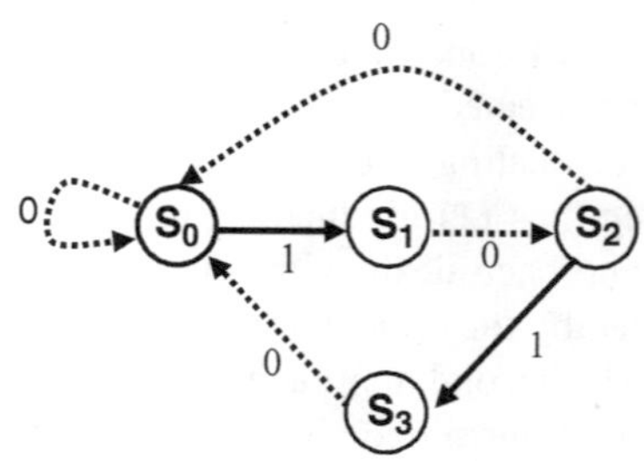

Fig. 8.14 FSTD of a constraint avoiding the resonance patterns 1101, 1011, and 11011

distinct), $c_k = c_l = c_m = 1$, then $c_{k+l-m} \neq 0$. If the above sequence is transmitted, then no ghost pulse will be created at any position (positions $\{0, 1, \ldots, L-1\}$). However, the amplitude jitter may be caused through the IFWM process. For $L = 4$, the constraint has the capacity of 0.8791 and may be described by an FSTD shown in Fig. 8.15. The high-speed applications require the simplest possible realization, and a simple block code of rate 4/5 is given in Table 8.2. The code is designed based on the approach described in Sect. 8.3 and has the efficiency (code rate) of 91%.

Constraint C

This constraint is based on the fact that the proper choice of initial pulse phases may result in complete cancelation of different ghost-pulse contributors at a zero-bit position in "resonance," which is illustrated in Fig. 8.16. A pseudoternary constrained code satisfying such property is shown in Fig. 8.17. The symmetry is imposed by a deliberate addition of a single pulse on the left or right side of the "asymmetric" sequence (e.g., the sequence 1, −1, 0, 1 is translated into 1, −1, 0, 1, −1). (Since 1 and −1 symbols in direct detection are considered as the same symbol the corresponding constraint is called here pseudoternary.)

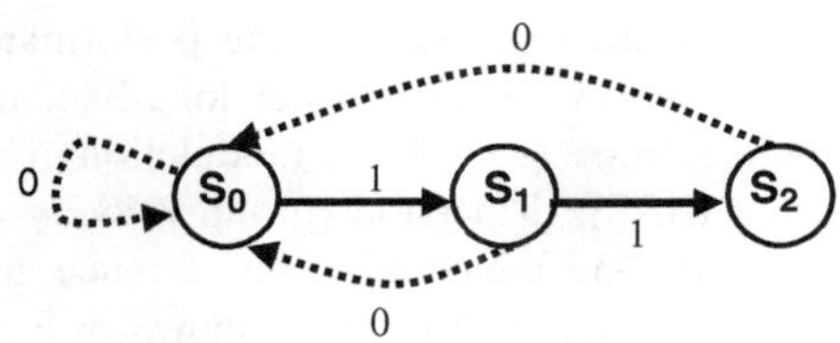

Fig. 8.15 FSTD of (0,2) RLL constraint

Table 8.2 Mapping between input and codewords for the rate 4/5 code

Input	Codeword	Input	Codeword
0000	00000	1000	00110
0001	00001	1001	01000
0010	00101	1010	01010
0011	01001	1011	01100
0100	10001	1100	10000
0101	10101	1101	10010
0110	00010	1110	10100
0111	00100	1111	10110

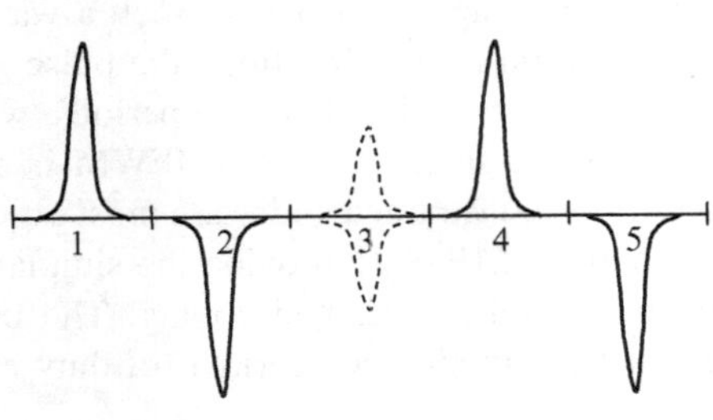

Fig. 8.16 Pseudoternary cancelation principle. Pulses at positions 1–4–2 and pulses at positions 2–5–4 creates two ghost pulses of the same amplitude but opposite signs at position 3

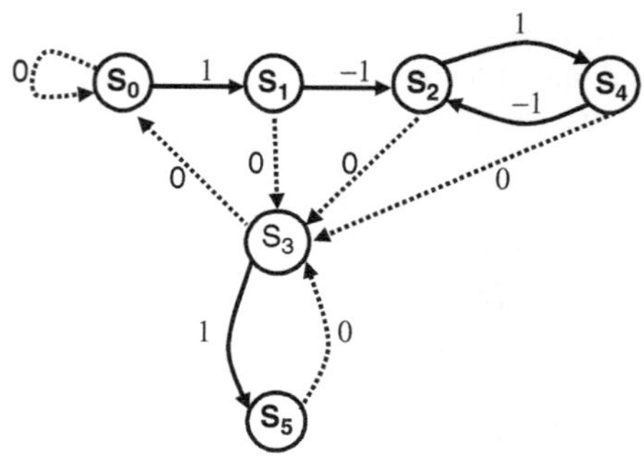

Fig. 8.17 Pseudoternary code of rate 0.9

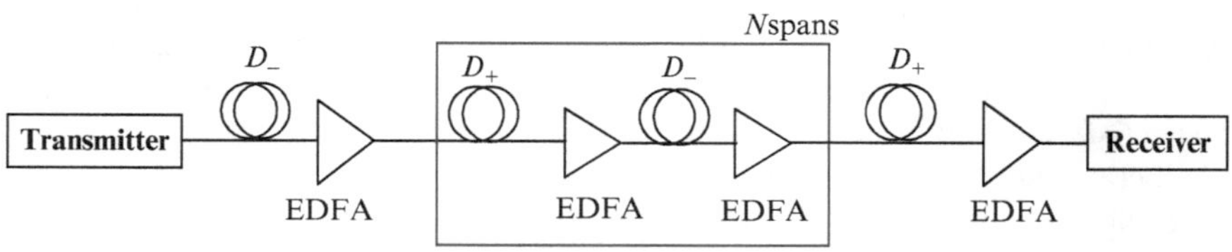

Fig. 8.18 Dispersion map used for simulation

We shall now illustrate the performance improvement of constrained coding on a 40-Gb/s single channel long-haul transmission system.The effects of Kerr nonlinearities (self-phase modulation, IXPM, IFWM), stimulated Raman scattering (SRS), dispersion [group-velocity dispersion (GVD), second-order GVD], crosstalk, and intersymbol interference are taken into account during simulation. The dispersion map used, as shown in Fig. 8.18, is composed of N (20–80) spans of length $L = 48$ km, each span consisting of $2L/3$ km of D_+ fiber followed by $L/3$ km of D_- fiber. D_+ fiber has a dispersion of 20 ps/(nm km), a dispersion slope of 0.06 ps/(nm^2 km), an effective area equal to 110 μm^2, and loss equal to 0.19 dB/km. D_- fiber has a dispersion of −40 ps/(nm km), a dispersion slope of −0.12 ps/(nm^2 km), an effective area equal to 30 μm^2, and loss equal to 0.25 dB/km. The nonlinear Kerr coefficient is set to 2.6×10^{-20} m^2/W in both types of fibers. The precompensation of −320 ps/nm and corresponding postcompensation are also employed. The erbium-doped fiber amplifiers (EDFAs) with a noise figure of 6 dB are deployed after every fiber section. The simulations are carried out with an average launched power of 0 dBm and a central wavelength of 1,552.524 nm. The dispersion map is selected in such a way that the IXPM is low. However, during transmission over a D_+ fiber, the pulse is spread over up to ten-bit periods rather than over tens or hundreds bit periods, which is typical for pseudolinear transmission. Under such conditions, IFWM is a predominant effect, and the previously described constrained codes are most efficient.

Figures 8.19–8.21 provide the simulation results of the specified optical transmission with constrained coding. The use of constrained encoding is compared against RZ modulation format (of duty cycle 33%). The pattern sequence used in

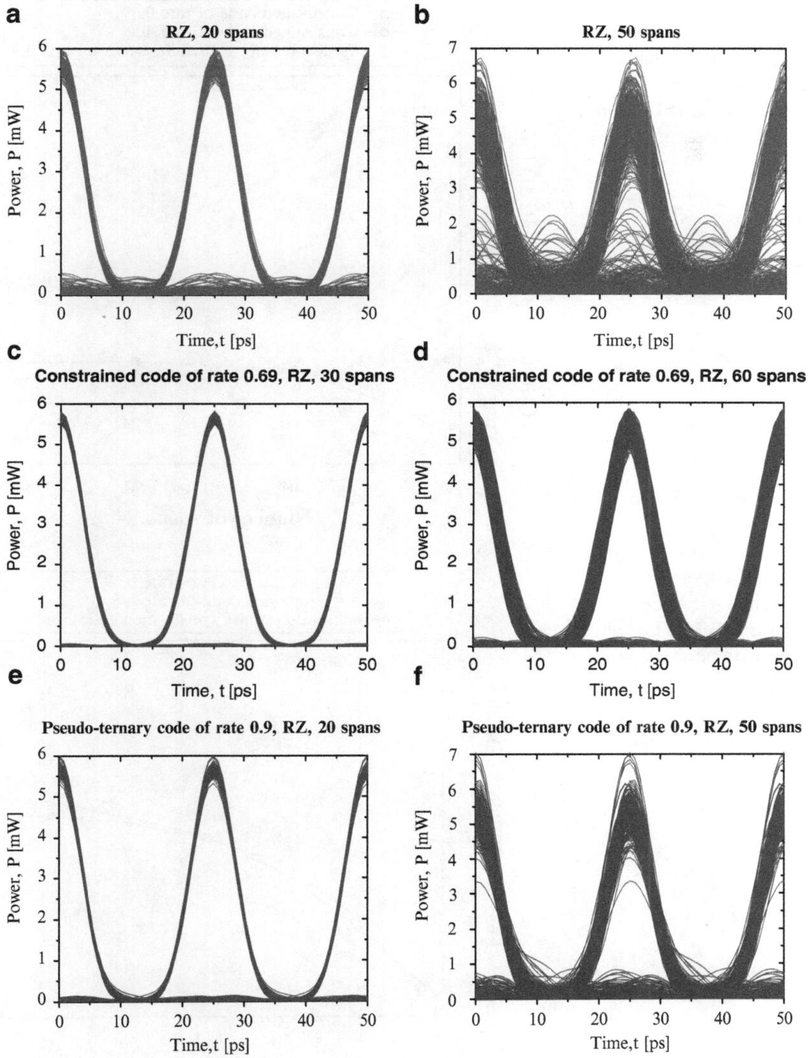

Fig. 8.19 Eye diagrams under constrained coding. (After ref. [6]; © IEEE 2006; reprinted with permission.)

simulations is of length $2^{15} - 1$. From Fig. 8.19, it is evident that the constrained codes are successful in suppressing both IFWM (ghost pulse and amplitude jitter) and IXPM (timing jitter). For the RZ modulation format, the eye diagram after 50 spans is closed, while the eye diagrams for constrained codes are widely open. The ability of constrained codes to substantially reduce the timing jitter due to IXPM is also important in enabling the clock recovery, which would be quite

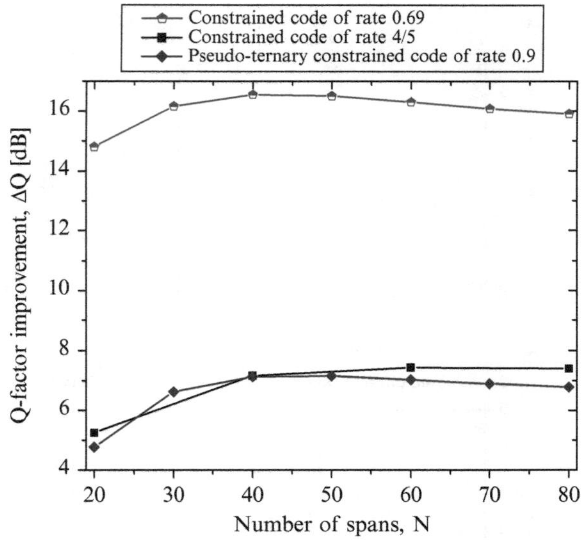

Fig. 8.20 Q-factor improvement over RZ for different number of spans in the absence of ASE noise. (After ref. [6]; © IEEE 2006; reprinted with permission.)

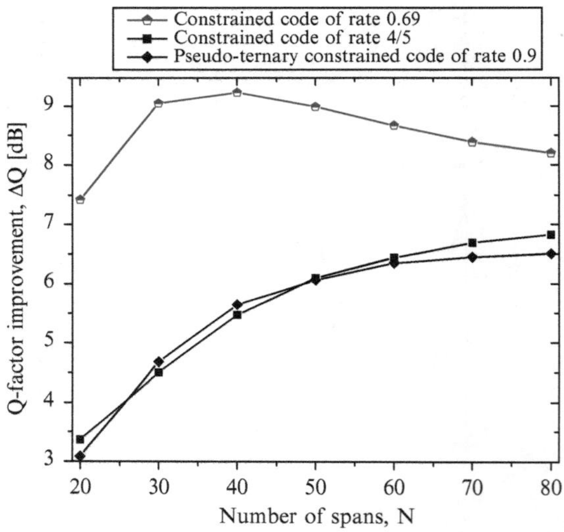

Fig. 8.21 Pseudoternary code of rate 0.9. (After ref. [6]; © IEEE 2006; reprinted with permission.)

challenging when the eye diagram is completely closed (see Fig. 8.19b). The Q-factor improvement against the number of spans is shown in Figs. 8.20 and 8.21. The Q-factor improvement is defined as

$$\triangle Q = 20 \log \left[Q_{\text{encoded}} / Q_{\text{uncoded}} \right] \text{ [dB]}.$$

Significant Q-factor improvement up to 16 dB is demonstrated depending on code rate and number of spans. Notice that in the presence of ASE noise, the eye diagrams are more closed due to noise, resulting in reduced Q-factor improvement.

8.5.2 Combined Constrained and Error Control Coding

Traditionally in coded systems, the inner code is a constrained code or modulation code and the outer code is an error correction code (ECC). That is, the message bits are first encoded by using an ECC and then the resultant code is encoded using a modulation code. Thus, at the receiver, the constrained code is decoded first followed by the decoder for the ECC. Even though the constrained decoding described previously is conceptually simple (it is based on sliding-block decoder), the main disadvantage with this scheme is that it decouples the channel and the error correction decoder since the decoder for the constrained code operates only on hard bits and does not produce soft information, which is necessary for iterative (LDPC/turbo) decoding that is typically used for the ECC. An alternative scheme that circumvents the above problem is known as *reverse concatenation* [13, 14], and it has been recently adapted for soft detectors. This scheme can operate in the presence of strong intrachannel nonlinearities when even advanced forward error correction (FEC) schemes would be overwhelmed with errors.

In the reverse concatenation scheme, the order of FEC encoder and constrained encoder is inverted, i.e., the modulation encoder is followed by ECC, and as a result, such a scheme facilitates the use of soft decoding. In order to realize reverse concatenated systems, it is typically required that the ECC be systematic. Figure 8.22a, b illustrates the standard concatenation scheme and the reverse concatenation scheme. In the scheme shown, an LDPC code of length n and dimension (number of message bits) k is used as the ECC. The message bits at the output of Constrained Encoder 1 pass through a systematic LDPC encoder unchanged. Parity check bits in general do not satisfy channel constraint and must be encoded using a separate constrained encoder (Constrained Encoder 2 in Fig. 8.22b). Encoder 2, of course, imposes the same constraint as Encoder 1. Soft constrained decoders 1 and 2 employ the Bahl,

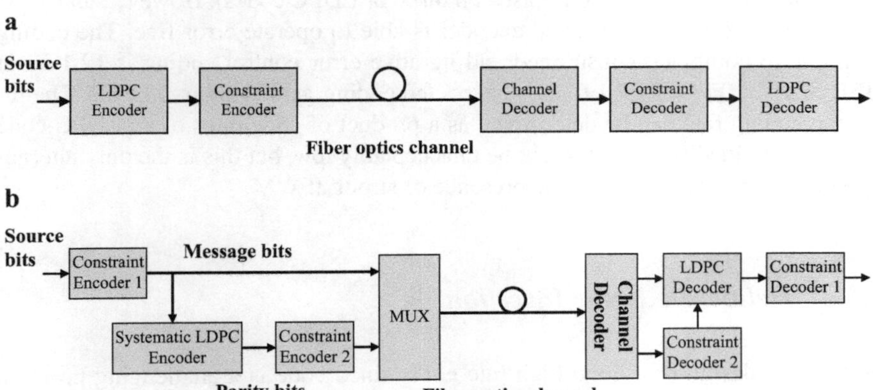

Fig. 8.22 (**a**) Standard concatenation scheme, (**b**) reverse concatenation scheme (optical transmitter and receiver parts not shown)

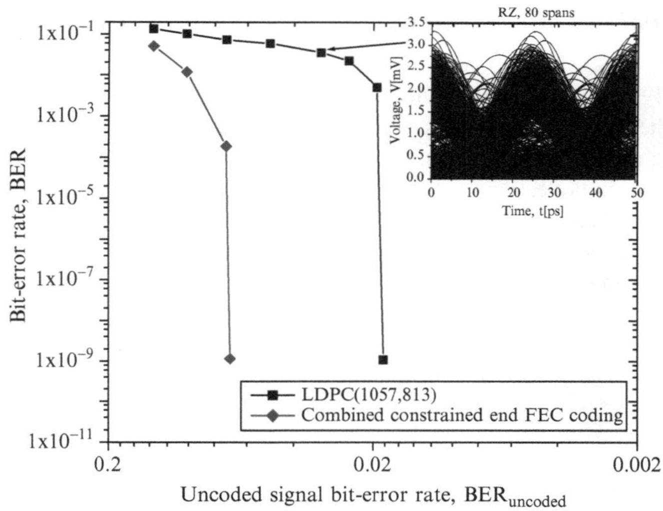

Fig. 8.23 BER performance of combined constrained and LDPC FEC scheme at 40 Gb/s. LDPC code belongs to the class of projective geometry codes. (After ref. [6]; © IEEE 2006; reprinted with permission.)

Cocke, Jelinek, and Raviv (BCJR) algorithm [15] performed on the trellis obtained by transforming the state-transition diagram in Fig. 8.15. Decoding for the LDPC code is carried out using the standard message-passing algorithm [16].

Figure 8.23 provides simulation results using this scheme for the same dispersion map from Fig. 8.18 wherein the EDFAs noise figure is set to 6 dB. After 80 spans, the electrical eye diagram of uncoded signal (shown in Fig. 8.23) is completely closed, and the input BER is too high for any FEC scheme to handle (even for the most advanced methods based on turbo or LDPC codes). However, an LDPC code combined with constrained encoder is able to operate error free. The coding gain due to combined constrained and iterative error control coding is 12.1 dB at BER of 10^{-9}. For a BER of 10^{-12}, expected coding gain is about 13.6 dB. The resulting coding rate can be determined as a product of code rates of an LDPC code and a constrained code, and might be unacceptably low, but this is the only alternative to enable transmission in the presence of strong IFWM.

8.5.3 Deliberate Error Insertion

Sometimes design of a good high-rate constrained code is a challenging problem when the desired constraint is complicated because the encoder and decoder might be unacceptably complex for high-speed applications, and large decoding window size may result into unacceptable error propagation [17]. The number of errors caused by error propagation due to increased decoding window size may result in exceeding error correction capability of the FEC code employed in the standard concatenation scheme (shown in Fig. 8.22a). Even though combined constrained and error correction scheme described previously provides an excellent coding gain, it requires using encoders and decoders of relatively high complexity.

The key idea behind deliberate error insertion [17], as illustrated in Fig. 8.24, is to completely remove the constrained encoder and instead impose the constraint by deliberate insertion of errors before the transmission. The redundancy to be used in imposing a given constraint might be used to employ a stronger FEC scheme capable of correcting both channel errors and deliberate errors. Constraint monitoring circuit, which replaces the constrained encoder, operates on the LDPC-encoded sequence. It monitors if the channel constraint is satisfied; if not, it introduces the smallest possible number of errors so that constraint of the channel is satisfied. The errors introduced by the constraint monitoring circuit are independent of channel conditions and may cause the *error floor* if the constraint is so strong that the number of introduced errors exceeds the error correction capability of LDPC code. In order to avoid the error floor, a weaker constraint such as a (0,3) constraint can be used as shown in Fig. 8.25. Moreover, instead of operating on a bit-by-bit basis it is suggested to operate on a block-by-block basis. The monitor circuit accepts K bits at a time from an LDPC encoder and checks if the (0,3) constraint is satisfied within the block of length K; if not, it flips the minimum number of bits such that the constraint is satisfied. By doing so, we allow for the violation of the constraint between two neighboring blocks, so that the loose (0,3) constraint is obtained. However, LDPC codes of higher code rates may be employed and the error floor may be avoided as shown in Fig. 8.26. At a BER of 10^{-9}, the deliberate error insertion technique provides a coding gain 10.17 dB, while the expected coding gain at BER of 10^{-12} is 11.56 dB.

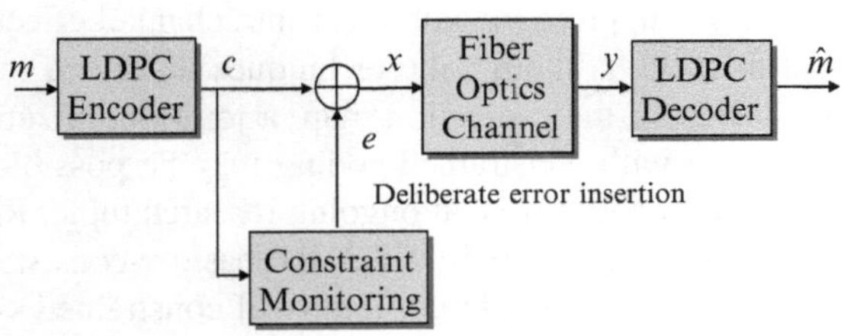

Fig. 8.24 Deliberate error insertion principle

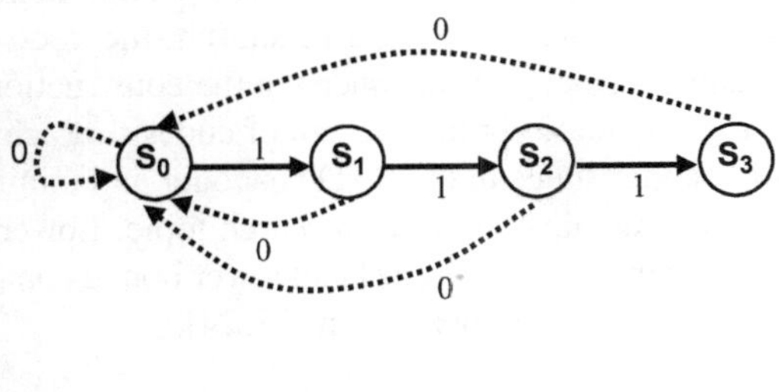

Fig. 8.25 FSTD of (0,3) RLL constraint

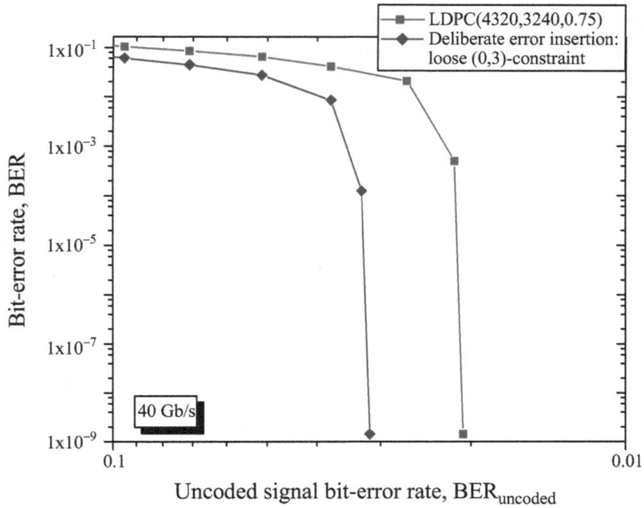

Fig. 8.26 BER performance of deliberate error insertion scheme for $K=4$. LDPC code that belongs to the class of block-circulant codes. (After ref. [6]; © IEEE 2006; reprinted with permission.)

8.6 Concluding Remarks

In this chapter, we covered the fundamental aspects of constrained coding and highlighted its usefulness in any communication system that requires the channel sequences to satisfy a given set of constraints in order to be valid. We then described in detail how constrained coding could be successfully employed in optical communications to suppress the nonlinear intrachannel effects present in high-speed optical transmission. Although the techniques described in Sect. 8.5 were independent of the design of the dispersion map, a joint optimization of both the dispersion map combined with constrained coding may be possible that could lead to further improvements and this is an ongoing research topic. Regarding construction of FSTD encoders described in Sect. 8.3, the basic **v**-consistent state-splitting algorithm can also be extended to a larger family of constrained systems called *almost finite-type* systems, such as the spectral null systems [18, 19]. Other types of state-splitting algorithms such as the generalized **v**-consistent state splitting and variable length state splitting can be used to shorten the code construction procedure [7, 20, 21]. Despite tremendous advances in the construction of codes for constrained systems, the problem of optimal design of codes that achieve the minimum possible number of encoder states in the FSTD encoder and minimum possible decoder window size W still remains an active research topic. Lower bounds on the number of encoder states were derived in [22] and lower bounds on the minimum sliding-block decoder window size were derived in [23, 24].

References

1. Essiambre R-J, Mikkelsen B, Raybon G (2004) Intra-channel cross-phase modulation and four-wave mixing in high-speed TDM systems. IEEE Electron Lett 35:1576–1578
2. Forzati M et al (2002) Reduction of intrachannel four-wave mixing using the alternate-phase RZ modulation format. IEEE Photon Technol Lett 14:1285–1287
3. Liu X et al (2002) Suppression of interchannel four-wave-mixing-induced ghost pulses in high-speed transmissions by phase inversion between adjacent marker blocks. Opt Lett 27: 1177–1179
4. Djordjevic IB, Vasic B (2006) Constrained coding techniques for the suppression of intra-channel nonlinear effects in high-speed optical transmission. IEEE/OSA J Lightwave Technol 24:411–419
5. Vasic B et al (2004) Ghost-pulse reduction in 40-Gb/s systems using line coding. IEEE Photon Lett 16:1784–1786
6. Immink KAS (1999) Codes for mass data storage systems. Shannon Foundation, Rotterdam, The Netherlands
7. Marcus BH, Siegel PH, Wolf JK (1992) Finite-state modulation codes for data storage. IEEE J Sel Areas Commun 10:5–37
8. Shannon C (1948) A mathematical theory of communication. Bell Syst Tech J 27:379–423, 623–656
9. Marcus B, Siegel P (1984) Constrained codes for PRML. IBM Res Rep RJ 4371
10. Marcus B, Siegel P (1988) Constrained codes for partial response channels. In: Proceedings of Beijing International Workshop on Information Theory, pp DI1.1–DI1.4
11. Khayrallah A, Neuhoff D (1989) Subshift models and finite-state codes for input-constrained noiseless channels: a tutorial. Based upon PhD dissertation by A Khayrallah, University of Mich
12. Djordjevic IB, Chilappagari SK, Vasic B (2006) Suppression of intrachannel nonlinear effects using pseudo-ternary constrained codes. IEEE/OSA J Lightwave Technol 24(2):769–774
13. Bliss WG (1981) Circuitry for performing error correction calculations on baseband encoded data to eliminate error propagation. IBM Tech Discl Bull 23:4633–4634
14. Mansuripur M (1991) Enumerative modulation coding with arbitrary constraints and post-modulation error correcting coding and data storage systems. Proc SPIE 1499:72–86
15. Bahl LR, Cocke J, Jelinek F, Raviv J (1991) Optimal decoding of linear codes for minimizing symbol error rate. IEEE Trans Inform Theory IT-20(2):284–287
16. Sankaranarayanan S, Vasic B (2005) Message-passing algorithm. In: Vasic B, Kurtas EM (eds.) Coding and signal processing for magnetic recording systems. CRC, Boca Raton, FL, pp 10–1–10–18
17. Vasic B, Pedagani K (2004) Run-length-limited low-density parity-check codes based on deliberate error insertion. IEEE Trans Magn 40:1738–1743
18. Karabed R, Marcus B (1988) Sliding-block coding for input-restricted channels. IEEE Trans Inform Theory 34(1):2–26
19. Marcus B (1985) Sofic systems and encoding data. IEEE Trans Inform Theory It-31(3): 366–377
20. Adler R, Friedman J, Kitchens B, Marcus B (1986) State splitting for variable length graphs. IEEE Trans Inform Theory IT-32(1):108–113
21. Heegard C, Marcus B, Siegel P (1991) Sliding-block coding for input-restricted channels. IEEE Trans Inform Theory 37(3):759–777
22. Marcus B, Roth R (1991) Bounds on number of states in ecnoder graphs for input-constrained channels. IEEE Trans Inform Theory 37(3):742–758
23. Ashley J (1988) A linear bound for sliding-block decoder with window size. IEEE Trans Inform Theory 34(3):389–399
24. Kamabe H (1989) Minimum scope for sliding-block decoder mappings. IEEE Trans Inform Theory 35(6):389–399

Chapter 9
Coding for Free-space Optical Channels

High bandwidth demands in metropolitan area networks (MANs) and requirements for flexible and cost-effective service cause the imbalance known as the "last mile bottleneck." Fiber-optics, RF, and copper/coaxial lines are the main state-of-the-art technologies to address the high bandwidth requirements. The incompatibility of RF/microwave and optical communication technologies due to large bandwidth mismatch between RF and optical channels is now widely believed to be the limiting factor in efforts to further increase the transport capabilities in the future. For this reason, RF/microwave–optical interface solutions that will enable aggregating multiple RF/microwave channels into an optical channel are becoming increasingly important. Free-space optical (FSO) communication is the technology that can address any connectivity needed in optical networks, be in a core, edge, or access [1,2]. In MANs, the FSO can be used to extend the existing MAN rings; in enterprise, the FSO can be used to enable local area network (LAN)-to-LAN connectivity and intercampus connectivity; and the FSO is an excellent candidate for the last-mile connectivity. However, an optical wave propagating through the air experiences fluctuations in amplitude and phase due to atmospheric turbulence. The intensity fluctuation, also known as the scintillation, is one of the most important factors that degrade the performance of an FSO communication link, even under the clear sky condition.

FSO links are considered as a viable solution for the last mile bottleneck problem because of the following properties: (1) high directivity of the optical beam provides high power efficiency and spatial isolation from other potential interferers, a property not inherent in RF/microwave communications, (2) the FSO transmission is unlicensed, (3) large fractional-bandwidth coupled with high optical gain using moderate powers permits very high-data rate transmission, (4) the state-of-the art fiber-optics-communications employ intensity modulation with direct detection (IM/DD), and the components for IM/DD are widely available, and (5) the FSO links are relatively easy to install and easily accessible for repositioning when necessary. The FSO communication may also be considered as an enabling technology to bring different technologies together and easy to integrate with variety of interfaces and network elements.

In this chapter, we describe several coding concepts enabling the communication over atmospheric turbulence channels: (1) coded MIMO, (2) raptor coding,

I. Djordjevic et al., *Coding for Optical Channels*,
DOI 10.1007/978-1-4419-5569-2_9,

and (3) coded OFDM. The chapter is organized as follows: the FSO channel model is introduced in Sect. 9.1; the concept of coded MIMO is introduced in Sect. 9.2; the concept of FSO-OFDM transmission is introduced in Sect. 9.3. In Sect. 9.4, we describe the use of coded OFDM in hybrid optical networks. The raptor coding concept for temporally FSO channels is described in Sect. 9.5. Finally, Sect. 9.6 summarizes this chapter.

9.1 Atmospheric Turbulence Channel Modeling

A commonly used turbulence model assumes that the variations of the medium can be understood as individual cells of air or eddies of different diameters and refractive indices. In the context of geometrical optics, these eddies may be observed as lenses that randomly refract the optical wave-front, generating a distorted intensity profile at the receiver of a communication system. The intensity fluctuation is known as scintillation and represents one of the most important factors that limit the performance of an atmospheric FSO communication link. The most widely accepted theory of turbulence is due to Kolmogorov [1–5]. This theory assumes that kinetic energy from large turbulent eddies, characterized by the parameter known as outer scale L_0, is transferred without loss to the eddies of decreasing size down to sizes of a few millimeters characterized by the inner scale parameter l_0. The inner scale represents the cell size at which energy is dissipated by viscosity. The refractive index varies randomly across the different turbulent eddies and causes phase and amplitude variations to the wave front. Turbulence can also cause the random drifts of optical beams – a phenomenon usually referred to as wandering – and can induce beam focusing.

Outer scale is assumed to be infinite in this chapter. We consider zero and nonzero inner scale conditions. Understanding the turbulence effects under zero inner scale is important as it represents a physical bound for the optical atmospheric channel and as such it has been of interest to researchers [1].

To account for the strength of the turbulence we use the unitless Rytov variance, given by [1–4]

$$\sigma_{\mathrm{R}}^2 = 1.23\, C_n^2\, k^{7/6} L^{11/6}, \tag{9.1}$$

where $k = 2\pi/\lambda$ is the wave number, λ is the wavelength, L is propagation distance, and C_n^2 denotes the refractive index structure parameter, which is constant for horizontal paths. Weak fluctuations are associated with $\sigma_{\mathrm{R}}^2 < 1$, the strong with $\sigma_{\mathrm{R}}^2 > 1$, and the saturation regime is defined by $\sigma_{\mathrm{R}}^2 \to \infty$ [1].

To characterize the FSO channel from a communication theory point of view, it is useful to give a statistical representation of the scintillation. The reliability of the communication link can be determined if we use a good probabilistic model for the turbulence. Several probability density functions (PDFs) have been proposed for the intensity variations at the receiver of an optical link [6–11]. Al-Habash et al.

[12] proposed a statistical model that factorizes the irradiance as the product of two independent random processes each with a Gamma PDF. The PDF of the intensity fluctuation is therefore

$$f(I) = \frac{2(\alpha\beta)^{(\alpha+\beta)/2}}{\Gamma(\alpha)\Gamma(\beta)} I^{(\alpha+\beta)/2-1} K_{\alpha-\beta}\left(2\sqrt{\alpha\beta I}\right), \quad I > 0, \tag{9.2}$$

where I is the signal intensity, α and β are parameters of the PDF, Γ is the gamma function, and $K_{\alpha-\beta}$ is the modified Bessel function of the second kind of order $\alpha - \beta$.

9.1.1 Zero Inner Scale

The parameters α and β of the PDF that predicts the scintillation experienced by plane waves in the case of $l_0 = 0$ are given by the expressions [4, 5]

$$\alpha = \left(\exp\left[\frac{0.49\sigma_R^2}{(1 + 1.11\sigma_R^{12/5})^{7/6}}\right] - 1\right)^{-1}, \quad \beta = \left(\exp\left[\frac{0.51\sigma_R^2}{(1 + 0.69\sigma_R^{12/5})^{5/6}}\right] - 1\right)^{-1}, \tag{9.3}$$

where σ_R^2 is the Rytov variance as given in (9.1). This is a very interesting expression, because the PDF of the intensity fluctuations at the receiver can be predicted from the physical turbulence conditions. The predicted distribution matches very well with the distributions obtained from numerical propagation simulations [1, 6].

9.1.2 Nonzero Inner Scale

In the presence of nonzero inner scale, the model must be modified to account for the change in the power spectrum of the refractive index variations. The PDF model is again a gamma–gamma distribution, but its parameters are now given by [1, 6]

$$\alpha = \left\{\exp\left[\sigma_{\ln X}^2\right] - 1\right\}^{-1}, \quad \beta = \left\{\exp\left[\frac{0.51\sigma_P^2}{(1 + 0.69\sigma_P^{12/5})^{5/6}}\right] - 1\right\}^{-1}, \tag{9.4}$$

where $\sigma_{\ln X}^2(l_0)$ is given by

$$\sigma_{\ln X}^2 = 0.16\sigma_1^2 \left(\frac{\eta_x Q}{\eta_x + Q}\right)^{7/6} \left[1 + 1.75\left(\frac{\eta_x}{\eta_x + Q}\right)^{1/2} - 0.25\left(\frac{\eta_x}{\eta_x + Q}\right)^{7/12}\right] \tag{9.5}$$

and

$$\eta_x = \frac{2.61}{1 + 0.45\sigma_1^2 Q^{1/6}}, \qquad Q = \frac{10.89L}{kl_0^2}. \tag{9.6}$$

The variance $\boldsymbol{\sigma}_{\mathrm{P}}^2$ parameter, from the second term in (9.4), is given by

$$\sigma_{\mathrm{P}}^2 = 3.86\sigma_1^2 \left\{ (1 + 1/Q^2)^{\frac{11}{12}} \left[\sin\left(\frac{11}{6}\tan^{-1} Q\right) + \frac{1.51}{(1+Q^2)^{\frac{1}{4}}} \sin\left(\frac{4}{3}\tan^{-1} Q\right) \right.\right.$$
$$\left.\left. - \frac{0.27}{(1+Q^2)^{\frac{7}{24}}} \sin\left(\frac{5}{4}\tan^{-1} Q\right) \right] - 3.5Q^{-\frac{5}{6}} \right\}. \tag{9.7}$$

The influence of both the atmospheric turbulence and electronic noise on QPSK and 16-QAM single side band (SSB) FSO-OFDM systems (and zero inner scale) is illustrated in Fig. 9.1. Results for an SSB OFDM system with 64 subcarriers

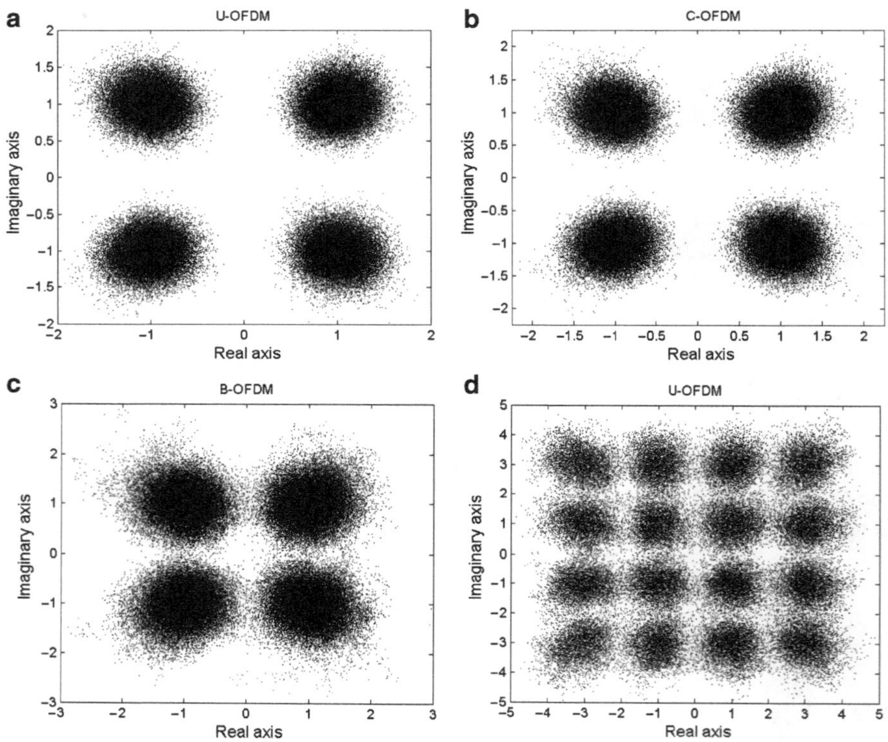

Fig. 9.1 Received constellation diagrams of QPSK (**a**)–(**c**) and 16-QAM (**d**) SSB FSO-OFDM systems with electrical SNR per bit of 18 dB under the weak turbulence ($\sigma_{\mathrm{R}} = 0.6$) for: (**a**), (**d**) U-OFDM scheme, (**b**) C-OFDM scheme, and (**c**) B-OFDM scheme. (After ref. [30]; © IEEE 2007; reprinted with permission.)

are shown. The average launched power is set to 0 dBm, the electrical signal-to-noise ratio (SNR) is set to 18 dB, and the received signal constellation diagrams are obtained assuming weak atmospheric turbulence ($\sigma_{\mathrm{R}} = 0.6$) and zero inner scale. The atmospheric turbulence changes the symmetry of clusters from circular for AWGN channel to elliptic (see Fig. 9.1). Both clipped (C)-OFDM and unclipped (U)-OFDM schemes seem to be more immune to the atmospheric turbulence than the biased (B)-OFDM scheme (different OFDM schemes with direct detection are discussed in Sect. 9.3).

9.1.3 Temporal Correlation FSO Channel Model

The channel model above ignores the temporal correlation. In the presence of temporal correlation the consecutive bits experience similar channel conditions. Here we describe the joint temporal correlative distribution model from [13], which describes the fading in an FSO channel at a single point of space at multiple instances of time. This method is based on the Rytov method to derive the normalized log-amplitude covariance function for two positions in a receiving plane perpendicular to the direction of propagation [13]:

$$b_X(d_{ij}) = \frac{B_X(P_i, P_j)}{B_X(P_i, P_i)}, \tag{9.8}$$

where d_{ij} is the distance between points P_i and P_j. B_X denotes the log-amplitude covariance function:

$$B_X(P_i, P_j) = E\left[X(P_i)X(P_j)\right] - E\left[X(P_i)\right] E\left[X(P_j)\right], \tag{9.9}$$

and X is the log-amplitude fluctuation.

In the weak turbulence regime, the Rytov method is commonly used to represent the field of electromagnetic wave as follows [1] (using cylindrical coordinates, $R = (r, L)$, L-the transmission distance)

$$U(\boldsymbol{R}) \equiv U(\boldsymbol{r}, L) = U_0(\boldsymbol{r}, L) \exp\left[\psi(\boldsymbol{r}, L)\right], \tag{9.10}$$

where $U_0(\boldsymbol{r}, L)$ is the electromagnetic field in the absence of turbulence, and ψ is the *complex phase perturbation* due to turbulence. The complex phase perturbation can be expressed as follows:

$$\psi(\boldsymbol{r}, L) = \log\left[\frac{U(\boldsymbol{r}, L)}{U_0(\boldsymbol{r}, L)}\right] = X + jY, \tag{9.11}$$

where X is the log-amplitude fluctuation and Y is the corresponding phase fluctuation. In the weak turbulence regime it is reasonable to assume X and

Y to be Gaussian random processes. To deal with phase fluctuations, which is important in coherent FSO communication systems, someone may use active modal-compensation of wave-front phase distortion [14]. The residual phase variance after modal-compensation can be described in terms of Zernike terms by [14]

$$\sigma_Y^2 = Z_J \left(\frac{D}{d_0} \right)^{5/3}, \tag{9.12}$$

where D is the aperture diameter, d_0 is the correlation length, and Z_J denotes the Jth Zenrike term not being compensated (commonly $J = 3, 6, 10, 20$).

The joint temporal distribution of n intensity samples $(I_1, I_2, \ldots, I_n)$ is given by [13]:

$$f_I(I_1, I_2, \ldots, I_n) = \frac{1}{2\pi \prod_{i=1}^{n} I_i} \frac{1}{(2\pi)^{n/2} |C_X|^{1/2}} \exp\left[-\frac{1}{8} \left(\ln \frac{I_1}{I_0} \ldots \ln \frac{I_n}{I_0} \right) \right], \tag{9.13}$$

where C_X is the covariance matrix of intensity samples:

$$C_X = \begin{bmatrix} \sigma_X^2 & \sigma_X^2 b_X\left(\frac{T}{\tau_0} d_0\right) & \cdots & \sigma_X^2 b_X\left(\frac{(n-1)T}{\tau_0} d_0\right) \\ \sigma_X^2 b_X\left(\frac{T}{\tau_0} d_0\right) & \sigma_X^2 & \cdots & \sigma_X^2 b_X\left(\frac{(n-2)T}{\tau_0} d_0\right) \\ \cdots & \cdots & \cdots & \cdots \\ \sigma_X^2 b_X\left(\frac{(n-1)T}{\tau_0} d_0\right) & \sigma_X^2 b_X\left(\frac{(n-2)T}{\tau_0} d_0\right) & \cdots & \sigma_X^2 \end{bmatrix}, \tag{9.14}$$

σ_X^2 denotes the variance of the log-normally distributed amplitude, which for plane wave can be approximated as [1]

$$\sigma_X^2 \cong 0.56 k^{7/6} \int_0^L C_n^2(x)(L-x)^{5/6} \mathrm{d}x, \tag{9.15}$$

where the wave number k, propagation length L, and the refractive index structure parameter C_n were introduced earlier. T is the time interval between observations, which corresponds to the symbol period; while τ_0 is the *coherence time*. Notice that expressions (9.14)–(9.15) are valid in the *weak* turbulence regime. In the same regime, the covariance function (9.14) is found to be exponential for both plane and spherical waves [1]:

$$b_X(\tau) = \exp\left(-\left(\frac{|\tau|}{\tau_0} \right)^{5/3} \right). \tag{9.16}$$

The typical values of coherence time τ_0 are in the range from 10 μs to 10 ms.

9.2 Coded MIMO FSO Communication

The performance of FSO communication systems can be improved by using MIMO communication techniques [15–21]. In the case of FSO communications, the MIMO concept is realized by employing *multiple optical sources at the transmitter side and multiple detectors at the receiver side*. In this section, two scenarios, both employing low-density parity-check (LDPC) codes, are described [22, 23]: (1) repetition MIMO and (2) space-time coding [20]. Although this concept is analogous to wireless MIMO concept, the underlying physics is different, and optimal and suboptimal configurations for this channel are needed.

9.2.1 LDPC-Coded MIMO Concept and Space-Time Coding

A block diagram of the coded MIMO scheme is shown in Fig. 9.2. M optical sources are all pointed toward the distant array of N photodetectors using an expanding telescope. We assume that the beam spots on the receiver side are sufficiently wide to illuminate a whole photodetector array. This approach might help in simplifying

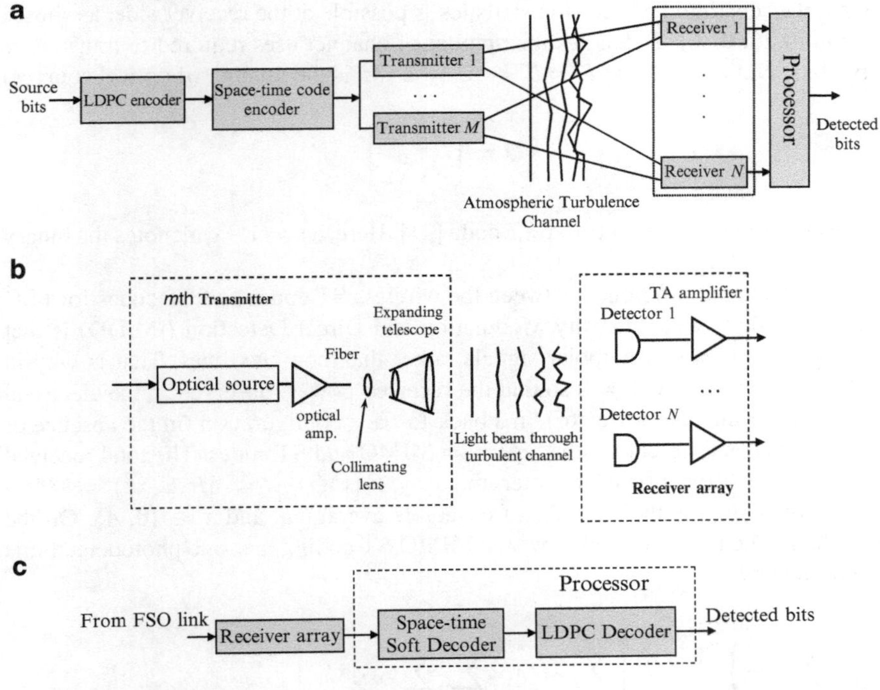

Fig. 9.2 (**a**) Atmospheric optical LDPC-coded MIMO system with space-time block codes, (**b**) mth transmitter and receiver array configurations, and (**c**) processor configuration

the transmitter–receiver pointing problem. We further assume that the receiver's implementation is based on a p.i.n. photodetector in a Trans-impedance Amplifier (TA) configuration. In the case of the *repetition* MIMO, the nth photodetector output scheme can be represented by

$$y_n(l) = x(l) \sum_{m=1}^{M} I'_{nm} + z_n(l), \quad n = 1, \ldots, N, \quad x(l) \in \{0, A\}, \tag{9.17}$$

where A denotes the intensity of the pulse in the absence of scintillation, $x(l)$ denotes data symbol at the lth time-slot. I'_{nm} represents the intensity channel coefficient between the nth photodetector ($n = 1, 2, \ldots, N$) and the mth ($m = 1, 2, \ldots, M$) optical source, which is described by the Gamma–Gamma PDF given by (9.2). The optical sources and photodetectors are positioned in such a way that different transmitted symbols experience different atmospheric turbulence conditions. z_n denotes the nth receiver TA thermal noise that is modeled as a zero-mean Gaussian process with double-side power spectral density $N_0/2$.

We assume an on–off keying (OOK) transmission over the atmospheric turbulence channel using incoherent light sources and direct detection. The information bearing signal is LDPC encoded. An ST encoder accepts K encoded bits x_k ($k = 1, 2, \ldots, K$) from an LDPC encoder. The ST encoder maps the input bits into the $T \times M$ matrix $\boldsymbol{O}$ whose entries are chosen from $\{x_1, x_2, \ldots, x_K, \overline{x}_1, \overline{x}_2, \ldots, \overline{x}_K\}$ so that the separation of decision statistics is possible at the receiver side, as shown later in (9.25, 9.26). T denotes the number of channel uses required to transmit K input bits. Notice that case $K = T = M = 2$ (M is the number of optical sources)

$$\boldsymbol{O} = \begin{bmatrix} x_1 & x_2 \\ x_2 & \overline{x}_1 \end{bmatrix}$$

corresponds to the Alamouti-like ST code [21]. Here, $\overline{x}_i = 1 - x_i$ denotes the binary complement of x_i.

A fundamental difference between the wireless ST codes and ST codes for FSO communications with Intensity Modulation and Direct Detection (IM/DD) is that the latter employs real unipolar signals rather than complex ones. That is why in IM/DD schemes, $E(x_k) \neq 0$ so that the received power (observed in the electrical domain, after the photodetector), in a back-to-back configuration (in the absence of scintillation), is different for the repetition MIMO and ST codes. The total received power in repetition MIMO is determined by $E\left[(Mx)^2\right] = M^2 E(x^2) = M^2 \cdot A^2/2$, where $E[\cdot]$ is the operator of ensemble averaging, and $x \in \{0, A\}$. On the other hand, the total received power in MIMO ST coding, per one photodetector, is determined by

$$\begin{aligned} E\left[\left(\sum_{m=1}^{M} x_m\right)^2\right] &= E\left[\sum_{m=1}^{M} x_i^2 + 2\sum_{m}\sum_{m'>m} x_m x_{m'}\right] \\ &= ME(x^2) + M(M-1)E^2(x) = \frac{1}{2}(M^2 + M) \cdot A^2/2. \end{aligned} \tag{9.18}$$

Therefore, for the same transmitted power, the received power per photododetector for repetition MIMO is $2M/(M+1)$ times larger. Therefore, we can impose two different comparison criteria: (1) to keep the received power upon photodetection, in the back-to-back configuration, constant, or (2) to keep transmitted power (observed in electrical domain) constant.

The received power (in electrical domain) per photodetector, for repetition MIMO transmission, is given by

$$E(y_n^2) = E\left[\left(x\sum_{m=1}^{M} I_{nm}\right)^2\right] = E\left[\left(\sum_{m=1}^{M} I_{nm}\right)^2 x^2\right]$$
$$= E\left(\sum_{m=1}^{M} I_{nm}^2 + 2\sum_{m}\sum_{m'>m} I_{nm}I_{nm'}\right)E(x^2) = \left[ME(I^2) + M(M-1)E[I]^2\right]A^2/2. \tag{9.19}$$

Assuming that the receiver TA thermal noise is white-Gaussian with a double-side-power spectral density $N_0/2$, the LLR of a symbol $x(l)$ (at the lth time-slot) for a binary repetition MIMO transmission is determined by

$$L(x(l)) = \log\left\{\prod_{n=1}^{N} \frac{\frac{1}{\sqrt{2\pi}\sqrt{N_0/2}}\exp\left[-\frac{y_n(l)^2}{2N_0/2}\right]}{\frac{1}{\sqrt{2\pi}\sqrt{N_0/2}}\exp\left[-\frac{\left(y_n(l)-\sum_{m=1}^{M} I_{nm}\right)^2}{N_0}\right]}\right\}$$
$$= \sum_{n=1}^{N}\left\{-\frac{y_n(l)^2}{N_0} + \frac{\left(y_n(l)-\sum_{m=1}^{M} I_{nm}\right)^2}{N_0}\right\}. \tag{9.20}$$

In the calculations of LLRs in (9.20), we assumed that the channel state information (CSI), denoted by I_{nm}, is known to the receiver, but not to the transmitter.

When the ST coding is used, it can easily be verified (by using $x_i^2 = x_i$ and $\overline{x}_i x_i = 0$) that for FSO systems, ST codes from orthogonal designs [20] do not satisfy the *orthogonality property*, $\det(\boldsymbol{O}\boldsymbol{O}^T) = \left[\sum_i x_i^2\right]^n$. Nevertheless the separation of decision statistics can be achieved, when the complements are properly chosen so that a simple a posteriori probability (APP) ST decoder still exists. Note that given the lack of orthogonality, $\boldsymbol{O}$ is not optimum. For the ST-coded MIMO scheme, the FSO channel can be described by

$$y = \begin{bmatrix} y_{11} & y_{21} & \cdots & y_{N1} \\ \cdots & \cdots & \cdots & \cdots \\ y_{1T} & y_{2T} & \cdots & y_{NT} \end{bmatrix} = \boldsymbol{O} \begin{bmatrix} I_{11} & I_{21} & \cdots & I_{N1} \\ \cdots & \cdots & \cdots & \cdots \\ I_{1M} & I_{2M} & \cdots & I_{NM} \end{bmatrix} + \begin{bmatrix} z_{11} & z_{21} & \cdots & z_{N1} \\ \cdots & \cdots & \cdots & \cdots \\ z_{1T} & z_{2T} & \cdots & z_{NT} \end{bmatrix}, \tag{9.21}$$

so that APP of x_k can be written as

$$\Pr(x_k|y) = Cf_k(x_k, y)\Pr(x_k), \tag{9.22}$$

where C is a constant and $f_k(x_k, \boldsymbol{y})$ is a function independent of other symbols (different from x_k). For example, for ST code $K = T = M = 4$ with an encoder mapper

$$\boldsymbol{O} = \begin{bmatrix} x_1 & x_2 & x_3 & x_4 \\ \overline{x}_2 & x_1 & \overline{x}_4 & x_3 \\ \overline{x}_3 & x_4 & x_1 & \overline{x}_2 \\ \overline{x}_4 & \overline{x}_3 & x_2 & x_1 \end{bmatrix} \tag{9.23}$$

and N photodetectors, the LLR of x_k, defined as

$$L(x_k) = \sum_{n=1}^{N} \log\left[\frac{p(x_k = 0|\boldsymbol{y}_n)}{p(x_k = 1|\boldsymbol{y}_n)}\right], \quad \boldsymbol{y}_n = [y_{n1}, \quad y_{n2}, \quad y_{n3}, \quad y_{n4}]^T \tag{9.24}$$

can be calculated by

$$L(x_k) = \sum_{n=1}^{N}\left[\frac{(\tilde{x}_k - 1)^2}{N_0} + \frac{I_{n1}^2 + I_{n2}^2 + I_{n3}^2 + I_{n4}^2 - 1}{N_0} - \frac{\tilde{x}_k^2}{N_0}\right], \quad k = 1, \ldots, 4 \tag{9.25}$$

where

$$\begin{aligned}
\tilde{x}_1 &= \sum_{n=1}^{N} (I_{n1}y_{n1} + I_{n2}y_{n2} + I_{n3}y_{n3} + I_{n4}y_{n4} - I_{n1}I_{n2} - I_{n2}I_{n3} - I_{n1}I_{n3} \\
&\quad - I_{n3}I_{n4} - I_{n1}I_{n4} - I_{n2}I_{n4}), \\
\tilde{x}_2 &= \sum_{n=1}^{N} \left(I_{n2}y_{n1} - I_{n1}y_{n2} - I_{n4}y_{n3} + I_{n3}y_4 + I_{n1}^2 + I_{n1}I_{n3} + I_{n1}I_{n4} + I_{n4}^2\right. \\
&\quad \left. - I_{n1}I_{n3} - I_{n2}I_{n3}\right), \\
\tilde{x}_3 &= \sum_{n=1}^{N} \left(I_{n3}y_{n1} + I_{n4}y_{n2} - I_{n1}y_{n3} - I_{n2}y_{n4} - I_{n1}I_{n4} - I_{n3}I_{n4} + I_{n1}^2\right. \\
&\quad \left. + I_{n1}I_{n4} + I_{n1}I_{n2} + I_{n2}^2\right), \\
\tilde{x}_4 &= \sum_{n=1}^{N} \left(I_{n4}y_{n1} - I_{n3}y_{n2} + I_{n2}y_{n3} - I_{n1}y_{n4} + I_{n1}I_{n3} + I_{n3}^2 - I_{n1}I_{n2}\right. \\
&\quad \left. - I_{n2}I_{n4} + I_{n1}^2 + I_{n1}I_{n2}\right).
\end{aligned} \tag{9.26}$$

The LLRs calculated in ST soft decoder (e.g., (9.25)) are forwarded to the LDPC decoder realized using an efficient implementation of sum-product algorithm. To derive (9.25), we used the Maximum-Likelihood (ML) detection rule as a starting point. Let $\boldsymbol{y}_n$ represent the nth column of $\boldsymbol{y}$ in (9.21), and $\boldsymbol{x}$ represent the sequence of transmitted bits $\boldsymbol{x} = [x_1 x_2 x_3 x_4]^{\mathrm{T}}$. The ML detection rule of the nth receiver can be formulated as follows: choose the estimate of $\boldsymbol{x}$ so that $\log[\mathrm{p}(\boldsymbol{y}_n|\boldsymbol{x})]$ is maximum. The $\log[\mathrm{p}(\boldsymbol{y}_n|\boldsymbol{x})]$ can be obtained by

$$\begin{aligned}\log\left[p(\boldsymbol{y}_n|\boldsymbol{x})\right] = &-\frac{[y_{n1} - (x_1 I_{n1} + x_2 I_{n2} + x_3 I_{n3} + x_4 I_{n4})]^2}{N_0}\\ &-\frac{[y_{n2} - (\overline{x}_2 I_{n1} + x_1 I_{n2} + \overline{x}_4 I_{n3} + x_3 I_{n4})]^2}{N_0}\\ &-\frac{[y_{n3} - (\overline{x}_3 I_{n1} + x_4 I_{n2} + x_1 I_{n3} + \overline{x}_2 I_{n4})]^2}{N_0}\\ &-\frac{[y_{n4} - (\overline{x}_4 I_{n1} + \overline{x}_3 I_{n2} + x_2 I_{n3} + x_1 I_{n4})]^2}{N_0}.\end{aligned}$$

By using $x_k^2 = x_k$ and $\overline{x}_k = 1 - x_k$ we were able to separate decisions for $x_k (k = 1, \ldots, 4)$, and derive (9.25).

The received power per photodetector (assuming $K = T = M$ and ST coding) is given by

$$\begin{aligned}E(y_n^2) &= E\left[\left(\sum_{m=1}^{M} I_{nm} x_m\right)^2\right] = E\left(\sum_{m=1}^{M} I_{nm}^2 x_m^2 + 2\sum_{m}\sum_{m'>m} I_{nm} I_{nm'} x_m x_{m'}\right)\\ &= \sum_{m=1}^{M} E(I_{nm}^2) E\{x_m^2\} + 2\sum_{m}\sum_{m'>m} E(I_{nm}) E(I_{nm'}) E\{x_m\} E\{x_{m'}\}\\ &= \left\{ME(I^2) + \frac{1}{2}M(M-1)E[I]^2\right\} E(x^2).\end{aligned} \tag{9.27}$$

Notice that the power efficiency of OOK-based MIMO is different from that of wireless communication PSK MIMO because $E[x] \neq 0$, and a part of the energy is used for transmission of the second term in (9.19) and (9.27). Notice that the second term in (9.19) is useful term because repletion MIMO is used, while the second term in (9.27), in the case of ST-coded MIMO, is in fact the interference term. To keep the total output power (observed in electrical domain) fixed, the parameter A is to be properly chosen (and it is different in repetition MIMO and ST-coded MIMO). For Gamma–Gamma distribution, $E[I] = 1$, and $E[I^2]$ is related to the Scintillation Index (SI) by [1]

$$\sigma_I^2 = \frac{E[I^2]}{E[I]^2} - 1 = \frac{1}{\alpha} + \frac{1}{\beta} + \frac{1}{\alpha\beta}, \tag{9.28}$$

where α and β have already been introduced in (9.3).

9.2.2 Bit-Interleaved LDPC-coded Pulse-Amplitude Modulation

The block scheme of the bit-interleaved (BI) LDPC-coded PAM technique, for a repetition MIMO transmission, is shown in Fig. 9.3. The source bit stream is encoded using an (n, k) LDPC code of the code rate $r = k/n(k$ - the number of information bits, $n-$ the codeword length). The $l \times L$ block-interleaver (L is an integer multiple of the codeword length n) collects l code-words written row-wise. The mapper accepts l bits at a time from the interleaver column-wise and determines the corresponding symbol for the Q-ary ($Q = 2^l$) PAM signaling using a *Gray mapping* rule. The number of columns in block-interlever L is determined by a data rate and temporal correlation of the channel. We assume that the interleaver size is sufficient to overcome a temporal correlation of the channel, so that the channel samples are uncorrelated. The basis function is given by

$$\phi_{\mathrm{PAM}}(t) = \frac{1}{\sqrt{T}}\mathrm{rect}(t/T), \quad \mathrm{rect}(t) = \begin{cases} 1, & 0 \leq t < 1, \\ 0, & \text{otherwise}, \end{cases}$$

while the signal constellation points by $A_q = qd(q = 0, 1, \ldots, Q - 1)$, where d is the separation between two neighboring points.

The average symbol energy is given by

$$E_{\mathrm{s}} = \frac{(Q-1)(2Q-1)}{6}d^2,$$

and it is related to the bit-energy E_{b} by $E_{\mathrm{s}} = E_{\mathrm{b}}\log_2 Q$. With this BI LDPC-coded modulation scheme, the neighboring information bits from the same source are allocated into different PAM symbols. The outputs of the N receivers in the repetition MIMO, denoted as $y_n(n = 1, 2, \ldots, N)$, are processed to determine the symbol reliabilities (log-likelihood ratios, LLRs) by

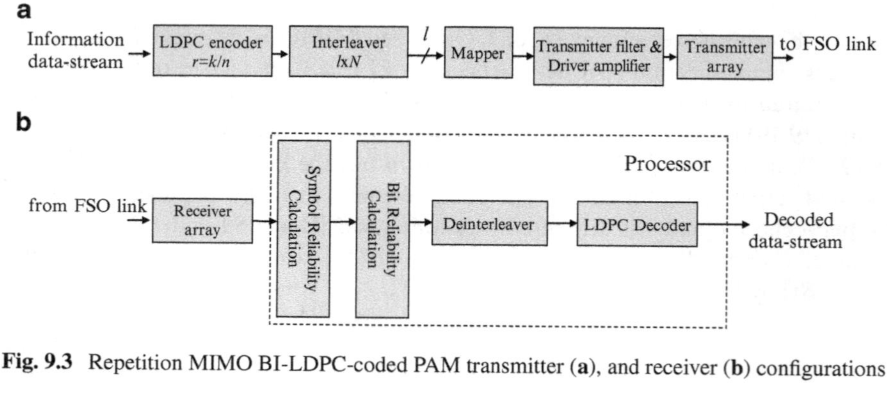

Fig. 9.3 Repetition MIMO BI-LDPC-coded PAM transmitter (**a**), and receiver (**b**) configurations

$$\lambda(q) = \log\left\{\frac{1}{\sqrt{2\pi}\sqrt{N_0/2}}\exp\left[-\frac{\left(y_n - \text{map}(q)d\sum_{m=1}^{M} I_{nm}\right)^2}{N_0}\right]\right\}$$
$$= -\sum_{n=1}^{N}\frac{\left(y_n - \text{map}(q)d\sum_{m=1}^{M} I_{nm}\right)^2}{N_0} + \log\left(\frac{1}{\sqrt{2\pi}\sqrt{N_0/2}}\right), \quad q = 0, 1, .. \tag{9.29}$$

where map(q) denotes a corresponding mapping rule. The second term in the second line of (9.29) can be neglected in simulations because it is constant for all symbols. Further, denote by c_j the jth bit in an observed symbol q binary representation $c = (c_1, c_2, \ldots, c_l)$. The bit reliabilities $L(c_j)$ are determined from symbol reliabilities by

$$L(c_i) = \log\frac{\sum_{c:c_i=0}\exp[\lambda(q)]\exp\left(\sum_{c:c_j=0, j\neq i} L_a(c_j)\right)}{\sum_{c:c_i=1}\exp[\lambda(q)]\exp\left(\sum_{c:c_j=0, j\neq i} L_a(c_j)\right)} \tag{9.30}$$

and forwarded to the LDPC decoder. Therefore, the ith bit reliability is calculated as the logarithm of the ratio of a probability that $c_i = 0$ and probability that $c_i = 1$. In the nominator, the summation is done over all symbols q having 0 at the position i, while in the denominator over all symbols q having 1 at the position i. With $L_a(c_j)$ we denoted a priori information determined from the LDPC decoder extrinsic LLRs. The inner summation in (9.30) is done over all bits of symbol q, selected in the outer summation, for which $c_j = 0$, $j \neq i$. By iterating the extrinsic reliabilities between APP demapper and LDPC decoder, the overall BER performance can be improved. The hard decisions from the LDPC decoder are delivered to the end user. Once more, in calculations of the symbol LLRs in (9.30), we assumed that CSI is known at the receiver side.

Other multilevel schemes, such as those based on Quadrature Amplitude-Modulation (QAM), are also applicable. However, the use of an additional DC bias is required because negative signals cannot be transmitted over an IM/DD system, and the power efficiency of such schemes is low.

To improve the BER performance, we perform the iteration of *extrinsic* information between APP PAM demapper and LDPC decoder. For a convergence behavior analysis, we perform the EXIT chart analysis [24]. To determine the Mutual Information (MI) transfer characteristics of the demapper, we model a priori input LLR, $L_{\text{M,a}}$, as a conditional Gaussian random variable. The MI between the bit c in a codeword and corresponding input LLR ($L_{\text{M,a}}$) is determined numerically. Similarly, the MI $I_{L_{\text{M,e}}}$ between c and $L_{\text{M,e}}$ is calculated numerically, but with the PDF of c and $L_{\text{M,e}}$ determined from the histogram obtained by Monte Carlo simulation.

By observing the $I_{\mathrm{LM,e}}$ as a function of the MI of $I_{\mathrm{LM,a}}$ and receiver SNR, E/N_0, in dB, the demapper EXIT characteristic (denoted as T_M) is given by

$$I_{\mathrm{LM,e}} = T_{\mathrm{M}}\left(I_{\mathrm{LM,a}}, E/N_0\right).$$

The EXIT characteristic of the LDPC decoder (denoted by T_{D}) is defined in a similar fashion as

$$I_{\mathrm{LD,e}} = T_{\mathrm{D}}\left(I_{\mathrm{LD,a}}\right).$$

The "turbo" demapping-based receiver operates by passing extrinsic LLRs between the demapper and LDPC decoder. The iterative process starts with an initial demapping in which $L_{\mathrm{M,a}}$ is set to zero having a consequence $I_{\mathrm{LM,a}} = 0$. The demapper output LLRs described by

$$I_{\mathrm{LM,e}} = I_{\mathrm{LD,a}}$$

are fed to the LDPC decoder. The LDPC decoder output LLRs described by

$$I_{\mathrm{LD,e}} = I_{\mathrm{LM,a}}$$

are fed to the APP demapper. The iterative procedure is repeated until the convergence or the maximum number of iterations has been reached. This procedure is illustrated in Fig. 9.4, where the APP demapper and LDPC decoder EXIT charts are shown together on the same graph. 4-PAM, 8-PAM, 16-PAM are observed, as well as the natural and Gray mapping. The EXIT curves have different slopes for different mappings. The existence of "tunnel" between corresponding demapping and decoder curves indicates that the iteration between demapper and decoder will be successful. The smallest SNR at which the iterative scheme starts to converge is known as the threshold (pinch-off) limit [24]. The threshold limit in the case of 16-PAM (Fig. 9.4b) is about 8 dB worse as compared with 4-PAM (Fig. 9.4a).

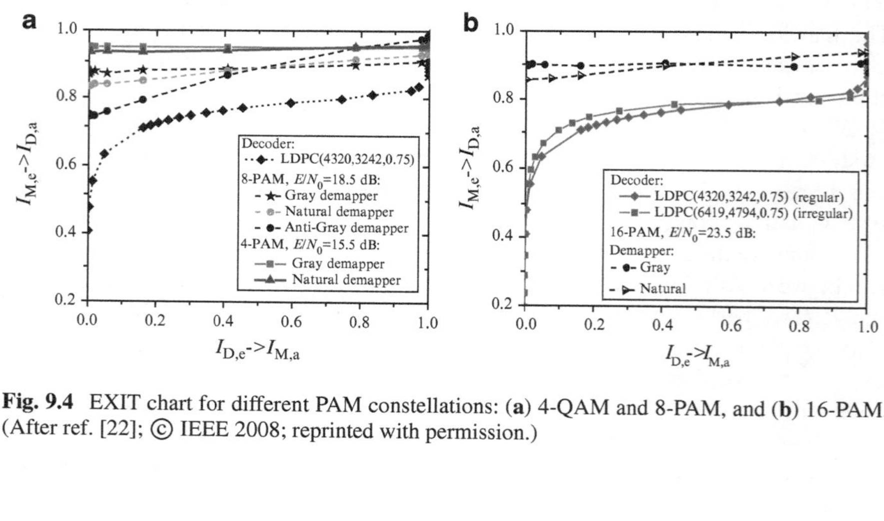

Fig. 9.4 EXIT chart for different PAM constellations: (**a**) 4-QAM and 8-PAM, and (**b**) 16-PAM. (After ref. [22]; © IEEE 2008; reprinted with permission.)

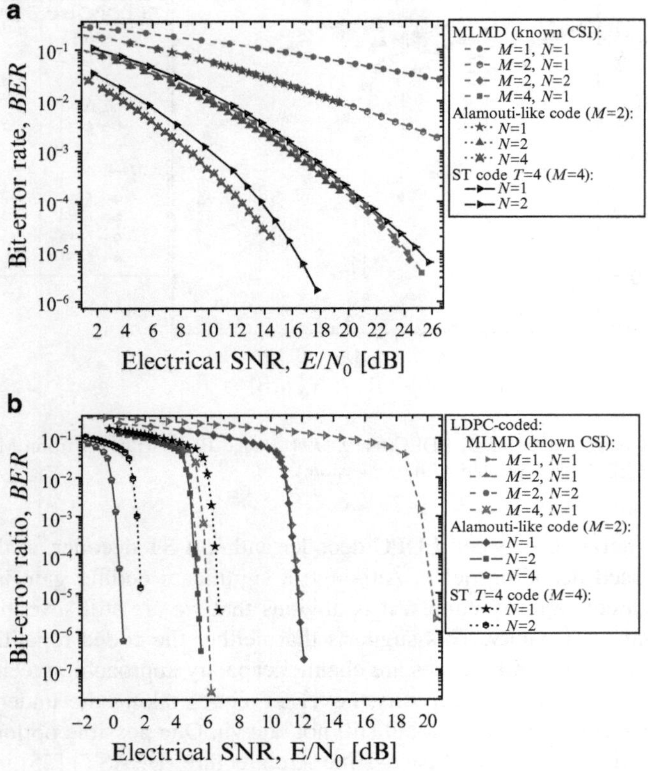

Fig. 9.5 BERs of binary LDPC(6419,4794)-coded MIMO ST coding scheme against LDPC-coded repetition MIMO: (**a**) uncoded case, and (**b**) coded case. (After ref. [22]; © IEEE 2008; reprinted with permission.)

The BER vs. electrical SNR in the presence of scintillation (per photodetector), for a strong turbulence regime ($\sigma_R = 3.0, \alpha = 5.485, \beta = 1.1156$), are shown in Fig. 9.5. The BER is shown for a different number of optical sources, and photodetectors, by employing an (6419,4794) irregular girth-6 LDPC code of a rate 0.747 designed using the concept of the PBD.

The Alamouti-like ST code performance is comparable to the repetition MIMO, while $T = 4$ ST performs worse than the corresponding repetition MIMO. The reason for such a behavior comes from the fact that we operate with nonnegative real signals rather than with complex, so that the space-time codes from orthogonal designs are not optimal in an FSO channel. Moreover, the second term in (9.27) behaves as interference, while in repetition MIMO it is a useful signal. The LDPC-coded MIMO with Alamouti-like code ($M = 2$) and $N = 4$ photodetectors provides about 20 dB improvement over LDPC-coded OOK with single optical source and single photodetector. Further performance improvements can be

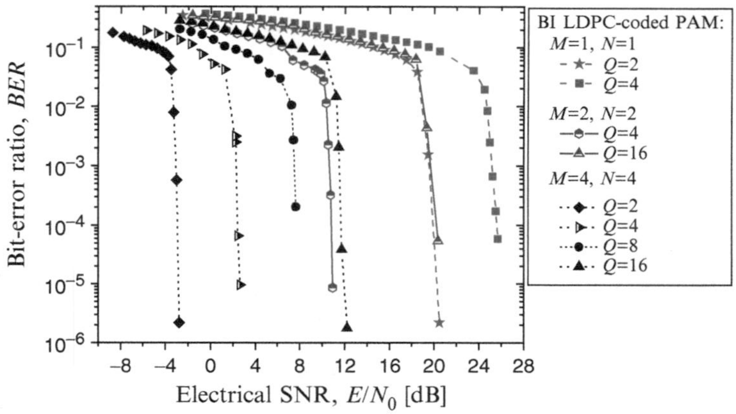

Fig. 9.6 BER performance of BI-LDPC(6419,4794)-coded PAM with repetition MIMO. (After ref. [22]; © IEEE 2008; reprinted with permission.)

obtained by iterating between LDPC decoder and soft ST decoder, at the expense of the increased decoding delay. Although a significant coding gain is obtained, from the channel capacity curves, it is obvious that we are still several dBs away from the channel capacity. This suggests that neither the coded repetition MIMO nor the wireless space-time codes are channel capacity approaching techniques. To come closer to the channel capacity, novel ST codes taking the underlying FSO physics into account are needed, but still not known. One possible option would be the use of Bell Labs Layered Space-Time Architecture (BLAST) [25] to deal with interference from (9.27), in combination with long LDPC codes.

The results of simulations for BI LDPC (6419,4794) -coded PAM are shown in Fig. 9.6 for different MIMO configurations and different number of signal constellation points employing the Gray mapping rule. Once more, although excellent BER performance improvement is obtained (about 23 dB for $M = N = 4, Q = 4$ over $M = N = 1$, $Q = 4$), there is still some space for improvement to come closer to the channel capacity, which was left for further research. The comparison for different component LDPC codes is given in Fig. 9.7. The scheme employing a girth-6 (g-6) irregular PBD-based LDPC code of rate 0.75 performs comparable to a girth-8 regular BC-LDPC code of the same rate. The scheme based on a girth-8 regular BIBD code of rate 0.81 performs worse than 0.75 codes. However, the difference is becoming less important as the constellation size grows.

9.2.3 Bit-Interleaved LDPC-coded Pulse-Position Modulation

The LDPC-coded MPAM scheme described in the previous section is not power efficient. In this section, we describe a power-efficient scheme based on Q-ary pulse-position modulation (PPM) [27]. The source bit streams originating from L different

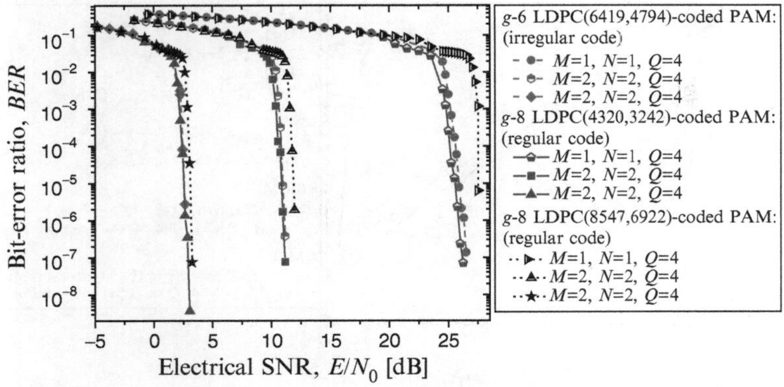

Fig. 9.7 BER performance of BI-LDPC-coded PAM with repetition MIMO for different LDPC component codes. (After ref. [22]; © IEEE 2008; reprinted with permission.)

RF/microwave sources are multiplexed together and encoded using an (n, k) LDPC code whose code rate is $r = k/n$, where k denotes the number of information bits, and n denotes the codeword length. The output of LDPC encoder is forwarded to an $m \times n$ block-interleaver, which collects m code-words in row-wise fashion. The mapper takes m bits at a time from the interleaver column-wise and determines the corresponding slot for Q-ary ($Q = 2^m$) PPM signaling based on a *Gray mapping* rule. Therefore, in this scheme, the neighboring information bits originating from the same source are allocated into different PPM symbols, which might improve the tolerance against atmospheric turbulence. In Q-ary PPM, at each signaling interval T_s a pulse of light of duration $T = T_s/Q$ is transmitted by a laser. Notice that the signaling interval T_s is subdivided into Q slots of duration T. The total transmitted power, denoted by P_{tot}, is the same regardless of the number of lasers, and the power per laser is P_{tot}/M. The ith ($i = 1, 2, \ldots, M$) laser modulated beam is projected toward the jth ($j = 1, 2, \ldots, N$) receiver using an expanding telescope. The receiver is based on a p.i.n. photodetector in a TA configuration.

We assume that the laser sources and p.i.n. photodetectors are positioned in such a way that different transmitted symbols experience different atmospheric turbulence conditions. The outputs of N receivers when the symbol q is transmitted, denoted as $Z_{n,q}(n = 1, 2, \ldots, N; q = 1, 2, \ldots, Q)$, are processed to determine the symbol reliabilities $\lambda(q)(q = 1, 2, \ldots, Q)$ by

$$\lambda(q) = -\frac{\sum_{n=1}^{N}\left(Z_{n,q} - \frac{\sqrt{E_s}}{M}\sum_{l=1}^{M} I_{n,l}\right)^2}{\sigma^2} - \frac{\sum_{n=1}^{N}\sum_{l=1,l\neq q}^{Q} Z_{n,l}}{\sigma^2}. \tag{9.31}$$

In (9.31), E_s denotes the symbol energy of uncoded symbol in electrical domain (in the absence of scintillation), which is related to the bit-energy E_b by $E_s = E_b \log_2 Q$. σ^2 denotes the variance of TA thermal noise (modeled as additive white Gaussian noise (AWGN)), and it is related to the double-side power spectral density

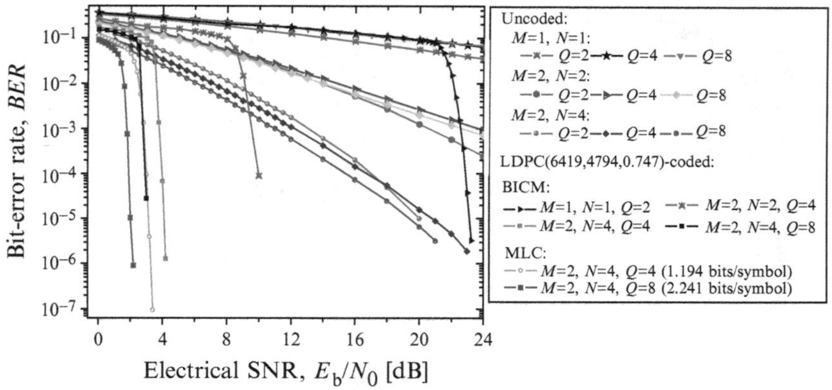

Fig. 9.8 BER performance of bit-interleaved LDPC-coded modulation against MLC for different MLMD configurations

N_0 by $\sigma^2 = N_0/2$. I_{nl} denotes the intensity of the light incident to nth photodetector $(n = 1, 2, \ldots, N)$, originating from lth $(l = 1, 2, \ldots, M)$ laser source.

The results of simulations assuming the operation under the strong turbulence regime ($\sigma_R = 3.0$) are shown in Fig. 9.8, for different number of lasers, photodetectors, and number of slots. The (6419,4794) irregular girth-6 LDPC code of rate 0.747 designed using the concept of pairwise balanced designs (PBD) [26] is employed. The MLC scheme with spectral efficiency of 2.241 bits/symbol combined with MLMD scheme employing two lasers and four photodetectors provides even 21 dB improvement over LDPC-coded binary PPM employing one laser and one photodetector. Corresponding BICM scheme of higher spectral efficiency (3 bits/symbol) performs about 1 dB worse. The number of inner iteration in sum-product LDPC decoder is set to 25 in both schemes, while the number of outer iterations in BICM scheme is set to 10. MLC employs a parallel-independent LDPC decoding. Therefore, the BI LDPC-coded modulation scheme, although simpler to implement than MLC scheme, performs slightly worse in BER performance, but provides higher spectral efficiency. Notice that simulation results presented in Fig. 9.8 are obtained assuming that receiver knows the CSI and TA thermal noise variance.

9.3 FSO-OFDM Transmission System

Optical OFDM systems can support high data rates by splitting a high-rate data-stream into a number of low-rate data-streams and transmitting these over a number of narrowband subcarriers [28–37]. The narrowband subcarrier data-streams experience smaller distortions than high-speed ones and require no equalization. Moreover, most of the required signal processing is performed in the RF domain.

This is advantageous because microwave devices are much more mature than their optical counterparts and because the frequency selectivity of microwave filters and the frequency stability of microwave oscillators are significantly better than that of corresponding optical devices. Furthermore, the phase noise levels of microwave oscillators are significantly lower than that of distributed feedback (DFB) laser diodes, which means that RF coherent detection is easier to implement than optical coherent detection. This, in turn, allows a system architect to directly apply the most advanced coherent modulation formats already developed for wireless communication.

The basic FSO-OFDM transmitter and receiver configurations are shown in Fig. 9.9a,b. The corresponding FSO link is shown in Fig. 9.9c. A 10-Gb/s information-bearing stream is demultiplexed into four 2.5 Gb/s streams, each encoded by identical LDPC encoders. The LDPC encoded outputs are further demultiplexed, and parsed into groups of B bits. The B bits in each group (frame) are subdivided into K subgroups with the ith subgroup containing b_i bits. The b_i

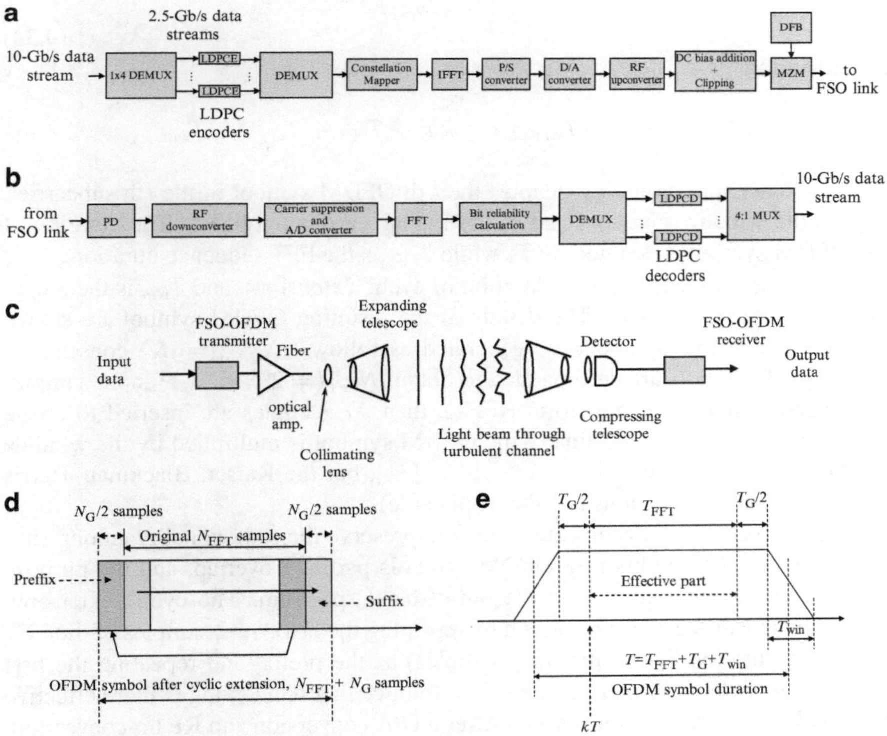

Fig. 9.9 FSO-OFDM system: (**a**) transmitter configuration, (**b**) receiver configuration, (**c**) FSO link, (**d**) OFDM symbol cyclic extension, (**e**) OFDM symbol after windowing. LDPCE-LDPC encoder, LPDCD-LDPC decoder, *DFB* distributed-feedback laser, *MZM* dual-drive Mach–Zehnder modulator, *P/S* parallel-to-serial

bits from the ith subgroup are mapped into a complex-valued signal from a 2^{b_i} -point signal constellation such as quadrature-amplitude modulation (QAM). The complex-valued signal points from K subchannels are considered to be the values of the discrete Fourier transform (DFT) of a multicarrier OFDM signal. Therefore, the symbol length (the time between two consecutive OFDM symbols) in an OFDM system is $T = KT_s$, where T_s is the symbol-interval length in an equivalent single-carrier system. By selecting K, the number of subchannels, sufficiently large, the OFDM symbol interval can be made significantly larger than the dispersed pulse-width in a single-carrier system, resulting in an arbitrarily small intersymbol interference. The transmitted OFDM signal can be written as

$$s(t) = s_{\mathrm{OFDM}}(t) + D, \tag{9.32}$$

where

$$s_{\mathrm{OFDM}}(t) = \mathrm{Re}\left\{ \sum_{k=-\infty}^{\infty} w(t-kT) \sum_{i=-N_{\mathrm{FFT}}/2}^{N_{\mathrm{FFT}}/2-1} X_{i,k} \cdot \mathrm{e}^{\mathrm{j}2\pi \frac{i}{T_{\mathrm{FFT}}}\cdot(t-kT)} \mathrm{e}^{\mathrm{j}2\pi f_{\mathrm{RF}}^{t}} \right\} \tag{9.33}$$

is defined for

$$kT - T_G/2 - T_{\mathrm{win}} \le t \le kT + T_{\mathrm{FFT}} + T_G/2 + T_{\mathrm{win}}.$$

In the above expression $X_{i,k}$ denotes the kth OFDM symbol in the ith subcarrier, $w(t)$ is the window function, and f_{RF} is the RF carrier frequency. The duration of the OFDM symbol is denoted by T, while T_{FFT} is the FFT sequence duration, T_{G} is the guard interval duration (the duration of cyclic extension), and T_{win} is the length of the windowing interval. The details of the resulting OFDM symbol are shown in Fig. 9.9d, e. The symbols are generated as follows: $N_{\mathrm{QAM}}(= K)$ consecutive input QAM symbols are zero-padded to obtain $N_{\mathrm{FFT}}(= 2^m, m > 1)$ input samples for inverse fast Fourier transform (IFFT), then N_{G} samples are inserted to create the guard interval T_{G} , and finally the OFDM symbol is multiplied by the window function (raised cosine function is used in [31], but the Kaiser, Blackman–Harris and other window functions are also applicable).

The purpose of the cyclic extension is to preserve the orthogonality among subcarriers when the neighboring OFDM symbols partially overlap, and the purpose of the windowing is to reduce the out-of-band spectrum. The cyclic extension, illustrated in Fig. 9.9d , is performed by repeating the last $N_{\mathrm{G}}/2$ samples of the FFT frame (of duration T_{FFT} with N_{FFT} samples) as the prefix, and repeating the first $N_{\mathrm{G}}/2$ samples (out of N_{FFT}) as the suffix. (Notice that windowing is more effective for smaller numbers of subcarriers.) After a D/A conversion and RF up-conversion, we convert the RF signal to the optical domain using one of two options: (1) for symbol rates up to 10 Gsymbols/s the OFDM signal directly modulates the DFB laser, and (2) for symbol rates above 10 Gsymbols/s the OFDM signal drives the dual-drive Mach–Zehnder modulator (MZM). The DC component (D in (9.32)) is

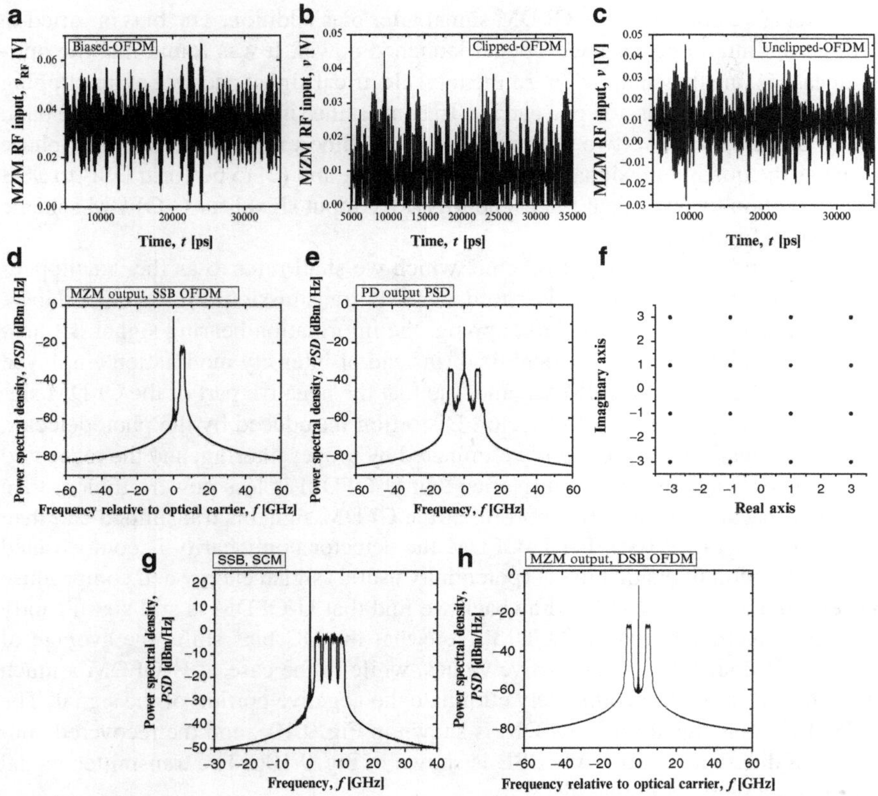

Fig. 9.10 Waveforms and power spectral densities of the SSB OFDM signal with 64 subcarriers at different points during transmission of an OFDM signal in a back-to-back configuration: (**a**) MZM RF input for B-OFDM, (**b**) MZM RF input for C-OFDM, (**c**) MZM RF input for U-OFDM, (**d**) PSD after MZM (U-OFDM), (**e**) photodectector (PD) output PSD (U-OFDM), (**f**) receiver constellation diagram for 16-QAM (U-OFDM), (**g**) PSD of SCM signal with four OFDM channels (U-OFDM), and (**h**) PSD of double-side band OFDM signal after MZM (U-OFDM). (After ref. [30]; © IEEE 2007; reprinted with permission.)

inserted to enable noncoherent recovery of the QAM symbols. In the remainder of this section, three different OFDM schemes are presented.

The first scheme is based on intensity modulation, and shall be referred to as the "biased-OFDM" (B-OFDM) scheme. Because bipolar signals cannot be transmitted over an IM/DD link, it is assumed that the bias component D is sufficiently large so that when added to $s_{\mathrm{OFDM}}(t)$ the resulting sum is nonnegative. For illustrative purposes the DFB laser driving signal is shown in Fig. 9.10a. The main disadvantage of the B-OFDM scheme is the poor power efficiency. To improve the power efficiency we present two alternative schemes. The first of these, which we shall refer to as the "clipped-OFDM" (C-OFDM) scheme, is based on SSB transmission, with clipping of the negative portion of the OFDM signal after bias addition. The bias is varied to find the optimum one for fixed optical launched power. It was found that the optimum case is one in which ~50% of the total electrical signal energy before clipping is allocated for transmission of a carrier. To convert the signal from double-side band (DSB) to SSB, we have two options: (1) to use Hilbert transformation of in-phase signal as the quadrature signal in electrical domain, and (2) to perform DSB to SSB transformation by an optical filter. The MZM RF input signal for C-OFDM scheme is shown in Fig. 9.10b.

The second power-efficient scheme, which we shall refer to as the "unclipped-OFDM" (U-OFDM) scheme, is based on SSB transmission employing $LiNbO_3$ MZM. To avoid distortion due to clipping, the information bearing signal is transmitted by modulating the electrical field (instead of intensity modulation employed in the B-OFDM and C-OFDM schemes) so that the negative part of the OFDM signal is transmitted to the photodetector. Distortion introduced by the photodetector, caused by squaring, is successfully eliminated by proper filtering, and the recovered signal distortion is insignificant. Notice that U-OFDM is less power efficient than C-OFDM because the negative portion of the OFDM signal is transmitted and then discarded (see Fig. 9.10c). For U-OFDM the detector nonlinearity is compensated by postdetection filters that reject (potentially useful) signal energy and compromise power efficiency. Despite this drawback we find that U-OFDM is still significantly more power efficient than B-OFDM. Note that the DC bias shifts the average of the C-OFDM signal toward positive values, while in the case of B-OFDM a much larger bias is needed to completely eliminate the negative portion of the signal. The MZM RF input signal for U-OFDM is shown in Fig. 9.10c, and the recovered constellation diagram for 16-QAM SSB is shown in Fig. 9.10g. The transmitted signal is recovered with negligible distortion.

The point-to-point FSO system considered here, shown in Fig. 9.9c, consists of an FSO-OFDM transmitter, propagation medium, and an FSO-OFDM receiver. The modulated beam is projected toward the receiver using the expanding telescope. At the receiver, an optical system collects the light, and focuses it onto a detector, which delivers an electrical signal proportional to the incoming optical power. The receiver commonly employs the transimpedance design, which is a good compromise between noise and bandwidth. A PIN photodiode plus preamplifier or an avalanche photodiode are typically used as optical detectors. During propagation through the air, the optical beam experiences amplitude and phase variations caused by scattering, refraction caused by atmospheric turbulence, absorption, and building sway. The photodiode output current can be written as

$$\begin{aligned} i(t) = R_{\mathrm{PD}}|a(t)s_{\mathrm{OFDM}}(t) + a(t)D|^2 = R_{\mathrm{PD}}\big[&|a(t)s_{\mathrm{OFDM}}(t)|^2 + |a(t)D|^2 \\ &+2R_{\mathrm{e}}\{a(t)s_{\mathrm{OFDM}}(t)a^*(t)D\}\big], \quad (9.34) \end{aligned}$$

where $|a(t)|^2$ denotes the intensity fluctuation due to atmospheric turbulence, and R_{PD} denotes the photodiode responsivity. The signal after RF down-conversion and appropriate filtering can be written as

$$r(t) = [i(t)k_{\mathrm{RF}}\cos(\omega_{\mathrm{RF}}t)]h_{\mathrm{e}}(\tau) + n(t), \tag{9.35}$$

where $h_{\mathrm{e}}(t)$ is the impulse response of the low-pass filter (having the transfer function $H_{\mathrm{e}}(\mathrm{j}\omega)), n(t)$ is electronic noise in the receiver, commonly modeled as a Gaussian process, k_{RF} denotes the RF downconversion factor, and the $*$ is the convolution operator. Finally, after the A/D conversion and cyclic extension removal, the transmitted signal is demodulated by the FFT algorithm. The soft outputs of the FFT demodulator are used to estimate the bit reliabilities that are fed to four identical LDPC iterative decoders based on the sum-product algorithm. The parameters of the overall OFDM-FSO system must be carefully chosen, as explained later in this section, so that the reconstructed sequence constellation diagram suffers minimal distortion in a back-to-back configuration.

For the sake of illustration, consider the signal waveforms and power-spectral densities (PSDs) at various points in the OFDM system given in Fig. 9.10. These examples were generated using SSB transmission in a back-to-back configuration. The bandwidth of the OFDM signal is set to 2.5 GHz, and the RF carrier to 7.5 GHz. The number of OFDM subchannels is set to 64. The OFDM sequence is zero-padded and the FFT is calculated using 128 points. The guard interval is obtained by a cyclic extension of 2×16 samples as explained above. The windowing (2×16 samples) is based on the Blackman–Harris windowing function. The average transmitted launched power is set to 0 dBm. The RF driver amplifier and MZM operate in linear regime (see Figs. 9.10a–c). The PSD for an SSB OFDM MZM output signal is shown in Fig. 9.10d, and the photodetector output signal (for SSB OFDM transmission) is shown in Fig. 9.10e. The OFDM term after beating in the photodetector (PD) (the third term in (9.34)), the low-pass term, and the squared OFDM term (the first term in (9.34)) can be easily identified.

If a 16-QAM OFDM system employing 64 subcarriers is used in combination with 39 Mb/s subchannels, the OFDM system proposed here allows transmitting a 10 Gb/s signal over a 2.5 GHz bandwidth, thereby increasing the spectral efficiency of OOK. To facilitate the implementation at higher speeds, OFDM may be combined with subcarrier multiplexing (SCM), the scheme is known as multiband OFDM in wireless literature. In this case (PSD shown in Fig. 9.10g), the spectral efficiency of $4 \times 10\mathrm{Gb/s}/11.25\,\mathrm{GHz} = 3.55$ bits/s/Hz is achieved, which is significantly better than that for OOK transmission over an FSO link. (For illustrative purposes, the PSD of DSB OFDM signal is also provided (see Fig. 9.10h).)

Before we are turning our attention to the BER performance evaluation, let us briefly explain how to interface RF/microwave and optical channels based on OFDM. In this section, we describe one possible interface between RF/microwave and optical channels, based on coded-OFDM. The block diagrams of the proposed transmitter and receiver configurations are shown in Fig. 9.11. The data streams from L different RF channels are combined using OFDM and encoded using an LDPC encoder. The LDPC encoded data stream is then parsed into groups of B bits, in a fashion similar to that explained above. The b_i bits from the ith subgroup are mapped into a complex-valued signal from a 2^{b_i} -point signal constellation such as QAM. The complex-valued signal points from all K subchannels are considered

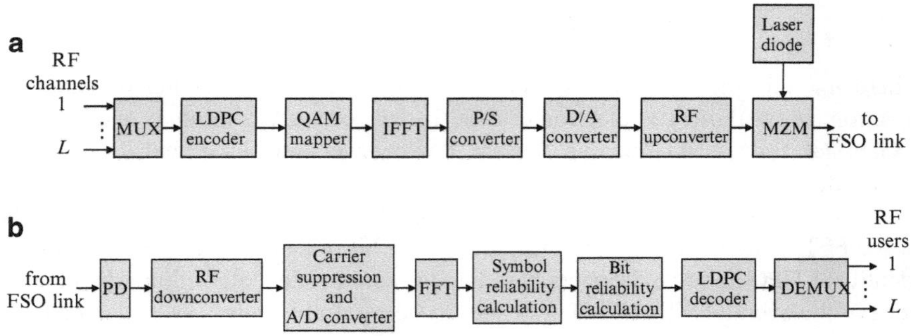

Fig. 9.11 The aggregation of RF/microwave channel by coded OFDM: (**a**) transmitter configuration, and (**b**) receiver configuration

as the values of the DFT of a multicarrier OFDM signal. After D/A conversion and RF up-conversion, the OFDM signal drives a MZM for transmission over the FSO link. The DC component facilitates recovering the QAM symbols incoherently. At the receiver, an optical system collects the light, and focuses it onto a detector, which delivers an electrical signal proportional to the incoming optical power. After the RF down-conversion, carrier suppression, A/D conversion, and cyclic extension removal, the transmitted signal is demodulated using the FFT algorithm. The soft outputs of the FFT demodulator are used to estimate the symbol reliabilities, which are converted to bit reliabilities, and provided as input to an LDPC iterative decoder. Therefore, the configuration of this scheme is similar to that of Fig. 9.9, from QAM block to bit-reliability calculation block. In this scheme, we used one LDPC code for all RF channels. Another option would be to use the LDPC encoder/decoder for every RF channel.

LDPC codes have been shown to significantly outperform turbo-product codes in bursty-error prone channels such as the fiber-optics channel in the presence of intrachannel nonlinear effects [38]. The quasi-cyclic LDPC codes similar to those described in Chap. 6 are employed here. In FSO communications, the receiver electronics noise is commonly modeled as a Gaussian noise (see e.g., [28–30, 39]). If r_{I} is the in-phase demodulator sample, and r_{Q} is the quadrature demodulator sample, then the *symbol* log-likelihood ratio (LLR) is calculated as

$$\lambda(s = (s_{\mathrm{I}}, s_{\mathrm{Q}})) = -\frac{(r_{\mathrm{I}} - s_{\mathrm{I}})^2}{2\sigma^2} - \frac{(r_{\mathrm{Q}} - s_{\mathrm{Q}})^2}{2\sigma^2}, \tag{9.36}$$

where s_{I} and s_{Q} are the coordinates of a transmitted signal constellation point and AWGN variance (σ^2) is determined from the required electrical SNR per bit $E_{\mathrm{b}}/N_{\mathrm{o}}$

$$\frac{E_{\mathrm{b}}}{N_{\mathrm{o}}} = \frac{E\left\{s_{i,k}\right\}}{\log_2 M}\frac{P_{\mathrm{o}}}{\sigma^2}. \tag{9.37}$$

P_o is the normalized received power, and $s_{i,k}$ denotes the QAM symbol in the kth subcarrier channel of the i-th OFDM frame. (With M we denote the number of points in the corresponding constellation diagram.) Notice that the definition of electrical SNR per bit, common in digital communications (see [31, 40, 41]), is different from that used in [1]. The initial *bit likelihoods*, provided to the iterative decoder, are calculated from the symbol LLRs, $\lambda(s)$, as

$$L(s_j) = \log \frac{\sum_{s:s_j=1} \exp[\lambda(s)]}{\sum_{s:s_j=0} \exp[\lambda(s)]}. \tag{9.38}$$

The Gaussian assumption in (9.36) may lead to BER performance degradation because the joint distribution is actually a convolution of the Gaussian and gamma–gamma PDFs. To reduce complexity, we use the Gaussian approximation in the calculation of symbol reliabilities. Nevertheless, dramatic performance improvement of an LDPC-coded FSO-OFDM system over an LDPC-coded FSO OOK system is obtained. In calculating bit reliabilities from symbol reliabilities (9.38) the following "max-star" operator, defined as $\max^*(x, y) = \log(e^x + e^y)$, is applied recursively $\max^*(x, y) = \max(x, y) + \log(1 + e^{-|x-y|})$.

Simulation results of an LDPC-coded SSB U-OFDM system for two different turbulence strengths, and zero inner scale, are shown in Fig. 9.12. For BPSK and QPSK, the coding gain improvement of an LDPC-encoded FSO-OFDM system over an LDPC-encoded FSO OOK system increases as the turbulence strength increases. However, the 16-QAM FSO-OFDM system is not able to operate in the regime of strong turbulence. For weak turbulence ($\sigma_R = 0.6$) (see Fig. 9.12a) the coding gain improvement of LDPC-coded FSO-OFDM system with 64 subcarriers over the LDPC-encoded FSO OOK system is 8.47 dB for QPSK and 9.66 dB for BPSK, at the BER of 10^{-5}. For strong turbulence ($\sigma_R = 3.0$) (see Fig. 9.12b) the coding gain improvement of the LDPC-coded FSO-OFDM system over the LDPC-coded FSO OOK system is 20.24 dB for QPSK and 23.38 dB for BPSK. In both cases, the block-circulant [42] LDPC code (4320,3242) of rate 0.75 is employed.

The comparison of different LDPC-coded SSB OFDM schemes, under the weak turbulence ($\sigma_R = 0.6$), is given in Fig. 9.13. The C-OFDM scheme slightly outperforms the U-OFDM scheme. Both C-OFDM and U-OFDM schemes outperform the B-OFDM scheme by approximately 1.5 dB at BER of 10^{-5}. For the results shown in Figs. 9.12–9.13, the received intensity samples are considered to be independent and uncorrelated. The results of simulations using the temporal correlation model described by (9.14) are shown in Fig. 9.14. The standard deviation σ_X is set to 0.6 (notice that σ_X is different from Rytov standard deviation σ_R used earlier, and for horizontal paths $\sigma_X \sim 0.498\sigma_R$). It is clear from Fig. 9.14 that LDPC-coded OFDM with or without interleaver provides excellent performance improvement even in the presence of temporal correlation. The BER performance can further be improved by using the interleaver with larger interleaving degree than that used in Fig. 9.14 (the star curve), at the expense of increasing encoder/decoder complexity. Notice the OOK scheme enters BER floor for this value of standard deviation ($\sigma_X = 0.6$), and even advanced FEC is not able to help too much. However, LDPC-coded OOK is

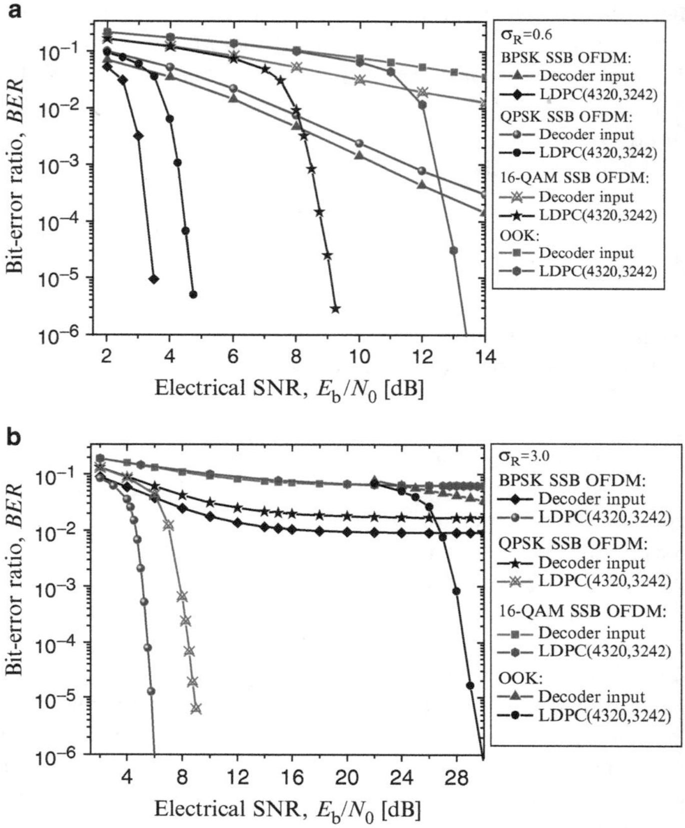

Fig. 9.12 BER performance of LDPC-coded SSB U-OFDM system with 64-subcarriers under: (**a**) the weak turbulence ($\sigma_R = 0.6$) and (**b**) strong turbulence ($\sigma_R = 3.0$). (After ref. [30]; © IEEE 2007; reprinted with permission.)

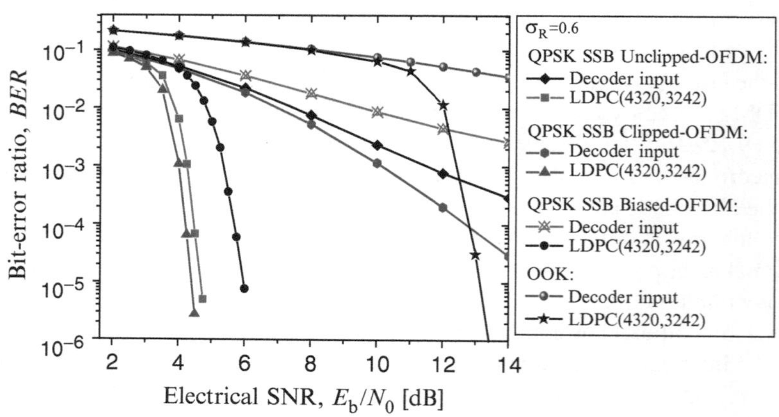

Fig. 9.13 Comparison of different LDPC-coded SSB FSO-OFDM systems with 64-subcarriers under the weak turbulence ($\sigma_R = 0.6$). (After ref. [30]; © IEEE 2007; reprinted with permission.)

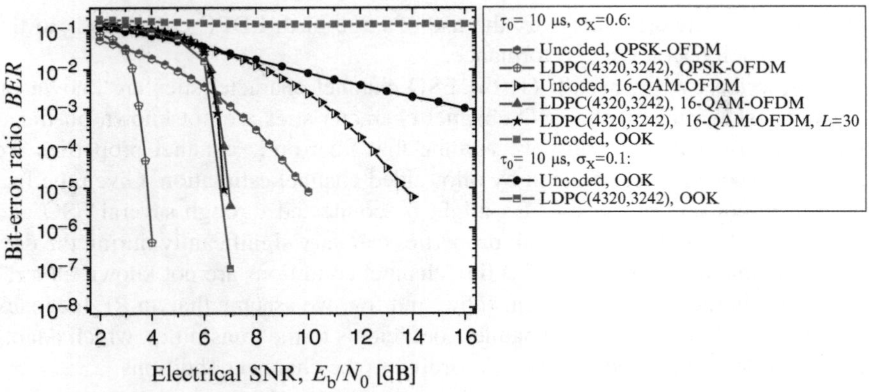

Fig. 9.14 BER performance of LDPC-coded OFDM in the presence of temporal correlation

able to operate properly at lower standard deviations σ_X. To generate temporally correlated samples we used two different methods, based on the Levinson–Durbin algorithm [43, 44] and the algorithm due to Wood and Chan [45], both of which provide the identical BER plots.

9.4 OFDM in Hybrid Optical Networks

The future optical networks will allow the integration of fiber-optics and FSO technologies, and will, therefore, have different portions of network being composed of fiber (either SMF or MMF) and FSO sections [46]. These hybrid optical networks might have significant impact for both military and commercial applications, when the pulling the ground fiber is expensive and takes a long time for deployment. Given the fact that hybrid optical networks will contain both FSO and fiber-optic sections, somebody has to study how to deal not only with atmospheric turbulence present in FSO portion of the network, but also to study the influence of fiber nonlinearities, polarization-mode dispersion (PMD), and chromatic dispersion in fiber-optic portion of the network.

In this section, we describe a coded-modulation scheme that is able simultaneously to deal with atmospheric turbulence, chromatic dispersion, and PMD in future hybrid optical networks. Moreover, the presented scheme supports 100 Gb/s per DWDM channel transmission and 100 Gb/s Ethernet, while employing the mature 10 Gb/s fiber-optics technology [47]. The described hybrid optical network scheme employs the OFDM as multiplexing and modulation technique and uses the LDPC codes as channel codes. With a proper design for 16-QAM-based polarization-multiplexed coded OFDM, the aggregate data rate of 100 Gb/s can be achieved for the OFDM signal bandwidth of only 12.5 GHz, which represents a scheme suitable for 100 Gb/s Ethernet. Notice that arbitrary FEC scheme can be used in proposed hybrid optical network. However, the use of large girth LDPC codes leads to the channel capacity achieving performance.

We describe two scenarios: (1) the FSO channel characteristics are known on transmitter side, and (2) the FSO channel characteristics are not known on transmitter side. In both scenarios, we assume that fiber-optic channel properties are known on receiver side, obtained by pilot-aided channel estimation. Given the fact that transmitter and receiver nodes might be connected through several FSO and fiber-optic links, and that FSO link properties can vary significantly during the day, it is reasonable to assume that FSO link channel conditions are not known on a receiver side. In the presence of rain, snow, and fog, we assume that an RF feedback channel is used to transmit the channel coefficients to the transmitter, which adapts the transmitted power and data rate according to the channel conditions.

The described scheme has many unique advantages: (1) demodulation, equalization, and decoding are jointly performed; (2) it is able to operate in the presence of channel impairments over different optical links, in SMF, MMF, and FSO; (3) it has high bandwidth efficiency (up to 10 bits/s/Hz); (4) it is compatible with future 100 Gb/s Ethernet technologies; and (5) the employed coded modulation provides excellent coding gains. We also describe how to determine the symbol reliabilities in the presence of laser phase noise and describe a particular channel inversion technique suitable to deal with PMD effects.

9.4.1 Hybrid Optical Networks

An example of a hybrid FSO – fiber-optic network is shown in Fig. 9.15a. This particular example includes intersatellite links and connection to aircrafts. The fiber-optic portion of network could be a part of an already installed MAN or a wide area network (WAN). The FSO network portion should be used whenever pulling the ground fiber is expensive and/or takes too long for deployment, such as urban and rural areas, where the optical fiber links are not already installed. The corresponding hybrid optical networking architecture is shown in Fig. 9.15b. We can identify three ellipses representing the core network, the edge network, and the access network. The FSO links can be used in both edge and access networks. The hybrid optical networks impose a big challenge to the engineers, because the novel signal processing techniques should be developed, which would be able to simultaneously deal atmospheric turbulence in FSO links; and with chromatic dispersion, PMD, and fiber nonlinearities in fiber-optic links. One such coded modulation technique is described in the rest of this section. By using the retro-reflectors the FSO systems can be applied even when there is no line of sight between transmitter and receiver.

The described coded modulation scheme employs the coded OFDM with coherent detection for the reasons given in introduction. The transmitter and receiver shown in Fig. 9.16, to be used in hybrid optical network from Fig. 9.15, are able to simultaneously deal with atmospheric turbulence, residual chromatic dispersion, and PMD. The bit streams originating from m different information sources are

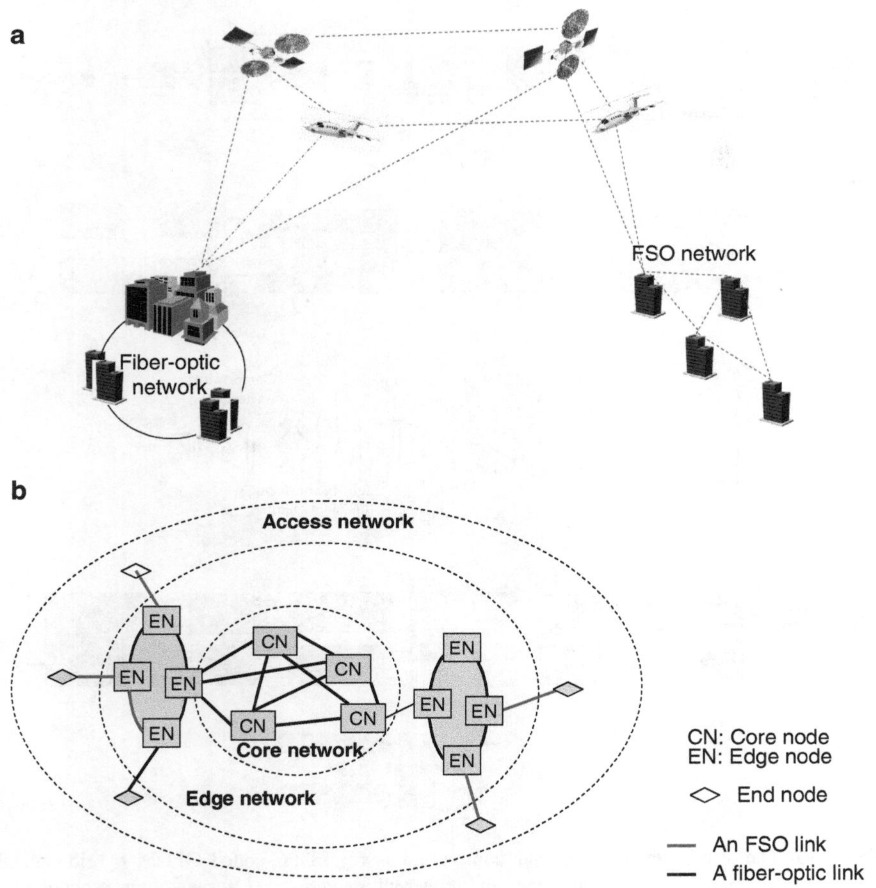

Fig. 9.15 (**a**) A hybrid FSO – fiber-optic network example, and (**b**) a hybrid optical networking architecture

encoded using different (n, k_i) LDPC codes of code rate $r_i = k_i/n.k_i$ denotes the number of information bits of ith $(i = 1, 2, \ldots, m)$ component LDPC code, and n denotes the codeword length, which is the same for all LDPC codes. The use of different LDPC codes allows us to optimally allocate the code rates. If all component LDPC codes are identical, the corresponding scheme is commonly referred to as the BI-coded modulation (BICM). The outputs of m LDPC encoders are written row-wise into a block-interleaver block. The mapper accepts m bits at time instance i from the $(m \times n)$ interleaver column-wise and determines the corresponding M-ary $(M = 2^m)$ signal constellation point $(\phi_{I,i}, \phi_Q, i)$ in two-dimensional (2D) constellation diagram such as M-ary phase-shift keying (PSK) or M-ary QAM. (The coordinates correspond to in-phase and quadrature components of M-ary two-dimensional constellation.) The operations of most of the blocks in

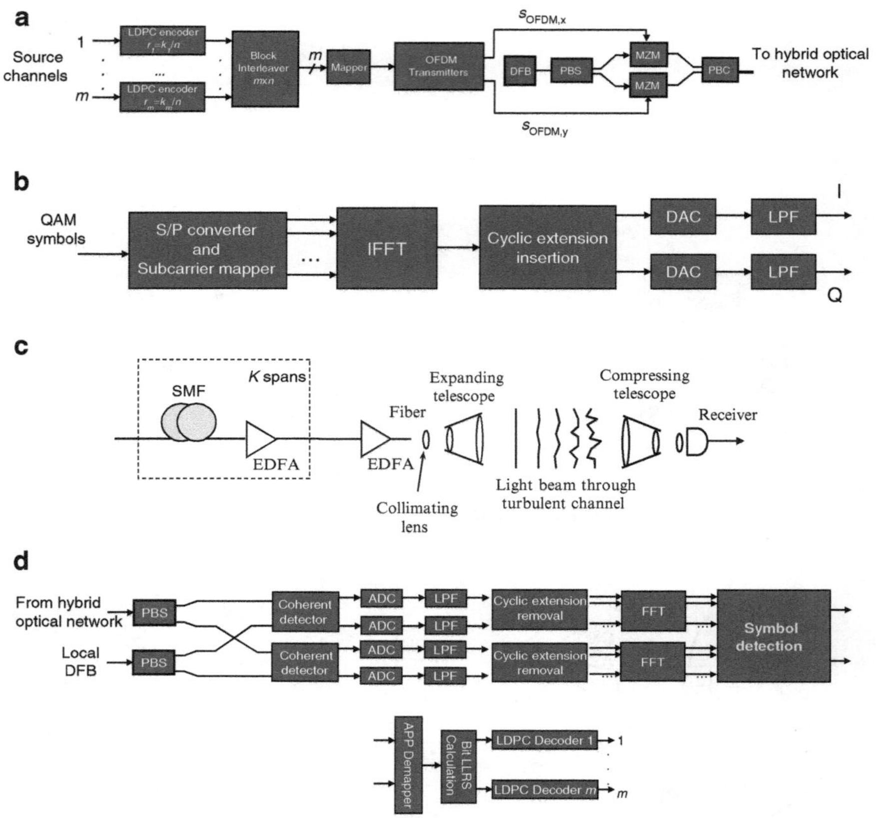

Fig. 9.16 The transmitter and receiver configurations for LDPC-coded OFDM hybrid optical system with polarization multiplexing and coherent detection: (**a**) transmitter architecture, (**b**) an OFDM transmitter architecture, (**c**) a hybrid optical link example, and (**d**) receiver architecture

transmitter and receiver are similar to those we presented in Chap. 6, here we provide only details relevant to hybrid optical networks.

The main motivation for use of OFDM in hybrid optical networks is the fact that the OFDM symbols with a large number of subcarriers (in order of thousands) have duration in the order of μs, and by means of interleavers in order of thousands we are able to overcome the atmospheric turbulence with temporal correlation in the order of 10 ms. For the OFDM scheme to be capable to simultaneously compensate for chromatic dispersion and PMD, in addition to the atmospheric turbulence, the cyclic extension guard interval should be longer that total delay spread due to chromatic dispersion and DGD, as indicated above. Because for high-speed signals a longer sequence of bits is affected by the deep fade in the ms range due to atmospheric turbulence, we propose to employ the polarization-multiplexing and large QAM constellations to achieve the aggregate data rate of $R_D = 100\,\text{Gb/s}$, while keeping the OFDM signal bandwidth in the order of 10 GHz. For example, by using the polarization-multiplexing and 16-QAM we can achieve $R_\mathrm{D} = 100\,\mathrm{Gb/s}$ for OFDM signal bandwidth of 12.5 GHz, resulting in bandwidth efficiency of 8 bits/s/Hz. Similarly, by using the polarization-multiplexing and 32-QAM we can achieve the same data rate ($R_\mathrm{D} = 100\,\mathrm{Gb/s}$) for OFDM signal bandwidth of 10 GHz, with bandwidth efficiency of 10 bits/s/Hz.

The receiver description requires certain knowledge of the channel. In what follows, we assume that fiber-optic channel characteristics are known on receiver side, because the fiber-optics channel coefficients can easily be determined by pilot-aided channel estimation. On the other hand, the hybrid optical network may contain different FSO and fiber-optic sections, while the channel characteristics of FSO link can change rapidly even during the day, so it is reasonable to assume that FSO channel characteristics are not known on receiver side. The FSO transmitter can use a retro-reflector and a training sequence to sense the FSO channel. We will further describe two concepts: (1) transmitter does not have any knowledge about the FSO link, and (2) transmitter knows the FSO link properties. When the transmitter knows the FSO link properties we can employ the transmitter diversity concept.

9.4.2 Description of Receiver and Transmission Diversity Scheme

The immunity to atmospheric turbulence can be improved by employing the diversity approaches. To maximize the receiver diversity, the multiple receivers should be separated enough so that independency condition is satisfied. Given the fact that the laser beam is getting expanded during propagation, it might not be possible always to separate the receivers sufficiently enough so that the independence condition is satisfied. On the other hand, by using transmission diversity instead, the independence condition is easier to satisfy. Moreover, it has been shown in [41] that transmitter diversity performs comparably to the maximum-ratio combining receiver diversity. In transmission diversity, the signal to be transmitted from ith transmitter, characterized by path gain $r_i \exp[-\mathrm{j}\theta_i]$, is premultiplied by complex gain $\alpha_i = a_i \exp[-\mathrm{j}\theta_i] (0 \leq a_i \leq 1)$. On the receiver side, the weight a_i that maximizes the SNR is chosen by [41]

$$a_i = r_i \Big/ \sqrt{\sum_{i=1}^{L} r_i^2},$$

where L is the number of transmitter branches. When the channel is not known on transmitter side we have to setup a_i to 1, and $\theta_i = 0$, and use Alamouti-type scheme instead [41]. Notice however, that Alamouti-type receiver requires the knowledge of FSO channel, and as such is not considered here.

The received complex symbols in x- and y-polarization, in the presence of PMD and scintillation, can be written by [46]

$$r_{x,i,k} = a_i(k)\alpha_i(k)e^{j\phi_Y}\left[h_{xx}(k)s_{x,i,k} + h_{xy}(k)s_{y,i,k}\right] + n_{x,i,k}, \tag{9.39}$$

$$r_{y,i,k} = a_i(k)\alpha_i(k)e^{j\phi_Y}\left[h_{yx}(k)s_{x,i,k} + h_{yy}(k)\,s_{y,i,k}\right] + n_{y,i,k}, \tag{9.40}$$

where we used the index k to denote the kth subcarrier, index i to denote the ith OFDM symbol, $h_{ij}(k)(i, j \in \{x, y\})$ are the channel coefficients due to PMD introduced in Chap. 6, $s_{x,i,k}$ and $s_{y,i,k}$ denote the transmitted symbols in x- and y-polarization, respectively; while corresponding received symbols are denoted by $r_{x,i,k}$ and $r_{y,i,k}$. The weight a_i is chosen in such a way to maximize the SNR, as explained above. In (9.39)–(9.40), $n_{x,i,k}$ and $n_{y,i,k}$ denote the ASE noise processes in x- and y-polarization. In the absence of ASE noise, (9.39) and (9.40) represent the system of linear equations with two unknowns $s_{x,i,k}$ and $s_{y,i,k}$, and upon solving we obtain:

$$\tilde{s}_{x,i,k} = \frac{\dfrac{h_{xx}^*(k)}{|h_{xx}(k)|^2}\left[r_{x,i,k} - \dfrac{h_{xy}(k)h_{yy}^*(k)}{|h_{yy}(k)|^2}r_{y,i,k}\right]}{1 - \dfrac{h_{xx}^*(k)h_{xy}(k)}{|h_{xx}(k)|^2}\dfrac{h_{yx}(k)h_{yy}^*(k)}{|h_{yy}(k)|^2}}, \tag{9.41}$$

$$\tilde{s}_{y,i,k} = \frac{h_{yy}^*(k)}{|h_{yy}(k)|^2}r_{y,i,k} - \frac{h_{yx}(k)h_{yy}^*(k)}{|h_{yy}(k)|^2}\tilde{s}_{x,i,k}. \tag{9.42}$$

where $\tilde{s}_{x,i,k}$ and $\tilde{s}_{y,i,k}$ denote the detector estimates of symbols $s_{x,i,k}$ and $s_{y,i,k}$ transmitted on kth subcarrier of ith OFDM symbol. Notice that the OFDM scheme with polarization diversity (described in Chap. 6), assuming that both polarizations are used on a transmitter side and equal-gain combining on a receiver side, is the special of case of symbol detector described by (9.41)–(9.42). By setting $s_{x,i,k} = s_{y,i,k} = s_{i,k}$ and using the symmetry of channel coefficients, the transmitted symbol can be estimated by:

$$\tilde{s}_{i,k} = \frac{h_{xx}^*(k)r_{x,i,k} + h_{xy}^*(k)r_{y,i,k}}{|h_{xx}(k)|^2 + |h_{xy}(k)|^2}.$$

In the presence of laser phase noise the symbols detector estimates are function of the laser phase noise process:

$$\tilde{s}_{x,i,k} = \frac{\dfrac{h_{xx}^*(k)}{|h_{xx}(k)|^2}e^{-j\phi_{PN}}\left[r_{x,i,k} - \dfrac{h_{xy}(k)h_{yy}^*(k)}{|h_{yy}(k)|^2}r_{y,i,k}\right]}{1 - \dfrac{h_{xx}^*(k)h_{xy}(k)}{|h_{xx}(k)|^2}\dfrac{h_{yx}(k)h_{yy}^*(k)}{|h_{yy}(k)|^2}}, \tag{9.43}$$

$$\tilde{s}_{y,i,k} = \frac{h_{yy}^*(k)e^{-j\phi_{PN}}}{|h_{yy}(k)|^2}r_{y,i,k} - \frac{h_{yx}(k)h_{yy}^*(k)}{|h_{yy}(k)|^2}\tilde{s}_{x,i,k}. \tag{9.44}$$

The detector soft estimates of symbols carried by kth subcarrier in ith OFDM symbol, $\tilde{s}_{x(y)i,k}$, are forwarded to the APP demapper, which determines the symbol LLRs $\lambda_{x(y)}(s)$ of x- (y-) polarization by

$$\lambda_{x(y)}(s|\phi_{PN}) = -\frac{(\mathrm{Re}[\tilde{s}_{i,k,x(y)}(\phi_{PN})] - \mathrm{Re}[\mathrm{QAM}(\mathrm{map}(s))])^2}{2\sigma^2} - \frac{(\mathrm{Im}[\tilde{s}_{i,k,x(y)}(\phi_{PN})] - \mathrm{Im}[\mathrm{QAM}(\mathrm{map}(s))])^2}{2\sigma^2}, \quad s = 0, 1, \ldots, 2^{n_b} - 1 \tag{9.45}$$

where Re[] and Im[] denote the real and imaginary part of a complex number, QAM denotes the QAM-constellation diagram, σ^2 denotes the variance of an equivalent Gaussian noise process originating from ASE noise, and map(s) denotes a corresponding mapping rule (Gray mapping rule is applied here). (n_b denotes the number of bits carried by symbol.) Notice that symbol LLRs in (9.45) are conditioned on the laser phase noise sample $\phi_\mathrm{PN} = \phi_\mathrm{T} - \phi_\mathrm{LO}$, which is a zero-mean Gaussian process (the Wiener–Lévy process) with variance $\sigma^2{}_\mathrm{PN} = 2\pi(\Delta\nu_\mathrm{T} + \Delta\nu_\mathrm{LO})|t|$ ($\Delta\nu_\mathrm{T}$ and $\Delta\nu_\mathrm{LO}$ are the corresponding laser linewidths introduced earlier). This comes from the fact that estimated symbols $\tilde{s}_{x(y)i,k}$ are functions of ϕ_PN. To remove the dependence on ϕ_PN we have to average the likelihood function, over all possible values of ϕ_PN:

$$\lambda_{x(y)}(s) = \log\left\{\int_{-\infty}^{\infty} \exp\left[\lambda_{x(y)}(s|\phi_\mathrm{PN})\right] \frac{1}{\sigma_\mathrm{PN}\sqrt{2\pi}} \exp\left(-\frac{\phi_\mathrm{PN}^2}{2\sigma_\mathrm{PN}^2}\right) \mathrm{d}\phi_\mathrm{PN}\right\}. \tag{9.46}$$

The calculation of LLRs in (9.46) can be performed by numerical integration. For the laser linewidths in order of tens of kHz it is sufficient to use the trapezoidal rule, with samples of ϕ_PN obtained by pilot-aided channel estimation. Let us denote by $b_{j,x(y)}$ the jth bit in an observed symbol s binary representation $\boldsymbol{b} = (b_1, b_2, \ldots, b_{nb})$ for x (y-) polarization. The bit LLRs required for LDPC decoding are calculated from symbol LLRs by

$$L\left(\hat{b}_{j,x(y)}\right) = \log\frac{\sum_{s:b_j=0} \exp\left[\lambda_{x(y)}(s)\right]}{\sum_{s:b_j=1} \exp\left[\lambda_{x(y)}(s)\right]}. \tag{9.47}$$

9.4.3 Performance Evaluation of Hybrid Optical Networks

We are turning our attention to the evaluation of the described hybrid optical network. In simulation results shown in Figs. 9.17–9.18 we assume that PMD channel coefficients are known at the receiver, because they can easily be determined by pilot-aided channel estimation. On the other hand, the FSO channel may change significantly during the day time, and as such is difficult to estimate. To illustrate the efficiency of this scheme, in Fig. 9.17a, b we show the constellation diagrams for aggregate rate of 100 Gb/s, corresponding to the $M = 16$ QAM and the OFDM signal bandwidth of 12.5 GHz in the presence of atmospheric turbulence ($\sigma_X = 0.1$ and $\sigma_Y = 0.05$), before (see Fig. 9.17a) and after (see Fig. 9.17b) PMD compensa-

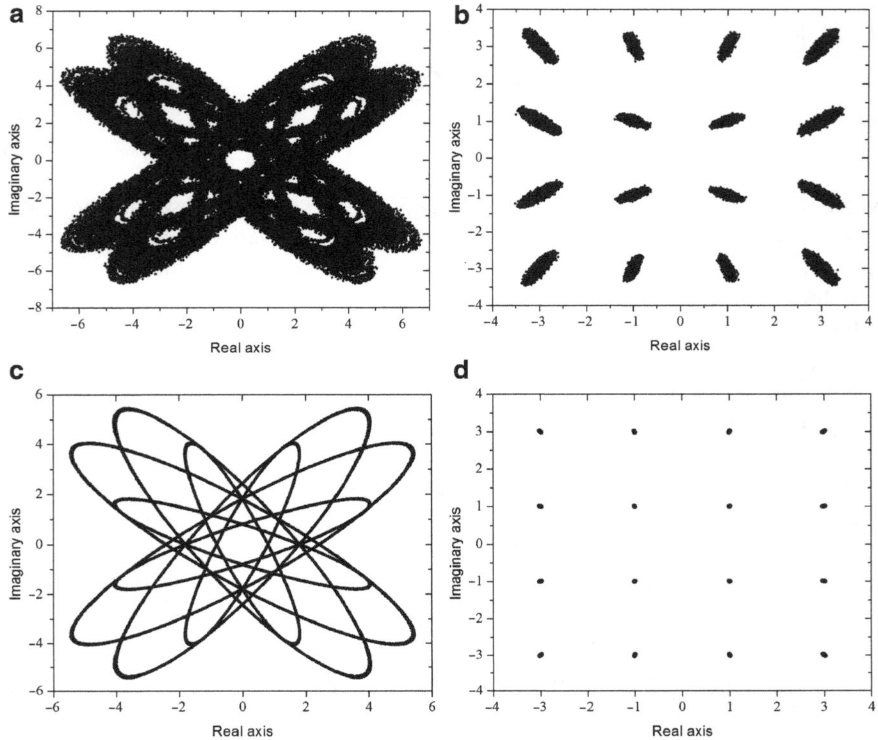

Fig. 9.17 The constellation diagrams for polarization-multiplexed 16-QAM (the aggregate data rate is 100 Gb/s) after 500 ps of DGD for $\sigma_X = 0.1$, $\sigma_Y = 0.05$, and $OSNR = 50\,\text{dB}$ observing the worst case scenario ($\theta = \pi/2$ and $\varepsilon = 0$) without transmission diversity: (**a**) before PMD compensation, and (**b**) after PMD compensation. The corresponding constellation diagrams in the presence of PMD only (for DGD of 500 ps): (**c**) before PMD compensation, and (**d**) after PMD compensation

tion, assuming the worst case scenario ($\theta = \pi/2$ and $\varepsilon = 0$). The corresponding constellation diagrams in the presence of PMD only are shown in Fig. 9.17c, d. The proposed coded modulation scheme is able to compensate for the PMD with DGD of up to 500 ps in the presence of atmospheric turbulence characterized by $\sigma_X = 0.1$ and $\sigma_Y = 0.01$.

In Fig. 9.18, we show the BER performance of the described scheme for both uncoded case (Fig. 9.18a) and LDPC-coded case (Fig. 9.18b). The OFDM system parameters were chosen as follows: the number of QAM symbols $N_{\text{QAM}} = 4096$, the oversampling is two times, OFDM signal bandwidth is set to either 10 GHz ($M = 32$) or 12.5 GHz ($M = 16$), and the number of samples used in cyclic extension $N_{\text{G}} = 64$. For the fair comparison of different M-ary schemes the OSNR on x-axis is given per information bit, which is also consistent with digital communication literature. The code rate influence is included in Fig. 9.11 so that the

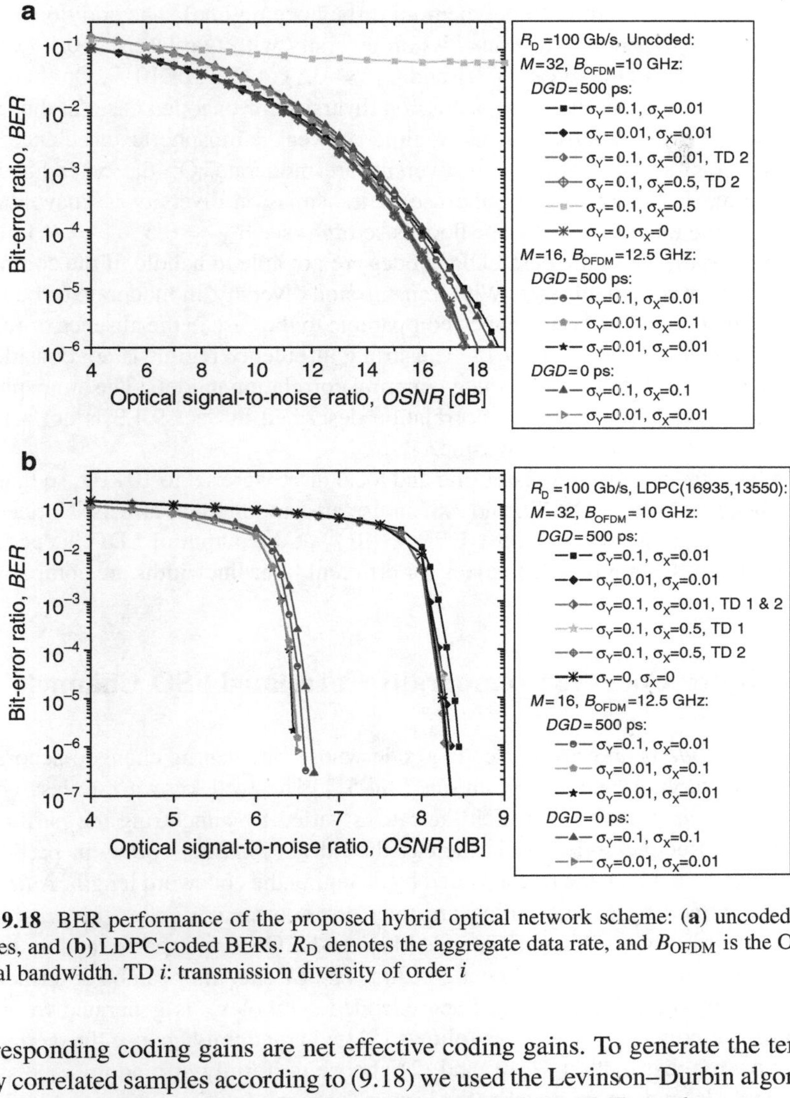

Fig. 9.18 BER performance of the proposed hybrid optical network scheme: (**a**) uncoded BER curves, and (**b**) LDPC-coded BERs. R_D denotes the aggregate data rate, and B_{OFDM} is the OFDM signal bandwidth. TD *i*: transmission diversity of order *i*

corresponding coding gains are net effective coding gains. To generate the temporally correlated samples according to (9.18) we used the Levinson–Durbin algorithm [43,44]. From Figs. 9.17 and 9.18 it can be concluded that PMD can be successfully compensated even in the presence of atmospheric turbulence. Most of the degradation is coming from FSO channel, as shown in Figs. 9.10 and 9.11. The 32-QAM case with aggregate data rate $R_D = 100$ Gb/s performs 1.9 dB (at BER $= 10^{-6}$) worse than 16-QAM (with the same aggregate rate) although the occupied bandwidth is smaller.

The net coded gain improvement (at BER of 10^{-6}) of LDPC-coded OFDM over uncoded-OFDM is between 11.05 dB ($M = 16$, $\sigma_X = 0.01$, $\sigma_Y = 0.01$, corresponding to the weak turbulence regime) and 11.19 dB ($M = 16$, $\sigma_X = 0.1$,

$\sigma_Y = 0.01$, corresponding to the medium turbulence regime). The additional coding gain improvement due to transmission diversity with two lasers is 0.19 dB for 32-QAM based OFDM ($\sigma_X = 0.01$ and $\sigma_Y = 0.1$) at BER of 10^{-6}. On the other hand, the improvement due to transmission diversity for uncoded case (at the same BER) is 1.26 dB. Therefore, in the regime of weak atmospheric turbulence, the improvements due to transmission diversity are moderate. On the other hand, in the moderate turbulence regime the use of transmission diversity is unavoidable. Otherwise, the uncoded BER error floor is so high (see $\sigma_X = 0.5$, $\sigma_Y = 0.1$ curve in Fig. 9.18a) that even the best LDPC codes are not able to handle, if the complexity is to be kept reasonably low. With transmission diversity, in moderate turbulence regime, we obtain BER performance comparable to the case in the absence of turbulence regime, as shown in Fig. 9.18. The strong turbulence regime is not considered here due to the lack of an appropriate temporal correlation model. (The atmospheric turbulence model with temporal correlation described in Sect. 9.1.3 is not a valid model in the strong turbulence regime.)

The laser linewidths of transmitting and local laser were set to 10 kHz, so that the atmospheric turbulence, PMD, and ASE noise are predominant effects. Notice that BER threshold required to achieve BER $= 10^{-6}$ at the output of LDPC decoder is 0.0196, and in this region BER values for different laser linewidths are comparable.

9.5 Raptor Codes for Temporally Correlated FSO Channels

A *rate-less code* is an error-correcting code whose rate can be changed according to the time-varying channel conditions [48–51]. The well known rate-less codes are: (1) *punctured codes*, in which the rate is varied by puncturing the parity bits so that the effective code rate is increased; and (2) *fountain codes*, in particular *Raptor codes*, in which the rate is varied by changing the codeword length. A *Raptor code* is obtained by concatenating an inner error-correcting code (the pre-code), with an outer Luby-transform (LT) code [49]. An LT Code is a sparse random linear code, with a very simple decoding algorithm. The *LT encoding* can be described by the following algorithm [49, 51]. Each encoded symbol $\boldsymbol{x}_n$ is generated from the message symbols $\boldsymbol{s}_1, \ldots, \boldsymbol{s}_{K,\mathrm{LT}}$ as follows [51]: (1) randomly choose the degree d_n from a degree distribution $\Omega(x)$, and (2) choose uniformly at random d_n distinct input symbols, and set $\boldsymbol{x}_n$ equal to the bitwise sum, modulo 2, of those d_n symbols. *LT decoding* can be described by the following algorithm. The decoder's task is to recover $\boldsymbol{s}$ from $\boldsymbol{x} = \boldsymbol{s}\boldsymbol{G}$, where $\boldsymbol{G}$ is the matrix associated with the graph (the pseudorandom matrix) by sum-product algorithm [51]:

1. Find a check node $\boldsymbol{x}_n$ that is connected to only one source symbol $\boldsymbol{s}_k$
   (a) Set $\boldsymbol{s}_k = \boldsymbol{x}_n$
   (b) Add $\boldsymbol{s}_k$ to all checks $\boldsymbol{x}'_n$ that are connected to $\boldsymbol{s}_k$: $\boldsymbol{x}'_n = \boldsymbol{x}_n + \boldsymbol{s}_k$ for all n' such that $\boldsymbol{G}_{nk} = 1$
   (c) Remove all the edges connected to the source symbol $\boldsymbol{s}_k$.

2. Repeat (1) until all s_k are determined.

LT codes, although simple to encode and decode, have the following two drawbacks: d_{n} can take any integer value up to the size of the input word, leading to a decoding complexity $O(K_{\mathrm{LT}} \log K_{\mathrm{LT}})$, and the error floor is observed at high SNRs. To avoid those problems the Raptor codes are advocated in [48, 50]. A Raptor code is formed by concatenating an inner error correcting code (the pre-code) with an outer LT code, as illustrated below, in Fig. 9.19. In this example, the inner code is a systematic code, so that we can collapse the Tanner graph as illustrated in Fig. 9.20. The joint Tanner graph is obtained by joining the inner code parity-check nodes with LT check nodes. We can use this graph to perform joint iterative decoding.

The LLR to be passed from check-node c to variable node v can be calculated as follows:

$$L\left(\mu_{c,v}^{(t)}\right) = \begin{cases} 2\tanh^{-1}\left[\tanh\left(\frac{L(y_c)}{2}\right)\prod_{w\in V_c, w\neq v}\tanh\left(\frac{L\left(\eta_{w,c}^{(t-1)}\right)}{2}\right)\right], & \forall c \in C_{\mathrm{LT}}, \\ 2\tanh^{-1}\left[\prod_{w\in V_c, w\neq v}\tanh\left(\frac{L\left(\eta_{w,c}^{(t-1)}\right)}{2}\right)\right], & \forall c \in C_{\mathrm{LDPC}}, \end{cases} \tag{9.48}$$

where V_c denotes the set of variable-nodes connected to check-node c, C_{LT} denotes the set of LT check nodes, and C_{LDPC} denotes the set of inner code check-nodes (in

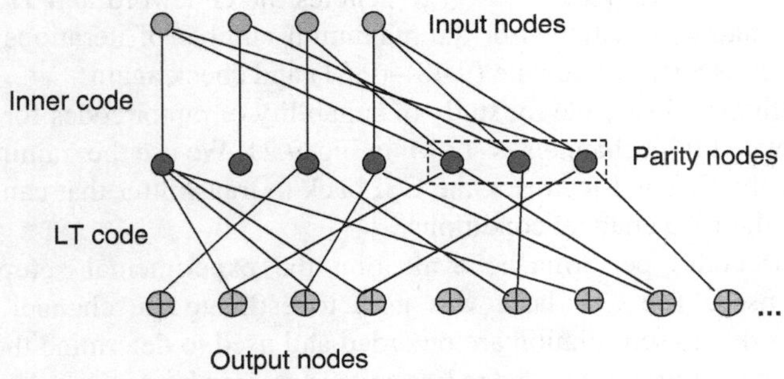

Fig. 9.19 Tanner graph representation of a raptor code

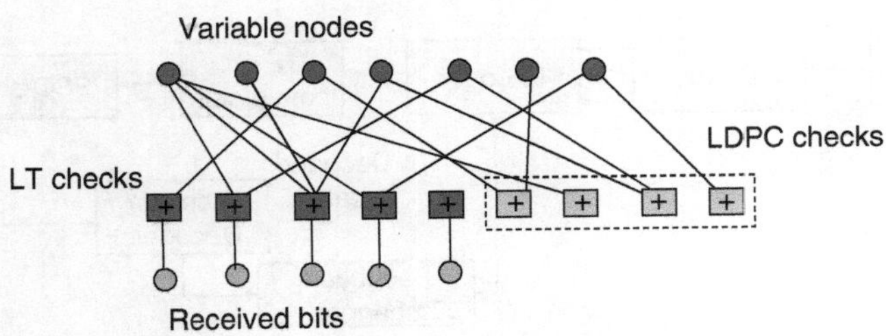

Fig. 9.20 Joint Tanner graph representation of a raptor code

example above the inner code is an LDPC code). In (9.48), we use the character μ to denote the passage of messages from c-nodes to v-nodes, and the character η to denote the passage of messages in opposite direction. The superscript t is used to denote the current iteration, and $(t-1)$ the previous one. y_c is the sample that corresponds to transmitted codeword bit x_v. The LLR to be passed from variable-node v to check-node c can be calculated as follows:

$$L\left(\eta_{v,c}^{(t)}\right) = \begin{cases} \sum_{d \in C_v, d \neq c} L\left(\mu_{d,v}^{(t-1)}\right), & \forall c \in C_{\mathrm{LT}}, \\ L\left(x_v^{(t)}\right) + \sum_{d \in C_v, d \neq c} L\left(\mu_{d,v}^{(t-1)}\right), & \forall c \in C_{\mathrm{LDPC}}. \end{cases} \tag{9.49}$$

The variable-node v LLR-update rule is given by

$$L\left(x_v^{(t)}\right) = \sum_{c \in C_v} L\left(\mu_{c,v}^{(t)}\right). \tag{9.50}$$

Finally, the decision step is as follows:

$$\hat{x}_v = \begin{cases} 1, & L\left(x_v^{(t)}\right) < 0, \\ 0, & \text{otherwise}. \end{cases} \tag{9.51}$$

If the syndrome equation $\hat{\boldsymbol{x}}\boldsymbol{H}^T = \mathbf{0}$($\boldsymbol{x}$ denotes the codeword and $\boldsymbol{H}$ is the joint parity-check matrix) is satisfied or the maximum number of iterations is reached, we stop, otherwise, we recalculate (9.48)–(9.51) and check again.

The system model suitable for study of suitability of raptor codes for use in temporally correlated FSO channels is shown in Fig. 9.21. We use the training sequence to determine the CSI and feedback the CSI back to transmitter that can be adapted according to the FSO channel conditions.

For raptor codes performance evaluation the experimental setup shown in Fig. 9.22 is used. The CW laser was used to estimate the channel conditions, the variations due to scintillation are recorded and used to determine the histogram and autocovariance functions. The raptor code is created by concatenating a regular LDPC(495,433) (MacKay) code of rate 0.87 and a LT code. The user bit sequence

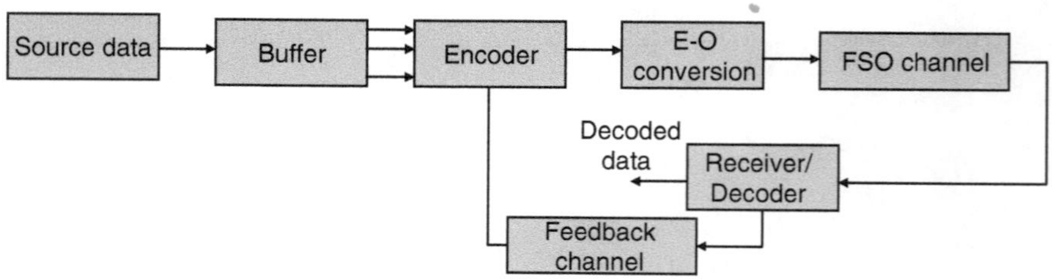

Fig. 9.21 FSO system with feedback channel

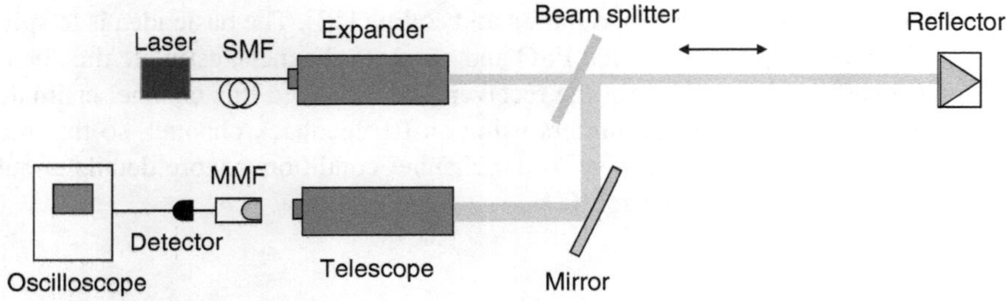

Fig. 9.22 Experimental setup. Total transmission link is 600 m (after [48])

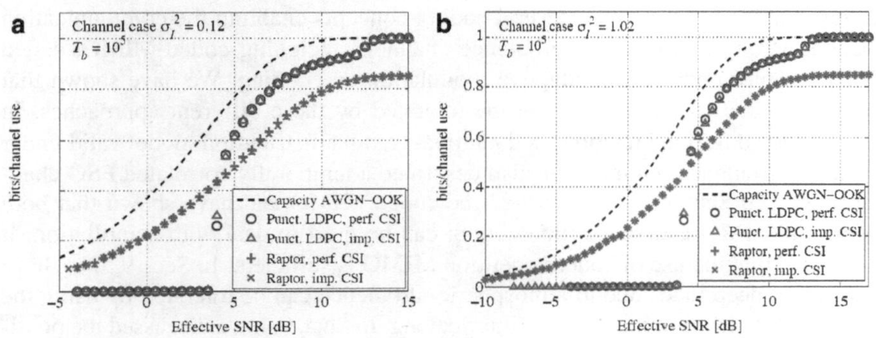

Fig. 9.23 Information rates at 1 Gb/s: (**a**) in the weak turbulence regime, and (**b**) in the strong turbulence regime (after [48])

is fed into a buffer that stores 40 words of $k_C = 433$ bits each, which are encoded using the LDPC code above. The LT number of information bits is therefore determined by $k_{LT} = 19800(= 40 \times 495)$. The results of Monte Carlo integration, used to determine the information rates as we described in [39], are shown in Fig. 9.23. The comparison is performed with respect to corresponding punctured LDPC code. We see that at high SNR the punctured LDPC codes outperform the raptor code described above. However, in the regime of low SNRs the punctured LDPC code cannot be used at all. Another interesting point to notice is that raptor code faces insignificant information rate degradation when the CSI is imperfect. The degradation of information rates in strong turbulence regime is small compared with the weak turbulence regime. Therefore, the raptor codes can be used to enable communication under strong turbulence regime.

Another interesting approach to enable communication under the strong turbulence regime is to use the adaptive modulation and coding [52]. The key idea behind this proposal is to estimate the channel conditions at the receiver side and feed this channel estimate back to the transmitter using an RF feedback channel, so that the transmitter can be adapted relative to the channel conditions. We demonstrate in [52] that deep fades in the order of 35 dB and above can be tolerated with this scheme. To improve further the tolerance to atmospheric turbulence we can use a hybrid

FSO-RF system with adaptive modulation and coding [53]. The basic idea is to split the LDPC-coded sequence between FSO and wireless channels, estimate the channel conditions in both channels at the receiver side, and feed this channel estimate back to both RF and FSO transmitters using an RF feedback channel, so that the transmitters can be adapted relative to the channel conditions. More details about this scheme can be found in Chap. 10.

9.6 Summary

In this chapter, we described several coding concepts enabling the communication over the strong atmospheric turbulence channels, including coded-MIMO, coded OFDM, raptor coding, and adaptive modulation and coding. We have shown that deep fades due to scintillation can be tolerated by these different approaches. In Sect.9.1, we described uncorrelated gamma–gamma distribution model valid under different turbulence regimes. We also described a temporally correlated FSO channel model. In Sect.9.2, we described the coded MIMO and have shown that both repetition MIMO and space-time coding can be used to deal with scintillation. It was found that the use of coded repetition MIMO is sufficient. In Sect. 9.3, we have shown that deep fades due to atmospheric turbulence can be tolerated by using the coded OFDM in combination with interleaving. In Sect. 9.4, we discussed the possibility of implementation of hybrid optical networks, in which the high-speed signal is transmitted over different fiber and FSO sections. We also discussed possible coding techniques suitable for compensation of channel impairments in both FSO and fiber links. In Sect.9.5, we described the use of rate-less codes as a means of enabling the communication over temporal correlated FSO channels. We also discussed the possibility of using the adaptive modulation and coding to tolerate deep fades due to scintillation.

References

1. Andrews LC, Philips RL (1998) Laser beam propagation through random media. SPIE Optical Engineering Press, Bellingham, WA
2. Willebrand H, Ghuman BS (2002) Free-space optics: enabling optical connectivity in today's networks. Sams Publishing, Indianapolis, Indiana
3. Ishimaru A (1978) Wave propagation and scattering in random media, vol 1–2. Academic, NY
4. Andrews LC, Phillips RL, Hopen CY (2001) Laser beam scintillation with applications. SPIE Press, Bellingham, WA
5. Karp S, Gagliardi R, Moran SE, Stotts LB (1988) Optical channels. Plenum, NY
6. Flatté SM, Bracher C, and Wang G-Y (1994) Probability-density functions of irradiance for waves in atmospheric turbulence calculated by numerical simulation. J Opt Soc Am A 11:2080–2092
7. Andrews LC, Phillips RL (1985) I-K distribution as a universal propagation model of laser beams in atmospheric turbulence. J Opt Soc Am 2:160–163

8. Andrews LC, Phillips RL (1986) Mathematical genesis of the I-K distribution for random optical fields. J Opt Soc Am 3:1912–1919
9. Churnside JH, Hill RJ (1987) Probability density of irradiance scintillations for strong path-integrated refractive turbulence. J Opt Soc Am A 4:727–733
10. Hill RJ, Frehlich RG (1997) Probability distribution of irradiance for the onset of strong scintillation. J Opt Soc Am A 14:1530–1540
11. Churnside JH, Frehlich RG (1989) Experimental evaluation of log-normally modulated Rician and IK models of optical scintillation in the atmosphere. J Opt Soc Am A 6:1760–1766
12. Al-Habash MA, Andrews LC, Phillips RL (2001) Mathematical model for the irradiance probability density function of a laser beam propagating through turbulent media. Opt Eng 40:1554–1562
13. Zhu X, Kahn JM (2002) Free-space optical communication through atmospheric turbulence channels. IEEE Trans Commun 50:1293–1300
14. Noll RJ (1976) Zernike polynomials and atmospheric turbulence. J Opt Soc Am 66:207–211
15. Wilson SG, Brandt-Pearce M, Cao Q, Leveque JJH, III (2005) Free-space optical MIMO transmission with Q-ary PPM. IEEE Trans Comm 53:1402–1412
16. Wilson SG, Brandt-Pearce M, Cao Q, Baedke M (2005) Optical repetition MIMO transmission with multipulse PPM. IEEE Selected Areas Comm 23:1901–1910
17. Simon K, Vilnrotter VA (2005) Alamouti-type space-time coding for free-space optical communication with direct detection. IEEE Trans Wireless Comm 4:35–39
18. Djordjevic IB, Vasic B, Neifeld MA (2006) Multilevel coding in free-space optical MIMO transmission with Q-ary PPM over the atmospheric turbulence channel. IEEE Photon Tehnol Lett 18(14):1491–1493
19. Cvijetic N, Wilson SG, Brandt-Pearce M (2007) Receiver optimization in turbulent free-space optical MIMO channels with APDs and Q-ary PPM. IEEE Photon Tehnol Lett 19(2): 1491–1493
20. Tarokh V, Jafarkani H, Calderbank AR (1999) Space-time block codes from orthogonal designs. IEEE Trans Inform Theory 45(5):1456–1467
21. Alamouti SM (1998) A simple transmit diversity technique for wireless communications. IEEE J Sel Areas Commun 16(8):1451–1458
22. Djordjevic IB, Denic S, Anguita J, Vasic B, Neifeld MA (2008) LDPC-coded MIMO optical communication over the atmospheric turbulence channel. IEEE/OSA J Lightwave Technol 26 (5):478–487
23. Denic S, Djordjevic IB, Anguita J, Vasic B, Neifeld MA (2008) Information theoretic limits for free-space optical channels with and without memory. IEEE/OSA J Lightwave Technol 26(19):3376–3384
24. ten Brink S (2000) Designing iterative decoding schemes with the extrinsic information transfer chart. AEÜ Int J Electron Comm 54:389–398
25. Foschini GJ (1996) Layered space-time architecture for wireless communication in a fading environment when using multi-element antennas. Bell Labs Tech J 1(2):41–59
26. Anderson I (1997) Combinatorial designs and tournaments. Oxford University Press, Oxford
27. Djordjevic IB (2007) LDPC-coded MIMO optical communication over the atmospheric turbulence channel using Q-ary pulse-position modulation. Opt Express 15(16):10026–10032
28. Djordjevic IB, Vasic B, Neifeld MA (2006) LDPC coded orthogonal frequency division multiplexing over the atmospheric turbulence channel. In: Proceedings of CLEO/QELS 2006, Paper no. CMDD5. (Invited Paper.), Long Beach, California, 21–26 May 2006
29. Djordjevic IB, Vasic B, Neifeld MA (2007) LDPC coded OFDM over the atmospheric turbulence channel. Optics Express 15:6332–6346
30. Djordjevic IB, Vasic B, Neifeld MA (2007) LDPC-coded OFDM for optical communication systems with direct detection. IEEE/LEOS J Sel Top Quantum Electron 13:1446–1454
31. Prasad R (2004) OFDM for wireless communications systems. Artech House, Boston
32. Shieh W, Djordjevic IB (2009) OFDM for optical communications. Elsevier and Academic Press, Amsterdam
33. Djordjevic IB (2008) LDPC coding for free-space optical communications. Rensselaer Polytechnic Institute, December 2008 (Invited Talk)

34. Djordjevic IB (2007) LDPC-coded optical communication over the atmospheric turbulence channel. In: Proceedings of the 41st Asilomar Conference on Signals, Systems, and Computers, Pacific Grove, California, 4–7 Nov 2007, pp 1903–1909, (Invited Paper)
35. Shieh W, Athaudage C (2006) Coherent optical frequency division multiplexing. Electron Lett 42:587–589
36. Lowery AJ, Du L, Armstrong J Orthogonal frequency division multiplexing for adaptive dispersion compensation in long haul WDM systems. In: Proceedings of The OFC Postdeadline Papers, Paper no. PDP39, Anaheim, CA, March 5–10, 2006
37. Jansen SL, Morita I, Takeda N, Tanaka H (2007) 20-Gb/s OFDM transmission over 4,160-km SSMF enabled by RF-pilot tone phase compensation. In: Proceedings of the OFC/NFOEC 2007 Postdeadline Papers, Paper no. PDP15, Anaheim, CA, USA, 25–29 Mar 2007
38. Djordjevic IB, Arabaci M, Minkov LL (2009) Next generation FEC for high-capacity communication in optical transport networks. IEEE/OSA J Lightw Technol 27(16):3518–3530
39. Anguita JA, Djordjevic IB, Neifeld MA, Vasic BV (2005) Shannon capacities and error–correction codes for optical atmospheric turbulent channels. IEEE/OSA J Opt Netw 4:586–601
40. Proakis JG (2001) Digital communications. McGraw Hill, Boston
41. Goldsmith A (2005) Wireless communications. Cambridge University Press, Cambridge
42. Milenkovic O, Djordjevic IB, Vasic B (2004) Block-circulant low-density parity-check codes for optical communication systems. IEEE J Sel Top Quantum Electron 10:294–299
43. Levinson N (1947) The Wiener RMS error criterion in filter design and prediction. J Math Phys 25:261–278
44. Durbin J (1959) Efficient estimation of parameters in moving-average models. Biometrica 46:306–316
45. Wood ATA, Chan G (1994) Simulation of stationary Gaussian processes in $[0,1]^d$. J Comp Graph Stat 3:409–432
46. Djordjevic IB Coded-OFDM in hybrid optical networks. IET Optoelectronics accepted for publication
47. Djordjevic IB, Xu L, Wang T (2009) Beyond 100 Gb/s optical transmission based on polarization multiplexed coded-OFDM with coherent detection. IEEE/OSA J Opt Commun Netw 1(1):50–56, June 2009
48. Anguita JA, Neifeld MA, Hildner B, Vasic B Rateless coding on experimental temporally correlated FSO channels. IEEE/OSA J Lightwave Technol, accepted for publication
49. Luby M (2002) LT codes. In: Proceeding of the 43rd Annual IEEE Symposium on Foundations of Computer Science, 271–280 November 2002
50. Shokrollahi A (2006) Raptor codes. IEEE Inf Theory 52:2551–2567
51. MacKay DJC (2005) Fountain codes. IEE Proc Commun 152:1062–1068
52. Djordjevic IB (2009) Adaptive modulation and coding for communication over the atmospheric turbulence channels. In: IEEE photonics society summer topicals, Newport Beach, California, Paper no. TuD3. 3, 20–22 July 2009
53. Djordjevic IB, Djordjevic GT (2009) On the communication over strong atmospheric turbulence channels by adaptive modulation and coding. Optics Express 17(20):18250–18262

Chapter 10
Optical Channel Capacity

Given the fact that LDPC-coded turbo equalizer described in Chap. 7 is an excellent candidate to deal with both linear and nonlinear channel impairments, there naturally raises the question about fundamental limits on channel capacity. There have been numerous attempts to determine the channel capacity of a nonlinear fiber optics communication channel [1–13]. The main approach, until recently, was to consider ASE noise as a predominant effect and to observe the fiber nonlinearities as the perturbation of linear case or as the multiplicative noise. In this chapter, which is based on a series of articles by authors [14–20], we describe how to determine the true fiber-optics channel capacity. Because in most of the practical applications the channel input distribution is uniform, we also describe how to determine the achievable information rates (AIRs) or uniform information capacity, which represent the lower bound on channel capacity [14–20]. This method consists of two steps (1) approximating probability density functions (PDFs) for energy of pulses, which is done by (a) evaluation of histograms [14–20], (b) instanton approach [15], or (c) edgeworth expansion [21], and (2) estimating AIRs by applying a method originally proposed by Arnold and Pfitser [22–24]. In most of the publications that are related to the channel capacity, it was assumed that fiber-optics channel is memoryless [1–13], while in reality the fiber-optics channel is the channel with memory [14–20].

An interesting approach to reduce the fiber-optics channel memory, introduced recently [25,26], is the backpropagation method. Namely, in point-to-point links, the receiver knows the dispersion map configuration and can propagate the received signal through the fictive dispersion map with fiber parameters [group velocity dispersion (GVD), second-order GVD, and nonlinearity coefficient] of opposite signs to that used in original map. However, the nonlinear interaction of ASE noise and Kerr nonlinearities cannot be compensated for and someone should use the method introduced in this chapter in information capacity calculation to account for this effect.

This chapter is organized as follows. In Sect. 10.1, we provide the basic information theory for both memoryless channel and channel with memory. In Sect. 10.2, we describe how to calculate the channel capacity of fiber-optic communication systems by modeling the optical channel as discrete dynamic nonlinear channel with memory. In Sect. 10.3, we provide both numerical and experimental results on calculation of channel capacity of fiber-optics communication systems with direct

I. Djordjevic et al., *Coding for Optical Channels*,
DOI 10.1007/978-1-4419-5569-2_10,

detection. The channel capacity of multilevel coherent detection systems is studied in Sect. 10.4. In Sect. 10.5, we report some results on channel capacity of optical OFDM systems for ASE noise-dominated scenario. In Sect. 10.6, we describe how to enable the communication over strong atmospheric turbulence channels by transmitting the encoding sequence partially over free-space optical (FSO) channel and partially over RF channel. We determine the optimum power adaptation that maximizes the total channel capacity. Finally, in Sect. 10.7, we describe how to improve the channel capacity by mode-multiplexing in multimode fiber (MMF) systems.

10.1 Channel Capacity Preliminaries

The concept of channel capacity for discrete memoryless channels has already been introduced in Chap. 4. Here we are concerned with channel capacity of continuous memoryless channels, such as the fiber-optics channel with backpropagation. In Fig. 10.1, we show the memoryless fiber-optics channel model for multicarrier transmission with coherent detection, such as orthogonal frequency division multiplexing (OFDM) [27–30]. The received symbol vector of kth subcarrier in ith OFDM symbol $\boldsymbol{r}_{i,k} = [r_{x,i,k}\ r_{y,i,k}]^{\mathrm{T}}$ can be represented by

$$\boldsymbol{r}_{i,k} = \boldsymbol{H}_k \boldsymbol{s}_{i,k} \mathrm{e}^{\mathrm{j}[\phi_{\mathrm{CD}}(k)+\phi_{\mathrm{T}}-\phi_{\mathrm{LO}}]} + \boldsymbol{n}_{i,k}, \tag{10.1}$$

where $\boldsymbol{s}_{i,k} = [s_{x\,i\,k}\ s_{y\,i\,k}]^{\mathrm{T}}$ denotes the transmitted symbol vector of kth subcarrier in ith OFDM symbol, for both polarizations, $\boldsymbol{n}_{i,k} = [n_{x\,i\,k}\ n_{y\,i\,k}]^{\mathrm{T}}$ denotes the noise vector dominantly determined by the amplified spontaneous emission (ASE) noise; ϕ_{T} and ϕ_{LO} denote the laser phase noise processes of transmitting and local lasers, $\phi_{\mathrm{CD}}(k)$ denotes the phase distortion of kth subcarrier due to chromatic dispersion, and the Jones matrix of kth subcarrier, denoted by $\boldsymbol{H}_k$, is given by

$$\boldsymbol{H}_k = \begin{bmatrix} H_{xx}(k) & H_{xy}(k) \\ H_{yx}(k) & H_{yy}(k) \end{bmatrix}. \tag{10.2}$$

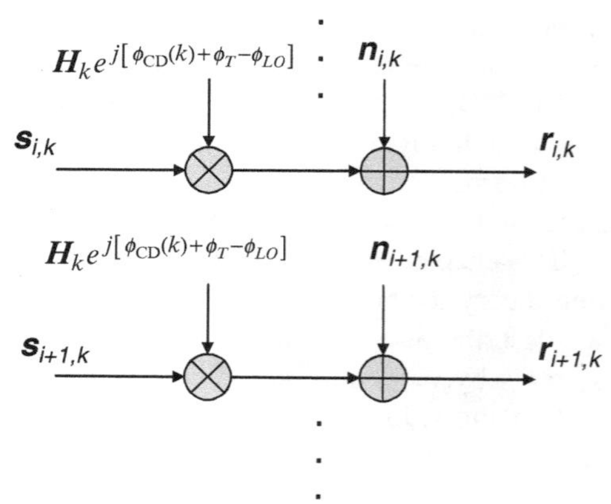

Fig. 10.1 Equivalent OFDM channel model

(In equations above, we use subscript x to denote x-polarization and y to denote y-polarization. We also assume, without loss of generality, that the photodiode responsivity is 1 A/W.) The single carrier system can be considered as special case of multicarrier model with only one subcarrier. Because the noise introduced by spontaneous emission has a flat frequency, as we have shown in Chap. 3, characterized by Gaussian PDF with (double-sided) power spectral density (PSD) in one state of polarization is given by

$$S_{\mathrm{sp}}(\nu) = (G-1)F_{\mathrm{no}}h\nu/2, \tag{10.3}$$

where G is the amplifier gain and $h\nu$ is the photon energy; we will introduce the concept of differential entropy associated with a multidimensional continuous random variable and determine the information capacity of additive Gaussian noise.

Let X denote an n-dimensional multivariate $X = [X_1, X_2, \ldots, X_n]$ with a PDF $p_1(x_1, x_2, \ldots, x_n)$, representing the channel input. The corresponding *differential entropy* is defined by [31, 32]

$$\begin{aligned} &h(X_1, X_2, \ldots, X_n) \\ &= \underbrace{-\int_{-\infty}^{\infty}\int_{-\infty}^{\infty}\ldots\int_{-\infty}^{\infty} p_1(x_1, x_2, \ldots, x_n)\log p_1\left(x_1, x_2, \ldots, x_n\right)}_{n} \mathrm{d}x_1\mathrm{d}x_2\cdots\mathrm{d}x_n \\ &= E\left[-\log p_1(x_1, x_2, \ldots, x_n)\right], \end{aligned} \tag{10.4}$$

where we use $E[\ ]$ to denote the expectation operator. (In this chapter, we will use logarithms to the base 2.) In order to simplify explanation and save some space, we will use the following symbolic notation of previous equation due to Reza [32]: $h(\tilde{X}) = E\left[-\log p_1(\tilde{X})\right]$. In similar fashion, the channel output can be represented as an m-dimensional random variable $Y = [Y_1, Y_2, \ldots, Y_m]$ with a PDF $p_2(y_1, y_2, \ldots, y_m)$, and corresponding differential entropy is defined by

$$\begin{aligned} &h\left(Y_1, Y_2, \ldots, Y_n\right) \\ &= \underbrace{-\int_{-\infty}^{\infty}\int_{-\infty}^{\infty}\int_{-\infty}^{\infty} p_2\left(y_1, y_2, \ldots, y_m\right)\log p_1\left(y_1, y_2, \ldots, y_m\right)}_{m} \mathrm{d}y_1\mathrm{d}y_2\cdots\mathrm{d}y_m \\ &= E\left[-\log p_2\left(y_1, y_2, \ldots, y_n\right)\right]. \end{aligned} \tag{10.5}$$

In compact form, the differential entropy of output can be written as $h(\tilde{Y}) = E[-\log p_1(\tilde{Y})]$. To account for the channel distortions and ASE noise influence, we can observe the corresponding conditional and joint PDFs:

$$\begin{aligned} &P\left(y_1 < Y_1 < y_1 + \mathrm{d}y_1, \ldots, y_m < Y_m < y_m + \mathrm{d}y_m | X_1 \right. \\ &\quad \left. = x_1, \ldots, X_n = x_n\right) = p(\tilde{y}|\tilde{x})\mathrm{d}\tilde{y} \\ &P\left(y_1 < Y_1 < y_1 + \mathrm{d}y_1, \ldots, y_m < Y_m < y_m + \mathrm{d}y_m \cap x_1 < X_1 < x_1 \right. \\ &\quad \left. + \mathrm{d}x_1, \ldots, x_n < X_n < x_n + \mathrm{d}x_n\right) = p(\tilde{x}, \tilde{y})\mathrm{d}\tilde{x}\,\mathrm{d}\tilde{y} \end{aligned} \tag{10.6}$$

The mutual information (also known as information rate) can be written in compact form as follows [32]:

$$I(\tilde{X};\tilde{Y}) = E\left[\log \frac{p(\tilde{X},\tilde{Y})}{p(\tilde{X})P(\tilde{Y})}\right]. \tag{10.7}$$

Note that various differential entropies $h(X)$, $h(Y)$, $h(Y|X)$ do not have direct interpretation as long as the information processed in the channel is concerned, compared to their discrete counterparts. Some authors, such as Gallager [33], prefer to define the mutual information directly by (10.7), without considering the differential entropies at all. The mutual information, however, has the information-theoretic meaning and represents the average information processed in the channel (or amount of information conveyed by the channel). The mutual information has the following important properties [31, 32] (1) the mutual information is symmetric: $I(X;Y) = I(Y;X)$; (2) it is a nonnegative; (3) the mutual information is finite; (4) mutual information is invariant under linear transformation; (5) it can be expressed in terms of the differential entropy of channel output by $I(X;Y) = h(Y) - h(Y|X)$, and (6) it is related to the channel input differential entropy by $I(X;Y) = h(X) - h(X|Y)$. The information capacity can be obtained by maximization of (7) under all possible input distributions:

$$C = \max I(\tilde{X};\tilde{Y}). \tag{10.8}$$

Let us now determine the mutual information of two random vectors $X = [X_1, X_2, \ldots, X_n]$ and $Y = [Y_1, Y_2, \ldots, Y_m]$, which are normally distributed. Let $Z = [X;Y]$ be the random vector describing the joint behavior. Without loss of generality we further assume: $\overline{X}_k = 0\,\forall\, k$ and $\overline{Y}_k = 0\,\forall\, k$. The corresponding PDF for X, Y, and Z are given, respectively, as [32]

$$p_1(\tilde{x}) = \frac{1}{(2\pi)^{n/2}(\det A)^{1/2}} \exp(-0.5(A^{-1}\tilde{x},\tilde{x})), \quad A = [a_{ij}],$$

$$a_{ij} = \int x_i x_j p_1(\tilde{x})\mathrm{d}\tilde{x}, \tag{10.9a}$$

$$p_2(\tilde{y}) = \frac{1}{(2\pi)^{n/2}(\det B)^{1/2}} \exp\left(-0.5\left(B^{-1}\tilde{y},\tilde{y}\right)\right), \quad B = [b_{ij}],$$

$$b_{ij} = \int y_i y_j p_2(\tilde{y})\mathrm{d}\tilde{y}, \tag{10.9b}$$

$$p_3(\tilde{z}) = \frac{1}{(2\pi)^{(n+m)/2}(\det C)^{1/2}} \exp\left(-0.5\left(C^{-1}\tilde{z},\tilde{z}\right)\right), \quad C = [c_{ij}],$$

$$c_{ij} = \int z_i z_j p_3(\tilde{z})\mathrm{d}\tilde{z}. \tag{10.9c}$$

By substitution of (10.9) into (10.7) we obtain [32]:

$$I(X;Y) = \frac{1}{2}\log\frac{\det A \det B}{\det C}. \tag{10.10}$$

The mutual information between two Gaussian random vectors can also be expressed in terms of their correlation coefficients [32]:

$$I(X;Y) = -\frac{1}{2}\log\left[\left(1-\rho_1^2\right)\cdots\left(1-\rho_l^2\right)\right], \quad l = \min(m,n), \tag{10.11}$$

where ρ is the correlation coefficient between X_j and Y_j.

In order to obtain the information capacity for additive Gaussian noise, we make the following assumptions (1) the input X, output Y, and noise Z are n-dimensional random variables; (2) $\overline{X}_k = 0, \overline{X_k^2} = \sigma_{x_k}^2\ \forall k$ and $\overline{Z}_k = 0, \overline{Z_k^2} = \sigma_{z_k}^2\ \forall k$; and (3) the noise is additive: $Y = X + Z$. Because

$$p_X(\tilde{y}|\tilde{z}) = p_X(\tilde{x}+\tilde{z}|\tilde{z}) = \prod_{k=1}^{n}\left[\frac{1}{(2\pi)^{1/2}\sigma_{z_k}}\mathrm{e}^{-z_k^2/2\sigma_{z_k}^2}\right] = p(\tilde{z}), \tag{10.12}$$

the conditional differential entropy can be obtained as

$$H(Y|X) = H(Z) = -\int_{-\infty}^{\infty} p(\tilde{z})\log p(\tilde{z})\mathrm{d}\tilde{z}. \tag{10.13}$$

The mutual information is then

$$I(X;Y) = h(Y) - h(Y|X) = h(Y) - h(z) = h(Y) - \frac{1}{2}\sum_{k=1}^{n}\log 2\pi e\sigma_{z_k}^2. \tag{10.14}$$

The information capacity, expressed in bits per channel use, is therefore obtained by maximizing $h(Y)$. Because the distribution maximizing the differential entropy is Gaussian, the information capacity is obtained as

$$\begin{aligned} C(X;Y) &= \frac{1}{2}\sum_{k=1}^{n}\log 2\pi e\sigma_{y_k}^2 - \frac{1}{2}\sum_{k=1}^{n}\log 2\pi e\sigma_{z_k}^2 = \frac{1}{2}\sum_{k=1}^{n}\log\left(\frac{\sigma_{y_k}^2}{\sigma_{z_k}^2}\right) \\ &= \frac{1}{2}\sum_{k=1}^{n}\log\left(\frac{\sigma_{x_k}^2+\sigma_{z_k}^2}{\sigma_{z_k}^2}\right) = \frac{1}{2}\sum_{k=1}^{n}\log\left(1+\frac{\sigma_x^2}{\sigma_z^2}\right), \end{aligned} \tag{10.15}$$

For $\sigma_{x_k}^2 = \sigma_x^2, \sigma_{z_k}^2 = \sigma_z^2$ we obtain the following expression for information capacity:

$$C(X;Y) = \frac{n}{2}\log\left(1+\frac{\sigma_x^2}{\sigma_z^2}\right), \tag{10.16}$$

where σ_x^2/σ_z^2 is the signal-to-noise ratio (SNR).

The expression (10.16) represents the maximum amount of information that can be transmitted per symbol. From practical point of view, it is important to determine the amount of information conveyed by the channel per second, that is, the information capacity per unit time, also known as channel capacity. For band-limited channels and Nyquist signaling, there will be $2W$ samples per second (W is the channel bandwidth) and the corresponding channel capacity is

$$C = W \log\left(1 + \frac{P}{N_0 W}\right) \text{[bits/s]}, \tag{10.17}$$

where P is the average transmitted power and $N_0/2$ is the noise PSD. This is the most famous Shannon's result in information theory [34].

The channel capacity per single polarization of the OFDM system shown in Fig. 10.1 with N independent subcarriers of bandwidth B_N can be evaluated by

$$C = \max_{P_i:\sum P_i = P} \sum_{i=0}^{N-1} W_N \log_2(1+\gamma_i), \gamma_i = |H_i|^2 P_i / N_0 W_N, \tag{10.18}$$

where P_i is the power allocated to ith subcarrier, γ_i is the SNR of ith subcarrier, and H_i is the fiber transfer function amplitude of ith subcarrier. It can be shown by using the Lagrangian method that the optimum power allocation policy is the water-filling over frequency [31]:

$$\frac{P_i}{P} = \begin{cases} 1/\gamma_{\text{tsh}} - 1/\gamma_i, & \gamma_i \geq \gamma_{\text{tsh}} \\ 0, & \text{otherwise} \end{cases}, \tag{10.19}$$

where γ_i is the SNR of ith subcarrier and γ_{tsh} is the threshold SNR. By substituting (10.19) into (10.18) the following channel capacity expression is obtained:

$$C = \sum_{i:\gamma_i > \gamma_{\text{tsh}}} B_N \log_2\left(\frac{\gamma_i}{\gamma_{\text{tsh}}}\right). \tag{10.20}$$

The ith subcarrier is used when corresponding SNR is above the threshold. The number of bits per ith subcarrier is determined by $m_i = \lfloor B_N \log_2(\gamma_i/\gamma_{\text{tsh}}) \rfloor$, where $\lfloor\rfloor$ denotes the largest integer smaller than enclosed number.

In digital optical communications, we are generally confronted with the time series. Let the input and output alphabets of the channel be finite and be denoted by $\{A\}$ and $\{B\}$, respectively; and the channel input and output are denoted by X and Y. For memoryless channels, the noise behavior is generally captured by a conditional probability matrix $P\{b_j|a_k\}$ for all $b_j \in B$ and $a_j \in A$. For the channels with finite memory, such as the optical channel, the transition probability is dependent on the transmitted sequences up to the certain prior finite instance of time. For example, for channel described by Markov process the transition matrix has the following form $P\{Y_k = b|\ldots, X_{-1}, X_0, X_1, \ldots, X_k\} = P\{Y_k = b|X_k\}$. We are interested into more general description, which is due to McMillan [35] and Khinchin [36] (see also [32]). Let us consider a member of input ensemble

x and its corresponding channel output y: $\{X\} = \{\ldots, x_{-2}, x_{-1}, x_0, x_1, \ldots\}$, $\{Y\} = \{\ldots, y_{-2}, y_{-1}, y_0, y_1, \ldots\}$. Let X' denote all possible input sequences and Y' denote all possible output sequences. By fixing a particular symbol at specific location, we obtain so-called *cylinder* [32]. For example, cylinder $x^{4,1}$ is obtained by fixing the symbol a_1 at position $x_4 : x^{4,1} = \ldots, x_{-1}, x_0, x_1, x_2, x_3, a_1, x_5, \ldots$. The output cylinder $y^{1,2}$ is obtained by fixing the output symbol b_2 at position 1: $y^{1,2} = \ldots, y_{-1}, y_0, b_2, y_2, y_3, \ldots$. To characterize the channel we have to determine the following transition probability $P(y^{1,2}|x^{4,1})$, that is the probability that cylinder $y^{1,2}$ was received given that cylinder $x^{4,1}$ was transmitted. Therefore, for all possible input cylinders $S_A \subset X'$ we have to determine the probability that cylinder $S_B \subset Y'$ was received given that S_A was transmitted. The channel is completely specified by (1) input alphabet A, (2) output alphabet B, and (3) transition probabilities $P\{S_B|S_A\} = v_x$ for all $S_A \in X'$ and $S_B \in Y'$. Thus the channel is specified by the triplet: $[A, v_x, B]$. If the transition probabilities are invariant with respect to time shift T, that is, $v_{Tx}(TS) = v_x(S)$, then the channel is said to be *stationary*. If the distribution of Y_k depends only on the statistical properties of the sequence $\ldots, x_{k-1}, x_k$, we say that the channel is without *anticipation*. If furthermore the distribution of Y_k depends only $x_{k-m}, \ldots, x_k$ we say that channel has the finite *memory* of m units.

The source and channel may be described as a new source $[C, \omega]$ with C being the product of input A and output B alphabets, namely $C = A \times B$, and ω is a corresponding probability measure. The joint probability of symbol $(x, y) \in C$, where $x \in A$ and $y \in B$, is obtained as the product of marginal and conditional probabilities: $P(x \cap y) = P\{x\}P\{y|x\}$.

Let us further assume that both source and channel are stationary. The following description due to Khinchin [32, 36] is useful in describing the concatenation of a stationary source and a stationary channel.

1. If the source $[A,\mu]$ (μ is the probability measure of the source alphabet) and the channel $[A, v_x, B]$ are stationary, the product source $[C, \omega]$ will also be stationary.
2. Each stationary source has an entropy, and therefore $[A, \mu]$, $[B, \eta]$ (η is the probability measure of the output alphabet), and $[C, \omega]$ each have the finite entropies.
3. These entropies can be determined for all n-term sequences $x_0, x_1, \ldots, x_{n-1}$ emitted by the source and transmitted over the channel as follows [32]:

$$\begin{aligned}
&H_n(X) \leftarrow \{x_0, x_1, \ldots, x_{n-1}\} \\
&H_n(Y) \leftarrow \{y_0, y_1, \ldots, y_{n-1}\} \\
&H_n(X, Y) \leftarrow \{(x_0, y_0), (x_1, y_1), \ldots, (x_{n-1}, y_{n-1})\} \\
&H_n(Y|X) \leftarrow \{(x_0|Y), (x_1|Y), \ldots, (x_{n-1}|Y)\} \\
&H_n(X|Y) \leftarrow \{(X|y_0), (X|y_1), \ldots, (X|y_{n-1})\}
\end{aligned} \tag{10.21}$$

It can be shown that the following is valid:

$$H_n(X, Y) = H_n(X) + H_n(Y|X); \quad H_n(X, Y) = H_n(Y) + H_n(X|Y). \tag{10.22}$$

Equation (10.22) can be rewritten in terms of entropies per symbol:

$$\frac{1}{n}H_n(X,Y) = \frac{1}{n}H_n(X)+\frac{1}{n}H_n(Y|X); \quad \frac{1}{n}H_n(X,Y) = \frac{1}{n}H_n(Y)+\frac{1}{n}H_n(X|Y). \tag{10.23}$$

For sufficiently long sequences the following channel entropies exist:

$$\lim_{n\to\infty}\frac{1}{n}H_n(X,Y) = H(X,Y); \quad \lim_{n\to\infty}\frac{1}{n}H_n(X) = H(X); \quad \lim_{n\to\infty}\frac{1}{n}H_n(Y) = H(Y);$$
$$\lim_{n\to\infty}\frac{1}{n}H_n(X|Y) = H(X|Y); \quad \lim_{n\to\infty}\frac{1}{n}H_n(Y|X) = H(Y|X). \tag{10.24}$$

The mutual information exists and it is defined as

$$I(X,Y) = H(X) + H(Y) - H(X,Y). \tag{10.25}$$

The *stationary information capacity* of the channel is obtained by maximization of mutual information over all possible information sources:

$$C(X,Y) = \max I(X,Y). \tag{10.26}$$

Equipped with this knowledge, in Sect. 10.2, we will discuss how to determine the information capacity of fiber-optics channel with memory.

10.2 Calculation of Information Capacity by Forward Recursion of BCJR Algorithm

Recent progress in DWDM transmission technology has lead to a fundamental and a very difficult question: "What is the Shannon's capacity of a nonlinear fiber optics communication channel?" There have been numerous attempts to tackle this problem [1–13]. The common approach, until recently, was to consider the ASE from in-line amplifiers as a predominant effect and include the effect of nonlinearities in an approximate fashion. The fiber nonlinearities are considered either as (1) the perturbation of a linear case [4, 12] or as (2) the multiplicative noise [6].

Although the perturbative methods, in general, may yield to reliable results in the domain of validity [4], they are applicable in the regime of relatively small nonlinearities, which is not applicable for high-speed transmission (at 40 Gb/s or higher) systems. Mitra and Stark [6] treat a nonlinear noisy channel as a linear one with an effective nonlinear noise. However, as indicated in [2], such an approximation needs to be carefully justified for each particular transmission system. In dispersion free transmission [7, 8], the nonlinear Schrödinger equation [37] can be solved analytically, but such a result is only of academic interest [7]. In [3], Tang determined the channel capacity of a multispan DWMD system employing dispersive nonlinear optical fibers and an ideal coherent optical receiver. The results obtained in [3] are based on solving the NLSE by Voltera series expansion up to the first order. Such a method is valid in systems for which the maximum nonlinear rotation is small compared to 2π [3]. The channel capacity is determined using the Pinsker's formula [38], which may lead to incorrect conclusions, as it was shown in [2], especially for high SNRs. The statistics of optical transmission in a noisy nonlinear channel with weak dispersion management and zero average dispersion is considered in [2], and the lower bounds for channel capacity are determined (although numerical results are not reported). However, this method is applicable only for weak dispersion management systems with zero average dispersion, as described in [39]. Some papers, [13], completely ignore the influence of dispersion and nonlinearities. The spectral efficiency limits in DWDM systems with coherent detection in nonlinear regime limited independently by cross-phase modulation or four-wave mixing are reported in [1]. As opposed to 10-Gb/s systems where the major nonlinearities are due to interchannel effects, in 40 Gb/s and above the major nonlinearities are due to intrachannel nonlinearities, such as intrachannel FWM, intrachannel XPM, and intrachannel SPM. Therefore, the weak nonlinearity assumption is not applicable, and the results on channel capacity are valid only if the combined effects of ASE noise, Kerr nonlinearities, dispersion and filtering effects are taken into account in calculations.

To calculate the channel capacity, we model the whole transmission system as the dynamical nonlinear ISI channel with memory, in which m previous and next m symbols influence the observed symbol. The optical communication system is characterized by the conditional PDF of the output complex vector of samples $\boldsymbol{y} = (y_1, \ldots, y_n, \ldots)$, where $y_i = (y_{i,\mathrm{I}}(x), y_{i,\mathrm{Q}}(x), y_{i,\mathrm{I}}(y), y_{i,\mathrm{Q}}(y)) \in \boldsymbol{Y}$ ($\boldsymbol{Y}$ represents the set of all possible output samples), given the source sequence of M-ary symbols $\boldsymbol{x} = (x_1, \ldots, x_n, \ldots)$, $x_i \in \boldsymbol{X} = \{0, 1, \ldots, M-1\}$. The $y_{i,I}(x)[y_{i,I}(y)]$ corresponds to the in-phase channel sample in x-polarization (y-polarization), and the $y_{i,\mathrm{Q}}(x)[y_{i,\mathrm{Q}}(y)]$ represents the quadrature channel sample in x-polarization (y-polarization).

An example of dynamical channel description by means of trellis diagram is shown in Fig. 10.2 for 4-level modulation formats (such as QPSK). This dynamical trellis is uniquely defined by the following triplet: the previous state, the next state, and the channel output. The state in the trellis is defined as $\boldsymbol{s}_j = (x_{j-m}, x_{j-m+1}, \ldots, x_j, x_{j+1}, \ldots, x_{j+m}) = \boldsymbol{x}[j-m, j+m]$, where x_k denotes the index of the symbol from the following set of possible indices $\boldsymbol{X} = \{0, 1, \ldots, M-1\}$. Every symbol carries $l = \log_2 M$ bits, using the appropriate mapping rule (natural, Gray, anti-Gray, etc.) The memory of the state is equal to $2m+1$, with $2m$ being the number of *symbols* that influence the observed symbol from both sides. The trellis has $M^{2m+1} = 64$ states $(\boldsymbol{s}_0, \boldsymbol{s}_1, \ldots, \boldsymbol{s}_{63})$, each of which corresponds to a different three-symbol patterns. The state index is determined by considering $(2m+1)$ symbols as digits in numerical system with the base M. For example, in Fig. 10.2, the quaternary numerical system (with the base 4) is used. (In this system 18 is represented by $(102)_4$.) The left column in the dynamic trellis represents the current states and the right column denotes the terminal states. The

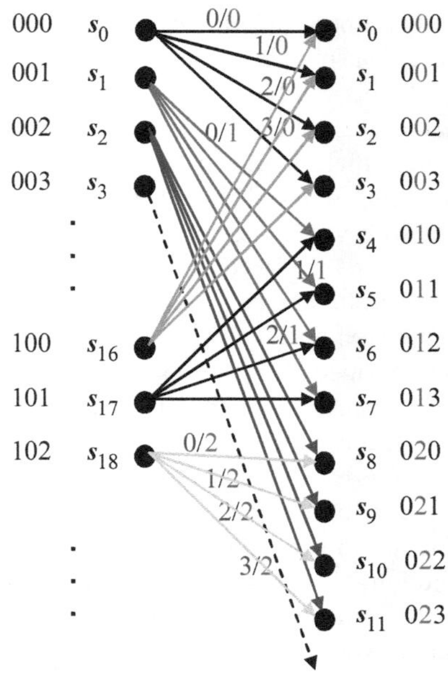

Fig. 10.2 A portion of trellis for four-level BCJR equalizer with memory $2m + 1 = 3$

branches are labeled by two symbols, the input symbol and the output symbol is the central symbol of terminal state. Therefore, the main difference with respect to other channels (such as magnetic recording channel), where the current symbol is affected only by the previous symbols, is that in optical channel current symbol is affected by both previous and incoming symbols. For the complete description of the dynamical trellis, the transition PDFs $p(y_j|x_j) = p(y_j|s)$, $s \in \boldsymbol{S}$ are needed; where $\boldsymbol{S}$ is the set of states in the trellis. The conditional PDFs can be determined by using *instanton-Edgeworth expansion* method, which we proposed in [21].

The information rate, expressed in bits/channel use, can be calculated by [31,32]:

$$I(\boldsymbol{Y};\boldsymbol{X}) = H(\boldsymbol{Y}) - H(\boldsymbol{Y}|\boldsymbol{X}), \tag{10.27}$$

where $H(\boldsymbol{U}) = E(\log_2 P(\boldsymbol{U}))$ denotes the entropy of a random variable $\boldsymbol{U}$ and $E(\cdot)$ denotes the mathematical expectation operator. By using the Shannon–McMillan–Brieman theorem that states [31,32]:

$$E(\log_2 P(\boldsymbol{Y})) = \lim_{n\to\infty} (1/n) \log_2 P(\boldsymbol{y}[1,n]), \tag{10.28}$$

the information rate can be determined by calculating $\log_2(P(\boldsymbol{y}[1,n]))$, by propagating the sufficiently long source sequence. Our model is a particular instance of discrete stationary channel model described in Sect. 10.1. The fiber-optics channel model described in this section can be considered stationary for sufficiently long information sequences and sufficiently long channel memories assumptions. Note that other papers on channel capacity [1–13, 25], also consider channel as stationary, because all calculations of channel capacity assume that corresponding transition PDFs are independent of initial moment. By substituting (10.28) into (10.27), we obtain the following expression suitable for practical calculation of independent identically distributed (i.i.d.) information capacity

$$I(Y;X) = \lim_{n\to\infty} \frac{1}{n}\left[\sum_{i=1}^{n} \log_2 P(y_i|y[1,i-1],x[1,n]) - \sum_{i=1}^{n} \log_2 P(y_i|y[1,i-1])\right]. \tag{10.29}$$

Based on Sect. 10.2, the conditional PDF $P(y_i|\boldsymbol{y}[1,i-1],\boldsymbol{x}[1,n])$ is related to $P(y_i|\boldsymbol{x}[i-m,i+m]) = P(y_i|\boldsymbol{s})$. Therefore, the first term in (10.29) can be straightforwardly estimated from transition PDFs $P(y_i|\boldsymbol{s})$. To calculate $\log_2 P(y_i|\boldsymbol{y}[1,i-1])$, we use the forward recursion of the multilevel BCJR algorithm that we described in Chap. 7, wherein the forward metric and backward metric are modified as follows:

$$\alpha_j(s) = \max_{s'}{}^{*}\left[\alpha_{j-1}(s') + \gamma_j(s',s) - \log_2 M\right] \quad \gamma_j(s',s) = \log\left[p(y_j|x[j-m,j+m])\right]. \tag{10.30}$$

In (10.23) $\alpha_j(\boldsymbol{s}) = \log\{p(\boldsymbol{s}_j = \boldsymbol{s}, \boldsymbol{y}[1,j])\}(j = 1,2,\ldots,n)$ denotes the forward metric and $\gamma_j(\boldsymbol{s}',\boldsymbol{s}) = \log[p(\boldsymbol{s}_j = \boldsymbol{s}, y_j, \boldsymbol{s}_{j-1} = \boldsymbol{s}')]$ is the branch metric. The ith term $\log_2 P(y_i|\boldsymbol{y}[1,i-1])$ can be calculated iteratively:

$$\log_2 P(y_i|y[1,i-1]) = \max_{s}{}^{*}\, \alpha_i(s), \tag{10.31}$$

where max*-operator was applied for all $\boldsymbol{s} \in \boldsymbol{S}$ ($\boldsymbol{S}$ denotes the set of states in the trellis shown in Fig. 10.2).

Information capacity is defined as

$$C = \max I(Y;X), \tag{10.32}$$

where the maximization is performed over all possible input distributions. Because the optical channel has the memory, it is natural to assume that optimum input distribution will be with memory as well. By considering the stationary input distributions of the form $p(x_i|x_{i-1},x_{i-2},\ldots) = p(x_i|x_{i-1},x_{i-2},\ldots,x_{i-k})$, we can determine the transition probabilities of corresponding Markov model that maximizes the information rate in (10.29) by nonlinear numerical optimization [40, 41].

This method is applicable to both memoryless channels and for channels with memory. In Fig. 10.3, we report the information capacities for different signal constellation sizes and two types of QAM constellations: square QAM and star QAM [42] (see also [25]), by observing a linear channel model. We also provide the information capacity for an optimum signal constellation, based on so-called iterative polarization quantization (IPQ) that we introduced in [43]. We can see that information capacity can be closely approached even with the independent identically distributed (i.i.d. or IID) information source providing that constellation size is suf-

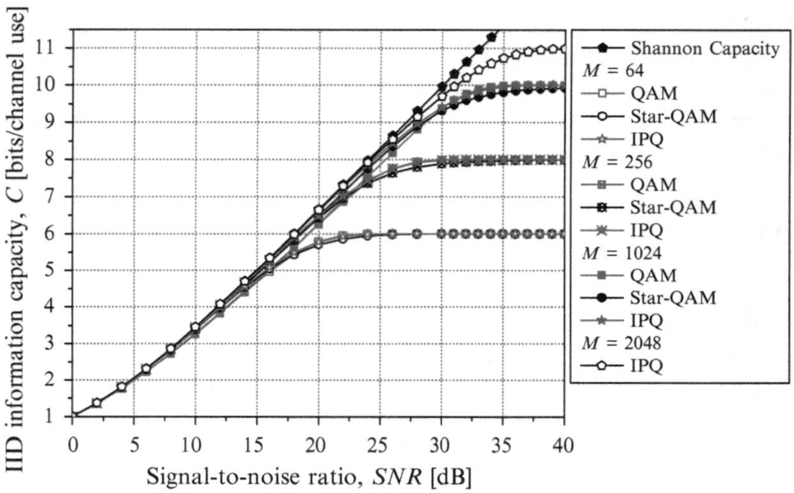

Fig. 10.3 i.i.d. information capacities for linear channel model and different signal constellation sizes. (64-star QAM contains 8 rings with 8 points each, 256-star QAM contains 16 rings with 16 points, and 1,024-star QAM contains 16 rings with 64 points.) SNR is defined as E_s/N_0, where E_s is the symbol energy and N_0 is the power spectral density

ficiently large. It is interesting to note that star QAM outperforms the corresponding square QAM for low and medium SNR, while for high SNRs square QAM outperforms star QAM. The IPQ significantly outperforms both square QAM and star QAM.

Given this description of i.i.d. information capacity calculation for fiber-optics channel in next two sections we study the information capacity of fiber-optics communication systems with direct detection (Sect. 10.3) and coherent detection (Sect. 10.4).

10.3 Information Capacity of Fiber-Optics Communication Systems with Direct Detection

In Fig. 10.4a, we show the i.i.d. channel capacity against optical SNR for different state memory assumptions, in the presence of PMD, for DGD of 100 ps, observing the RZ transmission at 10 Gb/s. As expected, the BCJR equalizer of memory $m = 1$ is sufficient to compensate for this level of DGD, while the memoryless assumption ($m = 0$) leads to significant i.i.d. information capacity loss.

In Fig. 10.4b, we show the i.i.d. channel capacity against optical SNR for different memory assumptions in BCJR equalizer, in the presence of residual chromatic dispersion of 11,200 ps/nm (which corresponds to 700 km of SMF with dispersion and dispersion slope being 16 ps/(nm km) and 0.08 ps/(nm^2 km), respectively). The NRZ transmission system operating at 10 Gb/s is observed. The BCJR equalizer

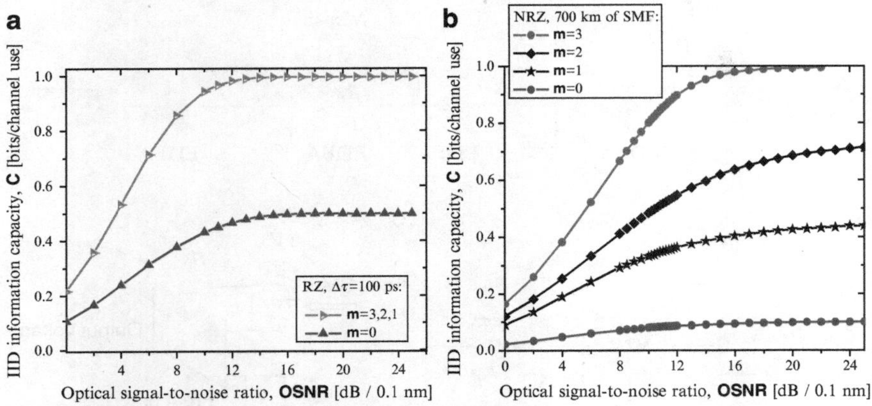

Fig. 10.4 (**a**) i.i.d. information capacity for DGD of 100 ps assuming RZ transmission at 10-Gb/s, (**b**) i.i.d. information capacity after 700 km of SMF assuming NRZ transmission at 10-Gb/s

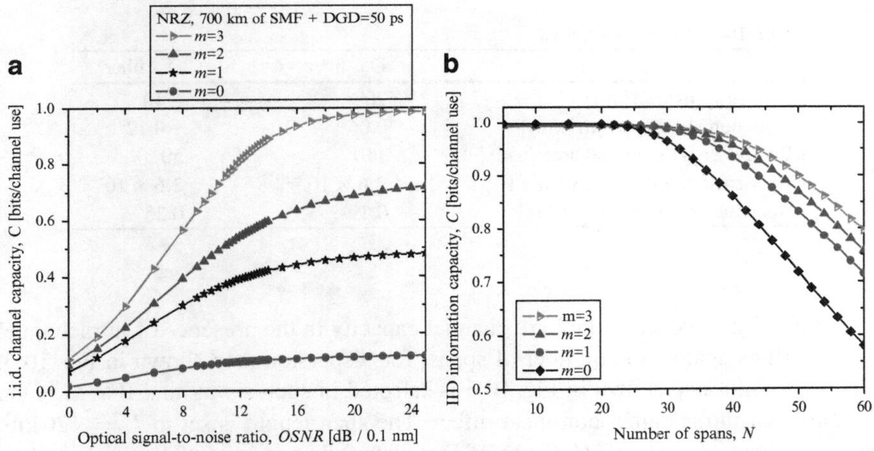

Fig. 10.5 (**a**) i.i.d. information capacity after 700 km of SMF and for DGD of 50 ps assuming NRZ transmission at 10 Gb/s, (**b**) influence of memory effects on i.i.d. information capacity vs. the number of spans for dispersion map shown in Fig. 10.6, assuming RZ transmission at 40 Gb/s

memory $m = 3$ is sufficient to compensate for this level of residual chromatic dispersion, while for the lower memories this effect cannot be eliminated at all by any equalization scheme alone, and FEC is unavoidable. To achieve the i.i.d. channel capacity of 1, for lower optical SNR values, we have to increase the state memory.

In Fig. 10.5a, we show the i.i.d. channel capacity against optical SNR for different state memories, in the presence of both residual chromatic dispersion (11,200 ps/nm) and PMD with DGD of 50 ps. The NRZ transmission system operating at 10 Gb/s is observed as well. For the memory $m = 3$, the unity information rate is achieved for higher OSNR values (about 4 dB higher).

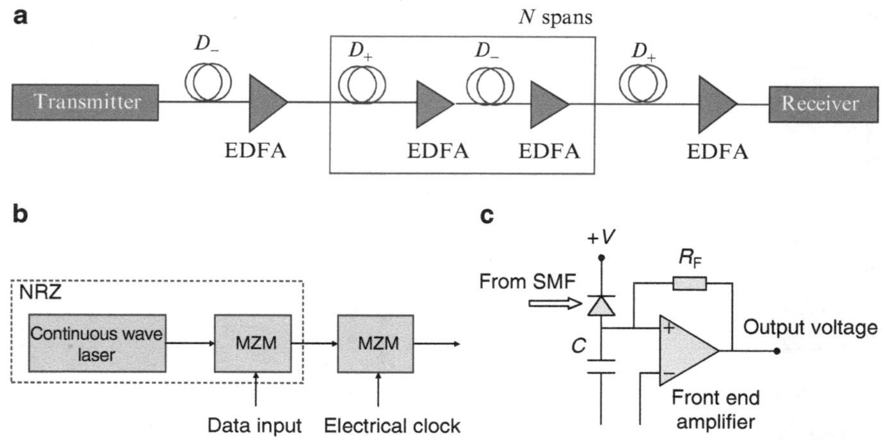

Fig. 10.6 (**a**) Dispersion map under study, (**b**) transmitter configuration, (**c**) receiver configuration

Table 10.1 Fiber parameters

	D_+ fiber	D_- fiber
Dispersion [ps/(nm km)]	20	−40
Dispersion slope [ps/(nm^2 km)]	0.06	−0.12
Effective cross-sectional area [μm^2]	110	50
Nonlinear refractive index [m^2/ W]	2.6×10^{-20}	2.6×10^{-20}
Attenuation coefficient [dB/km]	0.19	0.25

In Fig. 10.5b, we show the i.i.d. channel capacity in the presence of intrachannel nonlinearities against the number of spans, for dispersion map shown in Fig. 10.6. The dispersion map shown in Fig. 10.6 is selected in such a way that IFWM is the predominant intrachannel nonlinear effect. The span length is set to $L = 120$ km, and each span consists of $2L/3$ km of D_+ fiber followed by $L/3$ km of D_- fiber. Precompensation of $-1,600$ ps/nm and corresponding postcompensation are also applied. The parameters of D_+ and D_- fibers, used in simulations, are given in Table 10.1. RZ modulation format of a duty cycle of 33% is observed, the extinction ratio is 14 dB, and the launched power is set to 0 dBm. EDFAs with noise figure of 6 dB are deployed after every fiber section, the bandwidth of optical filter (modeled as super-Gaussian filter of eight order) is set to $3R_1$, and the bandwidth of electrical filter (modeled as Gaussian filter) to $0.7R_1$, with R_1 being the line rate [defined as the bit rate (40 Gb/s) divided by a code rate]. The line rate (defined above) is appropriately chosen so that the effective information rate is 40 Gb/s.

Figure 10.7 shows the conditional PDFs $p(y_j|\boldsymbol{x}[j-m, j+m])$ obtained for dispersion map described above for $m = 2$. As expected, by increasing the number of spans, the ghost pulse at the central bit position for the state $\boldsymbol{s} =$ "11011" grows, hence shifting the mean of the PDF to the right. After certain number of spans,

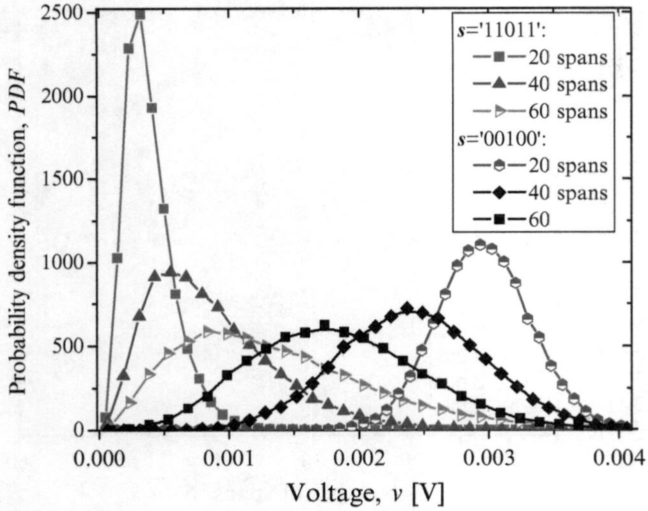

Fig. 10.7 Conditional PDFs $p(y_j|\boldsymbol{x}[j-m, j+m])$ against different number of spans estimated from simulator for RZ OOK

the mean of PDF exceeds the decision threshold resulting in an error. On the other hand, the mean of PDF for an isolated one (in state $\boldsymbol{s}$ = "00100") shifts to the left as the number of spans increases, suggesting that assumed memory $2m + 1 = 5$ is not sufficiently large to capture the effect of intersymbol interference completely. However, even the partial elimination of nonlinear ISI may lead to significant BER performance improvement, as shown later. To estimate the PDFs, the region of all possible samples is quantized in 64 bins, and the number of occurrences of samples in a given bin is counted and normalized with total number of samples in all bins.

For this particular dispersion map, it can be concluded from Fig. 10.5b, that if LDPC codes of rate 0.8 are used, and assuming that the BCJR equalizer operates with memory $m = 3$, 7,200 km can be achieved. To extend the transmission distance further, we have to increase the memory in trellis description of channel, employ an optimum dispersion map, and optimize the system parameters.

In Fig. 10.8, we show how the i.i.d. channel capacity degrades with quantization of LLRs, for the same system used in Fig. 10.6. The use of three bits to represent the LLRs results in moderate channel capacity loss, while the use of only 2 bits results in significant channel capacity loss.

We further perform the experimental evaluation of i.i.d. information capacity for OOK-NRZ systems with direct detection. Figure 10.9 shows the setup used for this experiment. A precoded test pattern was uploaded into a pulse generator (Anritsu MP1763C) via personal computer (PC) with GPIB interface. The 10-Gb/s NRZ signal was used to drive a zero-chirp Mach–Zehnder modulator (UTP). Then the signal passed through PMD emulator (JDSU PE3), where controlled amount of DGD was introduced. The distorted signal was mixed with controlled amount of ASE noise with 3-dB coupler.

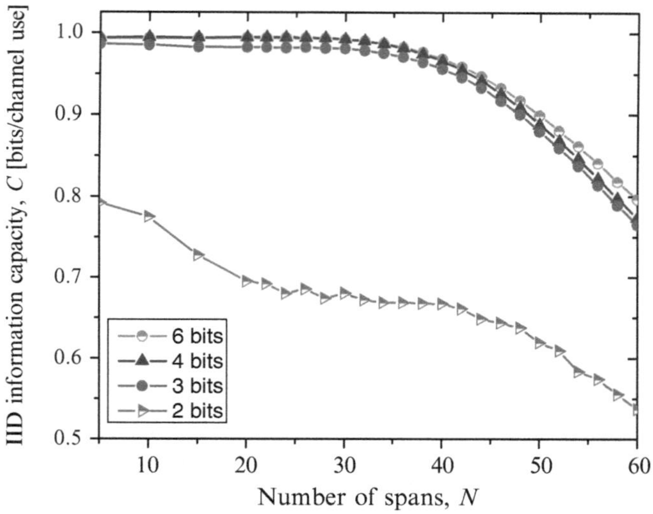

Fig. 10.8 i.i.d. information capacity loss due to quantization of LLRs vs. the number of spans for dispersion map shown in this figure, assuming RZ transmission at 40 Gb/s

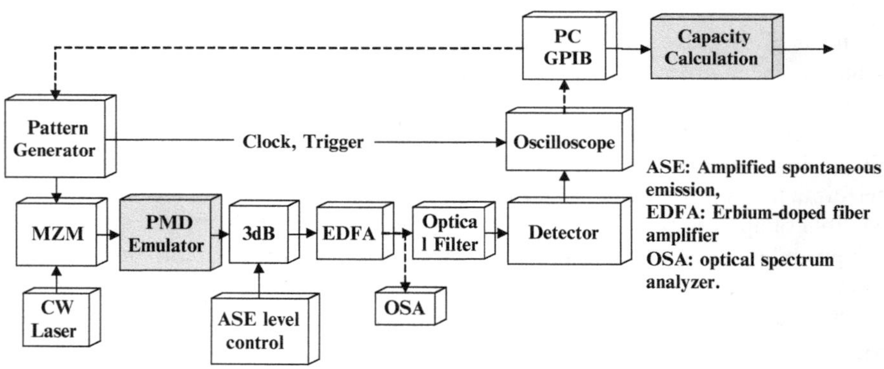

Fig. 10.9 Experimental setup

Modulated signal level was maintained at 0 dB while the ASE power level was changed to obtain different OSNRs. Next, the optical signal was preamplified (Optigain, Inc., 2000 series), filtered (JSDU 2-nm band-pass filter) followed by detection (Agilent 11982A). An oscilloscope (Agilent DCA 86105A) triggered by the data pattern was used to acquire the samples. To maintain constant power of −6 dBm at the detector, a variable attenuator was used. Data was transferred via GPIB back to the PC, which employed the forward step of the BCJR algorithm to calculate channel capacity.

Figure 10.10 shows the experimental results for the i.i.d. information for different values of the state trellis memory assumptions ($2m+1=5$ and $2m+1=1$). Unity

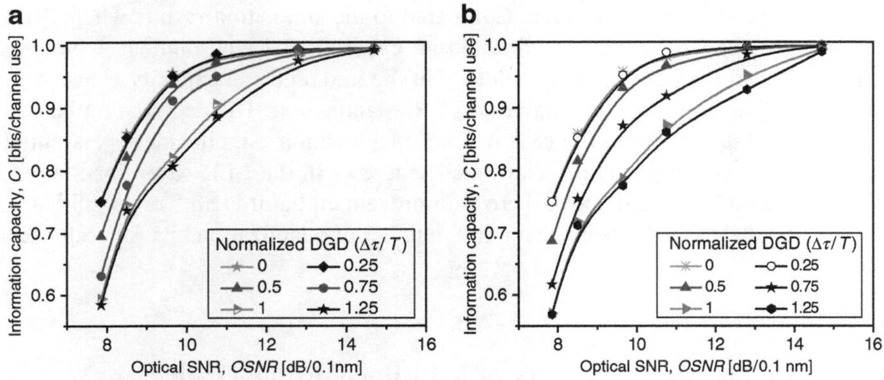

Fig. 10.10 Memory effects on i.i.d. information capacity (**a**) capacity plots at memory $2m + 1 = 5$, (**b**) capacity plots at memory $2m + 1 = 1$

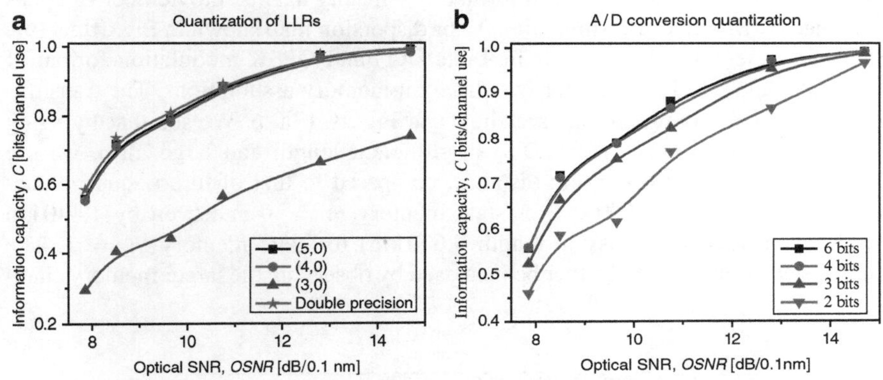

Fig. 10.11 Quantization effects influence on i.i.d. channel capacity (**a**) LLR quantization, (**b**) A/D conversion

capacity is achievable in both cases although the case of memory 0 and DGD of 1.25 indicates that for higher values of DGD capacity of 1 will be impossible to achieve. It is clear that the usage of dynamic discrete model for the optical channel is needed to effectively reduce PMD effects.

Figure 10.11 shows how the quantization effects influence the i.i.d. channel capacity loss. Figure 10.11a studies the quantization of the LLRs, while Fig. 10.11b studies the influence of A/D conversion on capacity. Let the real LLR be denoted by λ. Calculations were done with double precision numbers. A 2-tuple of integers (d_b, p_b) was used to represent the integer (with sign) and decimal part of λ, respectively [17, 18]. For example, in (5,0) representation, "5 denotes the number of bits used to represent the integer part of λ and 0 represents the number of bits used for decimal part." Figure 10.11a demonstrates that (5,0) and (4,0) representations are very close to the double precision one. Significant i.i.d. channel capacity degradation is observed for the (3,0) case. Compared to the simulation results where 3-bit quantization for LLRs was sufficient, in an experimental environment 4 bits are required. From Fig. 10.11b we conclude that the i.i.d. channel capacity is less sensitive to A/D quantization compared to LLR quantization. The degradation here is more gradual compared to the case of LLR quantization. Significant loss is noted for the (3,0) case, and it is not that much severe as in the LLR case. For number of quantization bits 4 and above there is improvement but it is not very significant. We can conclude that 4 bits is the optimal number of quantization bits for both LLR and A/D.

10.4 Information Capacity of Multilevel Fiber-Optics Communication Systems with Coherent Detection

In Fig. 10.12, we show the i.i.d. information capacity against the number of spans (obtained by Monte Carlo simulations), for dispersion map shown in Fig. 10.5a (the fiber parameters are the same as in Table 10.1) and QPSK modulation format of aggregate data rate 100 Gb/s, for two different memory assumptions. The transmitter and receiver configurations are shown in Fig. 10.13a, b. We see that by using the LDPC code (of rate $R = 0.8$) of sufficient length and large girth, we are able to extend the transmission distance, compared to that of turbo equalizer described in Chap. 7, by 600 km for state memory $m = 0$ and even by 1,440 km (resulting in total transmission length 9,600 km) for state memory $m = 1$. The transmission distance can further be increased by observing the larger memory chan-

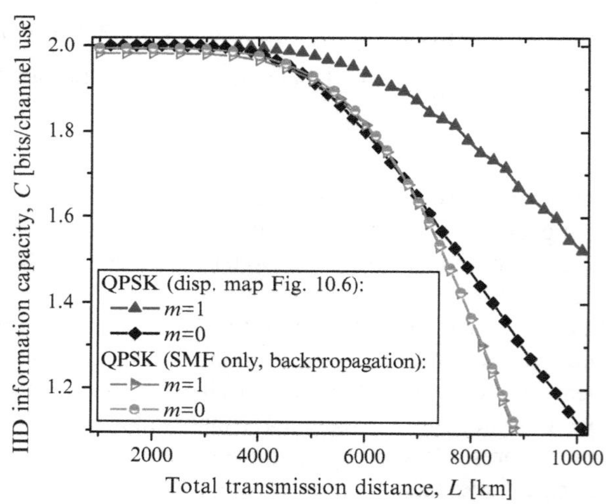

Fig. 10.12 i.i.d. information capacity for QPSK of aggregate data rate of 100 Gb/s against the transmission distance

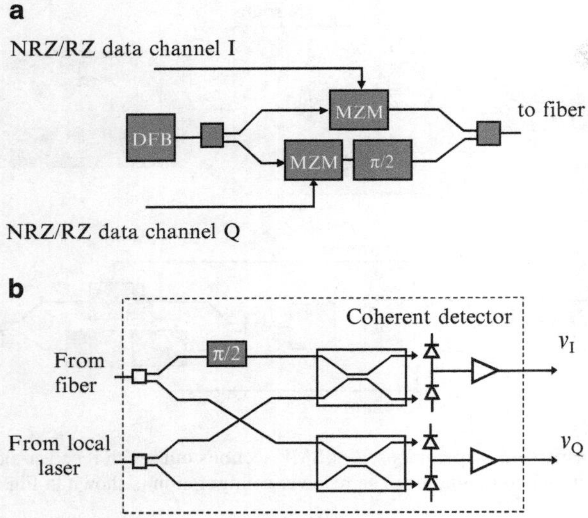

Fig. 10.13 (**a**) Transmitter and (**b**) receiver configurations for system shown in Fig. 10.5a. *DFB* distributed feedback laser, *MZM* Mach–Zehnder modulator

nel assumptions, which requires higher computational complexity for corresponding turbo equalizer. On the other hand, we can use backpropagation approach [25, 26] to keep the channel memory reasonably low and then apply the method described in this section. Note that backpropagation method cannot account for the nonlinear ASE noise–Kerr nonlinearities interaction, and someone should use the method introduced in Sect. 10.2 in information capacity calculation to account for this effect. In the same figure we show the i.i.d. information capacity, when backpropagation method is used, for dispersion map composed of standard SMF only with EDFAs of noise figure 6 dB being deployed every 100 km, as shown in Fig. 10.14. We see that backpropagation method helps reducing the channel memory, since the improvement for $m = 1$ over $m = 0$ case is small.

In Fig. 10.15, we show the i.i.d. information capacities for three different modulation formats (1) MPSK, (2) star QAM, and (3) IPQ; obtained by employing the dispersion map from Fig. 10.14. The symbol rate was 50 GS/s, and the launch power was set to 0 dBm. We see that IPQ outperforms star QAM and significantly outperforms MPSK. For transmission distance of 5,000 km, the i.i.d. information capacity is 2.72 bits/symbol (the aggregate rate is 136 Gb/s per wavelength), for 2,000 km it is 4.2 bits/symbol (210 Gb/s) and for 1,000 km the i.i.d. information capacity is 5.06 bits/symbol (253 Gb/s per wavelength).

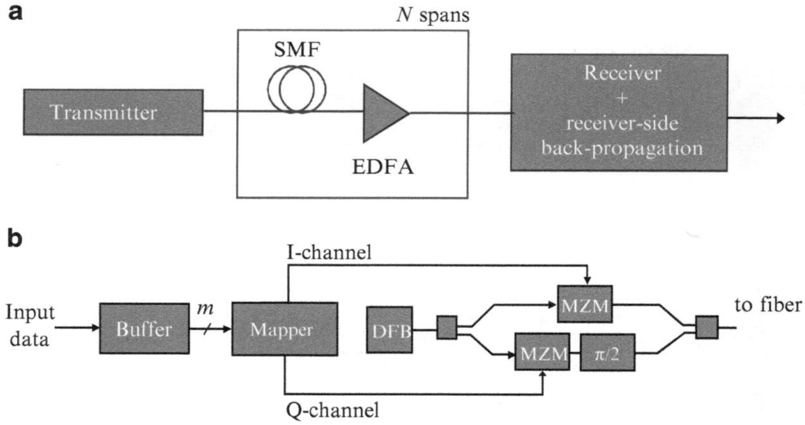

Fig. 10.14 (**a**) Dispersion map composed of SMF sections only with receiver-side backpropagation and (**b**) transmitter configuration. The receiver configuration is shown in Fig. 10.13b

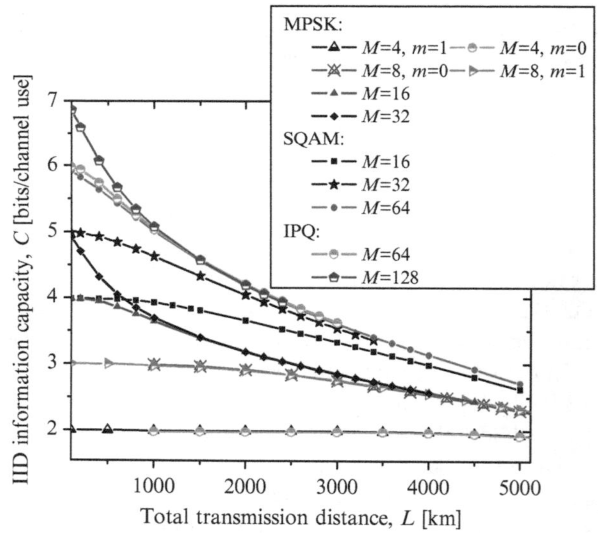

Fig. 10.15 i.i.d. information capacities for star-QAM (SQAM), MPSK, and IPQ for different constellation sizes and dispersion map from Fig. 10.14

10.5 On the Channel Capacity of Optical OFDM Systems for ASE-Noise-Dominated Scenario

In order to provide seamless integrated transport platforms, which can support heterogeneous networking, in this section, we describe the use of power-variable rate-adaptive low-density parity-check (LDPC)-coded polarization-multiplexed OFDM.

We describe the optimum power loading algorithm suitable to deal with bandwidth reduction due to concatenation of reconfigurable optical add–drop multiplexers (ROADMs) and wavelength cross-connects (WXCs), propose the rate adaptation scheme, determine the channel capacity, and show that the proposed scheme can closely approach the channel capacity. We show that similar scheme can be used to overcome the bandwidth limitation problem of GI-POFs. For RoF systems, the number of subcarriers is chosen in such a way that the bandwidth of the signal per subcarrier is smaller than the coherence bandwidth of wireless channel so that each subcarrier experiences the flat fading. To deal with flat fading, we show that the use of equal gain combining (EGC) diversity with two receive antennas is sufficient. In addition, we determine the channel capacity of this hybrid optical-wireless system and show that proposed scheme can closely approach the channel capacity.

This section is organized as follows. In Sect. 10.5.1, we describe a power-variable rate-adaptive polarization multiplexed coded OFDM scheme suitable for use in optically routed networks and determine corresponding channel capacities. In Sect. 10.5.2, we show that similar adaptive scheme can be used to overcome the limited bandwidth of GI-POFs and to enable beyond 100-Gb/s transmission. In Sect. 10.5.3, we describe an adaptive-coded OFDM scheme suitable for use in RoF systems and determine corresponding channel capacity.

10.5.1 Power-Variable Rate-Adaptive Polarization-Multiplexed Coherent-Coded OFDM

In an optically routed network, different signal transmission paths have different number of optical amplifiers, WXCs, and ROADMs. Different wavelength channels carrying the traffic to different destinations can have quite different SNRs and signal is differently impacted by various channel impairments including PMD, chromatic dispersion, fiber nonlinearities, and filtering effects due to concatenation of optical filters/ROADMs/WXCs. The optical networks should provide a target bit-error ratio (BER) performance regardless of the data destination. To address all these issues, it is possible use of power-variable rate-adaptive coded polarization-multiplexed optical OFDM system with coherent detection, which is shown in Fig. 10.16. The optimum power adaptation policy is already given by (10.19). The optimum threshold SNR is determined from condition that total power in all subcarriers cannot be larger than available power, $\sum_i P_i \leq P$, by

$$\gamma_{\text{tsh}} = N \Big/ \sum_{\gamma_i \geq \gamma_{\text{tsh}}} 1/\gamma_i, \qquad (10.33)$$

where N is the number of subcarriers with $\gamma_i \geq \gamma_{\text{tsh}}$, in an iterative procedure that converges fast (after few iterations).

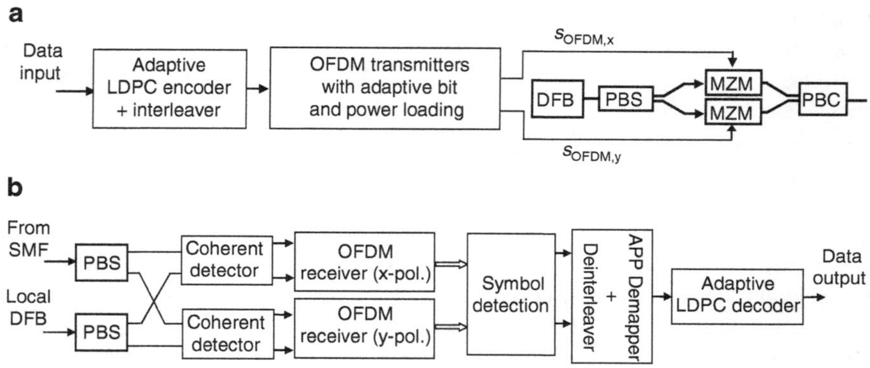

Fig. 10.16 Block diagram of the power-variable rate-adaptive LDPC-coded OFDM scheme (**a**) transmitter and (**b**) receiver configurations. *DFB* distributed feedback laser, *PBS/C* polarization-beam splitter/combiner, *MZM* dual-drive Mach–Zehnder modulator, *APP* a posteriori probability

The variable-rate adaptation is achieved by choosing the maximum product of integer m_i, corresponding to the number of bits per ith subcarrier, and corresponding code rate R_i as follows:

$$m_i R_i \leq C_i, \tag{10.34}$$

where C_i is the channel capacity of the ith subcarrier. The signal constellation size $M_i = 2^{m_i}$ per ith subcarrier and the corresponding code rate R_i of component LDPC code are chosen in accordance with the channel conditions. When the channel conditions are favorable (large SNR), the larger constellation sizes and higher code rate LDPC codes are employed. Among several candidate LDPC codes we employ one (based on subcarrier γ_i) which provides the largest product $m_i R_i$ closest to the subcarrier channel capacity (but lower than C_i). When channel conditions are poor, we reduce the code rate and signal constellation size according to (10.34).

The channel capacity of polarization-multiplexed optical coherent OFDM scheme (derived based on theory due to Bölcskei et al. [44]), with N subcarriers can be evaluated as follows:

$$C = E_H \left\{ \frac{1}{N} \max_{\mathrm{Tr}(\Sigma) \leq P} B_{\mathrm{sc}} \log_2 \left[\det \left(I_{2N} + \frac{1}{\sigma_n^2} \boldsymbol{H} \boldsymbol{\Sigma} \boldsymbol{H}^{\dagger} \right) \right] \right\}, \tag{10.35}$$

where $\boldsymbol{\Sigma}$ is the covariance matrix of Gaussian input vector, defined as

$$\Sigma = \mathrm{diag}\left\{ \Sigma_i \right\}_{i=0}^{N-1}, \quad \Sigma_i = \frac{P}{2N} I_2 \tag{10.36}$$

with P being the maximum overall transmit power, σ_n being the standard deviation of ASE noise process, and B_{sc} being the bandwidth of subcarrier channel. $\mathrm{Tr}(\Sigma)$ denotes the trace of matrix Σ, E_H denotes the expectation operator with respect to channel matrix $\boldsymbol{H}$, and $\boldsymbol{I}_{2N}$ is $2N \times 2N$ identity matrix. The channel matrix in (10.35) is a block diagonal matrix in (10.37) with ith block diagonal element corresponding to the 2×2 Jones matrix $\boldsymbol{H}(i)$ of ith subcarrier.

$$\boldsymbol{H} = \mathrm{diag}\{\boldsymbol{H}(i)\}_{i=0}^{N-1}, \quad \boldsymbol{H}(i) = \begin{bmatrix} H_{xx}(i) & H_{xy}(i) \\ H_{yx}(i) & H_{yy}(i) \end{bmatrix}, \qquad (10.37)$$

The chromatic dispersion effect and reduced bandwidth effect due to concatenation of ROADMs, as well as different linear channel imperfections, are all incorporated in $\boldsymbol{H}(i)$.

In Fig. 10.17, we show the channel capacity calculated based on (10.35)–(10.37) for polarization-multiplexed OFDM system when all order PMD with average DGD τ of 500 ps is observed, for 2,000 km of SMF, 128 subcarriers, and OFDM signal bandwidth BW of 25 GHz. The pilot-aided channel estimation is used to determine channel coefficients in (10.37). We see that when channel matrix (the channel state information, CSI) is perfectly known we can completely compensate for different linear channel impairments, which was expected because the Jones matrix is unitary. However, when the CSI is not ideal (the channel estimation is not perfect) we can have significant channel capacity degradation (see Fig. 10.17a). We see when channel coefficients are estimated to be 90% of nominal values we have small performance degradation. However, when the channel estimates are 50% of nominal value we have found significant channel capacity degradation. We also have found that for target BERs below 10^{-9} M-ary QAM (MQAM) is far away from the channel capacity. To reach the channel capacity, we have to invent better modulation formats and use them in combination with adaptive LDPC coding. In Fig. 10.17b, we study the efficiency of the optimum power adaptation [(10.19), (10.33)] in dealing with

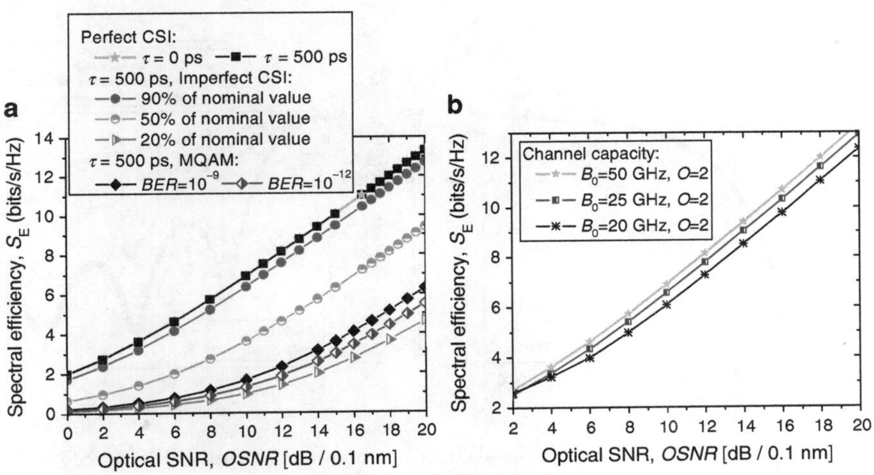

Fig. 10.17 Spectral efficiency limits of polarization-multiplexed coherent optical OFDM (with OFDM signal bandwidth BW = 25 GHz)

concatenation filtering problem (that is modeled as super-Gaussian filter of order $O = 2$ and bandwidth $B_0 = 20$, 25, and 50 GHz). We see that channel capacity is getting worse as the bandwidth of optical filter is getting smaller than OFDM signal bandwidth, but we are still quite close to the channel capacity.

10.5.2 Adaptive-Coded OFDM for Communication Over GI-POF Links

Because the transfer function of GI-POF is quasistatic as shown in [45, 46], we can quite easily implement the optimum power adaptation algorithm instead of sub-optimal algorithm used in [47]. For the verification of model proposed in [45] in Fig. 10.18a we report the refractive indices of core and cladding as well as the material dispersion for benzyl benzoate (BEN) doped poly(methyl methacrylate) (PMMA) GI-POF, while in Fig. 10.18b we provide frequency responses for a 100m long PMMA GI-POF for index exponent $g = 2.1$ at wavelength 850 nm. The corresponding frequency responses at 650 nm are already given in [45]. Both the uniform and Gaussian excitations are observed. We see that the Gaussian excitation provides higher bandwidth than uniform excitation, but the largest improvement in bandwidth is due to differential mode attenuation (DMA).

The corresponding information capacity expression is similar to (10.19). The variable-rate adaptation can be achieved, similarly as in previous case, by choosing the maximum product of integer m_i, corresponding to the number of bits per ith subcarrier, and code rate R of corresponding LDPC code as given by (10.34). The signal constellation size 2^{m_i} per ith subcarrier and the corresponding code rate R of LDPC code are chosen in accordance with the subcarrier SNR. When the subcarrier

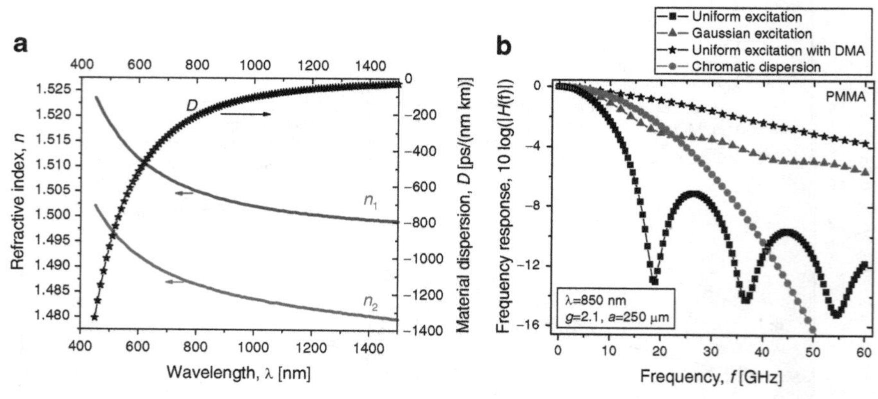

Fig. 10.18 (**a**) Refractive index and material dispersion for benzyl benzoate (BEN) doped PMMA GI-POF and (**b**) frequency responses for a 100-m long PMMA GI-POF for index exponent $g = 2.1$ at wavelength 850 nm

SNR γ_i is high, larger constellation size is employed. When γ_i is low, we reduce the signal constellation size according to (10.34) and do not transmit at all on subcarriers with subcarrier SNR below threshold γ_{tsh}.

In Fig. 10.19a, we report the channel capacity and spectral efficiencies for BEN-PMMA of 100 m in length, when OFDM of two different signal bandwidths (12.5 GHz and 25 GHz) is used, by employing the model and parameters reported in [45]. We see that spectral efficiency for MQAM at target BER of 10^{-6} is far away from channel capacity. However, when LDPC(16935,13550) code of large girth is used we are 1.97 dB away from channel capacity for aggregate rate of 100 Gb/s. By using longer LDPC(24015,19212) code we are getting only 0.4-dB improvement.

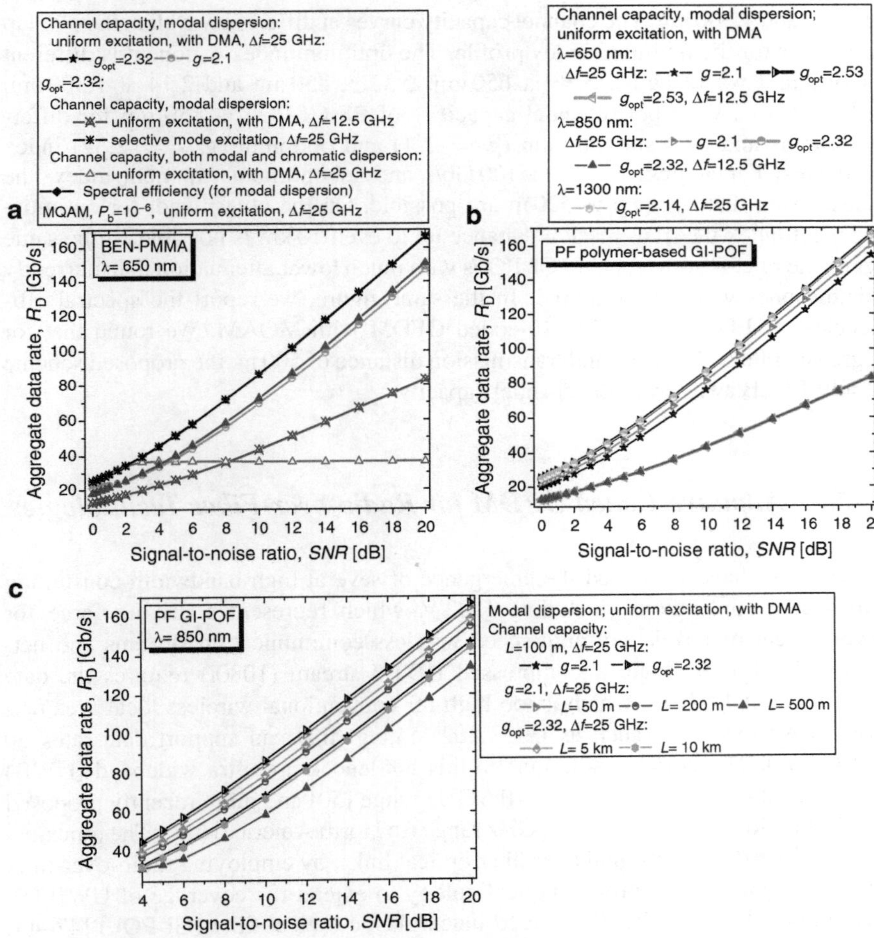

Fig. 10.19 Channel capacities for (**a**) 100-m long PMMA GI-POF at 650 nm and optimum index exponent, (**b**) 100 m of PF GI-POF at different wavelengths, and (**c**) PF GI-POF for different lengths

To come closer to the channel capacity, we have to invent better modulation formats. In the same figure, we study joint influence of chromatic and modal dispersion on channel capacity. We see that chromatic dispersion imposes serious limitations, and for this type of fiber, in addition to power and rate adaptation we have to perform chromatic dispersion compensation. In the same figure, we report the channel capacity for optimum index exponent ($g_{opt} = 2.32$). The improvement in SNR at 100 Gb/s due to optimum index profile is 2.37 dB over $g = 2.1$. The use of selective mode Gaussian excitation [the excited group modes are $(0.3\text{–}0.35)M_g$, where M_g is the total number of mode groups that can be excited] does not improve the channel capacity.

In Fig. 10.19b, we study the channel capacity of perfluorinated (PF) polymer-based GI-POF for parameters given in [46] for three different wavelengths: 650 nm, 850 nm, and 1,300 nm. The channel capacity curves at different wavelengths overlap each other for the optimum index profile. The optimum index exponent is different for different wavelengths: 2.53 at 650 nm, 2.32 at 850 nm and 2.14 at 1,300 nm. In Fig. 10.19c, we report channel capacities of PF GI-POF at 850 nm for different lengths and both suboptimum ($g = 2.1$) and optimum ($g_{opt} = 2.32$) index exponents. For aggregate rate of 100 Gb/s and suboptimum exponent index, the transmission distances up to 500 m are possible. On the other hand, for optimum exponent index the transmission distance up to even 10 km is possible, at the same aggregate rate. However, novel GI-POFs with much lower attenuation than currently existing ones would be required. In the same figure, we report the spectral efficiencies for LDPC(16935,13550)-coded OFDM with MQAM. We found that for aggregate rate of 100 Gb/s and transmission distance of 500 m, the proposed scheme is only 1.9 dB away from the channel capacity.

10.5.3 Adpative-Coded OFDM for Radio-over-Fiber Technologies

Recently we have witnessed the emergence of several high-bandwidth-consuming multimedia applications, including HDTV, which represent a driving force for development of novel very-high-speed wireless communication systems and networks [47–49]. A single uncompressed HDTV stream (1080i) requires the date rates of about 1.5 Gb/s, which is too high for conventional wireless local area network (LAN) systems, such as IEEE 802.11a,b,g (that can support data rates up to 54 Mb/s). The possible solutions to this bottleneck are ultra wideband (UWB) communications operating in 3.1–10.6 GHz range [50] and more recently proposed 60-GHz radio operating in 57–64 GHz range (in North America) [47]. The transmission of UWB/60-GHz signal over fiber optical links, by employing radio-over-fiber (RoF) technology, is a promising technology to extend the coverage of UWB/60-GHz radios. In particular, the graded index plastic optical fiber (GI-POF) [47–49] provides a cost-effective solution for UWB/60-GHz radios.

An adaptive RoF system architecture is shown in Fig. 10.20. The data generated in central station is first LDPC encoded. The LDPC-encoded data enters the

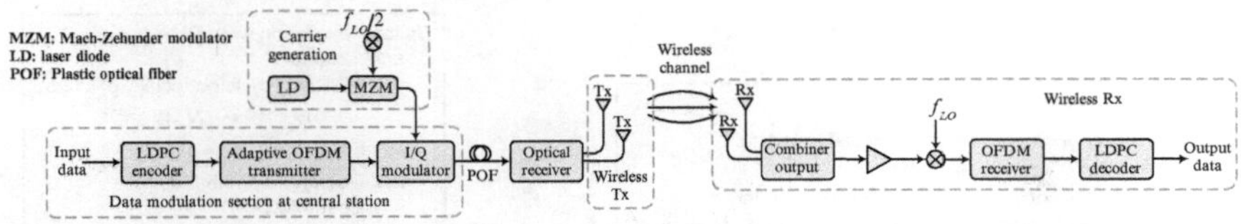

Fig. 10.20 Block diagram of an adaptive LDPC-coded OFDM scheme suitable for use in RoF systems. *MZM* Mach–Zehnder modulator, *LD* laser diode, *POF* plastic optical fiber

buffer in the adaptive OFDM block. Depending on ith subcarrier SNR, m_i bits are taken from the buffer. The mapper that follows buffer selects a corresponding constellation point from 2^{m_i}-QAM constellation diagram. After pilot insertion and serial-to-parallel (S/P) conversion, the inverse FFT (IFFT) is performed. The cyclic extension is performed but repeating the $N_G/2$ samples (of IFFT frame) as prefix and the first $N_G/2$ samples as suffix. The cyclic extension duration is chosen to be longer than the total delay spread due to dispersion in MMF and multipath fading in wireless channel.

To deal with multipath wireless fading, we propose the use of diversity principle. It will be shown later in the text that the use of two wireless receive antennas and EGC is sufficient to compensate for Rayleigh fading, which corresponds to the worst-case scenario (no line-of-site). To generate the optical 60-GHz carrier signal, the Mach–Zehnder modulator (MZM) is biased at null transmission point, and the local oscillator output signal, with frequency being half of that required in wireless domain, is used as RF input signal. To perform electrical-to-optical conversion, the I/Q modulator is used, driven by mm-wave OFDM signal. The signal is then transmitted over GI-POF link. At the end of GI-POF link, the signal is converted in electrical domain by a p.i.n. photodetector, amplified, and retransmitted by wireless antenna. On the wireless receiver end, the receiver diversity principle is employed. The signals from receiver wireless antennas are upon cophasing appropriately weighted, and after down-conversion we perform the reverse operations in OFDM receiver.

The corresponding channel capacity expression is derived to be

$$C = E_a\left[\sum_{i:\gamma_i>\gamma_{\text{tsh}}} B_{\text{sc}}\log_2\left(a^2\gamma_i/\gamma_{\text{tsh}}\right)\right], \tag{10.38}$$

where B_{sc} is the bandwidth of subcarrier channel, a is the Rayleigh fading wireless channel coefficient, and $E_a[]$ denotes the expectation operator with respect to a.

The variable-rate adaptation can be achieved by choosing the maximum product of integer m_i, corresponding to the number of bits per ith subcarrier, and code rate R of corresponding LDPC code as described by (10.34). The signal constellation size 2^{m_i} per ith subcarrier and the corresponding code rate R of LDPC code are chosen in accordance with the subcarrier SNR. When the subcarrier SNR γ_i is high, larger constellation size is employed. When γ_i is low, we reduce the signal constellation

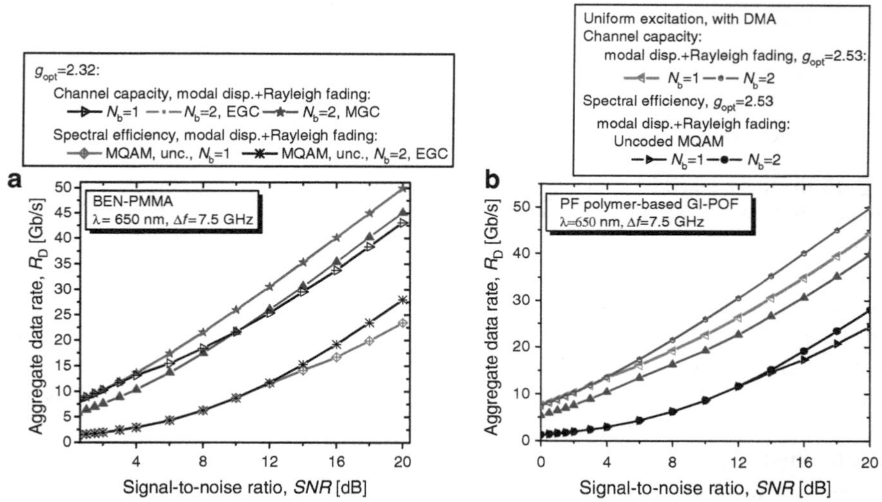

Fig. 10.21 Aggregate data rates for RoF systems of bandwidth 7.5 GHz for Rayleigh fading (**a**) 100-m long BEN-PMMA and (**b**) 100 m of PF GI-POF

size according to (10.34) and do not transmit at all on subcarriers with subcarrier SNR below threshold γ_{tsh}.

In Fig. 10.21, we show the channel capacities and spectral efficiencies of the proposed RoF system, corresponding to OFDM bandwidth of 7.5 GHz. In Fig. 10.21a, we report the corresponding plots for BEN-PMMA at 650 nm. We see that even 40-Gb/s transmission over 100 m of GI-POF and corresponding wireless link is possible by occupying whole UWB available bandwidth. When suboptimum exponent of index profile ($g = 2.1$) is used we have 0.65-dB degradation in channel capacity. The use of $N_b = 2$ receiver wireless antennas and EGC is sufficient to compensate for Rayleigh fading and achieve the full channel capacity. When the multilevel QAM (MQAM) is used without coding (observed at probability of error $P_b = 10^{-6}$), we face significant spectral efficiency degradation, and 40-Gb/s transmission is not possible at all. However, when LDPC(16935,13550)-coded OFDM is used in combination with MQAM we are able to achieve the 40-Gb/s transmission, but we are 2.72 dB away from channel capacity. The use of longer LDPC (24015,19212) code provides only small improvement. To come closer to the channel capacity, we have to invent a better modulation format designed for this particular RoF system. In Fig. 10.21b, we provide corresponding plots for perfluorinated (PF) polymer-based GI-POF.

In Fig. 10.22, we study the possibility of 100-Gb/s transmission and beyond in 60-GHz RoF systems based on PF GI-POFs (observing suboptimum exponent index $g = 2.3$) for different transmission distances. We see that even transmission distances of 2 km are possible at this aggregate rate. However, novel types of GI-POFs, with much lower attenuation than currently available ones, are to be fabricated.

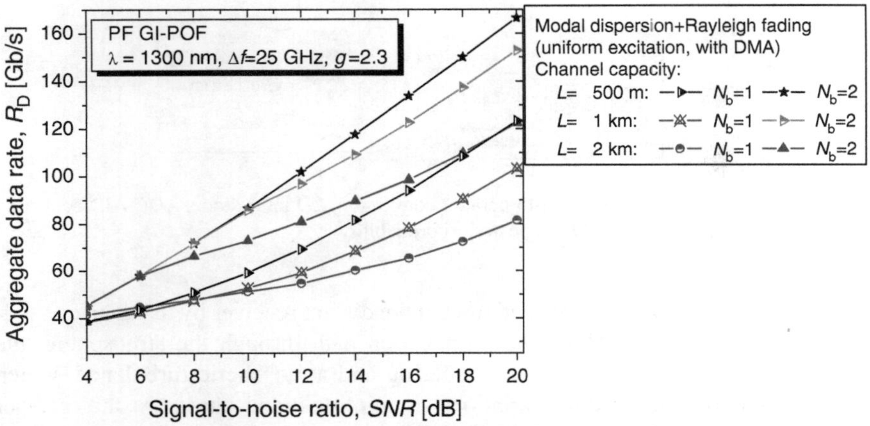

Fig. 10.22 Aggregate data rates for RoF systems of bandwidth 25 GHz for Rayleigh fading and different PF GI-POF lengths

10.6 On the Channel Capacity of Hybrid Free-Space Optical-Wireless Channels

In this section, we describe the use a hybrid FSO-RF system with adaptive modulation and coding as an efficient way to deal with strong atmospheric turbulence. Adaptive modulation and coding [51–53] can enable robust and spectrally efficient transmission over both $\alpha - \mu$ (or generalized Gamma) wireless fading channel and FSO channel. To enable high-speed communication over atmospheric turbulence channels, we describe a scheme in which an LDPC-coded sequence is partially over FSO channel and partially over wireless channel. The CSI of both channels is forwarded to the transmitters by an RF-feedback channel. The transmitters then adapt powers and rates so that total channel capacity is maximized.

10.6.1 Hybrid FSO-Wireless System Description

We first describe the hybrid system, wireless channel model, and FSO channel model. The adaptive hybrid FSO-RF communication system, shown in Fig. 10.23, consists of two parallel FSO and RF channels. The encoded data stream is partially transmitted over FSO and partially over RF channel. Operating symbol rate of FSO channel is commonly many times higher that that of RF channel. FSO channel comprises an FSO transmitter, propagation path through the atmosphere, and an FSO receiver. The optical transmitter includes a semiconductor laser of high launch power, adaptive mapper, and power control block. To reduce the system cost, the direct modulation of laser diode is used.

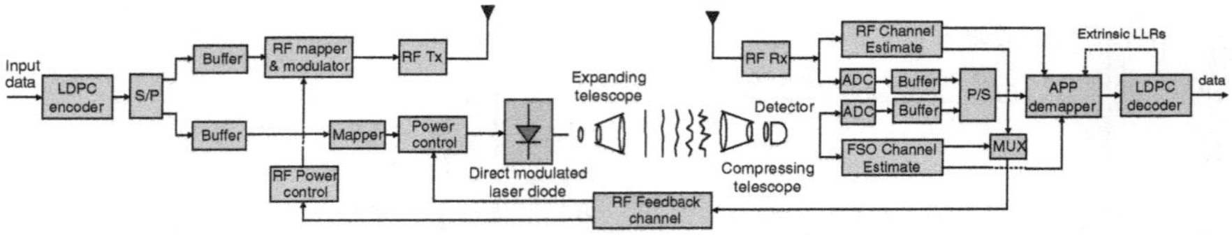

Fig. 10.23 System model. *S/P* serial-to-parallel conversion, *LD* laser diode, *ADC* A/D converter, *P/S* parallel-to-serial converter, *APP* a posteriori probability

The modulated beam is projected toward the distant receiver by using an expanding telescope assembly. Along the propagation path through the atmosphere, the light beam experiences absorption, scattering and atmospheric turbulence, which cause attenuation and random variations in amplitude and phase. At the receiver side, an optical system collects the incoming light and focuses it onto a detector, which generates an electrical current proportional to the incoming power. The RF channel comprises adaptive RF mapper, RF power control, RF transmitter (Tx), transmitting antenna, wireless propagation path, receiver antenna, and RF receiver (Rx).

The RF channel estimates and FSO irradiance estimates are transmitted back to transmitters using the same RF feedback channel. Because the atmospheric turbulence changes slowly, with correlation time ranging from 10 μs to 10 ms, this is a plausible scenario for FSO channels with data rates in the order of Gb/s. Note that erbiumdoped fiber amplifiers (EDFAs) cannot be used at all in this scenario because the fluorescence time is too long (about 10 ms). The semiconductor optical amplifiers (SOAs) should be used instead, if needed. The data rates and powers in both channels are varied in accordance with channel conditions. The symbol rates on both channels are kept fixed while the signal constellation diagrams sizes are varied based on channel conditions. When FSO (RF) channel condition is favorable larger constellation size is used, when FSO (RF) channel condition is poor smaller constellation size is used, and when the FSO (RF) channel SNR falls below threshold the signal is not transmitted at all. Both subsystems (FSO and RF) are designed to achieve the same target bit error probability (P_b). The RF subsystem employs M-ary quadrature amplitude modulation (MQAM), while FSO subsystem employs the M-ary pulse amplitude modulation (MPAM). MPAM is selected for FSO subsystem because negative amplitude signals cannot be transmitted over FSO channels with direct detection. These two modulation formats are selected as an illustrative example; the proposed scheme, however, is applicable to arbitrary multilevel modulations. The optimum variable-power variable-rate policy, maximizing total channel capacity, is described in next subsection. In the rest of this section, we describe the wireless and the FSO channel models. Both RF and FSO channels are modeled as *block-fading* channels, because symbol durations in both channels are much shorter than the corresponding coherence times. The duration of the block is proportional to the coherence time of the channel, which is from 10 μs to 10 ms for FSO channel and from 10 ms to 1 s for RF channels. We therefore assume the i.i.d. fading between the blocks.

The signal at the RF receiver antenna (see Fig. 10.23), denoted by s_k can be written as

$$s_k = r_k e^{j\gamma_k} \boldsymbol{c}_k + \boldsymbol{w}_k, \tag{10.39}$$

where r_k is the fading envelope for kth transmission block and γ_k is the random phase shift occurred during signal transmission of the same block of symbols over a fading channel. The fading at antenna is frequency nonselective; it does not change for the duration of block (slow fading) and it is independent from block to block. The $\boldsymbol{c}_k$ is the vector of transmitted symbols within the kth block. The $\boldsymbol{w}_k$ denotes the block of zero mean circular Gaussian noise samples corresponding to block k

The corresponding $\alpha - \mu$ PDF is given by [54, 55]

$$p\left(r_t\right) = \frac{\alpha \mu^{\mu} r_t^{\alpha\mu-1}}{\hat{r}^{\alpha\mu} \Gamma\left(\mu\right)} \exp\left(-\mu \frac{r_t^{\alpha}}{\hat{r}^{\alpha}}\right), \tag{10.40}$$

where $\Gamma(\,)$ is the Gamma function and $\mu > 0$ is the inverse of the normalized variance of r_t^{α}

$$\mu = E^2\left\{r_t^{\alpha}\right\} / \mathrm{Var}\left\{r_t^{\alpha}\right\}, \tag{10.41}$$

where $E\{\}$ is the expectation operator and $\hat{r}$ is a α-root mean value

$$\hat{r} = \sqrt[\alpha]{E\left\{r_t^{\alpha}\right\}}. \tag{10.42}$$

The $\alpha - \mu$ fading model is employed because Rayleigh, Nakagami-m, exponential, Weibull and one-sided Gaussian distribution functions are all special cases of this model. For example, by setting $\alpha = 2$ and $\mu = 1$ we obtain the Rayleigh distribution, while by setting $\alpha = 2$ and $\mu = 2$ we obtain Nakagami $m = 2$ distribution.

After the signal cophasing, the resulting signal can be written as

$$z_k = r_k \boldsymbol{x}_k + \boldsymbol{w}_k. \tag{10.43}$$

The FSO communication channel model is described by

$$\boldsymbol{y}_k = R i_k \boldsymbol{x}_k + \boldsymbol{n}_k, \tag{10.44}$$

where $\boldsymbol{x}_k$ is the kth transmitted block, i_k is the instantaneous intensity gain, $\boldsymbol{y}_k$ is the kth received block of symbols, $\boldsymbol{n}_k$ the vector of additive white Gaussian noise (AWGN) samples with a normal distribution $N(0, \sigma^2)$ representing the transimpedance amplifier thermal noise, and R denotes the photodiode responsivity. (Without the loss of generality in the rest of the section we will set $R = 1\mathrm{A/W}$.) [All signals in (10.44) are real valued.]

Several PDFs have been proposed for the intensity variations at the receiver side of an FSO link [56]. For example, Al-Habash et al. [56, 57] proposed a statistical model that factorizes the irradiance as the product of two independent random processes each with a Gamma PDF. It was shown in [57] that predicted distribution matches very well the distributions obtained from numerical propagation simulations and experiments and as such is adopted here. The PDF of the intensity fluctuation is given by [56]

$$p(i_k) = \frac{2(\alpha'\beta')^{(\alpha'+\beta')/2}}{\Gamma(\alpha')\Gamma(\beta')} i_k^{(\alpha'+\beta')/2-1} K_{\alpha'-\beta'}\left(2\sqrt{\alpha'\beta' i_k}\right), \quad i_k > 0, \quad (10.45)$$

where i_k $(k \geq 0)$ is the signal intensity, α' and β' are parameters of the PDF, and $K_{\alpha'-\beta'}$ is the modified Bessel function of the second kind of order $\alpha' - \beta'$. The parameters α' and β' are related to the scintillation, and in the case of zero inner scale ($l_0 = 0$) (for plane waves), they are given by [56, 57]

$$\alpha' = \frac{1}{\exp\left[\frac{0.49\sigma_R^2}{\left(1+1.11\sigma_R^{12/5}\right)^{7/6}}\right] - 1}, \quad \beta' = \frac{1}{\exp\left[\frac{0.51\sigma_R^2}{\left(1+0.69\sigma_R^{12/5}\right)^{5/6}}\right] - 1}, \quad (10.46)$$

where σ_R^2 is the Rytov variance [56]

$$\sigma_R^2 = 1.23 C_n^2 k^{7/6} L^{11/6}, \quad (10.47)$$

where $k = 2\pi/\lambda$ (λ is the wavelength), L denotes the propagation distance, and C_n^2 is refractive index structure parameter. Weak fluctuations are associated with $\sigma_R^2 < 1$, the moderate with $\sigma_R^2 \approx 1$, the strong with $\sigma_R^2 > 1$, and the saturation regime is defined by $\sigma_R^2 \to \infty$ [57].

10.6.2 Adaptive Modulation and Coding

There are many parameters that can be varied at the transmitter side relative to the FSO (RF) channel conditions including data rate, power, coding rate, and combinations of different adaptation parameters. The transmitter power adaptation can be used to compensate for SNRs variation due to atmospheric turbulence/fading, with the aim to maintain a desired BER in both FSO and RF channels. The power adaptation therefore "inverts" the FSO channel scintillation and fading in wireless channel so that both FSO and RF channels behave similarly as an AWGN channel. The FSO channel, upon channel inversion, appears to the receiver as standard AWGN channel with $\mathrm{SNR}^{\mathrm{FSO}} = \Gamma_0^{\mathrm{FSO}}/E\left[1/i_t^2\right]$, where $\Gamma_0^{\mathrm{FSO}} = E_s^{\mathrm{FSO}}/N_0$ is the SNR in the absence of scintillation, with E_s^{FSO} being the symbol energy and $N_0/2$ being the double-side PSD of AWGN related to variance by $\sigma^2 = N_0/2$. Note that this definition, commonly used in wireless communications [52–55], and also in [58], is different from [57] where SNR is defined as P_o/σ, where $E[i_t] \leq P_o$. The wireless fading channel, upon channel inversion, appears to the receiver as standard AWGN channel with $\mathrm{SNR}^{\mathrm{RF}} = \Gamma_0^{\mathrm{RF}}/E\left[1/h^2\right]$, where h is channel coefficient, $\Gamma_0^{\mathrm{RF}} = E_s^{\mathrm{RF}}/N_0$ is the SNR in the absence of fading, with E_s^{RF} being the symbol energy and $N_0/2$ being the double-side PSD of AWGN.

In the rest of this section, we derive the optimum power adaptation policy that maximizes the total channel capacity. We further derive the rate adaptation policy but assuming that symbol rates in both channels are fixed, and the symbol rate in RF channel is much smaller than that in FSO channel. As an illustrative example, we assume that MQAM is used in RF channel, while MPAM is used in FSO channel, and determine the spectral efficiency. MPAM is selected because negative amplitude signals cannot be transmitted over FSO channels with direct detection. The MQAM is not power efficient for transmission over FSO channel with direct detection because it requires the addition of DC bias to convert negative amplitudes to positive ones and as such is not considered here. Note that M-ary pulse-position modulation (MPPM) can also be used for transmission over FSO channel. Because MPPM is highly spectrally inefficient, we restrict our attention to MPAM instead.

Before we continue with the description of different adaptation policies, we have to derive a target bit-error probability equations P_b for MAPM and MQAM on an AWGN channel. In MPAM the transmitted signal x takes values from the discrete set $X = \{0, d, \ldots, (M-1)d\}$ $(M \geq 2)$, where d is the Euclidean distance between two neighboring points. If all signal constellation points are equiprobable, the average signal energy is given by $E_s = d^2(M-1)(2M-1)/6$, and it is related to the bit energy E_b by $E_s = E_b \log_2 M$, so that SNR per bit is defined by $\mathrm{SNR}_b = E_b/N_0 = E_s/(N_0 \log_2 M) = d^2(M-1)(2M-1)/(6N_0 \log_2 M)$. Following the derivation similar to that reported in [52], we derive the following expression for bit-error probability of FSO channel:

$$P_b^{\mathrm{MPAM}} \cong \frac{M-1}{M \log_2 M} \mathrm{erfc}\left(\sqrt{\frac{3\Gamma_0^{\mathrm{MPAM}}}{2(M-1)(2M-1)}}\right), \tag{10.48}$$

were the symbol SNR Γ_0^{MPAM} is the symbol SNR in the absence of scintillation, and erfc(z) function is defined by

$$\mathrm{erfc}(z) = \frac{2}{\sqrt{\pi}} \int_z^{\infty} \exp\left(-u^2\right) \mathrm{d}u.$$

Because (10.48) is not invertible, we derive the following empirical formula which is valid in the regime of medium and high SNRs:

$$P_b^{\mathrm{MPAM}} \cong 0.2 \exp\left[-\frac{1.85\Gamma_0^{\mathrm{MPAM}}}{2^{2.19 \log_2 M} - 1}\right]. \tag{10.49}$$

The corresponding expressions for MQAM are given in [52]. In derivations that follow we will assume that target bit-error probabilities in both channels are the same $P_b^{\mathrm{MPAM}} = P_b^{\mathrm{MQAM}} = P_b$. The total spectral efficiency R as the function of bit-error probability P_b can be found as

$$R = \frac{1}{2.19} B_{\mathrm{FSO}} \log_2\left(1 + K_{\mathrm{FSO}} \Gamma_0^{\mathrm{FSO}}\right) + B_{\mathrm{RF}} \log_2\left(1 + K_{\mathrm{RF}} \Gamma_0^{\mathrm{RF}}\right), \tag{10.50}$$

where $K_{\text{FSO}} = -1.85\backslash\ln(5P_{\text{b}})$ and $K_{\text{RF}} = -1.5\backslash\ln(5P_{\text{b}})$. Because the total spectral efficiency changes as the channel conditions in either channel change, the spectral efficiency is a function of FSO channel irradiance i_k and RF fading coefficient h as follows

$$R=\frac{1}{2.19}B_{\text{FSO}}\log_2\left(1+K_{\text{FSO}}\Gamma^{\text{FSO}}(i_k)\frac{P^{\text{FSO}}(i_k)}{P}\right)+B_{\text{RF}}\log_2\left(1+K_{\text{RF}}\Gamma^{\text{RF}}(h)\frac{P^{\text{RF}}(h)}{P}\right), \tag{10.51}$$

where $\Gamma^{\text{FSO}}(i_k) = i_k^2\Gamma_0^{\text{FSO}}$ and $\Gamma^{\text{RF}}(h) = h^2\Gamma_0^{\text{RF}}$. To derive the optimum power adaptation policy, subject to $P^{\text{FSO}}(i_k) + P^{\text{RF}}(h) \le P$, we have to define the corresponding Lagrangian, differentiate it with respect to $P^{\text{FSO}}(i_k)$ and $P^{\text{RF}}(h)$, and set corresponding derivatives to be equal to zero. The optimum power adaptation policy is obtained as the result of this derivation:

$$\frac{K_{\text{FSO}}P^{\text{FSO}}(i_k)}{P} = \begin{cases}\frac{1}{\Gamma_{\text{tsh}}} - \frac{1}{\Gamma^{\text{FSO}}}, & \Gamma^{\text{FSO}} \ge \Gamma_{\text{tsh}}\\ 0, & \Gamma^{\text{FSO}} < \Gamma_{\text{tsh}}\end{cases}, \quad \Gamma^{\text{FSO}} = \Gamma_0^{\text{FSO}}i_k^2,$$

$$\frac{K_{\text{RF}}P^{\text{RF}}(h)}{P} = \begin{cases}\frac{1}{\Gamma_{\text{tsh}}} - \frac{1}{\Gamma^{\text{RF}}}, & \Gamma^{\text{RF}} \ge \Gamma_{\text{tsh}}\\ 0, & \Gamma^{\text{RF}} < \Gamma_{\text{tsh}}\end{cases}, \quad \Gamma^{\text{RF}} = \Gamma_0^{\text{RF}}h^2, \tag{10.52}$$

where Γ_{tsh} is the threshold SNR, which is common to both channels. With this adaptation policy, more power and higher data rates are transmitted when the FSO (RF) channel conditions are good, less power and lower data rates are transmitted when FSO (RF) channel is bad, and nothing is transmitted when the SNR falls below the threshold Γ_{tsh}. The optimum threshold Γ_{tsh} can be obtained numerically by solving the following equation:

$$\int_{\sqrt{\Gamma_{\text{tsh}}/\Gamma_0^{\text{FSO}}}}^{\infty}\left(\frac{1}{K_{\text{FSO}}\Gamma_{\text{tsh}}} - \frac{1}{K_{\text{FSO}}\Gamma_0^{\text{FSO}}i_k^2}\right)p(i_k)\mathrm{d}i_k + b\int_{\sqrt{\Gamma_{\text{tsh}}/\Gamma_0^{\text{RF}}}}^{\infty}\left(\frac{1}{K_{\text{RF}}\Gamma_{\text{tsh}}} - \frac{1}{K_{\text{RF}}\Gamma_0^{\text{RF}}h^2}\right)p(h)\mathrm{d}h = 1, \tag{10.53}$$

where $b = B_{\text{RF}}/B_{\text{FSO}}$, $p(i_k)$ is the PDF of FSO irradiance i_k given by (10.45), and $p(h)$ is the PDF of RF channel coefficient h given by (10.40). The optimum spectral efficiency, defined as data rate R over channel bandwidth B, can be evaluated by substituting (10.52) into (10.51) to obtain:

$$\frac{R}{B} = \frac{1}{2.19}\int_{\sqrt{\Gamma_{\text{tsh}}/\Gamma_0^{\text{FSO}}}}^{\infty}\log_2\left(\frac{\Gamma_0^{\text{FSO}}i_k^2}{\Gamma_{\text{tsh}}}\right)p(i_k)\mathrm{d}i_k + b\int_{\sqrt{\Gamma_{\text{tsh}}/\Gamma_0^{\text{RF}}}}^{\infty}\log_2\left(\frac{\Gamma_0^{\text{RF}}h^2}{\Gamma_{\text{tsh}}}\right)p(h)\mathrm{d}h\ [\text{bits/s/Hz}] \tag{10.54}$$

Although this adaptive-rate adaptive-power scheme provides excellent spectral efficiencies, the optimum threshold computation in (10.54) is time extensive. Instead, we can perform the truncated channel inversion with fixed rate. The truncated channel inversion adaptation can be performed by

$$\frac{P^{\mathrm{FSO}}(i_k)}{P}=\begin{cases}\dfrac{1}{i_k^2 E_{\Gamma_{\mathrm{tsh}}}^{\mathrm{FSO}}\left[1/i_k^2\right]}, & \Gamma^{\mathrm{FSO}} \geq \Gamma_{\mathrm{tsh}} \\ 0, & \Gamma^{\mathrm{FSO}} < i_{\mathrm{tsh}}\end{cases};\ \frac{P^{\mathrm{RF}}(h)}{P}=\begin{cases}\dfrac{1}{h^2 E_{\Gamma_{\mathrm{tsh}}}^{\mathrm{RF}}\left[1/h^2\right]}, & \Gamma^{\mathrm{RF}} \geq \Gamma_{\mathrm{tsh}} \\ 0, & \Gamma^{\mathrm{RF}} < \Gamma_{\mathrm{tsh}}\end{cases}, \tag{10.55}$$

where

$$E_{\Gamma_{\mathrm{tsh}}}^{\mathrm{FSO}}\left[\frac{1}{i_k^2}\right]=\int_{\sqrt{\Gamma_{\mathrm{tsh}}/\Gamma_0^{\mathrm{FSO}}}}^{\infty}\frac{p(i_k)}{i_k^2}\mathrm{d}i_k,\quad E_{\Gamma_{\mathrm{tsh}}}^{\mathrm{RF}}\left[\frac{1}{h^2}\right]=\int_{\sqrt{\Gamma_{\mathrm{tsh}}/\Gamma_0^{\mathrm{RF}}}}^{\infty}\frac{p(h)}{h^2}\mathrm{d}h. \tag{10.56}$$

The threshold Γ_{tsh} in (10.55) is obtained by maximizing the spectral efficiency as given below:

$$\begin{aligned}R=\max_{\Gamma_{\mathrm{tsh}}}\Bigg\{&\frac{1}{2.19}B_{\mathrm{FSO}}\log_2\left(1+K_{\mathrm{FSO}}\Gamma_0^{\mathrm{FSO}}\frac{1}{E_{\Gamma_{\mathrm{tsh}}}^{\mathrm{FSO}}\left[1/i_k^2\right]}\right)P\left(i_k\geq\sqrt{\Gamma_{\mathrm{tsh}}/\Gamma_0^{\mathrm{FSO}}}\right)\\&+B_{\mathrm{RF}}\log_2\left(1+K_{\mathrm{RF}}\Gamma_0^{\mathrm{RF}}\frac{1}{E_{\Gamma_{\mathrm{tsh}}}^{\mathrm{RF}}\left[1/h^2\right]}\right)P\left(h\geq\sqrt{\Gamma_{\mathrm{tsh}}/\Gamma_0^{\mathrm{RF}}}\right)\Bigg\},\end{aligned} \tag{10.57}$$

where

$$\begin{aligned}P\left(i_k\geq\sqrt{\Gamma_{\mathrm{tsh}}/\Gamma_0^{\mathrm{FSO}}}\right)&=\int_{\sqrt{\Gamma_{\mathrm{tsh}}/\Gamma_0^{\mathrm{FSO}}}}^{\infty}p(i_k)\mathrm{d}i_k;\\P\left(h\geq\sqrt{\Gamma_{\mathrm{tsh}}/\Gamma_0^{\mathrm{RF}}}\right)&=\int_{\sqrt{\Gamma_{\mathrm{tsh}}/\Gamma_0^{\mathrm{RF}}}}^{\infty}p(h)\mathrm{d}h.\end{aligned} \tag{10.58}$$

In Fig. 10.24, we show the spectral efficiencies for FSO system only, which can be achieved using the optimum power and rate adaptation and MPAM for different target bit-error probabilities, and both (a) weak turbulence regime ($\sigma_R = 0.2$, $\alpha' = 51.913$, $\beta' = 49.113$) and (b) strong turbulence regime ($\sigma_R = 2$, $\alpha' = 4.3407$, $\beta' = 1.3088$). For example, the spectral efficiency R/B of 2 bits/s/Hz at $P_b = 10^{-9}$ is achieved for symbol SNR of 23.3 dB in weak turbulence regime and 26.2 dB in strong turbulence regime. In the same figure, we report the spectral efficiencies that can be achieved by both channel inversion ($\Gamma_{\mathrm{tsh}} = 0$) and truncated channel inversion ($\Gamma_{\mathrm{tsh}} > 0$). In the weak turbulence regime (see Fig. 10.24a), even simple channel inversion performs comparable to optimum adaptive-power adaptive-rate scheme. However, in the strong turbulence regime (see Fig. 10.24b), this scheme faces significant performance degradation. On the other hand, the truncated channel inversion scheme in strong turbulence regime faces moderate performance

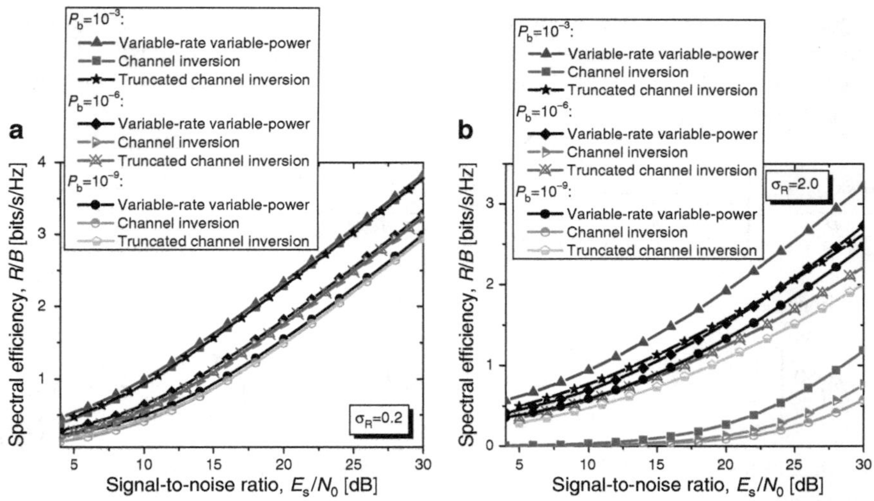

Fig. 10.24 Spectral efficiencies of FSO system against symbol SNR for different target bit probabilities of error (**a**) in weak turbulence regime and (**b**) in strong turbulence regime

degradation, about 3.7 dB at $P_b = 10^{-9}$ for spectral efficiency of 2 bits/s/Hz. The optimum adaptation policy for FSO channel at $P_b = 10^{-6}$ for spectral efficiency 4 bits/s/Hz provides moderate improvement of 3.3 dB in the weak turbulence regime over nonadaptive scheme, while the improvement in the strong turbulence regime is significant 31.7 dB.

In Fig. 10.25, we report the spectral efficiencies for hybrid FSO-RF system shown in Fig. 10.23, with RF subsystem fading parameters $\alpha = 3$, $\mu = 2$ in both weak turbulence regime (Fig. 10.25a) and strong turbulence regime (Fig. 10.25b). We assume that FSO subsystem symbol rate is ten times larger than RF subsystem data rate, that is $b = 0.1$. For spectral efficiency of 2 bits/s/Hz, the hybrid FSO-RF system outperforms the FSO system by 3.39 dB at BER of 10^{-6} and 3.49 dB at BER of 10^{-9}. It is interesting to note that even truncated channel inversion for hybrid system outperforms the optimum adaptation of FSO system by 0.8 dB at BER of 10^{-9} and spectral efficiency of 2 bits/s/Hz.

In Fig. 10.26, we report the spectral efficiencies for hybrid FSO-RF system, with RF subsystem fading parameters $\alpha = 2$, $\mu = 1$ (corresponding to Rayleigh fading) in both weak turbulence regime (Fig. 10.26a) and strong turbulence regime (Fig. 10.26b). This case corresponds to the situation where there is no line-of-site between transmit and receive antennas for RF subsystem. We again assume that $b = 0.1$. For spectral efficiency of 2 bits/s/Hz, the hybrid FSO-RF system outperforms the FSO system by 3.01 dB at BER of 10^{-6} and 3.11 dB at BER of 10^{-9}. The truncated channel inversion for hybrid system performs comparable to the optimum adaptation of FSO system.

In this section, we described two different adaptive modulation scenarios for both FSO system with RF feedback and hybrid FSO-RF system. Further improvements

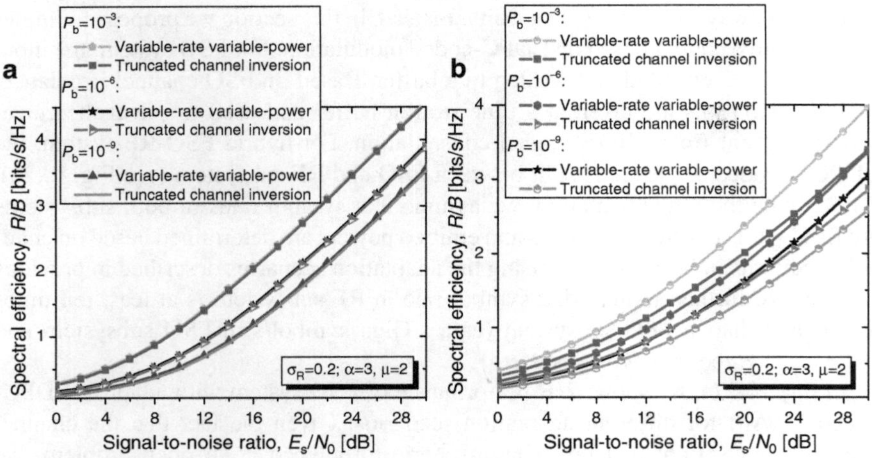

Fig. 10.25 Spectral efficiencies of hybrid FSO-RF system with $\alpha = 3$, $\mu = 2$ fading against symbol SNR for different target bit probabilities of error (**a**) in weak turbulence regime and (**b**) in strong turbulence regime

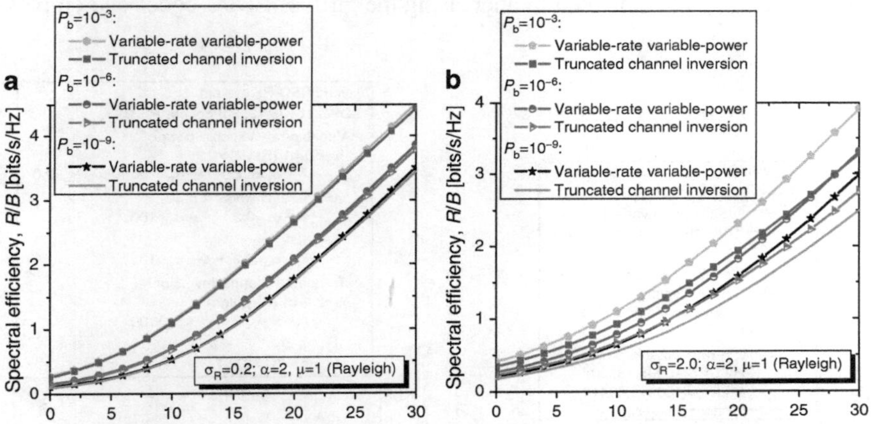

Fig. 10.26 Spectral efficiencies of hybrid FSO-RF system with $\alpha = 2$, $\mu = 1$ (Rayleigh) fading against symbol SNR for different target bit probabilities of error (**a**) in weak turbulence regime and (**b**) in strong turbulence regime

in spectral efficiency can be obtained by adaptive coding. By using the trellis coded modulation (TCM) or cosset codes, we can separate the encoding and modulation process (see [52, 53] for more details). However, to keep the complexity of this approach reasonably low the convolutional or block codes should be simple and short. Those codes are in principle of low rate and weak so that coding gains are moderate. For example, the adaptive coding scheme based on TCM proposed in [53] is about 5 dB away from channel capacity. Instead, in this section we propose to implement adaptive coding based on LDPC-coded modulation. For FSO system, the input data are LDPC encoded and written to a buffer. Based on FSO channel irradiance, i_t, $\log_2 M(i_t)$ bits are taken at a time from a buffer and used to select the corresponding point from MPAM signal constellation. For hybrid FSO-RF system, the LDPC-encoded sequence is split between FSO and RF subsystem (see Fig. 10.23). To facilitate the implementation, we assume that symbol rates in both subsystems are fixed while constellation sizes and emitted powers are determined based on channel conditions in both channels using the adaptation scenarios described in previous section. We further assume that symbol rate in RF subsystem is at least ten times lower than that in FSO subsystem (e.g., 1 Giga symbol/s in FSO subsystem and 100 Mega symbol/s in RF subsystem).

In Fig. 10.27a, we show R/B performance of FSO system with adaptive LDPC-coded MPAM for different adaptation scenarios. Given the fact that the channel capacity of FSO channel under atmospheric turbulence is an open problem, we show in the same figure an *upper bound* in the absence of atmospheric turbulence from [59]. The coding gain over adaptive modulation at $P_b = 10^{-6}$ for $R/B = 4$ bits/s/Hz is 7.2 dB in both (weak and strong) turbulence regimes. Larger coding gains are expected at lower BERs and for higher spectral efficiencies. Further improvements can be obtained by increasing the girth of LDPC codes and employ-

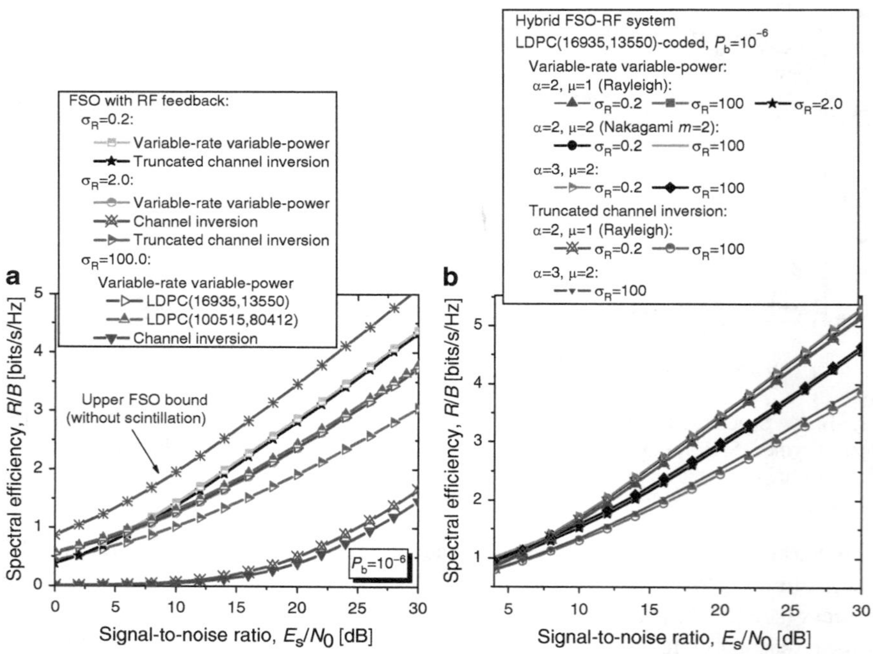

Fig. 10.27 Spectral efficiencies against symbol SNR for adaptive LDPC-coded modulation (**a**) FSO with RF feedback only and (**b**) hybrid FSO-RF system

ing better modulation formats. The increase in codeword length to 100,515 does not improve R/B performance that much as shown in Fig. 10.27a. It is interesting to note that by employing adaptive coding, the communication under saturation regime is possible, as shown in Fig. 10.27a. Moreover, for variable-rate variable-power scheme there is no degradation in saturation regime compared to strong turbulence regime. Overall improvement from adaptive modulation and coding for $R/B = 4$bits/s/Hz at $P_b = 10^{-6}$ over nonadaptive uncoded modulation ranges from 10.5 dB (3.3 dB from adaptive modulation and 7.2 dB from coding) in the weak turbulence regime to 38.9 dB in the strong turbulence regime (31.7 dB from adaptive modulation and 7.2 dB from coding).

In Fig. 10.27b, we show R/B performance of hybrid FSO-RF system with adaptive LDPC-coded modulation (MPAM is used in FSO subsystem and MQAM in RF subsystem) for different adaptation scenarios. The symbol rate in FSO subsystem is set to be ten times larger than that in RF subsystem ($b = 0.1$). For spectral efficiency of 4 bits/s/Hz at BER of 10^{-6}, the improvement of hybrid FSO-RF system over FSO system is 5.25 dB in Rayleigh fading ($\alpha = 2$, $\mu = 1$), 5.51 dB in Nakagami $m = 2$ fading ($\alpha = 2$, $\mu = 2$), and 5.63 dB in $\alpha = 3$, $\mu = 2$ fading. For spectral efficiency of 2 bits/s/Hz at the same BER, the improvement of hybrid FSO-RF system over FSO system is 3.32 dB in Rayleigh fading, 3.72 dB in Nakagami $m = 2$ fading, and 3.86 dB in $\alpha = 3$, $\mu = 2$ fading.

10.7 On the Channel Capacity of Optical MIMO MMF Systems

MMFs are widely used in LANs and optical interconnects. Many of the installed Ethernet backbones are based on MMFs and operate at data rates of about 1 Gb/s. Great deal of research activities is devoted to upgrade 1-Gb/s Ethernet backbones to 10 Gb/s. Nevertheless, it is important to note that MMFs has several advantages over SMFs (1) easy installation, maintenance, and handling, which leads to low cost; (2) the effective cross-sectional area of MMF is much larger than that of SMF, indicating much better tolerance to fiber nonlinearities; and (3) from information theory point of view the existence of many modes can be exploited in MIMO fashion, already in use in wireless communications [52], to improve the spectral efficiency. Recent experiment by Dr. Shieh's group [61], in which the 21.4-Gb/s polarization multiplexed coherent OFDM transmission over 200 km of MMF has been demonstrated, suggests that MMF may be used for beyond short reach applications.

Because OFDM has shown robustness in variety of wireless channels, it has been adopted in many communication standards [30] including wireless LAN (WiFi), digital-video and audio broadcasting standards, digital subscribed loop (DSL), and Worldwide Interoperability for Microwave Access (WiMaX, or IEEE 802.16). An example of OFDM-MIMO signaling scheme over MMF with coherent detection,

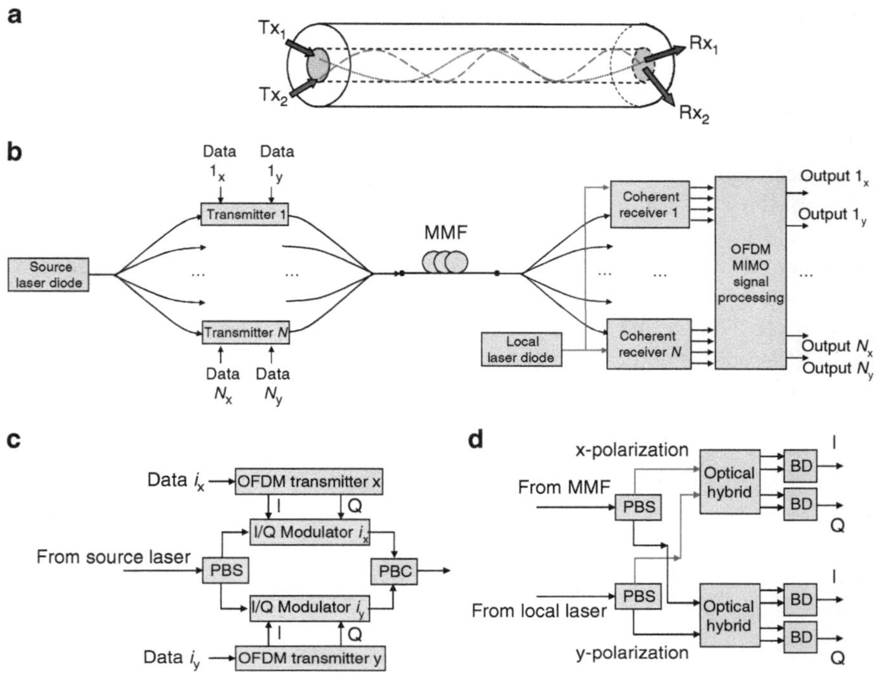

Fig. 10.28 Coherent optical OFDM-MIMO over MMF (**a**) 2 × 2 MIMO example, (**b**) a coherent optical MIMO MMF link, (**c**) *i*th transmitter configuration, and (**d**) coherent receiver configuration. *PBS* polarization beam splitter, *PBC* polarization beam combiner, *BD* balanced detector

employing both polarizations, is shown in Fig. 10.28 [30]. In Fig. 10.28a, 2×2 MIMO system is given along with propagation paths of two modes, which occupy a different cross section of the optical fiber core and take a slightly different path along the optical fiber. In Fig. 10.28b, an example of MIMO system with M_T transmitters and M_R receivers is given, for simplicity we set $M_T = M_R = N$. Because this system employs both polarizations, it is able to carry $2M_T$ independent data streams. The laser diode continuous wave output signal is split among M_R branches, each containing a transmitter, whose configuration is given in Fig. 10.28c. Two independent data streams are used as inputs to OFDM transmitter x, whose signal is transmitted over x-polarization, and OFDM transmitter y, whose signal is transmitted over y-polarization. To convert OFDM signals in optical domain, I/Q modulators are used. The optical OFDM signals from both polarizations are combined by polarization beam combiner. On receiver side, N different outputs from MMF are used as inputs to N coherent receivers, whose configuration is shown in Fig. 10.28d. The receiver configuration is typical for coherent OFDM systems, as it is already explained in previous chapters. One local laser is used for all coherent receivers to reduce the system cost. The in-phase (I) and quadrature (Q) channel outputs from both polarizations are used as inputs to OFDM-MIMO signal processing block. The use of OFDM facilitates multimode dispersion compensation. The scheme shown in Fig. 10.28 can be used in both broadcasting and multicasting applications [62, 63].

Let us now consider a MIMO system over MMF with M_T transmitter and M_R receivers, the corresponding input–output relation can be established as

$$y_i\,(t) = \sum_{j=1}^{M_\mathrm{T}} \sum_{k=1}^{P} h_{ji}\,(k)\, \mathrm{e}^{j\omega_\mathrm{c}(t-\tau_{p,k})} x_j\left(t - \tau_{g,k}\right) + w_i\,(t), \tag{10.59}$$

where $y_i(t)$ is the signal received by ith receiver ($i = 1, 2, \ldots, M_R$), while $h_{ji}(k)$ is the channel gain from jth transmitter through the kth mode. $\tau_{p,k}$ and $\tau_{g,k}$ denote the phase and group delay associated with kth mode, respectively. (ω_c is the carrier frequency and w_i is the noise signal.) $x_j(t)$ denotes the transmitted signal originating from jth transmitter and P denotes the number of modes. Note that expression (10.59) observes only one polarization. When the group delay spread $\Delta\tau_\mathrm{g} = \tau_\mathrm{g,P} - \tau_\mathrm{g,1}$ is small compared to symbol duration, the following is valid $x(t - \tau_{\mathrm{g},k}) \approx x(t - \tau_\mathrm{g})$, meaning that all paths arrive at approximately the same time, which is valid up the certain MMF length. Then the sampled baseband equivalent of (10.59) can be written in matrix form as follows:

$$\boldsymbol{y}\,(n) = \boldsymbol{H}x\,(n) + \boldsymbol{w}(n), \tag{10.60a}$$

where

$$\boldsymbol{y}(n) = \begin{bmatrix} y_1 & (nT_\mathrm{s}) \\ & \vdots \\ y_{M_\mathrm{R}} & (nT_\mathrm{s}) \end{bmatrix}, \quad \boldsymbol{H}_{ij} = \sum_{k=1}^{P} h_{ij}(k)\mathrm{e}^{-\mathrm{j}\omega_\mathrm{c}\tau_{p,k}}, \quad \boldsymbol{x}(n) = \begin{bmatrix} x_1 & (nT_\mathrm{s}) \\ & \vdots \\ x_{M_\mathrm{T}} & (nT_\mathrm{s}) \end{bmatrix},$$
$$\boldsymbol{w}(n) = \begin{bmatrix} w_1 & (nT_\mathrm{s}) \\ & \vdots \\ w_{M_\mathrm{R}} & (nT_\mathrm{s}) \end{bmatrix} \tag{10.60b}$$

When the number of modes is large so that each Tx and Rx launches and samples from sufficiently different groups of modes and when the product of carrier frequency and phase-delay spread $\omega_c\tau_{p,k} \gg 2\pi$ so that each element of $\boldsymbol{H}$ can be considered as uniformly distributed, then the elements of $\boldsymbol{H}$ will have complex Gaussian distribution with Rayleigh distribution for amplitudes [30, 52]. The simulations performed in [64] confirm validity of this assumption.

Because the MMF channel varies at relatively slower rate compared to data being transmitted, sending the estimated CSI back to transmitter is feasible. The channel capacity can be evaluated for two different scenarios (1) when CSI is available at transmitter side (CSIT) and (2) when CSI is not available at transmitter side.

The channel capacity for CSIT case can be obtained by averaging the capacities associated with each channel realization [52]:

$$C = E_H\left\{\max_{R_x:\,\mathrm{Tr}(R_x)=\rho} B\log_2\left[\det\left(\boldsymbol{I}_{M_R} + \boldsymbol{H}\boldsymbol{R}_x\boldsymbol{H}^\dagger\right)\right]\right\}, \quad \rho = \sum_{i=1}^{M_\mathrm{T}} E[\boldsymbol{x}_i\boldsymbol{x}_i^*], \tag{10.61}$$

where $\boldsymbol{R}_x$ is the covariance matrix of transmitted data and $\boldsymbol{I}$ is the identity matrix. We use Tr( ) to denote the trace of matrix (the sum of diagonal elements), $E[\,]$ to denote the expectation operator, and † to denote Hermitian operation (transposition and complex conjugate). By performing the singular value decomposition (SVD) of channel matrix

$$\begin{aligned} \boldsymbol{H} &= \boldsymbol{U}\boldsymbol{\Sigma}\boldsymbol{V}^\dagger; \quad \boldsymbol{U}^\dagger\boldsymbol{U} = \boldsymbol{I}_{M_\mathrm{R}}; \quad \boldsymbol{V}^\dagger\boldsymbol{V} = \boldsymbol{I}_{M_\mathrm{T}}; \quad \Sigma = \mathrm{diag}(\sigma_i), \\ \sigma_i &= \sqrt{\lambda_i}, \quad \lambda_i = \mathrm{eigenvalues}(\boldsymbol{H}\boldsymbol{H}^\dagger), \end{aligned} \tag{10.62}$$

the MIMO channel capacity can be written as

$$C = E_H\left\{\max_{P_i:\,\sum_i P_i \le \overline{P}} \sum_{i=1}^{R_H} B\log_2\left(1 + \frac{P_i\gamma_i}{\overline{P}}\right)\right\}, \quad \gamma_i = \frac{\sigma_i^2\overline{P}}{\sigma^2}, \quad \sigma^2 = E\left[w_i^2\right], \tag{10.63}$$

where $R_H = \mathrm{rank}(\boldsymbol{H}) \le \min(M_\mathrm{T}, M_\mathrm{R})$ is the number of nonzero singular values of $\boldsymbol{H}$.

When the CSI is not known on transmitter side the transmitter assumes zero-mean spatially white model for $\boldsymbol{H}$ so that corresponding ergodic channel capacity can be determined by [52]

$$C = \max_{R_x:\,\mathrm{Tr}(R_x)=\rho} E_H\left\{B\log_2\left[\det\left(\boldsymbol{I}_{M_\mathrm{T}} + \boldsymbol{H}\boldsymbol{R}_x\boldsymbol{H}^\dagger\right)\right]\right\}, \tag{10.64}$$

where the expectation is performed with respect to the distribution on the channel matrix. The optimum input covariance that maximizes ergodic channel capacity is the scaled identity matrix $\boldsymbol{R}_x = (\rho/M_\mathrm{T})\boldsymbol{I}_{M_\mathrm{T}}$, so that ergodic channel capacity can be written as

$$C = E_H\left\{B\log_2\left[\det\left(\boldsymbol{I}_{M_\mathrm{R}} + \frac{\rho}{M_\mathrm{T}}\boldsymbol{H}\boldsymbol{H}^\dagger\right)\right]\right\}. \tag{10.65}$$

The channel capacity of a 2 × 2 MIMO over MMF for both scenarios is shown in Fig. 10.29, from [64]. The results are obtained by propagating two signals over MMF link modeled using RSofts's LinkSIM. For comparison purposes, the ideal Gaussian MIMO case is provided as well. For high SNRs, we can see degradation of optical MIMO over MMF in comparison with ideal Gaussian MIMO. The significant spectral efficiency improvement was found for optical MIMO over optical single-input single-output (SISO). Note that those numerical results are obtained assuming that all modes arrive at the same time and only one polarization is observed. These results can be observed as upper bound for channel capacity of optical MIMO over MMF.

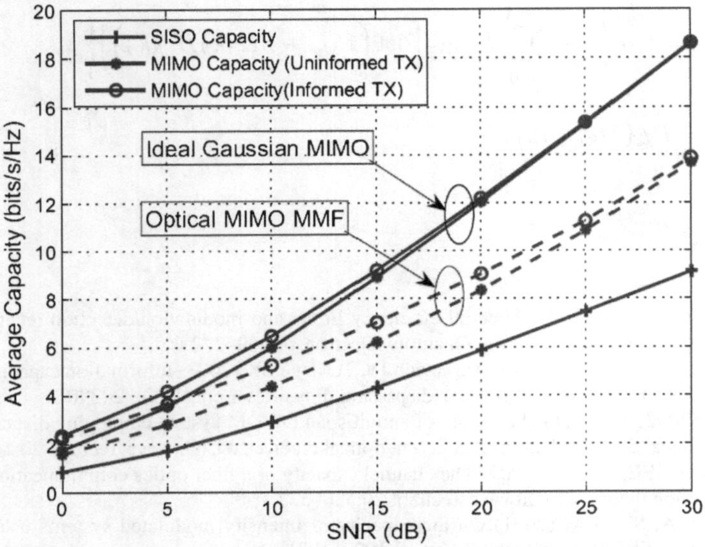

Fig. 10.29 Ergodic channel capacity against SNR for 2 × 2 MIMO over MMF. The number of modes was set to 500 (After ref. [64]; © IEEE 2009; reprinted with permission.)

The channel capacity discussed above is related to single-carrier coherent MIMO over MMF. The corresponding channel capacity expressions given above can be generalized to OFDM-MIMO systems with N_c subcarriers as follows [44, 65]

$$C = E_H \left\{ \frac{1}{N_c} \max_{\mathrm{Tr}(\Sigma)\le P} B \log_2 \left[\det \left(\boldsymbol{I}_{M_R N_c} + \frac{1}{\sigma_n^2} H \Sigma H^\dagger \right) \right] \right\}, \tag{10.66}$$

where Σ is the covariance matrix of Gaussian input vector, defined as

$$\Sigma = \mathrm{diag}\left\{\Sigma_k\right\}_{k=0}^{N_c-1}, \quad \Sigma_k = \frac{P}{M_R N_c} \boldsymbol{I}_{M_R}; \quad k = 0, 1, \ldots, N_c - 1 \tag{10.67}$$

with P being the maximum overall transmit power and σ_n being the standard deviation of noise process. The channel matrix in (10.66) is now a block diagonal matrix with ith block diagonal element corresponding to $M_T \times M_R$ MIMO over MMF channel:

$$\boldsymbol{H} = \begin{bmatrix} \boldsymbol{H}(0) & \ldots & 0 \\ \ldots & \ldots & \ldots \\ 0 & \ldots & \boldsymbol{H}(N_c - 1) \end{bmatrix} \tag{10.68}$$

With uniform power allocation, the OFDM MIMO over MMF channel capacity is given by

$$C = E_H \left\{ \frac{1}{N_c} \sum_{k=0}^{N_c-1} B \log_2 \left[\det \left(\boldsymbol{I}_{M_R} + \rho \boldsymbol{H}(k) \boldsymbol{H}^\dagger(k) \right) \right] \right\}, \tag{10.69}$$

where $\rho = P / \left(M_R N_c \sigma_n^2 \right)$.

References

1. Kahn JM, Ho KP (2004) Spectral efficiency limits and modulation/detection techniques for DWDM systems. IEEE Sel Top Quantum Electron 10:259–272
2. Turitsyn KS, Derevyanko SA, Yurkevich IV, Turitsyn SK (2003) Information capacity of optical fiber channels with zero average dispersion. Phys Rev Lett 91(20). 203901
3. Tang J (2002) The channel capacity of a multispan DWDM system employing dispersive nonlinear optical fibers and an ideal coherent optical receiver. J Lightwave Technol 20:1095–1101
4. Narimanov EE, Mitra P (2002) The channel capacity of a fiber optics communication system: perturbation theory. J Lightwave Technol 20:530–537
5. Mecozzi A, Shtaif M (2001) On the capacity of intensity modulated systems using optical amplifiers. IEEE Photon Technol Lett 13:1029–1031
6. Mitra PP, Stark JB (2001) Nonlinear limits to the information capacity of optical fiber communications. Nature 411:1027–1030
7. Tang J (2001) The Shannon channel capacity of dispersion-free nonlinear optical fiber transmission. J Lightwave Technol 19:1104–1109
8. Tang J (2001) The multispan effects of Kerr nonlinearity and amplifier noises on Shannon channel capacity for a dispersion-free nonlinear optical fiber. J Lightwave Technol 19: 1110–1115
9. Ho K-P, Kahn JM (2002) Channel capacity of WDM systems using constant-intensity modulation formats. In: Optical fiber communication conference (OFC'02), paper ThGG85
10. Kahn JM, Ho K-P (2002) Ultimate spectral efficiency limits in DWDM systems. In: Optoelectronics communication conference, Yokohama, Japan, July 8–12, 2002
11. Djordjevic IB, Vasic B (2004) Approaching Shannon's capacity limits of fiber optics communications channels using short LDPC codes. In: CLEO/IQEC 2004, paper CWA7
12. Narimanov E, Patel P. (2003) Channel capacity of fiber optics communications systems: WDM vs. TDM. In: Proceedings of conference on lasers and electro-optics (CLEO '03), pp 1666–1668
13. Li J (2003) On the achievable information rate of asymmetric optical fiber channels with amplifier spontaneous emission noise. In: Proceedings of IEEE military communication conference (MILCOM '03), Boston, MA, October 2003
14. Djordjevic IB, Vasic B, Ivkovic M, Gabitov I (2005) Achievable information rates for high-speed long-haul optical transmission. IEEE/OSA J Lightwave Technol 23:3755–3763
15. Ivkovic M, Djordjevic IB, Vasic B (2007) Calculation of achievable information rates of long-haul optical transmission systems using instanton approach. IEEE/OSA J Lightwave Technol 25:1163–1168
16. Djordjevic IB, Alic N, Papen G, Radic S (2007) Determination of achievable information rates (AIRs) of IM/DD systems and AIR loss due to chromatic dispersion and quantization. IEEE Photon Technol Lett 19(1):12–14
17. Minkov LL, Djordjevic IB, Batshon HG, Xu L, Wang T, Cvijetic M, Kueppers F (2007) Demonstration of PMD compensation by LDPC-coded turbo equalization and channel capacity loss characterization due to PMD and quantization. IEEE Photon Technol Lett 19(22): 1852–1854
18. Minkov LL, Djordjevic IB, Xu L, Wang T, Kueppers F (2008) Evaluation of large girth LDPC codes for PMD compensation by turbo equalization. Opt Express 16(17):13450–13455

19. Djordjevic IB, Minkov LL, Batshon HG (2008) Mitigation of linear and nonlinear impairments in high-speed optical networks by using LDPC-coded turbo equalization. IEEE J Sel Areas Commun Opt Commun Netw 26(6):73–83
20. Djordjevic IB, Minkov LL, Xu L, Wang T (2009) Suppression of fiber nonlinearities and PMD in coded-modulation schemes with coherent detection by using turbo equalization. IEEE/OSA J Opt Commun Netw 1(6):555–564
21. Ivkovic M, Djordjevic I, Rajkovic P, Vasic B (2007) Pulse energy probability density functions for long-haul optical fiber transmission systems by using instantons and edgeworth expansion. IEEE Photon Technol Lett 19(20):1604–1606
22. Arnold D, Kavcic A, Loeliger H-A, Vontobel PO, Zeng W (2003) Simulation-based computation of information rates: upper and lower bounds. In: Proceedings of IEEE international symposium on information theory (ISIT 2003), p 119
23. Arnold D, Loeliger H-A (2001) On the information rate of binary-input channels with memory. In: Proceedings of 2001 international conference on communications, Helsinki, Finland, June 11–14, 2001, pp 2692–2695
24. Pfitser HD, Soriaga JB, Siegel PH (2001) On the achievable information rates of finite state ISI channels. In: Proceedings of Globecom 2001, San Antonio, TX, Nov. 25–29, 2001, pp 2992–2996
25. Essiambre R-J, Foschini GJ, Kramer G, Winzer PJ (2008) Capacity limits of information transport in fiber-optic networks. Phys Rev Lett 101:163901-1–163901-4
26. Ip E, Kahn JM (in press) Nonlinear impairment compensation using backpropagation. In: Optical fibre, new developments. In-Tech, Vienna, Austria
27. Djordjevic IB, Vasic B (2006) Orthogonal frequency-division multiplexing for high-speed optical transmission. Opt Express 14:3767–3775
28. Shieh W, Athaudage C (2006) Coherent optical frequency division multiplexing. Electron Lett 42:587–589
29. Lowery AJ, Du L, Armstrong J (2006) Orthogonal frequency division multiplexing for adaptive dispersion compensation in long haul WDM systems. In: Proceedings of OFC postdeadline papers, Mar. 2006, paper no. PDP39
30. Shieh W, Djordjevic IB (2009) Optical orthogonal frequency division multiplexing. Elsevier, Amsterdam
31. Cover TM, Thomas JA (1991) Elements of information theory. Wiley, New York
32. Reza FM (1961) An introduction to information theory. McGraw-Hill, New York
33. Gallager RG (1968) Information theory and reliable communication. Wiley, New York
34. Shannon CE (1948) A mathematical theory of communication. Bell Syst Tech J 27:379–423, 623–656
35. McMillan B (1952) The basic theorems of information theory Ann Math Stat 24:196–219
36. Khinchin AI (1957) Mathematical foundations of information theory. Dover, New York
37. Agrawal GP (2001) Nonlinear fiber optics. Academic, San Diego, CA
38. Pinsker MS (1964) Information and information stability of random variables and processes. Holden Day, San Francisco, CA
39. Turritsyn SK et al (2000) Averaged model and integrable limits in nonlinear double-periodic Hamiltonian systems. Phys Rev E 61:3127–3132
40. Bertsekas DP (1999) Nonlinear programming, 2nd edn. Athena Scientific, Nashua, NH
41. Chong EKP Zak SH (2008) An introduction to optimization, 3rd edn. Wiley, New York
42. Webb WT, Steele R (1995) Variable rate QAM for mobile radio IEEE Trans Commun 43: 2223–2230
43. Peric ZH, Djordjevic IB, Bogosavljevic SM Stefanovic MC (1998) Design of signal constellations for Gaussian channel by iterative polar quantization In: Proceedings of 9th mediterranean electrotechnical conference, Tel-Aviv, Israel, 18–20 May 1998, vol 2, pp 866–869
44. Bölcskei H, Gesbert D, Paulraj AJ (2002) On the capacity of OFDM-based spatial multiplexing systems. IEEE Trans Commun 50:225–234
45. Yabre G (2000) Theoretical investigation on the dispersion of graded-index polymer optical fibers. J Lightwave Technol 18:869–877

46. Ishigure T, Koike Y, Fleming JW (2000) Optimum index profile of the perfluorinated polymer-based GI polymer optical fiber and its dispersion properties J Lightwave Technol 18:178–184
47. Yang H, Lee SCJ, Tangdiongga E, Breyer F, Randel S, Koonen AMJ (2009) 40-Gb/s transmission over 100 m graded-index plastic optical fiber based on discrete multitone modulation. In: Postdeadline papers, OFC/NFOEC 2009, Paper PDPD8
48. Stöhr A et al (2009) 60 GHz radio-over-fiber technologies for broadband wireless services. OSA J Opt Netw 8:471–487
49. Lee SCJ, Breyer F, Randel S, van den Boom HPA, Koonen AMJ (2008) High-speed transmission over multimode fiber using discrete multitone modulation [Invited]. J Opt Netw 7:183–196
50. Ghavami M, Michael LB, Kohno R (2007) Ultra wideband signals and systems in communication engineering. Wiley, New York
51. Djordjevic IB (2009) Adaptive modulation and coding for communication over the atmospheric turbulence channels. In: IEEE photonics society summer topicals 2009, 20–22 July 2009, Newport Beach, CA, Paper no. TuD3. 3
52. Goldsmith A (2005) Wireless communications. Cambridge University Press, Cambridge
53. Goldsmith A, Chua S-G (1998) Adaptive coded modulation for fading channels. IEEE Trans Commun 46:595–601
54. Yacoub MD (2002) The α–μ distribution: a general fading distribution In: Proceedings of IEEE international symposium PIMRC, September 2002, vol 2, pp 629–633
55. Yacoub MD (2007) The α–μ distribution: a physical fading model for the Stacy distribution IEEE Trans Veh Technol 56(1):27–34
56. Al-Habash MA, Andrews LC, Phillips RL (2001) Mathematical model for the irradiance probability density function of a laser beam propagating through turbulent media. Opt Eng 40:1554–1562
57. Andrews LC, Philips RL (2005) Laser beam propagation through random media. SPIE Press, Bellingham, WA
58. Djordjevic IB, Denic S, Anguita J, Vasic B, Neifeld MA (2008) LDPC-coded MIMO optical communication over the atmospheric turbulence channel. IEEE J Lightwave Technol 26(5):478–487
59. Farid AA, Hranilovic S (2007) Upper and lower bounds on the capacity of wireless optical intensity channels. In: Proceedings of ISIT 2007, Nice, France, June 24–29, 2007, pp 2416–2420.
60. Shah AR, Hsu RCJ, Sayed AH, Jalali B (2005) Coherent optical MIMO (COMIMO). J Lightwave Technol 23(8):2410–2419
61. Tong Z, Yang Q, Ma Y, Shieh W (2008) 21.4 Gb/s coherent optical OFDM transmission over multimode fiber In: Post-deadline papers technical digest, 13th Optoelectronics and communications conference (OECC) and 33rd Australian conference on optical fibre technology (ACOFT), Paper no. PDP-5
62. Bikhazi NW, Jensen MA, Anderson AL (2009) MIMO signaling over the MMF optical broadcast channel with square-law detection. IEEE Trans Commun 57(3):614–617
63. Agmon A, Nazarathy M (2007) Broadcast MIMO over multimode optical interconnects by modal beamforming. Opt Express 15(20):13123–13128
64. Hsu RCJ, Tarighat A, Shah A, Sayed AH, Jalali B (2005) Capacity enhancement in coherent optical MIMO (COMIMO) multimode fiber links. J Lightwave Technol 23(8):2410–2419
65. Wang J, Zhu S, Wang L (2005) On the channel capacity of MIMO-OFDM systems. In: Proceedings of international symposium on communications and information technologies 2005 (ISCIT 2005), Beijing, China, October 12–14, 2005, pp 1325–1328

Chapter 11
Future Research Directions in Coding for Optical Channels

Today's photonic infrastructure, whose foundations were established several decades ago, gradually extends from global backbone to access networks and even further, down to the curb, building, home, and desk. Recent studies indicate that each household in North America should be connected by at least 100 Mb/s (e.g., CANARIE [1]), which cannot be accommodated by the last century's technology. The 100 Gb/s Ethernet is currently under standardization, and according to industry experts 1 Tb/s Ethernet should be standardized by the year 2012–2013 [2–5]. Migrating to higher transmission rates comes along with certain challenges such as degradation in the signal quality due to different linear and nonlinear channel impairments and increased installation costs [5–9]. In addition to increased bandwidth, future optical networks will also require flexible wavelength management, the integration of transmission and switching, and optical signal processing functionality, while maintaining minimized operational and capital expenditures.

Current limitations of photonics-enabled networks also result from the heterogeneity of the infrastructure and consequential bottlenecks at different boundaries and interfaces. In optically routed networks, neighboring dense wavelength division multiplexing (DWDM) channels carry random traffic patterns in which different lightwave paths experience different penalties due to the deployment of reconfigurable optical add-drop multiplexers (ROADMs) and wavelength cross-connects (WXCs). Different wavelength channels carrying the traffic to different destinations can have quite different signal-to-noise ratios (SNRs) and spectral distortions due to cascaded filtering effects. The Internet of the future should be able to support a wide range of services containing a large amount of multimedia over different network types at high speed [10]. The future optical networks will allow the integration of fiber-optics and free-space optical (FSO) and RF and optical technologies [10–14].

It is interesting to note that the coding, modulation, detection, and signal-processing algorithms employed in today's fiber-optic communication channels are in many respects primitive when compared to the techniques used in wireless and storage (in particular magnetic recording) systems. This is mainly due to limitations imposed by the extremely high transmission speed and the lack of phase-preserving mechanism of direct detection, which is a prerequisite for many advanced techniques. On the other hand, the need to deal with fast fading in wireless channels has led to development of very sophisticated modulation, detection, and coding

I. Djordjevic et al., *Coding for Optical Channels*,
DOI 10.1007/978-1-4419-5569-2_11,

techniques that are also adaptive to time-varying channel conditions. The digital signal processing (DSP) has played a vital role in wireless communication and has enabled so-called software-defined radio (SDR) [15–17], once taught to be just a dream. Thanks to the recent resurgence of coherent detection [5] and recent drive toward dynamically reconfigurable optical networks [18] with transmission speeds beyond 100 Gb/s [5, 8, 19], DSP and FEC are becoming increasingly important. Because different signal transmission paths in an optically-routed network can pass different number of optical amplifiers, WXCs, and ROADMs they can have quite different SNRs at destination sides. Moreover, the corresponding signals can be differently impacted by various channel impairments including PMD, chromatic dispersion, fiber nonlinearities, and filtering effects due to concatenation of optical filters/ROADMs/WXCs. Regardless of the data destination, an optical transport system (OTS) must provide the predefined bit-error rate (BER) performance. To meet the needs of high-capacity, high-speed traffic, service providers are searching for plausible solutions that preferably work with the currently installed fiber-optic networks. To achieve target BER regardless of the data destination, we have recently proposed [20, 21, 30] to adjust the FEC strength according to the optical channel conditions. For long-haul transmission we employ stronger channel codes with lower code rate (higher overhead), while for the transmission over short and/or high-quality links we use the codes with high code rate (low overhead). Depending on the information obtained from the monitoring channels, we select the appropriate code rate matching to the SNR range that current channel SNR falls into, and to avoid frame synchronization issues we keep the codeword length fixed independent of the FEC code being used. Such an approach leads us toward the *software-defined optical transmission* (SDOT) [5] in which the transponder can be adapted or reconfigured to multiple standards, multiple modulation formats or code rates; the concept very similar to SDR. To achieve high-capacity, high-speed transmission with currently available technology we can reduce the symbol rate to 10, 25, or 40 Giga Symbols/s (GS/s) range and by using polarization-multiplexed, bandwidth-efficient multilevel modulation schemes based on LDPC codes we can achieve aggregate rate beyond 100 Gb/s per wavelength. The common denominator is the employment of quasi-cyclic (QC) LDPC codes in FEC.

In this chapter we study various FEC and coded modulation schemes suitable for use in next-generation optical transport networks. In Sect. 11.1 we describe binary QC-LDPC codes of large girth and their implementation in FPGA. The corresponding QC nonbinary LDPC codes are described in Sect. 11.2. Various binary-coded modulation schemes, including coded OFDM, are already introduced in Chap. 6. The corresponding nonbinary counterpart is described in Sect. 11.3. Next two sections are devoted to adaptive LDPC-coded modulation (Sect. 11.4) and adaptive LDPC-coded OFDM (Sect. 11.5). The overall decoding complexity of both binary and nonbinary LDPC codes can be reduced by employing so-called generalized LDPC (GLDPC) codes, which are described in Sect. 11.6. The information capacity formula for N-dimensional constellations, derived in Chap. 10, indicates that information capacity is logarithmic function of SNR, while linear function of dimension N. In Sect. 11.7, we study the possibility of improving spectral efficiency of optical transport networks by coded multidimensional hybrid frequency/amplitude/phase/polarization modulation. Finally, some important concluding remarks are provided in Sect. 11.8. This chap. is based on our several recent publications [20–30].

11.1 Binary QC-LDPC Codes of Large Girth

In this section, we describe a method for designing large-girth QC-LDPC codes and an FPGA implementation of decoders for QC codes.

11.1.1 Design of Large-Girth QC-LDPC Codes

Based on Tanner's bound for the minimum distance of an LDPC code [31]

$$d \geq \begin{cases} 1 + \frac{w_c}{w_c - 2}\left((w_c-1)^{\lfloor (g-2)/4 \rfloor} - 1\right), & g/2 = 2m+1, \\ 1 + \frac{w_c}{w_c - 2}\left((w_c-1)^{\lfloor (g-2)/4 \rfloor} - 1\right) + (w_c-1)^{\lfloor (g-2)/4 \rfloor}, & g/2 = 2m, \end{cases} \tag{11.1}$$

where g and w_c denote the girth of the code graph and the column weight, respectively, and where d stands for the minimum distance of the code; it follows that large girth leads to an exponential increase in the minimum distance, provided that the column weight is at least 3. ($\lfloor\rfloor$ denotes the largest integer less than or equal to the enclosed quantity.) For example, the minimum distance of girth-10 codes with column weight $r = 3$ is at least 10. The parity-check matrix of regular QC-LDPC codes [9, 32] can be represented by

$$H = \begin{bmatrix} I & I & I & \cdots & I \\ I & P^{S[1]} & P^{S[2]} & \cdots & P^{S[c-1]} \\ I & P^{2S[1]} & P^{2S[2]} & \cdots & P^{2S[c-1]} \\ \cdots & \cdots & \cdots & \cdots & \cdots \\ I & P^{(r-1)S[1]} & P^{(r-1)S[2]} & \cdots & P^{(r-1)S[c-1]} \end{bmatrix}, \tag{11.2}$$

where I is $p \times p$ (p is a prime number) identity matrix, P is $p \times p$ permutation matrix given by $P = (p_{ij})_{p \times p}$, $p_{i,i+1} = p_{p,1} = 1$ (zero otherwise), and where r and c represent the number of block-rows and block-columns in (11.2), respectively. The set of integers S are to be carefully chosen from the set $\{0, 1, \ldots, p-1\}$ so that the cycles of short length, in the corresponding Tanner (bipartite) graph representation of (11.2), are avoided. According to Theorem 2.1 in [32], we have to avoid the cycles of length $2k (k = 3 \text{ or } 4)$ defined by the following equation

$$S[i_1]j_1 + S[i_2]j_2 + \cdots + S[i_k]j_k = S[i_1]j_2 + S[i_2]j_3 + \cdots + S[i_k]j_1 \bmod p, \tag{11.3}$$

where the closed path is defined by $(i_1, j_1), (i_1, j_2), (i_2, j_2), (i_2, j_3), \ldots, (i_k, j_k), (i_k, j_1)$ with the pair of indices denoting row-column indices of permutation-blocks in (11.2) such that $l_m \neq l_{m+1}, l_k \neq l_1 (m = 1, 2, \ldots, k; l \in \{i, j\})$. Therefore, we have to identify the sequence of integers $S[i] \in \{0, 1, \ldots, p-1\} (i = 0, 1, \ldots, r-1; r < p)$ not satisfying (11.3), which can be done either by computer search or in a combinatorial fashion. For example, to design the QC-LDPC codes in [33], we introduced the concept of the cyclic-invariant difference set (CIDS). The CIDS-based codes come naturally as girth 6 codes, and to increase the girth we had to selectively remove certain elements from a CIDS. The design of LDPC codes of rate above 0.8, column weight 3, and girth 10 using the CIDS approach is a very challenging and still an open problem. Instead, in our recent paper [34], we solved this problem by developing an efficient computer search algorithm. We add an integer at a time from the set $\{0, 1, \ldots, p-1\}$ (not used before) to the initial set S and check if (11.3) is satisfied. If (11.3) is satisfied, we remove that integer from the set S and continue our search with another integer from set $\{0, 1, \ldots, p-1\}$ until we exploit all the elements from $\{0, 1, \ldots, p-1\}$. The code rate of these QC codes, R, is lower-bounded by

$$R \geq \frac{|S|p - rp}{|S|p} = \frac{1-r}{|S|}, \tag{11.4}$$

and the codeword length is $|S|p$, where $|S|$ denotes the cardinality of set S. For a given code rate R_0, the number of elements from S to be used is $\lfloor r/(1-R_0) \rfloor$. With this algorithm, LDPC codes of arbitrary rate can be designed.

Example 11.1. By setting $p = 2,311$, the set of integers to be used in (11.3) is obtained as $S = \{1, 2, 7, 14, 30, 51, 78, 104, 129, 212, 223, 318, 427, 600, 808\}$. The corresponding LDPC code has rate $R_0 = 1\text{–}3/15 = 0.8$, column weight 3, girth 10 and length $|S|p = 15 \cdot 2,311 = 34,665$. In the above example, the initial set of integers was $S = \{1, 2, 7\}$, and the set of row to be used in (11.3) is $\{1, 3, 6\}$. The use of a different initial set will result in a different set from that obtained above.

Example 11.2. By setting $p = 269$, the set S is obtained as $S = \{0, 2, 3, 5, 9, 11, 12, 14, 27, 29, 30, 32, 36, 38, 39, 41, 81, 83, 84, 86, 90, 92, 93, 95, 108, 110, 111, 113, 117, 119, 120, 122\}$. If 30 integers are used, the corresponding LDPC code has rate $R_0 = 1\text{–}3/30 = 0.9$, column weight 3, girth-8, and length $30 \cdot 269 = 8,070$.

11.1.2 FPGA Implementation of Large-Girth LDPC Codes

We use the min-sum algorithm, which is a simplified version of the min-sum-with-correction-term algorithm detailed in Chap. 5. Among various alternatives, we adopted a partially parallel architecture in our implementation since it is a natural choice for QC codes. In this architecture, a processing element (PE) is assigned to a group of nodes of the same kind instead of a single node. A PE mapped to a group of bit nodes is called a bit-processing element (BPE), and a PE mapped to a group

		BPEs				
		0	**1**	**2**	**…**	***c*-1**
	0	I	I	I	…	I
	1	I	$P^{S[1]}$	$P^{S[2]}$	…	$P^{S[c-1]}$
CPEs	**2**	I	$P^{2S[1]}$	$P^{2S[2]}$	…	$P^{2S[c-1]}$
	…	…	…	…	…	…
	***r*-1**	I	$P^{(r-1)S[1]}$	$P^{(r-1)S[2]}$	…	$P^{(r-1)S[c-1]}$

Fig. 11.1 The assignment of bit nodes and check nodes to BPEs and CPEs, respectively

Table 11.1 Memory allocation of the implementation

	MEM name				
	MEM B	MEM C	MEM E	MEM I	MEM R
Data word (bits)	8	11	1	8	32
Address word (bits)	16	16	15	15	10
Memory block size (words)	50805	50805	16935	16935	625

of check nodes is called a check-processing element (CPE). BPEs (CPEs) process the nodes assigned to them in a serial fashion. However, all BPEs (CPEs) carry out their tasks simultaneously. Thus, by changing the number of elements assigned to a single BPE and CPE, one can control the level of parallelism in the hardware. In Fig. 11.1, we depict a convenient method for assigning BPEs and CPEs to the nodes in a QC-LDPC code. This method is not only easy to implement but also advantageous since it simplifies the memory addressing.

The messages between BPEs and CPEs are exchanged via memory banks. In Table 11.1, we summarize the memory allocation in our implementation where we used the following notation: MEM B and MEM C denote the memories used to store bit node and check node edge values, respectively; MEM E stores the codeword estimate; MEM I stores the initial log-likelihood ratios; and finally, MEM R holds the state of the random number generator needed for AWGN source, which is based on Mersenne Twister algorithm.

In our initial design [35], we used the MitrionC hardware programming language, which is "an intrinsically parallel C-family language" developed by Mitrionics, Inc. [36]. Using MitrionC syntax, we provided a pseudo code in Fig. 11.2 showing how the data are transferred from MEM B to MEM C after being processed by BPEs. The code features three loop expressions of two types. The *for* loop sequentially executes its loop body for every bit node, i, in a BPE. On the contrary, the *for each* loop is a parallel loop, and hence, the operations in the loop body are applied to all the elements in its declaration simultaneously. To expatiate, due to the first *for each* loop, all BPEs perform their operations on their ith bit nodes in parallel. Since we are using a single memory in our implementation to store the edge values of all check nodes, the second *for each* loop causes a BPE to update its connections in

```
for i = 0 to p-1 do
    for each b = 0 to c-1 do
        - Read r data values from MEM B located in the range
          [b * p * r + i * r, b * p * r + i * r + r - 1].
        - Sum them up and store the sum.
        - Update MEM E at location (b * p + i).
        for each k = 0 to r-1 do
            - Subtract from the computed sum the value
            located at (b * p * r + i * r + k - 1) in MEM B.
            - Use this value to update MEM C at location
            (k * p + ((p - k * b) mod p)) * c + b).
        end
    end
end
```

Fig. 11.2 The pseudo code describing assignment of bit nodes and check nodes to BPEs and CPEs

MEM C in a pipelined fashion. As also shown in Fig. 11.1, we compute the memory addresses to read/write data from/to "on-the-fly" using the bit node ID (i), BPE ID (b) and CPE ID (k). This convenient calculation of addresses is possible because of the QC nature of the code and the way we assigned BPEs and CPEs.

We tested our design on the FPGA Subsystem located at the High Performance Computing (HPC) Center at The University of Arizona. This FPGA Subsystem consists of SGI RASC RC1000 Blade having two Virtex 4 LX2000 FPGAs. In Fig. 11.3, we present BER performance comparison of FPGA and software implementations for a girth-10 QC LDPC(16935,13550) code. We observe a close agreement between the two BER curves. Furthermore, the performance of the min-sum algorithm is only 0.2 dB worse than that of the min-sum-with-correction-term algorithm at the BER of 10^{-6} and the gap gets closer as the Q factor increases. The NECG of the min-sum algorithm for the same LDPC code at BER of 10^{-6} is found to be 10.3 dB.

The main problem in decoder implementation for large-girth binary LDPC codes is the excessive codeword length, and a fully parallel implementation on a single FPGA is quite a challenging problem. To solve this problem, in the next section, we will consider large-girth nonbinary LDPC codes over GF(2^m) [21, 23, 26, 37–39]. By designing codes over higher-order fields, we aim to achieve the coding gains comparable to binary LDPC codes but for shorter codeword lengths.

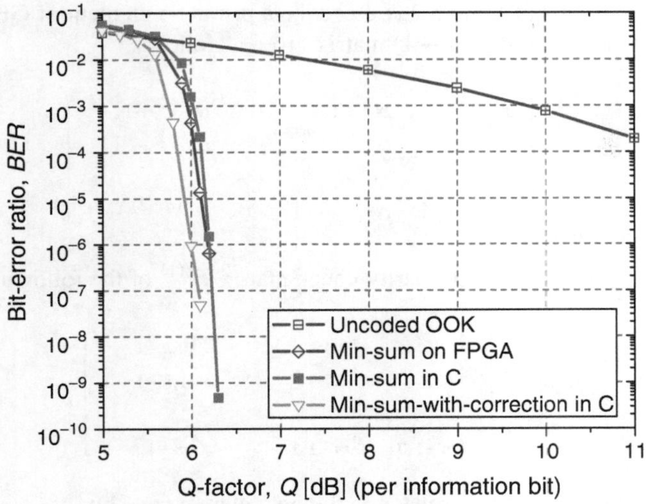

Fig. 11.3 BER performance comparison of FPGA and software implementations of the min-sum algorithm. (After ref. [9]; © IEEE 2009; reprinted with permission.)

11.2 Nonbinary QC-LDPC Codes

In this section, we describe a two-stage design technique for constructing nonbinary regular high-rate LDPC codes, we described in [22]. We show that the complexity of the nonbinary decoding algorithm over GF(4) used to decode this code is 1.1 times less complex compared to the min-sum-with-correction-term algorithm, described in Chap.5, used for decoding a bit-length-matched binary LDPC code. Furthermore, we demonstrate that by enforcing the nonbinary LDPC codes to have the same nonzero field element in a given column in their parity-check matrices, we can reduce the hardware implementation complexity of their decoders without incurring any degradation in the error-correction performance.

A q-ary LDPC code is a linear block code defined as the null space of a sparse parity-check matrix $\boldsymbol{H}$ over a finite field of q elements that is denoted by GF(q) where q is a power of a prime. Davey and MacKay [37] devised a q-ary sum–product algorithm (QSPA) to decode q-ary LDPC codes, where $q = 2^p$ and p is an integer. They also proposed an efficient way of conducting QSPA via fast Fourier transform (FFT-QSPA). FFT-QSPA is further analyzed and improved in [38]. A mixed-domain version of the FFT-QSPA (MD-FFT-QSPA) that reduces the computational complexity by transforming the multiplications into additions with the help of logarithm and exponentiation operations is proposed in [39]. Due to the availability of efficient decoding algorithms, we consider q-ary LDPC codes, where q is a power of two.

In the first step of our two-stage code design technique, we design binary QC-LDPC codes of girth-6 using the algebraic construction method based on the multiplicative groups of finite fields. Let α be a primitive element of GF(q) and let $\boldsymbol{W} = [\mathrm{w}_{i,j}]$ be a $(q-1)$-by-$(q-1)$ matrix given as follows:

$$W = \begin{bmatrix} \alpha^0 - 1 & \alpha - 1 & \cdots & \alpha^{q-2} - 1 \\ \alpha - 1 & \alpha^2 - 1 & \cdots & \alpha^{q-1} - 1 \\ \cdots & \cdots & \cdots & \cdots \\ \alpha^{q-2} - 1 & \alpha^{q-1} - 1 & \cdots & \alpha^{2(q-2)} - 1 \end{bmatrix}. \tag{11.5}$$

We can transform $\boldsymbol{W}$ into a QC parity-check matrix $\boldsymbol{H}^{(1)}$ of the following form:

$$H^{(1)} = \begin{bmatrix} A_{0,0} & A_{0,1} & \cdots & A_{0,n-1} \\ A_{1,0} & A_{1,1} & \cdots & A_{1,n-1} \\ \cdots & \cdots & \cdots & \cdots \\ A_{m-1,0} & A_{m-1,1} & \cdots & A_{m-1,n-1} \end{bmatrix}, \tag{11.6}$$

where every submatrix $\boldsymbol{A}_{i,j}$ is related to the field element $\mathrm{w}_{i,j}$ by

$$A_{i,j} = \left[z(w_{i,j})\ z(\alpha w_{i,j})\ z(\alpha^2 w_{i,j})\ \cdots\ z(\alpha^{q-2} w_{i,j})\right]^T, \tag{11.7}$$

where $z(\alpha^i) = (z_0, z_1, \ldots, z_{q-2})$ is a $(q-1)$-tuple over GF(2) whose ith component $z_i = 1$ and all other $q-2$ components are zero. Using Theorem 1 in [40], we can show that the parity-check matrix, $\boldsymbol{H}^{(1)}$, given in (11.6), which is a $(q-1)$-by-$(q-1)$ array of circulant permutation and zero matrices of size $(q-1)$-by-$(q-1)$, has a girth of at least six. We use this QC, girth-6 parity-check matrix $\boldsymbol{H}^{(1)}$ in the second stage.

If we simply choose γ rows and ρ columns from $\boldsymbol{H}^{(1)}$ while avoiding the zero matrices, we obtain a (γ, ρ)-regular parity-check matrix whose null space yields a (γ, ρ)-regular LDPC code with a rate of at least $(\rho\text{-}\gamma)/\rho$. Instead of a simple, random selection, however, if we choose rows and columns of $\boldsymbol{H}^{(1)}$ while avoiding performance-degrading short cycles, we can boost the performance of the resulting LDPC code. Hence, following the guidelines in [32], the first step in the second stage is to select γ rows and ρ columns from $\boldsymbol{H}^{(1)}$ in such a way that the resulting binary QC code has a girth of eight. In the second step, we replace the 1's in binary parity-check matrix with nonzero elements from GF(q) either by completely random selection or by enforcing each column to have the same nonzero element from GF(q) while letting the nonzero element of each column be determined again by a random selection. We denote the final q-ary (γ, ρ)-regular, girth-8 matrix by $\boldsymbol{H}^{(2)}$.

Following the two-stage design we had discussed above, we generated (3,15)-regular, girth-8 LDPC codes over the fields GF(2^p), where $0 \leq p \leq 7$. All the codes had a code rate (R) of at least 0.8 and hence an overhead OH $= (1/R - 1)$ of 25% or less. We compared the BER performances of these codes against each other and against some other well-known codes, namely the ITU-standard RS(255,239), RS(255,223), and RS(255,239) + RS(255,223) codes; and BCH(128,113) × BCH(256,239) turbo-product code (TPC). We used the binary

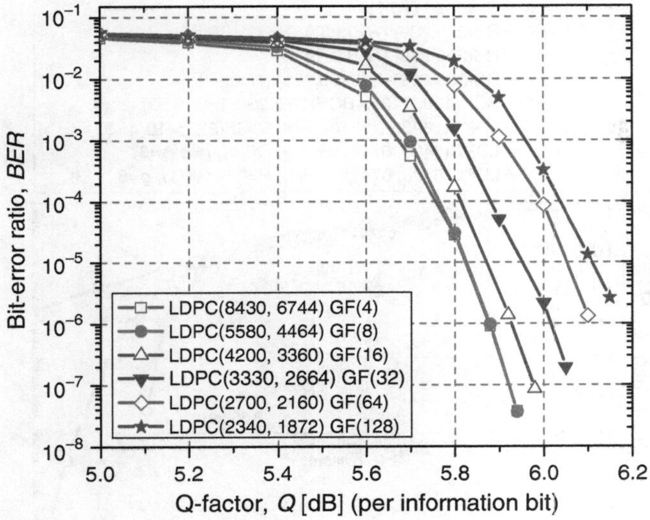

Fig. 11.4 Comparison of nonbinary, (3,15)-regular, girth-8 LDPC codes over BI-AWGN channel. (After ref. [9]; © IEEE 2009; reprinted with permission.)

AWGN (BI-AWGN) channel model in our simulations and set the maximum number of iterations to 50. In Fig. 11.4, we present the BER performances of the set of NB-LDPC codes discussed above. Using the figure, we can conclude that when we fix the girth of a nonbinary regular, rate-0.8 LDPC code at eight, increasing the field order above eight exacerbates the BER performance. In addition to having better BER performance than codes over higher-order fields, codes over GF(4) have smaller decoding complexities when decoded using MD-FFT-QSPA algorithm since the complexity of this algorithm is proportional to the field order. Thus, we focus our attention on nonbinary, regular, rate-0.8, girth-8 LDPC codes over GF(4) in the rest of the section.

In Fig. 11.5, we compare the BER performance of the LDPC(8430,6744) code over GF(4) discussed in Fig. 11.4 against that of the RS(255,239) code, RS(255,223) code, RS(255,239) + RS(255,223) concatenation code, and BCH(128, 113) $\times$ BCH(256, 239) TPC. We observe that the LDPC code over GF(4) outperform all of these codes with a significant margin. In particular, it provides an additional coding gain of 3.363 and 4.401 dB at BER of 10^{-7} when compared to the concatenation code RS(255,239) + RS(255,223) and the RS(255,239) code, respectively. Its coding gain improvement over BCH(128, 113) $\times$ BCH(256, 239) TPC is 0.886 dB at BER of 4×10^{-8}. Finally, we computed the NECG of the 4-ary, regular, rate-0.8, girth-8 LDPC code over GF(4) to be 10.784 dB at BER of 10^{-12}. We also presented in Fig. 11.5 a competitive, binary, (3,15)-regular, LDPC(16935,13550) code proposed in [34]. We can see that the 4-ary, (3,15)-regular, girth-8 LDPC(8430,6744) code beats the bit-length-matched binary LDPC code with a margin of 0.089 dB at BER of 10^{-7}. More importantly,

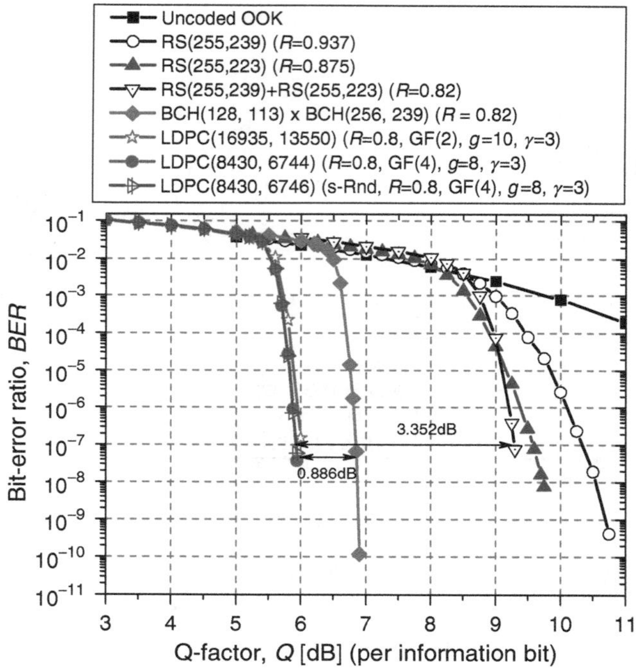

Fig. 11.5 Comparison of 4-ary (3,15)-regular, girth-8 LDPC codes; a binary, girth-10 LDPC code, three RS codes, and a TPC code. (After ref. [9]; © IEEE 2009; reprinted with permission.)

the complexity of the MD-FFT-QSPA used for decoding the NB-LDPC code is lower than the min-sum-with-correction-term algorithm [41] used for decoding the corresponding binary LDPC code. When the MD-FFT-QSPA is used for decoding a (γ, ρ)-regular q-ary LDPC$(N/\log q, K/\log q)$ code, which is bit-length-matched to a (γ, ρ)-regular binary LDPC(N, K) code, the complexity is given by $(M/\log q)2\rho q(\log q + 1\text{–}1/(2\rho))$ additions, where $M = N\text{–}K$ is the number of check nodes in the binary code. On the other hand, to decode the bit-length-matched binary counterpart using min-sum-with-correction-term algorithm [41], one needs $15M(\rho\text{-}2)$ additions. Thus, a (3,15)-regular 4-ary LDPC code requires 91.28% of the computational resources required for decoding a (3,15)-regular binary LDPC code of the same rate and bit length.

11.3 Nonbinary LDPC-Coded Modulation for Beyond 100 Gb/s Transmission

Compared to binary LDPC-coded modulation (BI-LDPC-CM) described in Chap. 6, nonbinary LDPC-coded modulation (NB-LDPC-CM) offers several advantages [23, 26] (1) m binary LDPC encoders/decoders needed for multilevel modulation are collapsed into a single 2^m-ary encoder/decoder (reducing the overall computational complexity of the system), (2) the use of block (de-) interleavers; binary-to-nonbinary and vice versa conversion interfaces are eliminated (reducing the complexity and latency in the system), (3) a posteriori probability (APP) demapper and LDPC decoder are integrated into a single block, and thus the need for iterating extrinsic information between APP demapper and LDPC decoder is eliminated (reducing the latency in the system). In addition to these advantages, we show that the proposed scheme provides strikingly higher coding gains than BI-LDPC-CM.

The transmitter and receiver configurations of the proposed NB-LDPC-CM scheme are shown in Fig. 11.6a, b, respectively. At the transmitter side, two independent data streams can be encoded using two different nonbinary LDPC codes of rates $R_i = K_i/N, i \in \{x, y\}$, where $K_x(K_y)$ denotes the number of information symbols used in the code corresponding to $x(y)$ polarization, and N denotes the codeword length, which is the same for both LDPC codes. The component LDPC codes operate over finite fields GF($q_i = 2^{m_i}$), $i \in \{x, y\}$, which can be different for each polarization. The Mapper $x(y)$ accepts a nonbinary symbol from the LDPC encoder $x(y)$ at time instance l and determines the corresponding q_x-ary (q_y-ary) constellation point $s_{l,x} = (I_{l,x}, Q_{l,x}) = |s_{l,x}| \exp(j\phi_{l,x})[s_{l,y} = |s_{l,y}| \exp(j\phi_{l,y})]$ whose coordinates are used as the inputs for a dual-drive Mach–Zehnder modulator $\mathrm{MZM}_x(\mathrm{MZM}_y)$. The outputs of the MZMs are combined using a polarization beam combiner (PBC). The same DFB laser is used as CW source with x- and y-polarizations being separated by a polarization beam splitter (PBS). Note that for

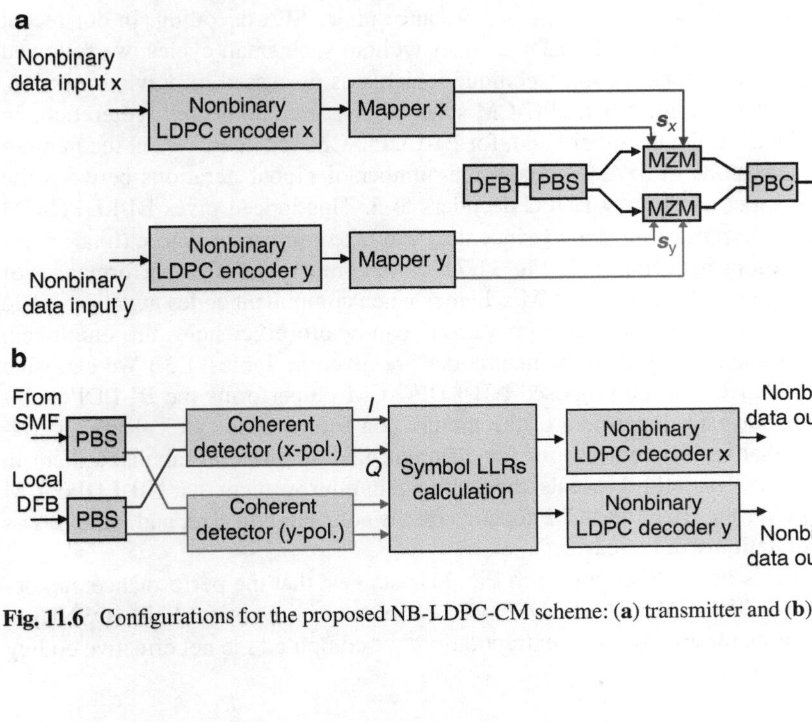

Fig. 11.6 Configurations for the proposed NB-LDPC-CM scheme: (**a**) transmitter and (**b**) receiver

different polarizations we can use different constellations and different LDPC codes, but we keep the codeword lengths the same to simplify the synchronization.

At the receiver side, the I- and Q-branches are sampled at the symbol rate. The symbol LLRs are then calculated as follows:

$$\lambda(\mathbf{s}) = \log \frac{P(\mathbf{s}|\,\mathbf{r})}{P(\mathbf{s}_0|\,\mathbf{r})} = \log \frac{P(\mathbf{r}|\,\mathbf{s})P(\mathbf{s})}{P(\mathbf{r}|\,\mathbf{s}_0)P(\mathbf{s}_0)}, \tag{11.8}$$

where $\mathbf{s} = (I_l, Q_l)$ and $\mathbf{r} = (r_\mathrm{I}, r_\mathrm{Q})$ denote the transmitted constellation point and the received symbol at time instance l (in either x- or y-polarization), and $\mathbf{s}_0$ represents the referent symbol. Note that the second equality follows from the Bayes' rule and that we determine $P(\mathbf{r}|\,\mathbf{s})$ via histograms.

If the symbol rate is set to R_s GigaSymbols/s (GS/s), we can achieve $2R_s \log_2(q)$ bits/s transmission using the proposed polarization-multiplexed NB-LDPC-CM scheme, where q signifies the finite field order which is also equal to the constellation size of the modulation. For example, using $R_s = 25$ GS/s, we can achieve 100, 150, and 200 Gb/s optical transmission using QPSK, 8-QAM and 16-QAM modulations, respectively.

We now compare the performance of NB-LDPC-CM and BI-LDPC-CM schemes assuming amplified spontaneous emission (ASE) noise-dominated scenario. To decode q-ary LDPC codes, the QSPA [37], and its FFT-based variants can be used [37–39]. When q is a power of 2, the complex arithmetic operations in FFT-based QSPA can be avoided; thus, we focus only on the cases where $q = 2^m$, for some positive integer m. Small cycles, particularly four cycles, in the Tanner graphs of LDPC codes deteriorate the code performance under SPA decoding. In our recent paper [22], to design NB-QC-LDPC codes without such small cycles, we followed a two-stage code construction technique which was discussed in previous section. In our simulations, for NB-LDPC-CM schemes, we used 50 decoding iterations in MD-FFT-QSPA. On the other hand, for BI-LDPC-CM schemes, we set the number of inner iterations in SPA to 25 and the number of global iterations between the APP demapper and binary LDPC decoders to 3. This indeed gives BI-LDPC-CM schemes an important advantage since they can take comparably more (total) number of iterations in decoding. In Fig. 11.7a–c, we compare the BER performances of BI-LDPC-CM and NB-LDPC-CM schemes when component codes at various code rates and modulation formats with various bandwidth efficiencies are employed. (Exact parameters of the component codes are given in Table 11.3.) We can state with confidence that the proposed NB-LDPC-CM outperforms the BI-LDPC-CM at all the code rates regardless of the modulation format used. We calculated coding gains that both schemes offer for different modulations and tabulated them in Table 11.2. As Table 11.2 reveals, the coding gain improvement that NB-LDPC-CM provides over the BI-LDPC-CM depends on the constellation size, and increases as the constellation size increases.

The slopes of the BER curves in Fig. 11.7 suggest that the performance gap between NB-LDPC-CM and BI-LDPC-CM increases as the optical SNR (OSNR) in the system increases. By using extrapolation, we computed the net effective coding

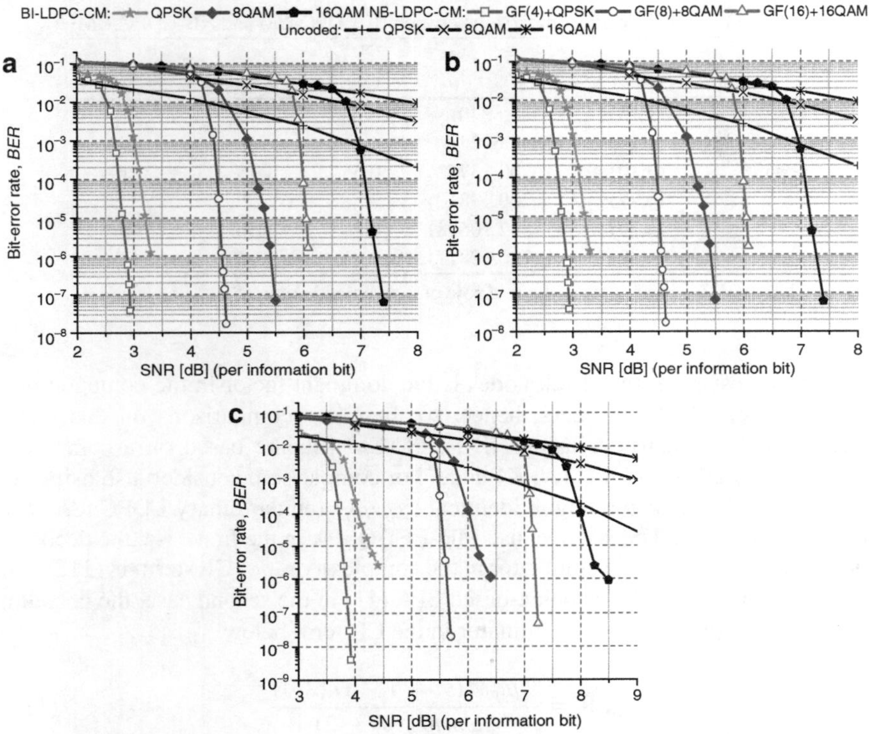

Fig. 11.7 BER performance comparison between corresponding BI-LDPC-CM and NB-LDPC-CM schemes based on component codes at various code rates, R: (**a**) $R = 0.8$, (**b**) $R = 0.85$, and (**c**) $R = 0.9$. Note that $R_s = 25$ GS/s

Table 11.2 Coding gains (dB) for BI-LDPC-CM and NB-LDPC-CM[a]

	Constellation size					
	4		8		16	
	BER		BER		BER	
Code rate	10^{-6}	10^{-12}	10^{-6}	10^{-12}	10^{-6}	10^{-12}
0.8	7.62	10.85	8.06	11.57	8.31	11.63
	7.19	10.22	7.21	10.50	7.11	10.16
0.85	7.27	10.57	7.64	11.19	7.87	11.22
	7.02	10.34	7.06	10.48	6.94	10.20
0.9	6.66	9.92	6.99	10.48	7.19	10.52
	6.00	–	6.24	–	5.96	–

[a] NB-LDPC-CM scheme is the first row of a given code rate

gain (NECG) that each scheme is expected to provide at the BER of 10^{-12} and presented them in Table 11.2. Comparing the NECG figures, we can clearly see the dramatic performance improvement that NB-LDPC-CM can bring to the FEC modules of future optical communication systems.

Table 11.3 Complexity comparison between BI-LDPC-CM and NB-LDPC-CM[a]

			CR		
Code rate	(γ, ρ)	(N, K)	QPSK	8-QAM	16-QAM
0.8	(3,15)	(8550,6840)	0.66	1.17	2.20
	(4,21)	(8547,6922)			
0.85	(3,20)	(16200,13770)	0.65	1.15	2.16
	(4,27)	(16200,13810)			
0.9	(3,29)	(11832,10608)	0.84	1.50	2.81
	(3,32)	(11936,10817)			

[a] NB-LDPC-CM scheme is the first row of a given code rate

Data processing at the check nodes is the dominant factor in the computational complexity of LDPC decoders; hence, we base our comparisons on this dominant factor. First, consider using NB-LDPC-CM scheme based on a component (γ, ρ)-regular 2^m-ary LDPC (N, K) code. For comparison, consider also using BI-LDPC-CM scheme employing m identical (γ_b, ρ_b)-regular binary LDPC (N_b, K_b) codes as component codes. If we use MD-FFT-QSPA in the first case, the decoding complexity is given by the numerator in the complexity ratio (CR) term in (11.9). On the other hand, if we use Jacobian-based SPA [41] in the second case, the decoding complexity is given by the denominator in the CR term below.

$$\mathrm{CR} = \frac{2\,\rho q M(m+1-1/(2\rho))}{15\,m M_b(\rho_b-2)}, \tag{11.9}$$

where $M = N - K$, $M_b = N_b - K_b$, and $q = 2^m$. In Table 11.3, we compare the complexity of corresponding NB-LDPC-CM and BI-LDPC-CM schemes employing component codes at various code rates and various modulation formats. When QPSK modulation is used, NB-LDPC-CM provides results in lower complexity systems but the additional coding gains that it offers over BI-LDPC-CM are not that high, i.e., in the ranges of 0.2–0.6 dB at the BER of 10^{-12} (see Table 11.2). If we use 8-QAM modulation, the complexities become comparable; however, the additional coding gain that NB-LDPC-CM offers over BI-LDPC-CM becomes quite significant, i.e., 0.7–1.0 dB at the BER of 10^{-12}. In the 16-QAM case, we observe that the complexity of NB-LDPC-CM is at least two times that of BI-LDPC-CM. On the other hand, NB-LDPC-CM provides impressive 1–1.5 dB coding gain improvement at the BER of 10^{-12} compared to BI-LDPC-CM, regardless of the component code rates. We should also note that error floors that appear in BI-LDPC-CM curves when component codes at the code rate $R = 0.9$ are used can be avoided using NB-LDPC-CM and nonbinary component codes. As opposed to doubling of the decoding complexity, reductions in latency, striking more than 1 dB improvement in coding gain and resilience to error floors can lead system designers to consider NB-LDPC-CM scheme even over much higher-order fields in order to balance their link budgets.

11.4 Adaptive Nonbinary LDPC-Coded Modulation

In order to have seamless integrated transport platforms, which can support heterogeneous networking, in this section, we describe the use of rate-adaptive nonbinary LDPC-coded polarization-multiplexed quadrature amplitude modulation (QAM) with coherent detection, we introduced in [21]. The use of different channel codes for different destinations would be costly to implement due to increased hardware complexity, unless unified encoding and decoding architectures can be used for all destinations. The structured QC-LDPC codes provide us with this unique feature. By using the multilevel modulation and polarization multiplexing, all related coding, signal processing, and transmission are performed at lower symbol rates (such as 25 GS/s), where dealing with fiber nonlinearities and PMD is more manageable, while the aggregate rate is kept at 100 Gb/s and above. The code rate of nonbinary LDPC (NB-LDPC) code, for a given constellation size, is chosen in accordance with the channel conditions. When the channel conditions are favorable (large SNR), higher code rate LDPC code is employed. Compared to the adaptive modulation approach, the proposed adaptive coding approach is very friendly from implementation point of view because the symbol rate is kept constant so that all synchronization issues related to the variable symbol rate modulation adaptation are avoided. The use of non-binary-coded multilevel modulation scheme is proposed because they offer lower decoding complexity and latency, and at the same time provide larger coding gains compared to their binary counterparts.

To demonstrate the potential of LDPC-coded modulation, we present in Fig. 11.8a the results of our comparison between BI-LDPC-CM with component codes of various code rates and two ITU-standard FEC codes. We should note that the symbol rate is $R_s = 25$ GS/s and we employ QPSK modulation, and hence, the aggregate data rate attained by the BI-LDPC-CM scheme is 100 Gb/s. As depicted in the Fig. 11.8a, regardless of the code rate of its component code, which ranges from 0.75 to 0.875, BI-LDPC-CM scheme outperforms the ITU-standard rate-0.82 BCH(128, 113) × BCH(256, 239) TPC. The coding gain improvement that BI-LDPC-CM scheme provides over another ITU-standard rate-0.82 RS(255,239)+ RS(255,223) concatenation code is much larger, i.e., 2.5 dB or more.

Table 11.4 presents the parameters of the component NB-QC-LDPC codes of three different code rates employed in our rate-adaptive scheme. In order to keep the (transmission) data rate constant, we keep the codeword length of component codes, N, the same. In addition, by keeping the codeword length the same over the family of component codes, one keeps the frame synchronization circuitry intact and independent of the component code rate adjustments. As shown in Table 11.4, the column weight (γ) is the same for all component codes. Hence, changing the code rate is tantamount to changing the subblock size (B) or the row weight (ρ).

As Fig. 11.8b confirms, LDPC-coded modulation offers even a higher potential to be exploited if it is used with NB-LDPC component codes rather than component codes over the binary field. We use component codes of code rate 0.875 as the base of our comparison between BI-LDPC-CM and NB-LDPC-CM. Also, we consider three different modulation formats, namely QPSK, 8-QAM, and 16-QAM.

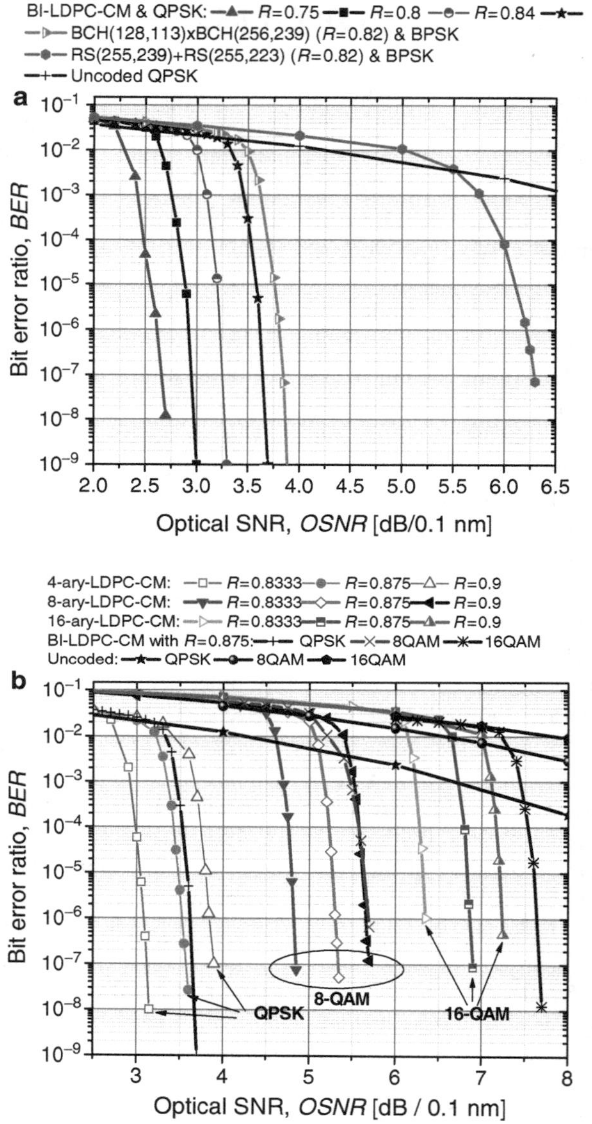

Fig. 11.8 Bit-error ratio performance curves (**a**) for the B-QC-LDPC codes employed in the BI-LDPC-CM scheme, (**b**) for the NB-QC-LDPC codes employed in the NB-LDPC-CM scheme

Table 11.4 Component nonbinary LDPC code parameters

Code rate	N	K	M	γ	ρ	B
0.833	13680	11400	2280	3	18	760
0.875	13680	11970	1710	3	24	570
0.9	13680	12312	1368	3	30	456

Table 11.5 Coding gains at the BERs of 10^{-7} and 10^{-12} for different NB-LDPC-CM schemes employing different code rates and modulations

	QPSK		8-QAM		16-QAM	
Code rate	10^{-7}	10^{-12}	10^{-7}	10^{-12}	10^{-7}	10^{-12}
0.833	8.13	10.70	8.57	11.36	8.80	11.41
0.875	7.67	10.21	8.08	10.85	8.28	10.87
0.9	7.35	9.88	7.72	10.48	7.91	10.50

We can observe from Fig. 11.8b that NB-LDPC-CM outperforms BI-LDPC-CM regardless of the modulation format used. A close look at Fig. 11.8b reveals that the coding gain improvement NB-LDPC-CM provides over the corresponding BI-LDPC-CM increases when we increase the constellation size of the modulation format. More specifically, NB-LDPC-CM with a component (3,24)-regular q-ary QC-LDPC(13680,11970) code provides 0.4 and 0.78 dB additional coding gain at the BER of 10^{-7} compared to the corresponding BI-LDPC-CM with component (4,32)-regular binary QC-LDPC(28800,25200) codes when 8-QAM, i.e., $q = 8$, and 16-QAM, i.e., $q = 16$, modulations are used, respectively. (Note that both binary and nonbinary component codes are of girth 8.) Table 11.5 presents coding gains at the BER of 10^{-7} along with the NECGs at the BER of 10^{-12}, obtained via extrapolation, for NB-LDPC-CM schemes depicted in Fig. 11.8b.

From Fig. 11.8b and Table 11.5, we can conclude that in order to maximize the coding gain improvement offered by NB-LDPC-CM schemes, one can consider adapting not only the component code rates but also the employed modulation format. NB-LDPC-CM offers much higher coding gains than BI-LDPC-CM if modulation formats with high bandwidth efficiencies are used. However, this appealing benefit does not come without an expense since the complexity at the receiver front ends might increase in order to support different modulation formats.

Until now, we focused our attention only on the error correction performance of the corresponding NB-LDPC-CM and BI-LDPC-CM schemes. As the previous section also alluded to, computational complexity is another important aspect apart from the error correction performance. Using the tools developed in previous section and the code parameters presented in Table 11.4, we can also compare the corresponding NB-LDPC-CM and BI-LDPC-CM schemes from the computational complexity standpoint. We use the complexity ratio figure of merit, we introduced by (11.9). Because $\rho = 24, M = 1,710$ for NB-LDPC-CM while $\gamma_b = 4, \rho_b = 32, M_b = 3,600$ for BI-LDPC-CM, the complexity ratio as the function of constellation size $q = 2^m$ can be written as [30]:

$$\mathrm{CR} = \frac{2\,\rho q M\left(m+1-{}^{1}\!/_{(2\rho)}\right)}{15\,m M_b(\rho-2)} = \frac{19}{375}\frac{2^m\left(m+1-{}^{1}\!/_{48}\right)}{m} \cong \frac{19}{375}\frac{2^m(m+1)}{m}.$$

When $m = 2$, i.e., QPSK modulation, CR is equal to 30.4%. That is, NB-LDPC-CM, which performs slightly better than BI-LDPC-CM for QPSK modulation, can achieve this performance with less than one-third of the complexity of

BI-LDPC-CM. Setting $m = 3$ for 8-QAM, we see that CR = 54%. Thus, by using NB-LDPC-CM, we obtain 0.4 dB additional coding gain at the BER of 10^{-7} over BI-LDPC-CM while reducing the computational complexity by almost 50%. In the 16-QAM case, where $m = 4$, CR = 101.3%, i.e., the computational complexities of the two schemes are comparable. However, we should stress here that for the same computational load, NB-LDPC-CM offers 0.78 dB more coding gain at the BER of 10^{-7} compared to BI-LDPC-CM in addition to reduced latency due to structural simplifications. These are indeed important assets in link budget analysis.

We demonstrate in Fig. 11.8b our simulation results for the NB-LDPC-CM scheme based on 4-ary QC-LDPC components codes modulated with the QPSK modulation and 8-ary QC-LDPC codes modulated with 8QAM modulation. The coding gain improvements obtained using the NB-LDPC-CM scheme over the ITU-standard rate-0.82 RS(255,239) + RS(255,223) and BCH $(128, 113) \times$ BCH$(256, 239)$ schemes are outstanding. Table 11.5 also presents the NECG figures observed at different BERs for the NB-LDPC-CM schemes discussed above. To the best of our knowledge, these are the highest NECG figures reported in the literature for multilevel modulations. A fair performance comparison between BI-LDPC-CM and NB-LDPC-CM schemes can be carried out when they both employ the same rate component codes with almost equal codeword lengths in bits. Thus, we chose the rate-0.875 codes as the basis for our comparison. In Fig. 11.8b, we can see that the NB-LDPC-CM scheme employing a (3,24)-regular, girth-8 4-ary QC-LDPC(13680,11970) outperforms the BI-LDPC-CM scheme employing a (4,32)-regular, girth-8 binary QC-LDPC(28800, 25200) by almost 0.1 dB at the BER of 10^{-6}.

11.5 Adaptive LDPC-Coded OFDM

In an optically-routed network, different signal transmission paths have different number of optical amplifiers, WXCs, and ROADMs. Different wavelength channels carrying the traffic to different destinations can have quite different SNRs and signal is differently impacted by various channel impairments. The optical networks should provide a target BER performance regardless of the data destination. To address all these issues, here we describe the use of power-variable rate-adaptive LDPC-coded polarization-multiplexed optical OFDM system with coherent detection, which is shown in Fig. 11.9.

To deal with concatenated ROADMs/WXCs and bandwidth-limitation problems, and imperfections of various devices, we employ the adaptive water-filling algorithm [42] to determine the optimum power to be allocated to ith subcarrier P_i as follows:

$$P_i/P = \begin{cases} 1/\gamma_{\text{tsh}} - 1/\gamma_i, & \gamma_i \geq \gamma_{\text{tsh}}, \\ 0, & \text{otherwise}, \end{cases} \tag{11.10}$$

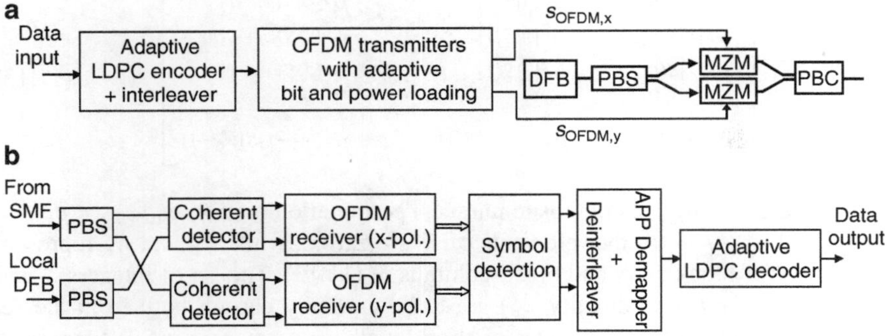

Fig. 11.9 Block diagram of proposed power-variable rate-adaptive LDPC-coded OFDM scheme: (**a**) transmitter, and (**b**) receiver configurations. *DFB* distributed feedback laser, *PBS/C* polarization beam splitter/combiner, *MZM* dual-drive Mach–Zehnder modulator, *APP* a posteriori probability

where P is the total available power, and γ_i is the SNR in ith subcarrier. The optimum threshold SNR is determined from condition that total power in all subcarriers cannot be larger than available power, $\sum_i P_i \le P$, by

$$\gamma_{\text{tsh}} = N/\left[1 + \sum_{\gamma_i \ge \gamma_{\text{tsh}}} 1/\gamma_i\right] \qquad (11.11)$$

where N is the number of subcarriers with $\gamma_i \ge \gamma_{\text{tsh}}$, in an iterative procedure that converges fast (after few iterations).

The variable-rate adaptation is achieved by choosing the maximum product of integer m_i, corresponding to the number of bits per ith subcarrier, and corresponding code rate R_i as follows

$$m_i R_i \le C_i, \qquad (11.12)$$

where C_i is the channel capacity of the ith subcarrier. The signal constellation size $M_i = 2^{m_i}$ per ith subcarrier and the corresponding code rate R_i of component LDPC code are chosen in accordance with the channel conditions. When the channel conditions are favorable (large SNR) the larger constellation sizes and higher code rate LDPC codes are employed. Among several candidate LDPC codes we employ one (based on subcarrier γ_i) which provides the largest product $m_i R_i$ closest to the subcarrier channel capacity (but lower than C_i). When channel conditions are poor we reduce the code rate and signal constellation size according to (11.12). The use of different channel codes for different destinations would be costly to implement due to increased hardware complexity, unless a unified encoding and decoding architectures can be used for all destinations. The structured QC-LDPC codes [32] provide us with this unique feature. Their parity-check matrices can be written in the following form (that represents a generalization of Fossorier's design [32] and (11.2))

$$H = \begin{bmatrix} P^{r[0]S[0]} & P^{r[0]S[1]} & \ldots & P^{r[0]S[w_r-1]} \\ P^{r[1]S[0]} & P^{r[1]S[1]} & \ldots & P^{r[1]S[w_r-1]} \\ P^{r[2]S[0]} & P^{r[2]S[1]} & \ldots & P^{r[2]S[w_r-1]} \\ \ldots & \ldots & \ldots & \ldots \\ P^{r[w_c-1]S[0]} & P^{r[w_c-1]S[1]} & \ldots & P^{r[w_c-1]S[w_r-1]} \end{bmatrix}, \qquad (11.13)$$

where P is $q \times q$ (q is a composite number) permutation matrix ($p_{i,i+1} = p_{q,1} = 1, i = 1, 2, \ldots, q-1$; other elements of P are zeros), while w_c and w_r represent the number of block-rows and block-columns in (11.13). The set of integers S and set of integers r (of cardinality w_c) are to be carefully chosen both from the set $\{0, 1, \ldots, q-1\}$; so that the cycles of short length, in corresponding Tanner graph representation of (11.13) are avoided, as explained earlier. The code rate adaptation can be achieved by varying different parameters in (11.13). To simplify implementation and resolve the synchronization problems related to the variable symbol rate transmission, similarly as previous section we keep the codeword length (symbol rate) fixed, but vary the permutation-block P size and row weight in (11.13) instead.

The channel capacity of polarization-multiplexed optical coherent OFDM scheme (derived based on theory due to Bölcskei *et al.* [43]), with N subcarriers can be evaluated as follows:

$$C = E_H \left\{ \frac{1}{N} \max_{Tr(\Sigma) \le P} B_{sc} \log_2 \left[\det \left(\boldsymbol{I}_{2N} + \frac{1}{\sigma_n^2} \boldsymbol{H\Sigma H}^{\dagger} \right) \right] \right\}, \qquad (11.14)$$

where Σ is the covariance matrix of Gaussian input vector, defined as

$$\Sigma = \mathrm{diag}\,\{\Sigma_i\}_{i=0}^{N-1}, \quad \Sigma_i = \frac{P}{2N} I_2 \qquad (11.15)$$

with P being the maximum overall transmit power, and σ_n being the standard deviation of ASE noise process, and B_{sc} being the bandwidth of subcarrier channel. $\mathrm{Tr}(\Sigma)$ denotes the trace of matrix Σ, E_H denotes the expectation operator with respect to channel matrix $\boldsymbol{H}$, and $\boldsymbol{I}_{2N}$ is $2N \times 2N$ identity matrix. The channel matrix in (11.14) is a block diagonal matrix in (11.16) with ith block diagonal element corresponding to the 2×2 Jones matrix $\boldsymbol{H}(i)$ of ith subcarrier.

$$\boldsymbol{H} = \mathrm{diag}\{\boldsymbol{H}(i)\}_{i=0}^{N-1}, \quad \boldsymbol{H}(i) = \begin{bmatrix} H_{xx}(i) & H_{xy}(i) \\ H_{yx}(i) & H_{yy}(i) \end{bmatrix}. \qquad (11.16)$$

The chromatic dispersion effect and reduced bandwidth effect due to concatenation of ROADMs, as well as different linear channel imperfections, are all incorporated in $\boldsymbol{H}(i)$.

We designed rate-adaptive QC-LDPC code of codeword length 28,800 (that is shorter than TBC proposed by Mizuochi *et al.* [44]), with possible rates {0.875, 0.84, 0.8, and 0.75}, whose BER performance has already been shown in Fig. 11.8a. Even highest rate code (0.875) outperforms the TBC of rate 0.82, and significantly

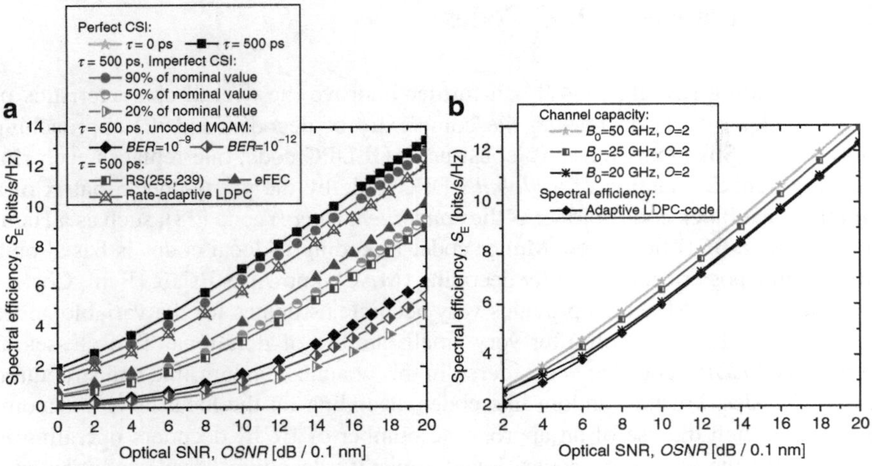

Fig. 11.10 Spectral efficiency limits of polarization-multiplexed coherent optical OFDM (with OFDM signal bandwidth $BW = 25\,\text{GHz}$)

outperforms the concatenated RS code. In Fig. 11.10, we show the channel capacity calculated based on (11.14)–(11.16) for polarization-multiplexed OFDM system when all order PMD with average DGD τ of 500 ps is observed, for 2,000 km of SMF, 128 subcarriers, and OFDM signal bandwidth BW of 25 GHz. The pilot-aided channel estimation is used to determine channel coefficients in (11.16).

We see that when channel matrix (the channel state information, CSI) is perfectly known we can completely compensate for different linear channel impairments, which was expected because the Jones matrix is unitary. However, when the CSI is not ideal (the channel estimation is not perfect) we can have significant channel capacity degradation (see Fig. 11.10a). We see when channel coefficients are estimated to be 90% of nominal values we have small performance degradation. However, when the channel estimates are 50% of nominal value we have found significant channel capacity degradation. We also have found that for target BERs below 10^{-9} M-ary QAM (MQAM) is far away from the channel capacity. However, with adaptive LDPC coding we can closely approach the channel capacity. To reach the channel capacity we have to invent better modulation formats and use them in combination with adaptive LDPC coding. In Fig. 11.10b we study the efficiency of the optimum power-adaptation (11.10)–(11.12) in dealing with concatenation filtering problem (that is modeled as super-Gaussian filter of order $O = 2$ and bandwidth $B_0 = 20, 25$, and 50 GHz). We see that channel capacity is getting worse as the bandwidth of optical filter is getting smaller than OFDM signal bandwidth, but we are still quite close to the channel capacity.

11.6 Generalized LDPC Codes

GLDPC coding [27, 28, 45–47] can further improve the overall characteristics of LDPC codes by: (1) decreasing the complexity of decoder, and (2) approaching closer to the Shannon's limit. To construct a GLDPC code, one replaces each single parity-check equation of a *global* LDPC code by the parity-check matrix of a simple linear block code, known as the *constituent* (*local*) code [28], such as a Hamming code, a BCH or a Reed–Muller code. Decoding of local codes is based on a maximum a posteriori probability decoding (MAP), known as BCJR (Bahl–Cocke–Jelinek–Raviv) [48], which provides very accurate estimates for the variable nodes in the global LDPC graph after very small number of iterations. Two classes of generalized LDPC codes are considered in this chapter: random-like and structured GLDPC codes. For the random-like codes, decoding on the local level is accomplished through the use of an appropriate number of BCJR decoders operating in parallel; for the structured codes, low-complexity decoding based on suboptimal message-passing algorithm is used instead.

In general, one seeks GLDPC codes for which the local codes have large minimum distance and a rate as high as possible, and for which the global code has very large girth. This is a consequence of the following lower bound on minimum distance D of a GLDPC code [31]:

$$D \geq \begin{cases} d\dfrac{[(d-1)(J-1)]^{(g-2)/4}-1}{(d-1)(J-1)-1} + \dfrac{d}{J}[(d-1)(J-1)]^{(g-2)/4}, & \text{for } g/2 \text{ odd}, \\ d\dfrac{[(d-1)(J-1)]^{g/4}-1}{(d-1)(J-1)-1}, & \text{for } g/2 \text{ even}, \end{cases} \quad (11.17)$$

where g and J denote the girth (shortest cycle length) of the global code graph and the column weight of the global code, respectively, and where d stands for the minimum distance of the local code. Obviously, a large girth leads to an exponential increase in the minimum distance, while large values of d lead to an increase of the bases of this exponent. GLDPC codes can be designed in such a way that the minimum distance D grows linearly with the code length [46].

Depending on the structure of the local components, GLDPC codes may be classified as follows [28]:

- GLDPC codes with algebraic local codes of short length, such as Hamming codes, BCH codes, RS codes, or Reed–Muller codes [27]
- GLDPC codes for which the local codes are high-rate regular or irregular LDPC codes with large minimum distance
- Fractal GLDPC codes in which the local code is in fact another GLDPC code

Each of these three constructions offers some specific advantages. For example, in the first scenario, the MAP decoders of the local codes allow for fast propagation of accurate estimates of the LDPC variables. The second construction is very well suited for longer-length codes. Since the local components in this case can have large length and therefore large minimum distance as well, the minimum distance of the resulting GLDPC codes is very high. The third approach is an attractive idea in terms of possible practical partly-parallel VLSI implementations, since the same code structure can be reused on different levels.

To construct a GLDPC code, one can replace each single parity-check equation of a *global* LDPC code by the parity-check matrix of a simple linear block code, known as the *constituent* (*local*) code, and this construction is proposed by Lentmaier and Zigangirov [46], and we will refer to this construction as LZ-GLDPC code construction. In another construction proposed by Boutros et al. in [45], referred here as B-GLDPC code construction, the parity-check matrix, H, is a sparse matrix partitioned into W submatrices $H_1, \ldots, H_W$. H_1 is a block-diagonal matrix generated from an identity matrix by replacing the ones by a parity-check matrix of a local code of codeword-length n and dimension k. Each submatrix H_j is derived from H_1 by random column permutations. For more details on LZ-GLDPC and B-GLDPC-like codes, and their generalization-fractal GLDPC codes (in which a local code is another GLDPC code) an interested reader is referred to [28]. The code rate of a GLDPC code is lower bounded by $R = K/N \geq 1 - W(1 - k/n)$, where K and N denote the dimension and the codeword-length of a GLDPC code, W is the column weight of a global LDPC code, and k/n is the code rate of a local code (k and n denote the dimension and the codeword-length of a local code). The GLDPC codes with component Hamming codes are considered in [28], here we are concerned with GDLPC codes with component codes based on RM or BCH codes [27]. A Reed–Muller code RM(r, m) of order r and codeword length $n = 2^m$ [49] is the set of all binary vectors associated with coefficients of Boolean polynomials of degree at most r in m variables. The minimum distance of RM(r, m) code is 2^{m-r}, and the dimension is determined as [49]

$$k = 1 + \binom{m}{1} + \cdots + \binom{m}{r}.$$

An interesting property of RM codes is that they can be defined recursively [49]: $\mathrm{RM}(r, m) = \{(a|a + b) : a \in \mathrm{RM}(r, m - 1), b \in \mathrm{RM}(r - 1, m - 1)\}$, where $(x|y)$ denotes the concatenation operation. The generator matrix of RM(r, m) code, denoted as $G(r, m)$, can be therefore defined recursively by

$$G(r, m) = \begin{bmatrix} G(r, m-1) & G(r, m-1) \\ 0 & G(r-1, m-1) \end{bmatrix}, \tag{11.18}$$

and can be observed as two-level generalized concatenated code [49]. RM$(0, m)$ is a repetition code, RM$(m - 1, m)$ is a parity-check code, and RM(m, m) corresponds to 2^m-tuples of a vector space. The generator matrix of RM(m, m) can be represented by

$$G(m, m) = \begin{bmatrix} G(m-1, m) \\ 0\,0 \cdots 0\,1 \end{bmatrix}. \tag{11.19}$$

Another interesting property of RM codes is that the dual of RM(r,m) code is another RM$(m-r-1,m)$ code. Therefore, the generator matrix of RM$(m-r-1,m)$ code can be used as the parity-check matrix of RM(r,m) code. If the recursion (11.18) is applied successively several times the RM(r,m) can be decomposed into several parity-check codes RM$(m'-1,m')$, repetition codes RM$(0,m')$, and the first-order RM$(1,m')$ codes. The MAP decoding of parity-check or repetition codes is trivial, while the first-order RM$(1,m')$ codes can be decoded using an efficient MAP-decoding algorithm proposed in [50], based on fast Hadamard–Walsh transform (FHWT). The overall complexity of that algorithm is in order $n'\log_2 n'$ (where $n' = 2^{m'}$), which is significantly lower than complexity of the BCJR algorithm that requires about n^{n-k+1} operations [50]. Therefore, the complexity of GLDPC codes with RM component codes is of order $N\ \log_2 n$. Since the complexity of SPA is of order $(N_{\mathrm{LDPC}} - K_{\mathrm{LDPC}})w_{\mathrm{r}}$, with w_{r} being the row weight of LDPC code parity check matrix, by proper selection of global LDPC code length N and local RM code length n, the complexity of GLDPC codes is about $(N_{\mathrm{LDPC}} - K_{\mathrm{LDPC}})w_{\mathrm{r}}/[(N/n)\sum(n'\log_2 n')](n' < n)$ times lower. For example, RM(4,6) code can be decomposed using (11.18) on RM(1,2), RM(1,3), RM(2,2), RM(3,3), and RM(4,4) component codes. Decoding of RM$(m',m')(m' = 1,\ 2,\ 3,\ 4)$ is trivial while the complexity of RM(4,6) is dominated by complexity of RM(1,3) decoding block and three RM(1,2) blocks, which is of order $\sum(n'\log_2 n') = 8\ \log_2 8 + 3 \cdot 4\log_2 4 = 48$. B-GLDPC code with $W = 2$ and length $N = 4,096$ based on RM(4,6) code has therefore complexity 11 times lower than girth-8 column-weight-4 LDPC code of length 8,547 (and row weight 21). Also note that GLDPC decoder for 4,096 code contains $4,096/64 = 64$ decoder blocks [composed of RM(1,2), RM(2,3) and RM$(m',m')(m' = 1,\ldots,4)$ decoders], operating in parallel, and this structure is suitable for FPGA or VLSI implementation. In order to keep the code rate high, we select the column-weight of a global code to be $W = 2$. In that case, if the girth of global code is $g = 8$, the minimum distance of GLDPC code is simply $D \geq d^2$.

The BCH codes are already introduced in Chap. 4. Note that BCH codes can be decoded using an efficient MAP algorithm proposed in [50] with complexity $n/[n-k)\log_2 n]$ times lower than that of BCJR algorithm (see [48] and Chap. 5 for more details about BCJR algorithm). The GLDPC codes described here can be put in systematic form, so that the efficient encoding algorithm due to Zhang and Parhi [51] can be employed. This algorithm can efficiently be performed in general purpose digital signal processors, as shown in [51].

The results of simulation for the AWGN channel model are shown in Fig. 11.11, and are obtained by maintaining the double-precision. GLDPC codes based on BCH(63,57) and RM(4,6) component codes for $W = 2$ perform comparably. The RM(4,6)-based GLDPC code outperforms the BCH(128,113) $\times$ BCH(256,239) TPC based with Chase II decoding algorithm on $p = 3$ least reliable bit positions by 0.93dB at BER of 10^{-9} (see Fig. 11.11a). Note that similar TPC was implemented in LSI technology (see [44] for more details). The TPC codeword is significantly longer, and the decoding complexity of GLDPC code based on RM(4,6) is at

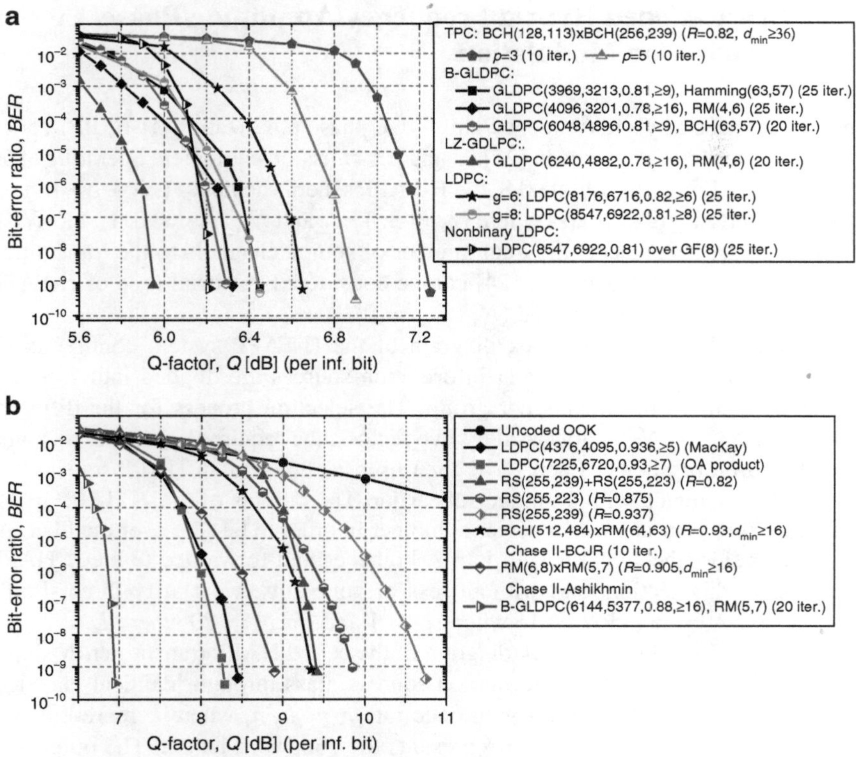

Fig. 11.11 BER performance on an AWGN channel: (**a**) GLDPC codes against TPCs and LDPC codes, (**b**) high-rate codes. (After ref. [27]; © IEEE 2009; reprinted with permission.)

least ten times lower. During decoding, TPC decoder employs 239 Chase II blocks operating in parallel, while GLDPC code on RM(4,6) code requires only 64 low-complexity MAP decoders as explained above. In simulations presented here we have employed an efficient realization of Chase II algorithm described in Chap. 5. In Fig. 11.11b BER performance of several classes of iteratively decodable codes (TPCs, LDPC, and GLDPC codes) of high code rate are compared against conventional RS, and concatenated RS codes. B-GLDPC code of rate 0.88 outperforms concatenated RS code of rate 0.82 by 2.47 dB (also at BER $= 10^{-9}$). RM(6,8) $\times$ RM(5,7) TPC of rate 0.905 outperforms concatenated RS code ($R = 0.82$) by 0.53 dB at BER of 10^{-9}. LDPC code of rate 0.93, designed using the concept of product of orthogonal arrays (OAs), outperforms the same RS concatenated code by 1.15 dB at BER of 10^{-9}. $R = 0.93$ LDPC code outperforms RS code of rate 0.937 by 2.44 dB at BER of 10^{-9}.

11.7 LDPC-Coded Hybrid Frequency/Amplitude/Phase/Polarization Modulation

Here we describe a hybrid frequency/amplitude/phase/polarization (H-FAPP)-coded modulation scheme, proposed in [25]. Using transmission equipment operating at 50 GS/s, and an LDPC code of rate 0.8, the H-FAPP scheme achieves $(N_1 + N_2 + \ldots + N_L) \times 40$ Gbit/s aggregate rates where $N_1 = 3, 4, \ldots$ and $N_{L \geq l \geq 2} = 0, 1, \ldots$. $(N_1 + N_2 + \ldots + N_L)$ represents the total number of source channels to the transmitter. HAPP scheme we introduced in [24] can be considered a special case of H-FAPP where only one subcarrier is used ($N_{L \geq l \geq 2} = 0$).

Figure 11.12a shows the block diagram of the H-FAPP system configuration. N input bit streams from different information sources are divided into Lgroups variable in number of streams per group. The selection process for the different groups $N_1, N_2, \ldots, N_L$ is governed by two factors, the required aggregate rate, and the polyhedron of choice. Each N_l is then used as input to an HAPP transmitter, where it is modulated with a unique subcarrier. The outputs of the L HAPP transmitters are then forwarded to a power combiner in order to be sent over the fiber. At the receiver side, the signal is split into L branches and forwarded to the L HAPP receivers. In this section, and without loss of generality, we clarify three simple examples for $N = 8$ and $N = 16$ where $L = 1$ and for $N = 20$ where $L = 2$.

Figure 11.12b shows the block diagram of the coded HAPP transmitter. N_l input bit streams from l different information sources, pass through identical encoders that use structured LDPC codes with code rate $r = k/n$, where k represents the number of information bits, and n represents the codeword length. The outputs of the encoders are then interleaved by an $N_l \times n$ bit-interleaver where the sequences are written row-wise and read column-wise. The output of the interleaver is sent in one bitstream, N_l bits at a time instant i, to a mapper. The mapper maps each N_l bits into a2^{N_l}-ary signal constellation point on a vertex of a polyhedron inscribed in a Poincaré sphere based on a lookup table. (Please note that the vertices of all the L polyhedrons define a regular polyhedron inscribed in the Poincaré sphere). The signal is then modulated by the HAPP modulator.

The HAPP modulator, shown in Fig. 11.12c, is composed of three simpler modulators, two amplitude modulators (AM) and one phase modulator (PM). Therefore, the lookup table maps each N_l bits into a set of three voltages $(f_{1,i}, f_{2,i}, f_{3,i})$ needed to control the set of modulators. As, the polyhedrons used are inscribed in a Poincaré sphere, Stokes parameters are used for the design of the polyhedron. Stokes parameters shown in (11.20) [25], are then converted into amplitude and phase parameters according to (11.21).

$$\begin{aligned} s_1 &= a_x^2 - a_y^2, \qquad s_2 = 2a_x a_y \cos(\delta), \\ s_3 &= 2a_x a_y \sin(\delta), \quad \delta = \phi_x - \phi_y, \end{aligned} \tag{11.20}$$

where

$$E_x = a_x(t)e^{\mathrm{j}(wt+\phi_x(t))}, \quad E_y = a_y(t)e^{\mathrm{j}(wt+\phi_y(t))} \tag{11.21}$$

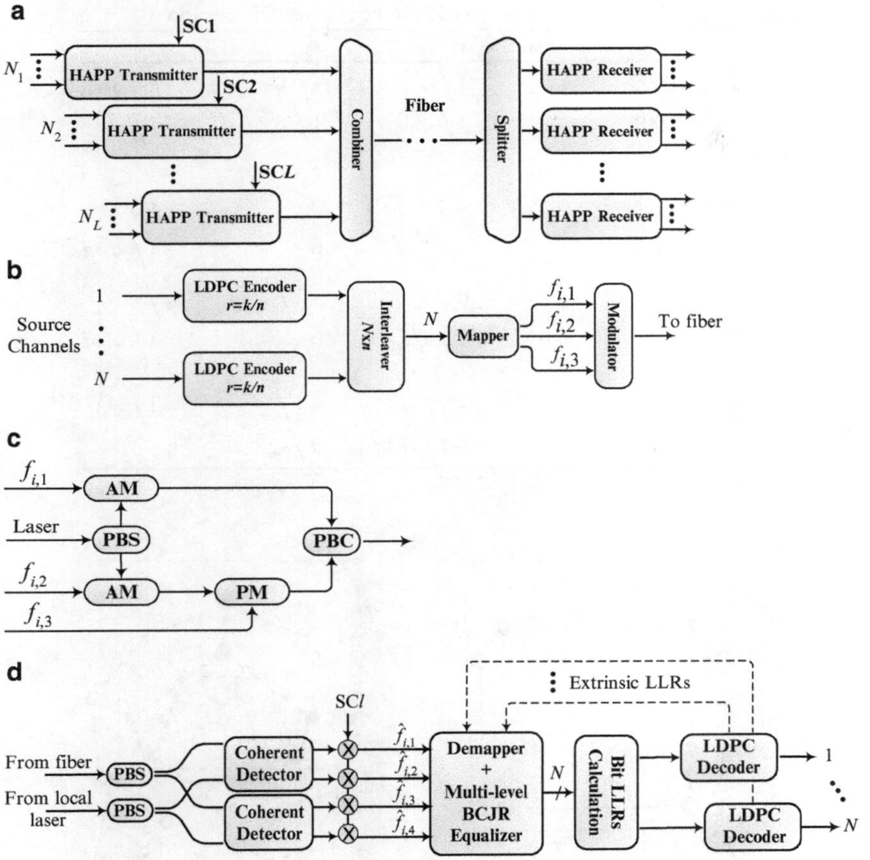

Fig. 11.12 H-FAPP bit-interleaved LDPC-coded modulation block diagrams: (**a**) H-FAPP system, (**b**) HAPP transmitter (**c**) HAPP modulator and (**d**) HAPP receiver configurations. *SC* subcarrier, *AM* amplitude modulator, *PM* phase modulator, *PBS* polarization beam splitter, *PBC* polarization beam combiner

Without loss of generality, we can assume that $\phi_x = 0$ at all times, hence $\delta = -\phi_y$. This yields a system of three equations with three unknowns that can easily be solved. Using symmetrical geometric shapes results in closed form numbers for the voltages as shown in Table 11.6. The corresponding lookup table for the example where $N_1 = 3$ is already given in Chap. 6. The constellation forms a cube inscribed inside the Poincaré sphere as $2^3 = 8$. On the other hand, the second scenario is for $N_1 = 4$ and $N_2 = 2$, the constellation contains $16 + 4 = 20$ points, hence the polyhedron of choice is the dodecahedron. For the latter case, $L = 2$, all the vertices of the dodecahedron expect four that can be chosen to be equidistant and furthest from each other are used for the polyhedron defined by N_1, and the tetrahedron that is formed by the remaining four points is used for N_2. Both constellations are shown in Fig. 11.13.

Table 11.6 Mapping rule lookup table for the H-FAPP-20 scenario

Group	Interleaver output	s_1	s_2	s_3
N_1	0000	$1/\sqrt{3}$	$1/\sqrt{3}$	$1/\sqrt{3}$
	0001	$1/\sqrt{3}$	$1/\sqrt{3}$	$-1/\sqrt{3}$
	⋮			
	1110	$-d/\sqrt{3}$	0	$1/\sqrt{3}d$ [a]
	1111	$-d/\sqrt{3}$	0	$-1/\sqrt{3}d$
N_2	00	0	$-1/\sqrt{3}d$	$d/\sqrt{3}$
	01	$d/\sqrt{3}$	0	$-1/\sqrt{3}d$
	10	$-1/\sqrt{3}$	$-1/\sqrt{3}$	$-1/\sqrt{3}$
	11	$-1/\sqrt{3}d$	$d/\sqrt{3}$	0

[a] d is the golden ratio: $(1+\sqrt{5})/2$

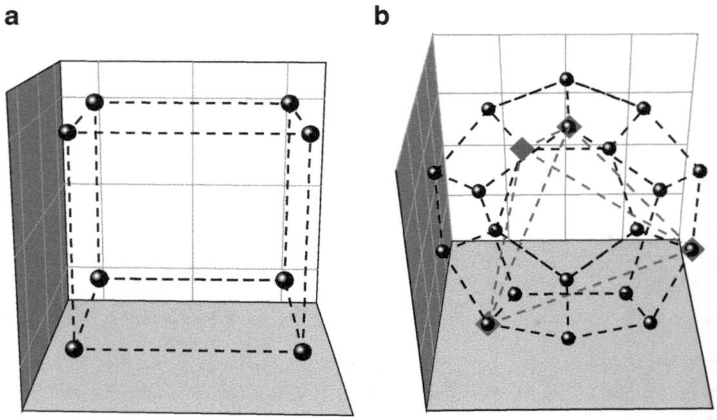

Fig. 11.13 Signal constellations in Stokes coordinates for: (**a**) $N_1 = 3$, $N_2 = 0$ and (**b**) $N_1 = 4$, $N_2 = 2$

Figure 11.13 shows the case for which (a) $N_1 = 3$, and (b) $N_1 = 4$ and $N_2 = 2$, with the diamond representing the vertices of the tetrahedron of N_2. Figure 11.12d shows the block diagram of the HAPP receiver. The signal from fiber is passed into two coherent detectors then to four branches. The output of each branch is demodulated by the subcarrier specified for the corresponding HAPP receiver, then sampled at the symbol rate then forwarded to the demapper and the multilevel BCJR algorithm-based equalizer (BCJR equalizer). The multilevel BCJR equalizer [9] used in this scheme is a generalization of the original version of the BCJR algorithm presented in [7]. The output of the equalizer is then forwarded to the bit log-likelihood ratios (LLRs) calculator which provides the LLRs required for the

LDPC-decoding process. The LDPC decoder forwards the extrinsic LLRs to the BCJR equalizer, and the extrinsic information is iterated back and forth between the decoder and the equalizer until convergence is achieved unless the predefined maximum number of iterations is reached. This process is denoted by *outer iterations*, as opposed to the *inner iterations* within the LDPC decoder itself. The outer iterations help in reducing the BER at the input of the LDPC decoder so it can efficiently decode the data within a small predefined number of inner iterations, without increasing the complexity of the system.

The proposed scheme is tested using VPITransmisionMaker [52], for a symbol rate of 50 GS/s, for 20 iterations of SPA for the LDPC decoder, and 3 outer iterations between the LDPC decoder and the multilevel BCJR equalizer. The simulations are done assuming an ASE noise-dominated channel scenario for both, a pseudo random bit sequence (PRBS) and an LDPC-coded bit sequence. The coded bit sequence uses LDPC(16935,13550) code of rate 0.8, which yields an actual effective information rate of the system of $3 \times 50 \times 0.8 = 120$, 160, and 240 Gb/s for the cases where $N_1 = 3$; $N_1 = 4$; and $N_1 = 4$, and $N_2 = 2$, respectively. Utilizing higher rate codes allows a higher actual transmission rate, or allows transmission components of lower speeds to achieve the current transmission rate.

The results of these simulations are summarized in Fig. 11.14. We show the uncoded and coded BER performance versus the OSNR per information bit. As noted from the figure, for the ASE noise-dominated scenario, the 8-HAPP scheme outperforms its QAM counterpart by 2 dB, while outperforms the PSK counterpart by 4 dB at BER of 10^{-6}. Moreover, the 16-HAPP outperforms its QAM counterpart by 1.1 dB at BER of 10^{-6}. On the other hand, the proposed scheme of H-FAPP that utilizes the 3D-space more efficiently increases the aggregate transmission rate by 80 Gb/s in comparison with 16-HAPP, and improves the performance by 1.75 dB at

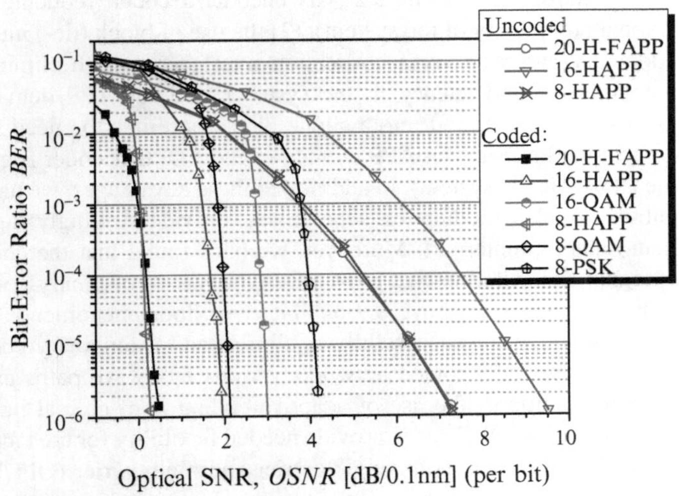

Fig. 11.14 BER performance versus the OSNR per bit for both uncoded and LDPC-coded data

BER of 10^{-6}. Furthermore, 20-H-FAPP doubles the aggregate transmission rate of 8-HAPP while keeping the BER performance of the system almost intact.

This approach can be further modified in the future to increase the number of constellation points and provide higher aggregate rates. One simple modification to the dodecahedron discussed in this paper that allows us to increase the total transmission per subcarrier to 200 Gbit/s is to exploit the fact that a dodecahedron has 12 faces. By doing so, we can add 12 points to the constellation at the projection of the center of each face on the Poincaré sphere. This way the total number of constellation points increases to $32 = 2^5$. Applying this configuration and utilizing five different orthogonal subcarriers allow a total aggregate rate of 1 Tb/s per wavelength that can be used as enabling technology for future 1 Tb/s Ethernet.

11.8 Concluding Remarks

In this chapter, we discussed various FEC and coded modulation schemes that are currently in research-and-development phase. We first described the binary QC-LDPC codes of large girth and their implementation in FPGA. The large-girth LDPC codes provide large coding gains and do not exhibit error floor phenomenon, when properly designed, in the region of interest for optical communications. Because large-girth LDPC codes of high-rate usually come with excessive lengths, we also discuss the possibility of using LDPC codes over higher-order fields. We have shown that (bit-length-matched) NB-LDPC codes outperform corresponding binary counterparts while offering comparable or lower decoding complexity. Compared to binary-coded modulation schemes, the non-binary-coded modulation offers several advantages: (1) m binary LDPC encoders/decoders needed for multilevel modulation are collapsed into a single 2^m-ary encoder/decoder (reducing the overall computational complexity of the system), (2) the use of block (de-) interleavers; binary-to-nonbinary and vice versa conversion interfaces can be eliminated (reducing the complexity and latency in the system), and (3) APP demapper and LDPC decoder can be integrated into a single block, and thus the need for iterating extrinsic information between APP demapper and LDPC decoder is eliminated (reducing the latency in the system). In addition to these advantages, we have shown that the nonbinary-coded modulation scheme can provide strikingly higher coding gains than binary counterpart. Moreover, we have found that the coding gain increases as signal constellation size grows. Certain classes of binary-coded modulation schemes of code rate above 0.9 exhibit error floor phenomenon for large constellation sizes, which can successfully be eliminated by nonbinary-coded modulation. Because in optically routed networks various ligthwave paths experience different channel conditions, the use of adaptive coding is of crucial importance. We have shown that QC-LDPC codes provide needed flexibility for the employment in this scenario. We discussed both single-carrier and multicarrier (OFDM)-based adaptive LDPC-coded modulation.

We demonstrated that overall decoding complexity of both binary and NB-LDPC codes can be reduced by using the GLDPC codes. The GLDPC codes offer numerous advantages for hardware implementation, because the same structures can be reused at different levels, and are very friendly to parallelization. The main problem, however, arises in the design of strong GLDPC codes of high-rate, in particularly above 0.9. Despite the Internet "bubble," which occurred in the early 2000s, the Internet traffic has been growing at an astonishing rate ranging from 75 to 125% per year [2–5]. Recent growth of Internet usage, IPTV, VoIP, and YouTube impose very stringent bandwidth requirements that optical networks must satisfy in near future. As the response to these bandwidth demands, we discussed the possibility of improving spectral efficiency of optical transport networks by using multidimensional coded hybrid frequency/amplitude/phase/polarization modulation. Because the information capacity is a logarithmic function of SNR but linear in number of dimensions, this approach seems to provide a promising alternative to new devices development.

References

1. http://www.canarie.ca/
2. McDonough J (2007) Moving standards to 100 GbE and beyond. IEEE Appl Pract 45(11):6–9
3. Melle S, Jaeger J, Perkins D, Vusirikala V (2007) Market drivers and implementation options for 100-GbE transport over the WAN. IEEE Appl Pract 45:18–24
4. Internet2 and level 3 communications to deploy next generation nationwide research network. Internet2 press release. Available at http://news.internet2.edu/index.cfm
5. Shieh W, Djordjevic I (2009) OFDM for optical communications. Elsevier/Academic, Burlington, MA
6. Essiambre E-J, Raybon G, Mikkelsen B (2002) Pseudo-linear transmission of high-speed TDM signals at 40 and 160 Gb/s. In: Kaminow IP Li T (eds) Optical fiber telecommunications IVB. Academic, San Diego, CA, pp 233–304
7. Djordjevic IB, Minkov LL, Batshon HG (2008) Mitigation of linear and nonlinear impairments in high-speed optical networks by using LDPC-coded turbo equalization. IEEE J Sel Areas Comm, Opt Comm Netw 26(6):73–83
8. Djordjevic IB, Cvijetic M, Xu L, Wang T (2007) Using LDPC-coded modulation and coherent detection for ultra high-speed optical transmission. IEEE/OSA J Lightwave Technol 25: 3619–3625
9. Djordjevic IB, Arabaci M, Minkov L (2009) Next generation FEC for high-capacity communication in optical transport networks. IEEE/OSA J Lightwave Technol 27(16):3518–3530 (invited paper)
10. Djordjevic IB Coded-OFDM in hybrid optical networks. IET Optoelectron (accepted for publication)
11. Djordjevic IB, Vasic B, Neifeld MA (2007) LDPC coded OFDM over the atmospheric turbulence channel. Opt Express 15:6332–6346
12. Djordjevic IB, Denic S, Anguita J, Vasic B, Neifeld MA (2008) LDPC-coded MIMO optical communication over the atmospheric turbulence channel. IEEE/OSA J Lightwave Technol 26(5):478–487
13. Anguita JA, Neifeld MA, Hildner B, Vasic B Rateless coding on experimental temporally correlated FSO channels. IEEE/OSA J Lightwave Technol, accepted for publication
14. Djordjevic IB Djordjevic GT (2009) On the communication over strong atmospheric turbulence channels by adaptive modulation and coding. Opt Express 17(20):18250–18262

15. Burns P (2003) Software defined radio for 3G. Artech House, Boston, MA
16. Kenningotn PB (2005) RF and baseband techniques for software defined radio. Artech House, Boston, MA
17. Mitola J (1995) The software radio architecture. IEEE Commun Magazine 33(5):26–38
18. Basch EB, Egorov R, Gringeri S, Elby S (2006) Architectural tradeoffs for reconfigurable dense wavelength division multiplexing systems. IEEE/LEOS J Sel Top Quantum Electron 12:615–626
19. Winzer PJ, Raybon G, Duelk M (2005) 107-Gb/s optical ETDM transmitter for 100 G Ethernet transport. In: Proceedings of the European Conference on Optical Communication, Glasgow, Scotland paper no. Th4.1.1
20. Djordjevic IB (2009) Adaptive LDPC-coded multilevel modulation with coherent detection for high-speed optical transmission. In: Proceedings of the IEEE Photonics Society Summer Topicals, 20–22 July 2009, Newport Beach, CA, paper no. WC1. 2
21. Arabaci M, Djordjevic IB, Saunders R, Marcoccia RM (2009) Rate-adaptive non-binary-LDPC-coded polarization-multiplexed multilevel modulation with coherent detection for optically-routed networks. In: Proceedings of the 11th international conference on transparent optical networks (ICTON 2009), Island of São Miguel, Azores, Portugal, paper no. Tu.B2.2
22. Arabaci M, Djordjevic IB, Saunders R, Marcoccia RM (2009) High-rate non-binary regular quasi-cyclic LDPC codes for optical communications IEEE/OSA J Lightwave Technol 27(23):5261–5267, December 1, 2009
23. Arabaci M Djordjevic IB, Saunders R, Marcoccia R (2009) Non-binary quasi-cyclic LDPC-coded modulation for beyond 100 Gb/s optical transmission. In: Proceedings of the 22nd annual meeting of the IEEE photonics society, 4–8 October 2009 Belek-Antalya, Turkey, paper WE 4
24. Batshon HG Djordjevic IB (2009) Hybrid amplitude/phase/polarization coded modulation for 100 Gb/s optical transmission and beyond. In: Proceedings of the 22nd annual meeting of the IEEE photonics society 4–8 October 2009, Belek-Antalya, Turkey
25. Batshon HG Djordjevic IB Beyond 240 Gb/s per wavelength optical transmission using coded hybrid subcarrier/amplitude/phase/polarization modulation. IEEE Photon Technol Lett, accepted for publication
26. Arabaci M, Djordjevic IB, Saunders R, Marcoccia RM Non-binary quasi-cyclic LDPC based coded modulation for beyond 100 Gb/s transmission. IEEE Photon Technol Lett (submitted)
27. Djordjevic IB, Xu L, Wang T Cvijetic M (2008) GLDPC codes with Reed-Muller component codes suitable for optical communications. IEEE Comm Lett 12:684–686
28. Djordjevic IB, Milenkovic O, Vasic B (2005) Generalized low-density parity-check codes for optical communication systems. IEEE/OSA J Lightwave Technol 23:1939–1946
29. Batshon HG, Djordjevic IB, Xu L, Wang T (2009) Multidimensional LDPC-coded modulation for beyond 400 Gb/s per wavelength transmission. IEEE Photon Technol Lett 21(16): 1139–1141
30. Arabaci M, Djordjevic IB, Saunders R, Marcoccia RM Polarization-multiplexed rate-adaptive non-binary-LDPC-coded multilevel modulation with coherent detection for optical transport networks. Opt Express (submitted)
31. Tanner RM (1981) A recursive approach to low complexity codes. IEEE Trans Inform Theory IT-27:533–547
32. Fossorier MPC (2004) Quasi-cyclic low-density parity-check codes from circulant permutation matrices. IEEE Trans Inform Theory 50:1788–1793
33. Milenkovic O, Djordjevic IB, Vasic B (2004) Block-circulant low-density parity-check codes for optical communication systems. IEEE/LEOS J Sel Top Quantum Electron 10:294–299
34. Djordjevic IB, Xu L, Wang T, Cvijetic M (2008) Large girth low-density parity-check codes for long-haul high-speed optical communications. In: Proceedings of the OFC/NFOEC. 24–28 February 2008 San Diego, CA, paper no. JWA53
35. Arabaci M Djordjevic IB (2008) An alternative FPGA implementation of decoders for quasi-cyclic LDPC Codes. In: Proceedings of the TELFOR 2008, 25–27 November 2008, Belgrade, Serbia, pp 351–354

36. Mitrionics (2008) Mitrion users guide, Mitrionics Inc., v1.5.0–001
37. Davey MC (1999) Error-correction using low-density parity-check codes. PhD. Thesis, University of Cambridge
38. Spagnol C, Marnane W, Popovici E (2007) FPGA implementations of LDPC over GF(2^m) decoders. In: Proceedings of the IEEE workshop on signal processing systems, Shanghai, China, pp 273–278
39. Voicila A, Verdier F, Declercq D, Fossorier M, Urard P (2007) Architecture of a low-complexity non-binary LDPC decoder for high order fields. In: Proceedings of the ISIT, Nice, France, pp 1201–1206
40. Lan L, Zeng L, Tai YY, Chen L, Lin S, Abdel-Ghaffar K (2007) Construction of quasi-cyclic LDPC codes for AWGN and binary erasure channels: a finite field approach. IEEE Trans Inform Theory 53:2429–2458
41. Chen J, Dholakia A, Eleftheriou E, Fossorier M Hu X-Y (2005) Reduced-complexity decoding of LDPC codes. IEEE Trans Comm 53:1288–1299
42. Cover TM, Thomas JA (1991) Elements of information theory. Wiley, New York
43. Bölcskei H, Gesbert D, Paulraj AJ (2002) On the capacity of OFDM-based spatial multiplexing systems. IEEE Trans Comm 50:225–234
44. Mizuochi T et al (2004) Forward error correction based on block turbo code with 3-bit soft decision for 10-Gb/s optical communication systems. IEEE/LEOS J Sel Top Quantum Electron 10(2):376–386
45. Boutros J, Pothier O, Zemor G (1999) Generalized low density (Tanner) codes. In: Proceedings of the 1999 IEEE international conference on communication (ICC'99), 6–10 June 1999, Vancouver, BC, Canada, vol 1, pp 441–445
46. Lentmaier M, Zigangirov KSh (1999) On generalized low-density parity-check codes based on Hamming component codes. IEEE Comm Lett 3:248–250
47. Parhi T, Zhang KK (2001) A class of efficient-encoding generalized low-density parity-check codes. In: Proceedings of the IEEE international conference on acoustics, speech, and signal processing (ICASSP '01), vol 4, pp 2477–2480
48. Bahl LR, Cocke J, Jelinek F, Raviv J (1974) Optimal decoding of linear codes for minimizing symbol error rate. IEEE Trans Inform Theory IT-20(2):284–287
49. Morelos-Zaragoza RH (2002) The art of error correcting coding. Wiley, Boston, MA
50. Ashikhmin A, Lytsin S (2004) Simple MAP decoding of first-order Reed-Muller and Hamming codes. IEEE Trans Inform Theory 50:1812–1818
51. Zhang T, Parhi KK (2001) A class of efficient-encoding generalized low-density parity-check codes. In: Proceedings of the ICASSP '01, vol 4, pp 2477–2480
52. VPITransmisionMaker. Available at http://www.vpiphotonics.com

Index

C

D

E

N

O

P

T

FSC
www.fsc.org
MIX
Papier | Fördert gute Waldnutzung
FSC® C083411

Kolibri 360 GmbH
Ferdinand-Jühlke-Straße 7
99095 Erfurt, Deutschland
produktsicherheit@kolibri360.de